PRACTICAL ENGINEERING DRAWING

GITIN M. MAITRA

NEW AGE INTERNATIONAL (P) LIMITED, PUBLISHERS
New Delhi • Bangalore • Chennai • Cochin • Guwahati • Hyderabad
Jalandhar • Kolkata • Lucknow • Mumbai • Ranchi
PUBLISHING FOR ONE WORLD
Visit us at www.newagepublishers.com

Published by New Age International (P) Ltd., Publishers
First Editon: 2000 (HB)
First Edtion: 2009 (PB)

Branches:

- 36, Malikarjuna Temple Street, Opp. ICWA, Basavanagudi, **Bangalore**. ✆ (080) 26677815
- 26, Damodaran Street, T. Nagar, **Chennai**. ✆ (044) 24353401
- Hemsen Complex, Mohd. Shah Road, Paltan Bazar, Near Starline Hotel, **Guwahati**. ✆ (0361) 2543669
- No. 105, 1st Floor, Madhiray Kaveri Tower, 3-2-19, Azam Jahi Road, Nimboliadda, **Hyderabad**. ✆ (040) 24652456
- RDB Chambers (Formerly Lotus Cinema) 106A,1st Floor, S.N. Banerjee Road, **Kolkata**. ✆ (033) 22275247
- 18, Madan Mohan Malviya Marg, **Lucknow**. ✆ (0522) 2209578
- 142C, Victor House, Ground Floor, N.M. Joshi Marg, Lower Parel, **Mumbai**. ✆ (022) 24927869
- 22, Golden House, Daryaganj, **New Delhi**. ✆ (011) 23262370, 23262368

ISBN : 978-81-224-2690-8

Rs. 250.00

C-09-01-3323

Printed in India at Mohan Lal Printers, Delhi.
Typeset at EPTECH, Delhi.

PUBLISHING FOR ONE WORLD
NEW AGE INTERNATIONAL (P) LIMITED, PUBLISHERS
4835/24, Ansari Road, Daryaganj, New Delhi-110002
Visit us at **www.newagepublishers.com**

To
JAYATI and INDRANI
and their mother ANJANA
with love and affection

PREFACE

It is imperative for anyone working in the technical field to have a thorough grasp of the principles which govern engineering drawing procedures. An engineering drawing is executed according to certain rules and codes, and is meant to illustrate the form and dimensions of an object in such a way that it can be generally made by the conventional methods of manufacturing processes. In other words, such drawing chiefly aims at utility and usefulness, so that the job can actually be implemented by a correct interpretation of the drawing.

Several books have been written on engineering drawing, but virtually all of these tend to be mainly theoretical in nature. That is, these books cover such aspects as the pictorial part of drawing making and allied features thereof, but they do not usually dwell on the practical implications of engineering drawings nor do they contain enough technical data to enable a drawing maker to produce manufacturing drawings of engineering components or equipment. Such books may be useful in educational institutions for passing examinations, but they are hardly adequate for making practical, industry-oriented, manufacturing drawings. Hence, the mere graphical representation of an object on a drawing sheet is not good enough for industrial purposes. It has to be supplemented by sufficient engineering data warranted for manufacturing the part.

The purpose or scope of this book is chiefly twofold: to familiarise the reader with the essential and advanced features of modern draughting practice so that he can understand the conceptual ideas as well as the underlying principles of geometrical drawing making, and to enable him to produce industrial drawings, i.e., drawings which are working drawings complete with the relevant information and data that are required for actual manufacturing of the parts. This book is, therefore, pragmatic in nature and there is a dearth of such books on engineering drawing which take congnizance of such practical needs as correct material specifications of common engineering components, assignment of proper fits and tolerances, incorporation of machining symbols, heat-treatment and hardness required for the purposes, and other allied data and instructions which comprise a full-fledged industrial drawing. Coping with this lacuna has been the main impetus to me to write this book, coupled with the fact that my years of working experiences in industries at home and abroad have made me fully conversant with the nature of problems the technical personnel encounter while working in the design and drawing office of industries. This book is intended to be a guide-book for these people. Most of the new engineers or draughtsmen lack the requisite knowledge for making or "reading" an industrial drawing when they leave the educational institutions and enter the practical field. This book will help them to do that plus provide them with the technique of the conventional methods of drawing making along with the knowledge of engineering standards, components and items, so that they need not resort to other manuals, Standards, etc. in short, this is a complete professional book to help make a complete professional man, so far as technical drawing is concerned. Besides, he can use this as a reference book throughout his professional career.

Since this book is written with a bias towards practical application, special emphasis has been laid on fits and tolerances, material selections commensurate with the applications to which the parts are subjected, proper heat-treatment, treatises on manufacturing processes, mensuration to calculate volume and mass, practical guidelines for methodical checking of working drawings, proper procedures and codes for dimensioning in drawing, common mistakes and their rectifications, common engineering components like fasteners, transmission elements like gears, belts, chains, etc., surface development of pieces generally required in industries, and data as regards pipes and accessories, jig bushes, etc. Written completely in SI units, this book quotes copiously from the latest versions of the relevant Indian Standards and ISO Specifications, and contains comprehensive Appendices.

Since the main objective is to provide the reader with as much practical engineering data and technical information as possible, certain aspects like the uses of drawing instruments, lettering, types of lines, etc., are not elaborately dealt with as these are detailed in conventional books on engineering drawing. Though primarily meant for draughting purposes, simple elementary engineering formulae and calculation procedures involving fits, gears, shafts, etc., have been included, so that the draughtsman may become familiar with the basic design concepts and considerations involved in arriving at the optimum dimensions and allied parameters of the engineering components, and can work independently, if the need arises.

I hope the book will serve the purpose for which it is meant. I shall gratefully receive suggestions aimed towards improvement of the book, and also corrections of any errors which might have inadvertently crept in.

I am thankful to M/s Tata McGraw Hill, New Delhi for granting me permission to produce certain drawings and Tables from *Handbook of Gear Designs*. These are duly acknowledged wherever they appear.

GITIN M. MAITRA

Contents

CHAPTER 1

FUNDAMENTALS OF ENGINEERING DRAWING

1.1 INTRODUCTION

From time immemorial, man has been expressing his thoughts in pictorial representation of objects and of abstract ideas. This graphic language is universal, and is found in one form or another from the cave paintings of primitive people made thousands of years ago to the modern Chinese and Japanese scripts which are basically of pictorial nature. The ancient Egyptians made their hieroglyphic writings in a systematic and methodical manner and these were also of pictorial script. The various modern alphabets used now all over the world can trace back their origin to some form of pictorial base.

Pictures of ancient times were of artistic nature. Later on, when man began to build places of worship, monuments, aqueducts, bridges and other structures, the architects and other concerned persons must have used some sort of drawings on which they laid down their ideas about the objects they were going to build and these drawings, which were presumably true to some scale to effect a faithfully conceived representation of the objects, are the forerunners of what is now known as "technical drawing". Viewed from modern perspective, these may seem crude attempts, but some of the amazing and colossal structures of ancient times are so accurate even in ruins, it is quite logical to presume that they could not have been possibly constructed unless some sort of accurate technical drawings were made beforehand to carry out the execution of these huge projects. Technical knowledge could not have developed to its highly sophisticated and intricate form of present day standard unless there has been a simultaneous growth of systematised technical drawing making procedures. And these procedures developed through many stages - from Vitruvius of ancient Rome through Leonardo da Vinci of Middle Ages to the comparatively recent Gaspard Monge who is credited with the invention of modern methods of descriptive geometry. This is but to name a few as the modern methods of technical draughting are the brain children of a host of highly technical, innovative and original minds.

While in fine art, the artist using his creative imagination aims to produce a picture which will convey to the observer the same visualisation of the object as it is in the artist's mind but which may or may not have a true semblance to the object concerned, the technical draughtsman can take no such liberties and is supposed to produce a picture which is an accurate representation of the object, giving precise details of every facets which are required from the engineering point of view for manufacture or repair of the part for maintenance and other purposes. The artist can impart his impression on the canvas by using different colours, tonal graduation and other means, and his end product is completely individualistic. The technical man follows rigid rules laid down for technical draughting so that there is no scope for ambiguity in interpreting his end product. Besides, an artist's picture can be more or less understood by a layman (barring the so-called modern art), a technical drawing can be understood only by the trained personnel.

To use a hackneyed expression : "Drawing is the language of engineers". While designing a machine or a part thereof, the design engineer first takes recourse to a preliminary sketch to express his still nebulous ideas in a more or less concrete shape. When his ideas are finalised, he or the draughtsman works on the drawing board and produces a manufacturing or "shop" drawing. The shop engineer then interprets or "reads" the drawing and produces the actual part accordingly.

The free-hand sketch helps the designer to visualise his technical problems, such as, space restrictions, vulnerability to failure at weak points, machining problems that may arise when the part actually goes into production and other allied problems and pitfalls. It also enables him to plan and to formulate the possible solutions. However, a sketch has its limitations. Since it is not drawn to scale, little or no idea can be formed as to the size and proportion of the part. Only a picture drawn to scale can show the exact shape, size and other details. When such a picture is drawn according to the rules of the orthographic projections, contains the relevant informations, such as, the necessary dimensions, tolerances, machining symbols, material specifications and other necessary instructions, *e.g.*, casting or welding details, heat treatment procedures, (if any), than a manufacturing or engineering drawing ensues.

Before the manufacturing drawing is made, the designer must arrive at the dimensions, shape, size and other peculiarities of the part after considering several factors. Beside strength, reliability and durability, the cost involved is the single most important factor. In this respect, Value Engineering techniques are of great help. They enable the designer to determine, for example, whether it is justified to cast a part (involving pattern making expenses) or a welded one will serve the purpose. The designer must also consider safety, ease of regular maintenance, operational conveniences, lubrication, number of parts required and sometimes, an aesthetic appeal of the part. While he is supposed to think for himself when developing the drawing, it is not always imperative that his ideas should be entirely original. In many cases, time tested ideas which are known to be sound in theory and practice have been incorporated in new designs. The designer can, of course, improve upon these ideas, but the fact remains that there is hardly any engineering appliance which is absolutely original or is the work of one man only. Thus it is always advisable to study many aspects which others have already developed. Often it has so happened that the underlying principle of a contrivance is successfully deployed in another one which is meant for an entirely different purpose.

Since drawing is the very basis of engineering, a thorough and comprehensive knowledge of engineering drawing is essential for all engineering personnel. If such a person shirks or does not possess the working knowledge of drawings, then he cannot be considered as professionally literate. An engineer or a draughtsman must realise the fact that a drawing serves many purposes in the engineering field. It serves to develop a fundamental plan, helps to convey the ideas to the relevant and concerned people for manufacturing, assembly and other purposes, and afterwards it is kept as a record in the archives for future reference. Besides, a basic knowledge of different manufacturing methods, heat treatment and other engineering processes is also required. A competent man will also take cognizance of the facilities available within his own manufacturing unit or factory where he works so that the product for which he has made the drawing can be made there itself to reduce cost. Often it pays to consult the shop people. This way, the draughtsman can hit upon better and innovative ideas. Also, many a time the machinist or the foundry man or some other specialist may have a final say in drawing making by pointing out some intrinsic design defect of a product or constraints inherent in its actual manufacture.

This book is essentially meant for mechanical drawing and its ramifications. Mechanical drawings are either outline drawings or working drawings. In an outline drawing, only the general appearance, some guiding dimensions and other informations are given so that it can be used in catalogues, brochures, *etc.*, to give the prospective buyer an idea of the essential features of the product. Working drawings may be assembly drawings or detail drawings. These are used for shop purposes. While the assembly drawings contain only the principal dimensions and instructions to the erector, the detail drawing must contain all the dimensions and complete information to facilitate the manufacture of the part. As pointed out earlier, a technical drawing can be read only by trained personnel. Though orthographic projection is mainly used, sometimes perspective drawing is also drawn though its use is very rare as it does not help the manufacturing processes involved. Main emphasis, therefore, has been given to orthographic projection methods such as multi-view projection and axonometric projection.

Guidelines for engineering drawings are laid down in the latest version of IS: 696 and in a few other Indian Standards. These guidelines as well as some foreign drawing standards such as DIN have been mainly followed in preparing this book. For further perusal of the subject, the reader may refer to the Bibliography in which copious references have been given as regards these standards and books on engineering drawing and allied subjects.

1.2 GENERAL REQUIREMENTS IN DRAUGHTING

It is a common experience the world over that most of the engineering graduates are under the impression that once they graduate, they are not expected to do any draughting work - that being entirely the domain of the professional draughtsmen. Nothing can be far from the actual practice in working life. An engineer is often required to work on the drawing board. Besides attending to his usual responsibilities, such as, site selection and evaluation, costing and estimating, equipment selection, engineering calculations pertaining to design, making of feasibility report and tender specifications, tender evaluation, selection of relevant engineering specifications and other allied tasks, an engineer is supposed to chalk out preliminary layouts and make complete drawings with or without the help of draughtsmen. He is also sometimes called upon to assist in construction and installation of plants and equipments for which a thorough knowledge of engineering drawing is imperative. It is, therefore, to his advantage that an engineering graduate starts his career at the bottom in a design and drawing office and gradually works up because there is so much to learn in the practical working life in the engineering field. It goes without saying that training and experience in the drawing office are primary prerequisites for a successful career. In short, the ability to make and "read" a drawing is of vital importance to an engineering personnel - be he an engineering graduate or a technical draughtsman.

An engineering drawing is a graphic language. It enables a trained person to visualise a solid object from views and projections which are drawn on papers which are two dimensional. In fact, to "think in three dimensions" from the two-dimensional views, projections and sections is the main objective of an engineering drawing. These views, projections and sections are drawn as per certain rules and norms. A thorough grasp over the subject, a mastery of these rules and norms, knowledge about the common engineering standards, and familiarity with standard engineering components - all these are attributes to give a practical shape to constructive imagination.

Before we proceed further to discuss about these rules, norms, standards and drawing practices in subsequent chapters and sections in this book, it is pertinent to dwell upon certain general and relevant aspects in drawing making procedures. These are dealt with below :

Accuracy : Accuracy is the most important factor in an engineering drawing. An inaccurate drawing is of little use, no matter how pleasing it may look apparently, and such a drawing often leads to futile exercise, entailing wastage of man-power, time, energy and money. An inaccurate drawing may also result in dangerous repercussions if components are actually made as per such a drawing and if these components eventually fail in operation leading to far-reaching consequences. It is, therefore, of utmost importance that an engineering drawing be made with utmost care and accuracy. Besides, an accurate drawing made by an engineer or a draughtsman brings him self-confidence which in turn acts as an impetus to take up more complicated and challenging jobs. This not only helps to further his career, it also helps him to acquire job satisfaction for doing something worthwhile.

Speed : In the modern day world of high technology, the technological products and manufacturing processes are rapidly changing, leading to obsolescence in many products. In this ever changing industrial world, speed is, therefore, an essential factor if a manufacturing industry wants to keep pace in this fiercely competitive market. And since everything begins at the drawing board stage, speed is required not only in all spheres of manufacturing activities including shop practices, it is equally essential in design and drawing offices. Therefore, beside skill in draughting and other desirable qualities, a draughtsman must acquire speed in his work. Speed in draughting comes with study and practice, intelligent thinking and continuous work.

Neatness : A drawing must be neat and clean. Just as it is a pleasure for the examiner to read a neat answer paper, so it is with a technical man who reads a neat drawing. Some draughtsmen are notorious in turning out dirty drawings which are vulnerable to misinterpretations and consequences thereof, beside being an eyesore. Every draughtsman, therefore, should always endeavour to acquire the habit of neatness and cleanliness.

If a draughtsman follows a few simple procedures with conscious effort, his drawing should be reasonably clean. He should always keep his hands clean and free from oil or perspiration. The draughting machine and drawing tools, instruments, set-squares, scale, T-square and other items used for drawing making should be kept clean all the time. Graphos, nibs and other inking appliances should be immediately cleaned with water after final use during the day lest they become clogged with dried ink. Drawings become dirty mainly by the graphite from the drawing pencils. Pencils should not be sharpened over the drawing board. Surroundings should be frequently dusted to keep them free from loose graphite particles and pencil chippings. Arms or palms should not rest on the drawing unless a piece of paper is kept in-between. For removing eraser particles, dust and similar objects, a brush or duster should be used. At the end of the working day, the drawing sheet should be covered by paper or plastic.

Legibility : Since technical drawing is a means of communication between the drawing maker and the drawing reader, it is obvious that this graphic language should be intelligible and meaningful to the user. Views and sections should be drawn as per the conventional rules, and they should be so placed that they are neither too crowded nor wide apart. Dimension-lines should have optimum gaps between them (usually 8 to 10 mm). Beside, particular attention should be given to the legibility of the lettering portions, executed either by hand or (preferably) by stencil, and to the dimensional and other numerals, so that there is no scope for ambiguity. Proper thickness should be maintained for different types of lining work as laid down in the Section dealing with lines. The whole drawing should be made in such a way that when a print is taken from the tracing, an overall impression of uniformity is imparted to the user.

These and other allied aspects in this connection will be treated at relevant Sections in the book.

Reliability : A technical drawing is a valuable document, and a considerable amount of money is at stake since parts will be made as per the drawing in a manufacturing concern. Hence the accuracy of the drawing which has already been emphasized.

Though a drawing is checked and approved by a checker or a supervisor after its completion, the main responsibility of producing an accurate drawing lies on the person who draughts it. The draughtsman, therefore, must be reliable, have self confidence, deliver the goods in time to meet the termin or time schedule and enjoy the overall credibility of the management, beside having the usual attributes of a good draughtsman, such as, technical skill, clear conception of the problem, relevant knowledge of standards, engineering materials, standard components, machining, tolerance system, weight calculations, strength properties, and manufacturing facilities available inside the concern where he works. In short, he should strive to equip himself in such a way that he can work independently with confidence and sufficient knowledge.

1.3 DRAWING INSTRUMENTS AND ACCESSORIES

Drawing instruments and accessories are necessary implements for making technical drawings. Whether regular drawings are made in drawing and design offices of a firm or freehand sketches are made at the site or elsewhere, some drawing equipments and allied items will always be required. In this Section, these equipments and their uses will be briefly discussed.

For general use in a drawing office, the following items are usually required.

1. Draughting machine : Nowadays, all modern drawing offices are equipped with draughting machines. These contrivances combine the functions of drawing board, T-square, set-squares, scale and angle protractors. Since around 35% of time in mechanical drawing is saved, draughting machines are being extensively used in commercial and other concerns.

2. Drawing boards : Where no draughting machine facilities are provided, drawings are usually made by fixing the drawing paper on drawing boards. These are generally made of some sort of soft wood so that thumb tacks can be easily inserted for fixing the drawing paper on to the board. However, the modern trend is now to use materials with hard surface since drawing sheets are now generally fixed on to the board by means of adhesive tape (Scotch tape or cello tape) and the tacks have become outmoded. The left-hand edge of the board along which T-square head slides while draughting is known as the working edge. It is to be ensured that this edge is straight and it should be tested with a steel straightedge.

3. T-square : The T-square used for drawing parallel lines on the drawing sheet, and it consists of a long arm called the blade to which a shorter piece called the head is rigidly fixed at right angles. The upper edge of the blade or the working edge, and the inner edge of the head must be straight and free from defects. The blade should preferably have transparent plastic edges to permit the drawing maker to see the lines in the vicinity.

4. Set-squares : The set-squares, also called triangles, are generally made of transparent plastic having standard angles 90° × 45° × 45° and 90° × 60° × 30°. These are made to ply to and fro on the working edge of the T-square for drawing verticals and inclined lines.

5. Protractors : Protractors are used for setting off or measuring angles other than the standard angles of set squares. They can be metallic or made of plastic material. Drawing protractors should never be used for measuring angles of a part lest they become dirty or damaged.

6. Scales : Different types of scales are used in a drawing office for different purposes. They may be flat type or of triangular cross-section. A set of scales are used to permit the draughtsman to draw parts in larger or reduced sizes in standard ratios. Scales may be made of wood, rigid or flexible plastics or cardboard.

7. Irregular curves : Irregular curves, also known as French curves, are used to draw curves other than circular area. A box of such curves, which are generally made of some plastic material, contains many pieces of different sizes, shapes and forms of curves. For drawing graphs, *etc.*, from points located at different places on a sheet, one or more French curves can be used. Portions of these curves are placed beside the relevant points, and these points are then joined together by pencil or ink so that a smooth curve ensues.

8. Drawing instruments box : A box containing instruments needed for draughting is by far the most important set of items. A beginner should be careful in selecting such a box inasmuch as sub-standard drawing instruments are vulnerable to defects and these in turn are liable to affect the quality of drawings. It is, therefore, prudent to take the opinion of experienced people before purchasing such a set.

These instruments are sold in a box containing the usual items required for draughting or they can be bought separately. The essential instruments are :

(i) The compass : A compass is required for drawing circles and circular arcs. The compasses come in different sizes. A standard instrument box normally contains a large compass and a smaller one called spring compass. For drawing big circles, lengthening bars are provided which are fixed to the large compass to facilitate drawing of such circles by increasing the radius. For extra large circles, beam compasses are used, but these are not supplied in a standard instrument box.

(ii) The dividers: These are similar to compass in construction and are used for dividing lines, *etc.*, into equal parts or to transfer distances from one place to another on a sheet. Like compasses, dividers are also available in large and small sizes. There are some adjustable types in which by replacing a leg, the same instrument can be used as a divider or a pencil compass or an ink compass by inserting the appropriate leg on to the instrument.

Beside the above, the box may include a container for pencil leads, thumb-tacks and a few other items. Some of the items already referred to, *viz.*, protractors, small scales, pair of small set squares may also be included in an instrument box.

9. Inking set : Nowadays, almost all drawings are inked. Beside the consideration for aesthetic values, an inked tracing produces better and more legible prints. The ink used for drawing is made from finally ground carbon in colloidal suspension. To make it waterproof, material such as natural or synthetic gum is added. The ink is supplied in handy bottles which are provided with droppers.

For inking purposes, penholders (popularly known as graphos pen) and various types of detachable nibs are used. Nibs may be meant for drawing lines or for lettering in conjunction with stencils. There are also composite pens of various grades, from coarse to fine.

10. Stencils : These are essentially meant for inked drawings. Stencils are generally made of plastics and are meant for lettering, writing numerals, drawing standard radii, machining symbols, ellipses and other assorted figures. Stenciling makes the drawing of uniform character, pleasant to look at and it also saves time. Alphabets and numerals in stencils come in different sizes, viz., 2 mm, 3 mm, 5 mm, *etc.*, and these characters may be of vertical or inclined types.

11. Erasers : Erasers (popularly known as rubbers) are used for removing pencil marks or ink-lines, usually pencil-lines. They are available in various degrees of hardness to suit different purposes. Erasers made from glass fibres and meant for removing ink-lines are also used, but they are costly. Erasing shield is used to protect the lines in the neighbourhood of those being erased. To remove particles after erasing, a duster or a dusting brush should be used.

12. Pencils : For technical drawing, high quality pencils should be used. They come in various grades - from 7 B (softest) to 9 H (hardest), but for ordinary drawing office use, 2 H, H and HB are sufficient for all practical purposes. Beside wood pencils which are commonly used, there are also holders available in the market in which leads (which are actually graphites) of suitable grade can be inserted. These holders are provided with a collet arrangement at the front end to grasp the lead firmly and a spring-loaded button at the other end to control the movement of the collet.

13. Drawing paper : Drawings can be made on opaque paper or tracing paper. Almost all industrial drawings are now made on transparent tracing papers from which blue prints, ammonia prints or similar contact prints can be made. Their sizes have been standardized (see Section 1.6). Tracing papers also come in rolls from which suitable portion can be cut out for special drawings or for non-standard sizes. Tracing cloths are not much used these days.

14. Pencil sharpener : Pencils can be sharpened by knives or sharpening machines. The use of razor blades for sharpening or erasing purposes is not recommended.

Besides, other useful items are sandpaper pad, templates, proportional dividers, thumbtacks, steel tape, *etc.*

1.4 LINES USED IN DRAWING

Since in a technical drawing, the basis of drawing is the line, every line on such a drawing carries a definite meaning and significance. A set of conventional lines has been standardised in IS: 10714 - 1983, covering the characteristics of the lines normally needed for different purposes. The following guidelines are generally followed :

1. All industrial drawings are now executed on transparent tracing papers as prints are taken from these transparencies by various processes. To obtain good reproduction, all lines should be sharply defined and dense.
2. The types and fields of applications of different lines are shown in Table 1.1. These should be adhered to unless other types are used for special purposes, *e.g.*, electrical or pipe-line diagrams, etc., or for applications other than those detailed in Table 1.1. If some other conventions conforming to other International Standards are adopted as regards lining, then this should be indicated by relevant notes on the drawing.

Table 1.1 Types and fields of application of lines

Line	*Description*	*General application*
A	Continuous thick	A1 Visible outlines A2 Visible edges
B	Continuous thin (straight or curved)	B1 Imaginary lines of intersection B2 Dimension lines B3 Projection lines B4 Leader lines B5 Hatching B6 Outlines of revolved sections in place B7 Short centre lines B8 Root line of thread.
C	Continuous thin freehand	C1 Limits of partial or interrupted views, sections, if the limit is not a chain thin line.
D	Continuous thin (straight) with zigzags	D1 Limit line
E	Dashed thick	E1 Hidden outlines E2 Hidden edges
F	Dashed thin	F1 Hidden outlines F2 Hidden edges
G	Chain thin	G1 Centre lines G2 Lines of symmetry G3 Trajectories
J	Chain thick	J1 Indication of lines or surfaces to which a special requirement applies. J2 Parting line of section.

K	——-- ——-- ——-- ——	Chain thin double dashed	K1 Outlines of adjacent parts K2 Alternative and extreme positions of movable part K3 Centroidal lines K4 Initial outlines prior to forming (see figure) K5 Parts situated in front of the cutting plane

In Fig. 1.1(a) and (b), typical applications of different types of lines are shown. This figure is to be read in conjunction with Table 1.1.

3. In technical drawings, two thicknesses of lines are generally used—thick and thin. The ratio of thick to thin line is around 2 : 1. According to the size and the type of the drawing, the thickness of the lines should be selected from the following range :

   0.18, 0.25, 0.35, 0.5, 0.7, 1 , 1.4 and 2 mm.

   The use of line of thickness 0.18 is not encouraged as it poses problem in certain methods of reproduction.

   It will be noticed that the thickness steps are more or less in the ratio of $1 : \sqrt{2}$, *i.e.,* 1 : 1.414. The significance of this lies in the fact that after micro-filming the original tracing, when later on enlarged prints are taken on sheets of standard formats, the thicknesses of the lines of the ensuing standard sheets (AO, A1, *etc.*) will be of the relevant standard thicknesses, because of the same relation (*viz,* $1 : \sqrt{2}$) existing between the different standard formats (see Section 1.6). From Table 1.1, we can see that the types A, E and J belong to the thick category and B, C, D, F, G, and K belong to the thin category. The designation of a particular line-group is determined by its thick line. Thus, if in a line-group, the thick line is of 2 mm thickness and the thin one is of 1 mm thickness, - (the thick line being 2 times as thick as the thin one, as discussed earlier)—the line-group will belong to the 2 mm category. Usually, in industrial drawings, the groups 0.5 and 0.7 are preferred.
4. When draughting views of one part using the same scale, the same thicknesses of lines should be maintained.
5. As regards the minimum space between parallel lines (including hatching lines), this should never be less than twice the thickness of the thickest line. A minimum space of 0.7 mm should be maintained.
6. Often in technical drawings, two or more lines of different types coincide or superimposed on one another. In such cases, the following order of priority should be observed (see Fig.1.2):
   (i) Visible outlines and edges (continuous thick line, type A),
   (ii) Hidden outlines and edges (dashed or dotted line, type E or F),
   (iii) Cutting planes (chain thin line type J),
   (iv) Centre lines and lines of symmetry (chain thin line, type G)
   (v) Centroidal lines (chain thin double - dashed line, type K),
   (vi) Projection lines (continuous thin line, type B).

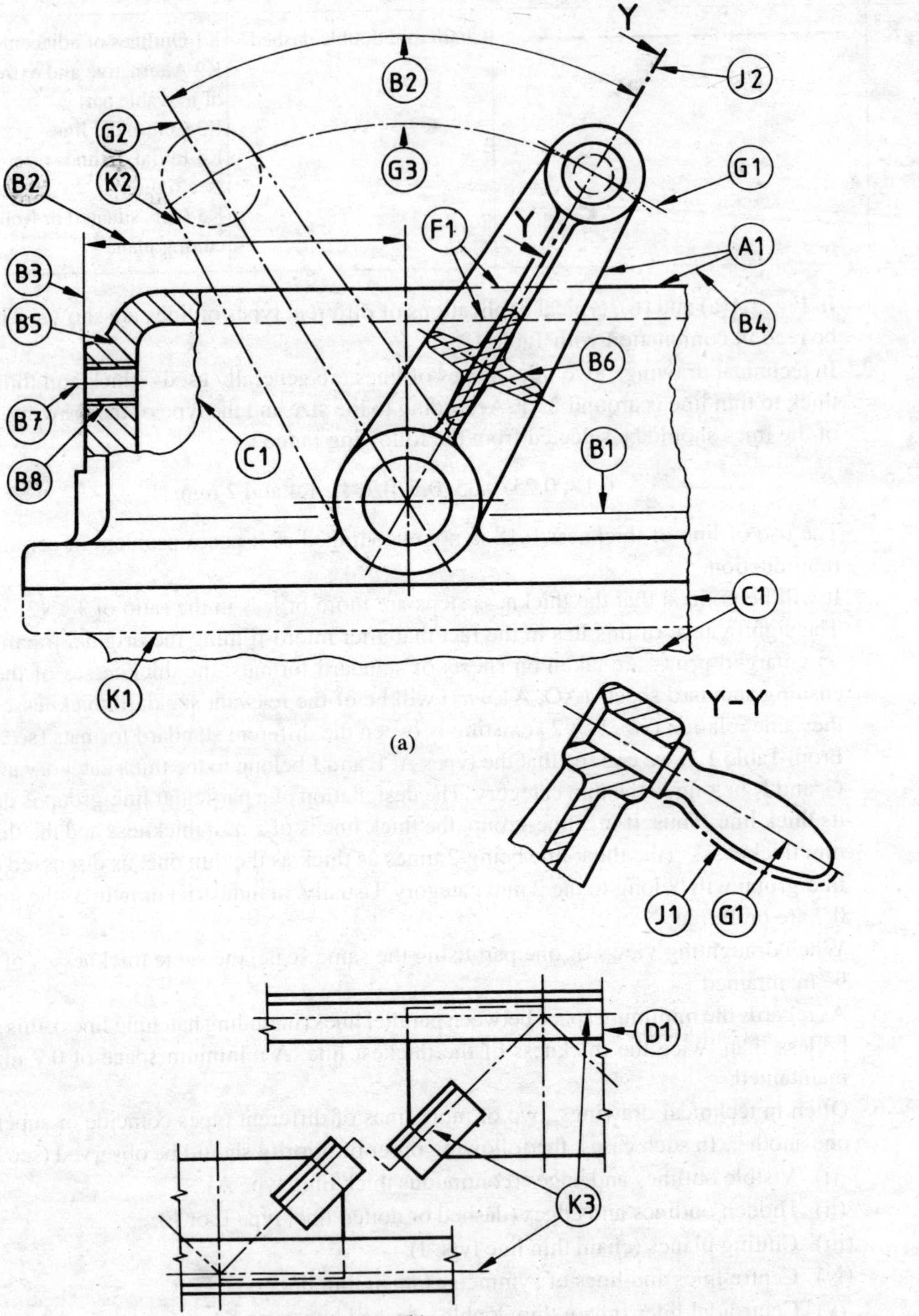

Fig. 1.1. Typical applications of different types of lines

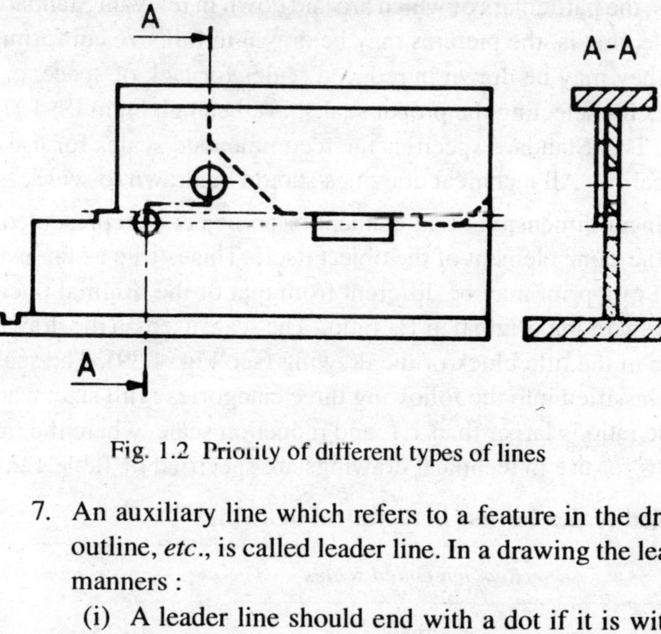

Fig. 1.2 Priority of different types of lines

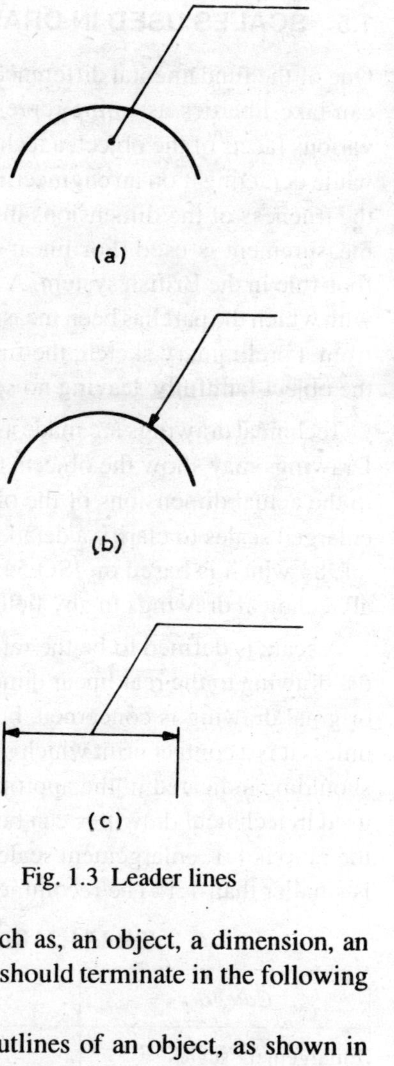

Fig. 1.3 Leader lines

7. An auxiliary line which refers to a feature in the drawing, such as, an object, a dimension, an outline, *etc*., is called leader line. In a drawing the leader lines should terminate in the following manners :
   (i) A leader line should end with a dot if it is within the outlines of an object, as shown in Fig.1.3(a),
   (ii) It should end with an arrow head, if it just touches the outline of an object, as shown in Fig. 1.3(b),
   (iii) It should end on a dimension line, without any dot or arrowhead, as shown in Fig. 1.3(c).
8. Hidden lines showing some interior or hidden surfaces, - (dashed thick or thin), - should be included only where these lines definitely help in the interpretation of the drawing.
9. Centre lines should normally project out for a short distance beyond the outline to which they refer. They may, however, be extended when it is necessary to aid dimensioning or to correlate views or sections.
10. If a drawing is to be reproduced to a smaller size by photographic process, the thickness of lines in the original should be enhanced accordingly so that legibility and clarity are ensured in the reproduction.

1.5 SCALES USED IN DRAWING

One of the fundamental differences between artistic drawing and technical drawing is that while an artist can take liberties as to the representation of the different proportions of the object while drawing the various facets of the object, a technical drawing maker must adhere to the actual dimensions of the object while depicting it on an engineering drawing. This aspect has been emphasized in Section 1.1. Obviously, the trueness of the dimensions in a drawing can be achieved only if some sort of standard yardstick for measurement is used. For linear measurement, such a yardstick is a meter scale in metric system or a foot-rule in the British system. A technical drawing is dimensioned as per the reading of these yardsticks with which the part has been measured in case of spare part drawing. In case of a design drawing developed from a preliminary sketch, the final drawing must show the dimensions in true proportions to represent the object faithfully, leaving no scope for imagination or ambiguity to the reader of the drawing.

Technical drawings are made as per scales, the particulars of which are laid down in relevant Standards. Drawings may show the objects true to scale, that is, the pictures may be drawn in full size conforming to the actual dimensions of the objects, or they may be drawn in reduced scales for lack of space, or in enlarged scales to clarify a detail. Guidelines for selecting the proper scale have been given in IS: 10713 - 1983 which is based on ISO 5455 - 1979. This Standard specifies the recommended scales for use on all technical drawings in any field of engineering. All technical drawings should be drawn to scale.

A scale is defined to be the ratio of the linear dimension of an element or an object as represented in the drawing to the real linear dimension of the same element of the object itself. This is true as far as the original drawing is concerned, but the scale of a print may be different from that of the original tracing unless it is a contact print which shows the print of the original in 1:1 ratio. The scale used on the drawing should be indicated in the appropriate place in the title block of the drawing (see Fig. 4.39). The scales used in technical drawings can be broadly classified into the following three categories : full size, where the ratio is 1:1, enlargement scale, where the ratio is larger than 1:1, and reduction scale, where the ratio is smaller than 1:1. The recommended scales for use in technical drawings are specified in Table 1.2.

Table 1.2 Recommended scales for use in technical drawings

Category	*Recommended scales*		
Enlargement scales	50:1 5:1	20:1 2:1	10:1
Full size		1:1	
Reduction scales	1:1 1:20 1:200 1:2000	1:5 1:50 1:500 1:5000	1:10 1:100 1:1000 1:10000

Besides the above scales, a ratio of 1 : 2.5 is also sometimes used, particularly in drawings conforming to the older DIN Standards. While entering in the drawing, the relevant scale can be written as for example,

SCALE 1 : 5

If, however, there is no scope for misunderstanding, the word "SCALE" may be omitted and only the ratio is written. If the drawing contains the detail view or section of a portion of the part separately, then the size of the scale - (usually enlargement scale) - should be written below the detail or somewhere adjacent to the item reference number. If situation warrants it, then scales besides those shown in Table 1.2 can also be used, provided that the required scale should be derived from a recommended scale by multiplying by whole number power of 10. However, in exceptional cases, intermediate scales may also be chosen.

As regards the proper selection of scales for drawing making, it entirely depends on the discretion of the drawing maker, the guiding factors being the complexity of the object to be depicted and the purpose of representation. In any case, the chosen scale should be large enough for easy, and clear interpretation of the part. On the other hand, it should not be too large to avoid unnecessary wastage of space and material of the drawing paper. Since the size of the drawing depends upon the size of the object and the scale chosen, the size of the scale should be judiciously decided upon.

IS: 10713 recommends that when a small object is drawn to a large scale, a full size view should also be added in the drawing for information. It will help the reader of the drawing to have a clear conception as to the actual size of the part. This full size view need not be detailed, and may be simplified by showing only the outlines of the part.

Due to lack of space or some other reason, drawings are sometimes made not to any scale. In such cases, the experiment: "N.T.S.", meaning 'not to scale', should be written in the box provided for writing scale in the title block. Such practice, however, should be avoided, if possible, because the reader of drawing may get confused as to the real shape and form of the object if it is drawn not to scale. It is also liable to misinterpretation. However, if it is unavoidable, dimensions which are not drawn as per scale in an otherwise true to scale drawing should be underlined. This is the conventional practice to draw the attention of the reader to such dimensions.

1.6 DRAWING SHEETS AND THEIR SIZES

Engineering drawings are executed on drawing sheets which may be transparent for taking prints out of them by means of various methods of reproduction which are in practice or the drawings may also be made on opaque papers for display or for some other purposes.

In IS: 10711-1983, the size of drawing sheets and their destinations have been standardised. This IS is in conformity with ISO 5457-1980 and it supersedes certain relevant subject matters covered in IS: 696-1972. The present IS is made with a view to meeting the needs of the common methods of reproduction for prints, handling of drawings, and also to cope with the modern methods, namely, Micro-filming, automatic trimming, *etc*. The Standard is applicable for all technical drawings in any field of engineering, and it specifies sizes of blanks and preprinted drawing sheets.

Sizes of sheets : In Tables 1.3, 1.4, and 1.5, the preferred sizes of the trimmed sheets have been given, the preference or choice being in that order. Drawing sheets may be used with the longer sides positioned either horizontally or vertically, depending on the orientation of the objects for which the views, sections, *etc*., are drawn, so that the purpose of conveying the impressions of the objects depicted to the reader of the drawing is suitably served.

The sizes given in Table 1.3 are the preferred sizes of the trimmed sheets. They are selected from the main ISO - A series.

Table 1.3 Sizes of drawing sheets, series ISO-A

Designation	*Dimensions (mm)*
A0	841 × 1189
A1	594 × 841
A2	420 × 594
A4	210 × 297

If a sheet of length greater than the normal length is required for any special purposes, then it can be chosen from sizes given in Table 1.4. The sizes in Table 1.4 are arrived at by extending the shorter sides of the formats given in ISO-A series (Table 1.3) to lengths which are integral multiples of the shorter side of the basic format chosen.

Table 1.4 Sizes of drawing sheets, special elongated sizes

Designation	*Dimensions (mm)*
A3 × 3	420 × 891
A3 × 4	420 × 1189
A4 × 3	297 × 630
A4 × 4	297 × 841
A4 × 5	297 × 1051

When still larger or longer sheets are required, then suitable sizes should be selected from Table 1.5.

Table 1.5 Sizes of drawing sheets, exceptional elongated sizes

Designation	*Dimensions (mm)*
A0 × 2	1189 × 1682
A0 × 3	1189 × 2523
A1 × 3	841 × 1783
A1 × 4	841 × 2378
A2 × 3	594 × 1261
A2 × 4	594 × 1682
A2 × 5	594 × 2102
A3 × 5	420 × 1486
A3 × 6	420 × 1783
A3 × 7	420 × 2080
A4 × 6	297 × 1261
A4 × 7	297 × 1471
A4 × 8	297 × 1682
A4 × 9	297 × 1892

Besides the above sizes, the size A5 of which the dimensions are 148 × 210 is also still in vogue, particularly as compartments of bigger sheets in which drawings for small or simple objections are contained.

It can be seen that the different formats of preferred sizes (Table 1.3) bear geometrical similarity to one another. The sides of each size are virtually in the ratio of 1 : $\sqrt{2}$ or 1 : 1.414, *e.g.*, for A0 size, 841 × 1.414 ≈ 1189, for A4 size, 210 × 1.414 ≈ 297, *etc*. This has, among others, important significance on the thickness of the lines used in the drawings when prints of different standard sizes, - (larger or smaller) - are taken out from an original of a particular standard size. This has been discussed in Section 1.4. Areas of successive sheets are in the ratio of 2 : 1. Thus, A0 = 2 × A1 and A1 = 2 × A2, *etc*.

The frame limiting the drawing space (which is meant for containing the drawing) is surrounded by borders on all the four sides. These borders are enclosed by the edges of the trimmed sheet. The recommended minimum widths of these borders are : 20 mm for A0 and A1 sizes, and 10 mm for A2, A3 and A4 sizes. Normally, these values are sufficient for gripping purposes during printing. For some types of reproduction machines, however, the minimum values may be reduced to 10 mm and 7 mm, respectively.

The title block of the drawing should be within the drawing space. To facilitate correct positioning of the drawing when reproduced or micro-filmed, drawing sheet should be provided with four centring marks which are placed at the ends of the two axes of symmetry of the trimmed sheet.

For easy location of a particular detail or aspect on a drawing sheet, a sort of co-ordinate system, called grid reference system, should be provided for all sizes of drawing sheets. The number of rectangles comprising the grid inside the border should be even, and the length of each such rectangle should lie between 25 mm and 75 mm. For reference, the rectangles of the grid should contain capital letters along two opposite sides and numerals along the other two opposite sides. Letters and numerals should start at the sheet corner which is diagonally opposite to the title block corner. The letters and numerals, which should be placed close to the frame, should be at a minimum distance of 5 mm from the edges of the trimmed sheet and they should be within the rectangles in symmetrical fashion.

All the features of the drawing sheets discussed in the foregoing paragraphs have been illustrated in Fig. 1.4.

1.7 TITLE BLOCKS OF DRAWINGS

Every engineering drawing normally contains a title block. It is an important feature in a drawing and is generally placed at the bottom right-hand corner of the drawing sheet. Prints taken from the original tracing are folded in such a manner that the title blocks can be readily seen with the prints in the folded condition.

In IS: 10711-1983 which is as per ISO 5457-1980, instructions for the orientation as regards the placement of the title block in a drawing sheet have been laid down. Title block should be within the drawing space and positioned in such a manner that the portion of the title block containing the identification of the drawing is situated in the bottom right-hand corner of the drawing space. The identification portion of the title block includes the title of the drawing, registration number, *etc*., and it shall have a maximum length of 170 mm. Though the title block is supposed to be placed at the bottom right-hand corner of the drawing sheet, it may so happen that for a special reason, *e.g.*, economisation of

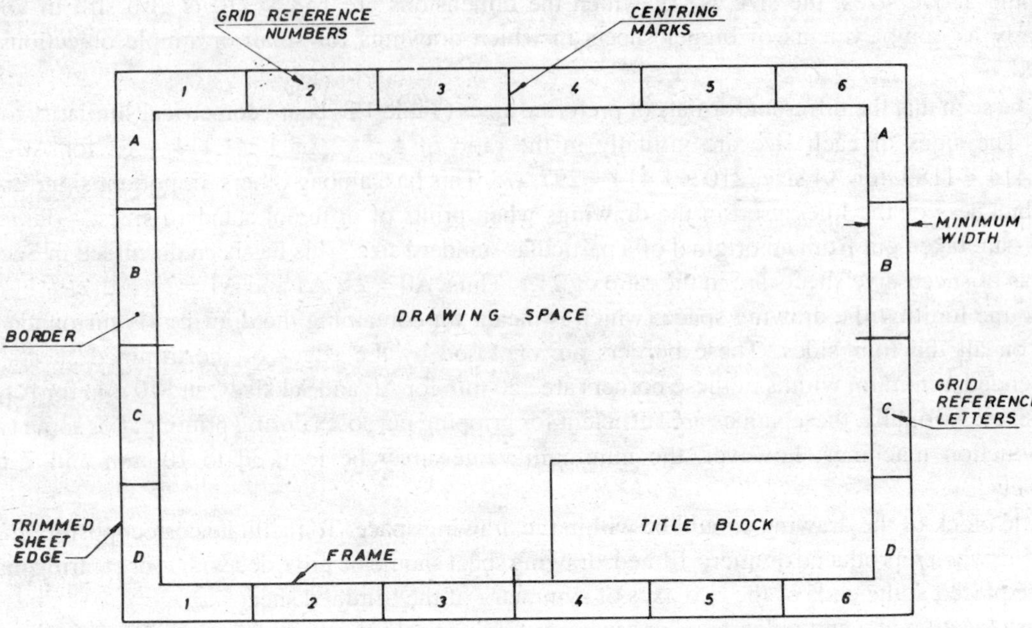

Fig. 1.4 A typical drawing sheet

space, *etc.*, the drawing has to be executed on a preprinted sheet after fixing it on the board in such a manner that the vertical axis of the sheet is thereby rotated by 90°. This is permitted, and in such cases the title block obviously occupies a position at the top right-hand corner. The title block now can be read when viewed from the right.

The title block should contain certain basic informations as regards the drawing to which it belongs. These are :

1. Name or title of the drawing: This should include the department, unit and description of the equipment or parts thereof for which the drawings have been made on the sheet. Some firms have systematic nomenclatures for their departments and units, and these are strictly followed while entering them on the drawing. This alleviates confusion and promotes easy location of the drawing. Also, such practice helps in methodical registration of the drawings in the drawing office and in maintaining systematized catalogues.

2. Drawing number: Every drawing must be assigned with an identification number. The numbering systems vary in different companies. Some are simple while some firms follow quite complex numbering system. In simple numbering system, a drawing number may consist of just a serial number or a serial number with a prefix or suffix to indicate the sheet size, *viz.*, B/50946. In a complex numbering scheme, a drawing number may include codified symbols denoting department, unit, modal number of machine or any other relevant information. With the advent and increasing use of computer, some companies have evolved fully computerised systems of numbering which may run into seven or eleven digits or any other forms, depending on the system used. When a set comprising several drawings is required for an equipment or a part, then each of these drawings are to be correlated. Thus, for example, if the set consists of eight separate drawing sheets, then the drawings should carry the information:

SHEET 1 OF 8, SHEET 2 OF 8, *etc.*,

and these entries should be made below the main drawing number in the box provided in the title block.

3. Scale: The scale used in the drawing should be indicated in the box provided in the title block.

4. Symbol denoting the method of projection: This is now optional since IS recommends the use of first angle projection only (see Section 1.8).

5. Name of the firm: The name and symbol (if any) of the company should be clearly displayed.

6. Signatures: The signatures with dates of all persons concerned with the drawing, *viz.*, the designer, the draughtsman, the checker, the approving authority, should be entered.

Besides the main features enumerated above, the following information should also be provided:

7. Key to machining and other symbols: Different firms use different machining symbols which belong to various series or degree of fineness. A legend pertaining to the particular surface finishes used by the firm and specifying their numerical values should be indicated.

8. Tolerance: The specification number and grade of tolerance should be given in the block.

9. Supersession: Any supersession of previous drawing should be mentioned.

10. Job or Order number: This should be entered in the title block.

11. Reference drawing numbers: If the drawing relates to old drawings, this should be given.

12. Revision: Each drawing carries a revision panel by the side of the main title block. It is imperative that any revision of any kind must be properly recorded in this panel. Brief description of revision, revision number, zone (*e.g.*, B - 3, C - 2, *etc.*), date and signature of the person approving the change should be contained in the revision panel. In keeping with the number of revision, the main drawing number should be suitably suffixed to indicate the revision number, namely, 6051 - R3, C/43528/R1, B - 25764/R2, etc. Also, the revision carried out should be indicated on the body of the drawing by means of proper symbols, such as,

255 △1
~~260~~

which means that the dimension has been changed from 260 mm to 255 mm, and this revision has been indicated by the symbol: △1 denoting first revision (see Fig. 4.39). The same symbol is carried over to the revision pannel (see Fig. 4.39). Since others may be affected by the revision, it is advisable to send a note or - better still - to send prints of revised tracings to agencies who are directly involved in the manufacture or repair of the part, thus, altered. The revised dimension should not be erased but simply struck off, as shown. In compartmentalised drawings, each compartment should preferably contain a miniaturised title block. If part number 2 is depicted in a compartment, this small block can be numbered as B/50946/2 if the main reference number is B/50946 (see Fig. 4.39).

The bill of materials or the parts list is generally placed above the main title block. The reader is advised to refer Sections 4.6 and 4.7 in which examples of the topics discussed in the foregoing paragraphs of this Section are illustrated with appropriate figures. In IS: 696 - 1972, the overall size of the title block recommended was 185 mm × 65 mm. This clause has since been superseded in the latest standard, *i.e.*, IS: 10711 - 1983 which deals with a part of the subjects covered in IS: 696.

1.8 TYPES OF PROJECTION

Engineering drawing is a graphic language by means of which the conception about the shape of an object is imparted to the technical person. This objective can be achieved by taking recourse to various methods of representation which are in vogue. In olden days, this graphic language mainly consisted of pictorial representation of the object, and it largely depended on the maker of the picture. In other words, it was mainly individualistic as no systematised pictorial method of drawing making was developed in those days. In modern times, the technical graphic language has evolved into a highly systematised and methodical way of representing a part, so that there is no scope for any ambiguity as regards its correct shape and form. The various types of representation are briefly discussed here, and these will be elaborately dealt with in Chapter 3.

To describe the shape of a part in graphic language, help is taken of a systematised method of what is known as the "projection" drawing. To make such a drawing, four aspect are involved, namely,

(i) the object of which the drawing is to be made;

(ii) the observer who sees the object from a place called the "station point":

(iii) the plane on which the (imaginary) image of the object is projected, and

(iv) the lines of sight, called the "projectors" or "projection line", which connect the station point with the object.

Thus, the process of projection drawing essentially consists of effecting a picture of the object to be formed on a plane when the object is viewed by the observer in a particular direction. By varying the station points and the rays of sight (*i.e.*, the projectors), different images of the object can be projected on to different planes. These images constitute the shape of the object. For the size of the object, the images are to be properly dimensioned. A complete technical drawing which is suitable for the construction of the part ensues only when the images describe the shape and contain the necessary dimensions, relevant notes, and other information.

Depending upon the direction and manner in which the rays of sight or projectors are oriented, the methods of projection are broadly classified into the following two categories :

1. Parallel projection, and
2. Perspective projection.

When the observer or the station point is imagined to be situated at an infinite distance from the object, then projectors falling on the plane of projection will be parallel as shown in Fig. 1.5(a). This type of projection is called parallel projection. When the station point is imagined to be situated at a finite distance, a perspective projection results. An image formed on the retina of the eye, that is, the way the object is presented as it would appear to the eye, and a photograph taken by a camera are examples of perspective projection. See Fig.1.5(b).

Parallel projection can be sub-divided into the following two classes:

1. Orthographic projection, and
2. Oblique projection.

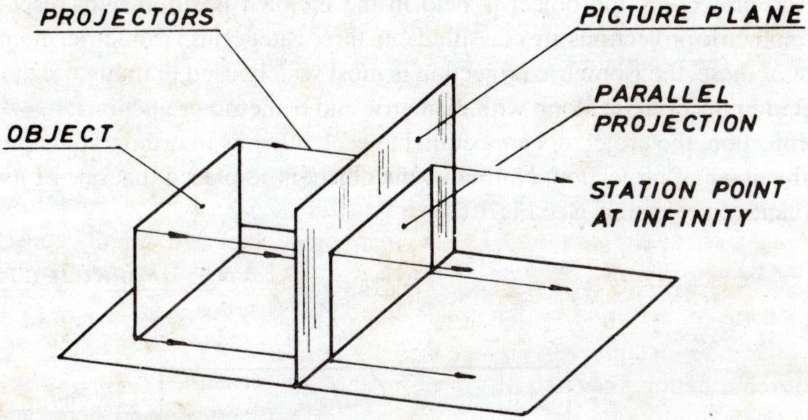

(a) PARALLEL PROJECTION

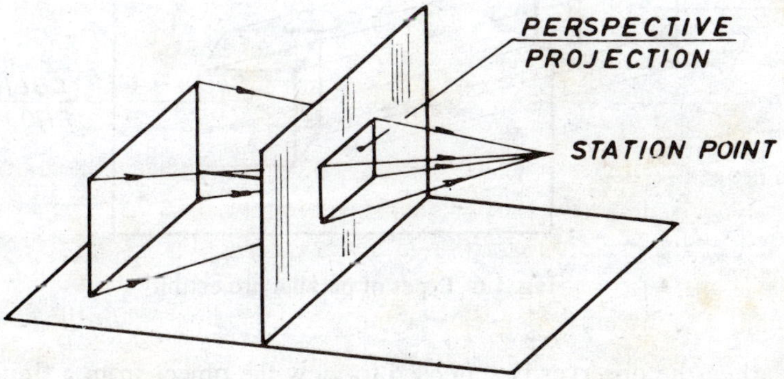

(b) PERSPECTIVE PROJECTION

Fig. 1.5 Types of projection
(a) Parallel projection (b) Perspective projection

In an orthographic projection, the projectors are parallel to each other and are perpendicular to the plane of projection. (The word "orthographic" means "to write or to draw at right angles"). An oblique projection results when the projectors are parallel to each other, but they are so oriented that they are oblique to the plane of projection.

Orthographic projections are of two types, namely, multi-view projection and axonometric projection. It is the multi-view projection which is most commonly used in engineering practice and as such, this method has become synonymous with orthographic projection in usual technical parlance. Depending on the quadrant in which the object is situated, multi-view projection can be termed as first, second, third or fourth angle projection. This is elaborated in Section 3.1.

In axonometric projection, the object is held in the inclined position with respect to the plane of projection. Axonometric projections are classified into three categories : isometric, diametric and trimetric projections. Out of these, the isometric projection is most widely used in industrial drawings and will be discussed in detail in Section 3.2 along with diametric and trimetric projections.

In oblique projection, the projectors are parallel to each other as in orthographic projections, but they are oblique to the plane of projection. Normally, the object is so placed that one of its principal faces is parallel to the plane of projection. See Fig.1.6.

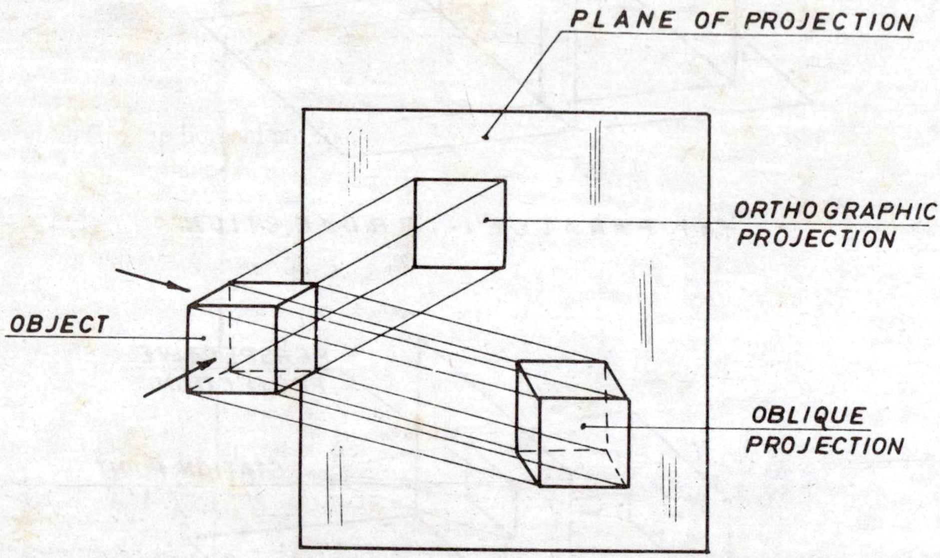

Fig. 1.6 Types of parallel projection

As indicated earlier, the observer is supposed to view the object from a finite distance in case of a perspective projection. Rays from all points on the object converge on to the observer's eye or the station point, so that a "cone of rays" result.

Perspective, oblique and the various axonometric projections (isometric, etc.) are also known as "pictorial drawing". Although the multi-view projection method makes it possible to accurately represent a part by means of a series of views, sections, *etc.*, and is the method most commonly used in industry to make manufacturing drawings, one of its drawbacks lies in the fact that to execute and to interpret such a drawing, one has to have specialised knowledge or training in the subject. Then only one can visualise the object from the views given in the drawing.

However, it is sometimes necessary to prepare drawings which can be easily understood by persons who need not have any technical training. In such cases, pictorial drawings are resorted to. In catalogues depicting buildings, perspective drawings are often provided because in perspective projection the object is represented as it appears to the eye. In piping diagrams and layouts, drawings are often given in isometric projection. Many mechanical and structural details which are not clear enough in conventional multi-view projections become readily understandable when drawn in oblique or any other form of pictorial representation. Technical illustrations in seminars, Patent Office drawings, sales literature diagrams, *etc.*, are normally made in pictorial methods, the main reason being that these methods enable the drawing maker to present a three-dimensional visulisation of an object on a two-dimensional paper, thereby making the shape of the object understandable even to a layman. A trained designer, too, often takes the help of pictorial methods as he goes through successive stages of design procedures till he develops an optimum and satisfactory design of the part.

PREFERENCE OF ADOPTION OF THE FIRST ANGLE PROJECTION SYSTEM

It has been stated earlier that the multi-view projection method is the most commonly adopted system in industrial drawings. IS: 10714 - 1983 lays down the code of practice for general principles of presentation in technical drawings. In the original version of the Standard, both the first angle and the third angle projection methods have been explained, and no clear-out distinction has been made as to the preference for adoption of any of the two systems involved, the selection of the method of projection being left to the discretion of the relevant organisation. Each of the two methods could equally have been used without prejudice to the principle involved.

In a subsequent Amendment to the above IS (which conforms to ISO 128-1982 T on Technical Drawings), it has been decided that as a basic requirement, the use of the first angle projection method only is to be followed. This need for standardisation of a single method of projection in engineering drawing for the whole country and the benefit which ensues by the adoption of the first angle projection stem from the following considerations :

1. If the industries use both the systems, the inter-communication between industries becomes more difficult, leading to confusion. Adoption of one system at the national level will eliminate this possibility and harmonize communication.
2. It has been found that the majority of the industries are in favour of first angle projection within the country.
3. These industries have mostly long association and collaboration with industries in other countries using the first angle as their code of practice. Hence, better communication and transfer of technology have resulted, alleviating delay and extra cost.
4. The Standards issued by the International Organisation for Standardisation (ISO) are all in first angle projection. By following the first angle projection, adoption of ISO Standards at the national level will be simpler. Besides, this will ensure international communication if only one set Standard is envisaged and the adoption will also promote the economical benefit thereof.
5. It will be desirable to follow the first angle projection system in view of the fact that trade interests with the countries subject to GATT agreement will be augmented.
6. Technical communication and trade interests will also be easier and enhanced when dealing with the European countries since the majority of these countries follow the first angle projection system in technical drawings.

In view of the foregoing, all the drawings and illustrations in this book have been executed in the first angle projection, unless otherwise stated.

CHAPTER 2

GEOMETRICAL CONSTRUCTIONS

2.1 COMMON GEOMETRICAL CONSTRUCTIONS

While draughting an engineering drawing, constructions based on plane geometry are often resorted to. Hence, the drawing maker should be proficient in his knowledge about the principles involved in these types of pictorial representations. This knowledge is imperative for solving problems which arise while making a technical drawing. Ordinarily, common drawing instruments such as the scale, compass, set squares, dividers and protractors are sufficient to make these constructions. Special instruments are required for draughting uncommon figure. Some of the common geometrical constructions which are often encountered while making engineering drawings will be discussed in this section.

Bisection of a line : Line *AB* is to be bisected. Draw arcs 1 and 2 with *r* as radius ($r > 1/2\ AB$) with *A* and *B* as centres respectively. Join the points of intersection by line 3 which is the perpendicular bisector of *AB* (Fig. 2.1).

Perpendicular from a point : With point *P* as centre, draw arc 1. Draw arcs 2 and 3 with *r* as radius ($r > 1/2\ AB$). Line 4 joining *P* with the point of intersection of the arcs is the required perpendicular. (Fig. 2.2).

Erection of a perpendicular from a point on a line : To erect a perpendicular from the point *P* on the line *LP*, draw arc 1 with any radius *r* and *P* as centre. Draw arc 2 with the same radius and the centre at *A* which is the point of intersection of arc 1 and the line *LP*. Arc 2 cuts arc 1 at *B*. Join *A* with *B* and extend it upto *C*, making $BC = AB$. Line 4, joining *C* and *P*, is the required perpendicular. (Fig. 2.3).

Bisector of an angle : To bisect angle α, draw arc 1 with *S* as centre, cutting the legs constituting the angle at *A* and *B*. Draw arc 2 and 3 with radius *r* from centre *A* and *B* respectively ($r > 1/2\ AB$). Line 4 joining *S* with the point of intersection of arcs 2 and 3 is the bisector of the angle (Fig. 2.4).

Tangent from a point on a circle : To draw a tangent at *P* on the circumference of the circle of radius *r*, draw arc 1 with radius *r* and centre *P*, cutting the circle at *A*. Join *M* with *A* and extend it upto *E*, making $MA = AE$. Line *EP* is the required tangent (Fig. 2.5).

Tangent to a circle from a point : To draw a tangent to a circle from the point *P* outside the circle, join *P* with the centre *M* and bisect *PM*. With the middle point *A* as centre, draw a semi-circle with radius *AM*, cutting the circle at *T*. Line *PT* is the required tangent (Fig. 2.6).

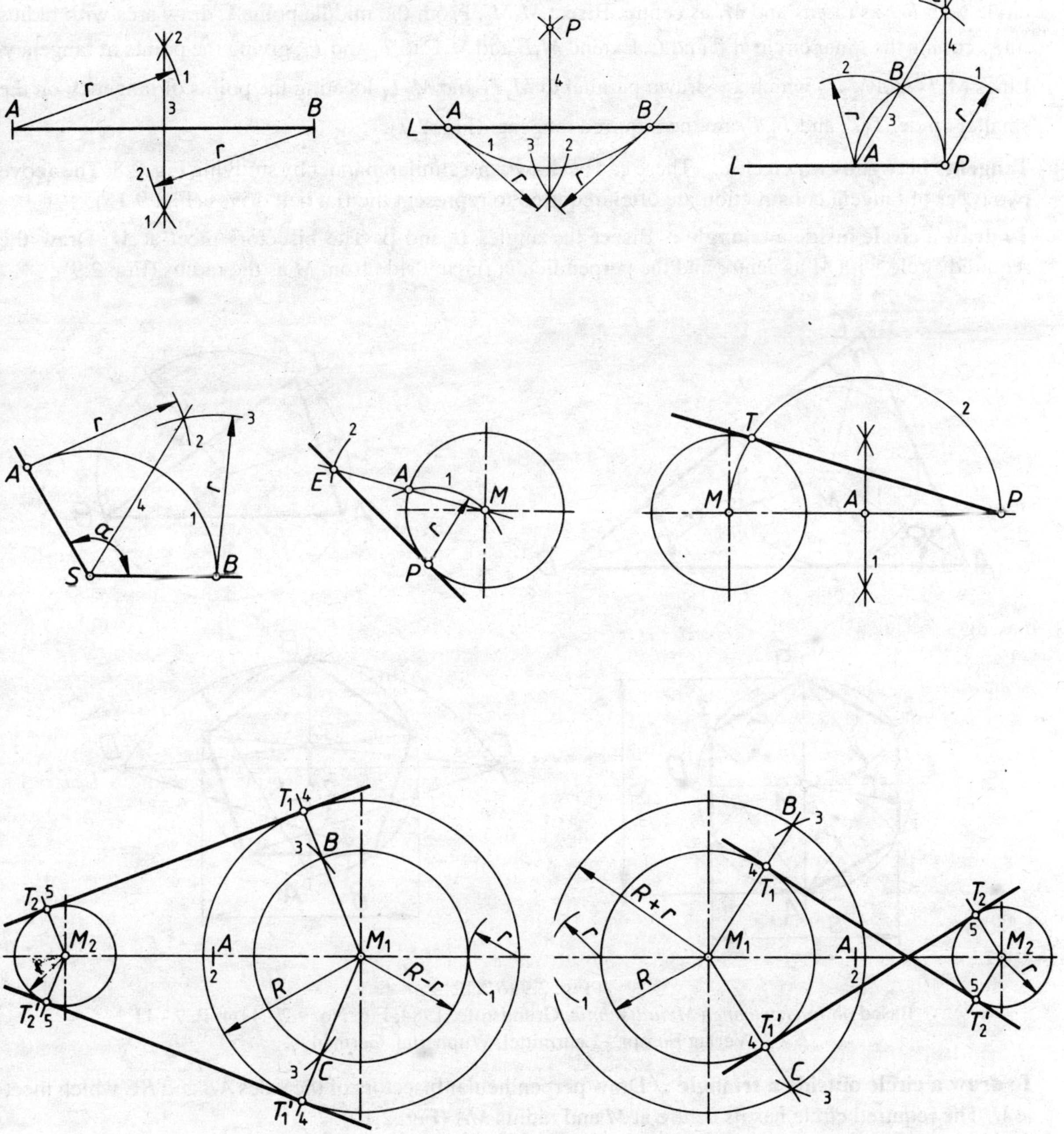

Fig. 2.1 to 2.8

Tangents on two circles : The circles with radii R and r with centres at M_1 and M_2 are given, Draw a circle with R-r as radius and M_1 as centre. Bisect M_1M_2. From the middle point A, draw arcs with radius AM_1, cutting the inner circle at B and C. Extend M_1B and M_1C to T_1 and T_1', giving the points of tangency. Lines M_2T_2 and M_2T_2' which are drawn parallel to M_1T_1 and M_1T_1' locating the points of tangency on the smaller circle. T_1T_2 and $T_1\ T_2'$ are the required tangents (Fig. 2.7).

Tangents between two circles : These can be drawn in a similar manner by studying Fig. 2.8. The above two types of tangent construction are often required to represent the flat belt drives (Fig. 9.15).

To draw a circle inside a triangle : Bisect the angles α and β. The bisectors meet at M. Draw the required circle with M as centre and the perpendicular on any side from M as the radius (Fig. 2.9).

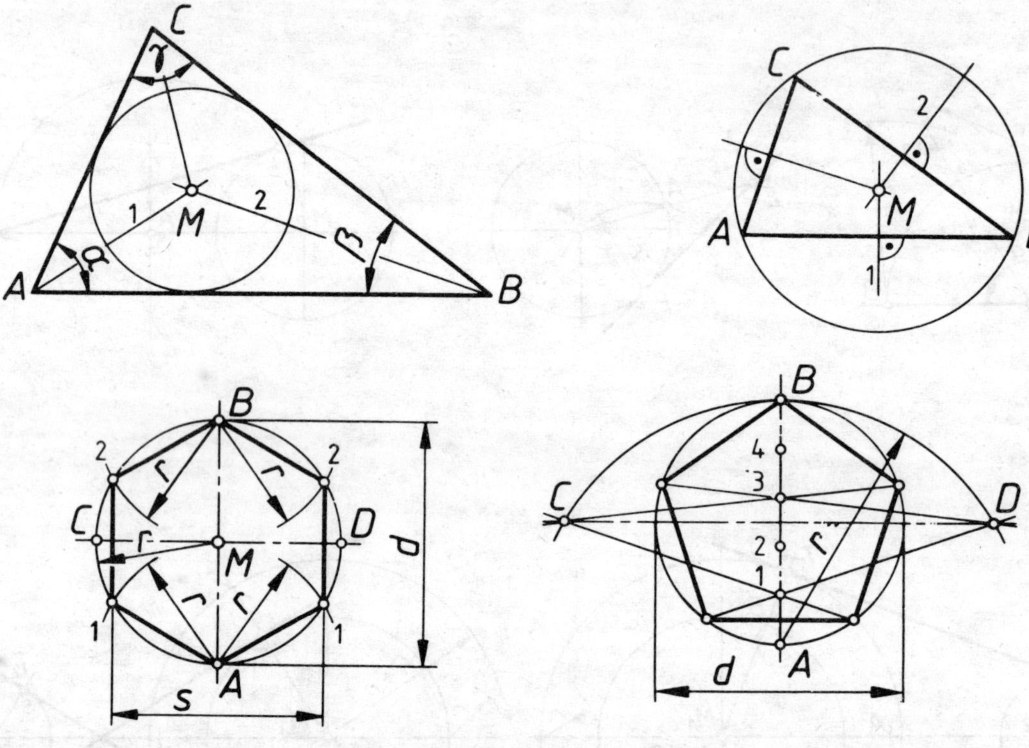

Fig. 2.9 to 2.12
Based on *Fachzeichnen Metalltechnik,* Grundstufe, 1984, Fig. No. 9/2 - 11/4, P. 9 - 11, Verlag Europa - Lehrmittel, Wuppertal, Germany.

To draw a circle outside a triangle : Draw perpendicular bisectors of the sides AB and BC which meet at M. The required circle has its centre at M and radius MA (Fig. 2.10).

To draw a hexagon inside a circle : Draw arcs 1 and 2 with centres A and B and radius $r = d/2$, where d is the diameter of the circle. The points of intersection of these arcs with the circumference of the circle are joined as shown in Fig. 2.11 to complete the hexagon. The nuts and bolt heads are usually hexagons. The following relations are useful in this context (Fig. 2.11):

Corner dimension, $e = d$. Across flat dimension, $s = 0.866d$, $d = 1.155s$.

To draw a regular polygon inside a circle : To draw a pentagon, divide AB into five equal parts. Draw arc CD with A as centre and AB as radius. Join C and D with points on AB bearing odd numbers, i.e., 1 and 3, and extend the lines to meet the circle. Join the points as shown in Fig. 2.12 to obtain the pentagon. For drawing a polygon with even number of sides, the lines should go through the even points, *i.e.*, 2, 4, 6, etc. (Fig. 2.12).

2.2 CONSTRUCTION OF COMMON CURVES

In a design and drawing office, it is often required to draw curves which follow some mathematical relations between their relevant parameters. Curves which are collectively known as "conic sections" appear frequently in engineering drawings. These are dealt with separately in Section 2.3. Curves which belong to the "spiral" category, including the involute, are described in Section 2.4. In this section, some common trigonometrical curves as well as cycloidal curves are discussed. Trigonometrical curves are often required to be drawn to depict the characteristic motions of certain mechanisms. Cycloidal curves are required for drawing the teeth profile of cycloidal gears. In all the cases of drawing regular curves, the best and the most accurate results are obtained by plotting the curves by using their mathematical equations. These equations are given here as well as elsewhere along with the pictorial representations of the curves.

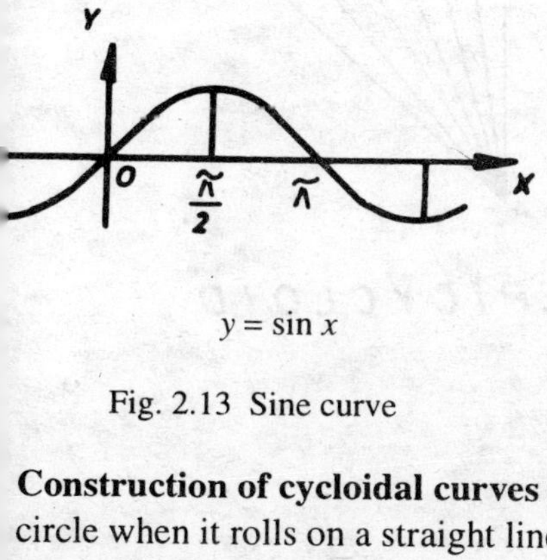

$y = \sin x$

Fig. 2.13 Sine curve

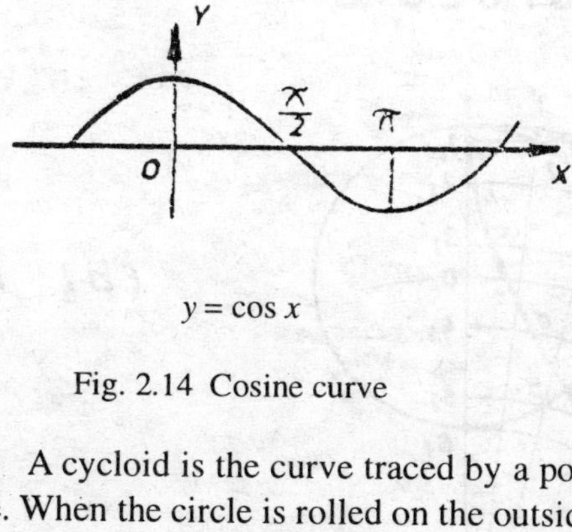

$y = \cos x$

Fig. 2.14 Cosine curve

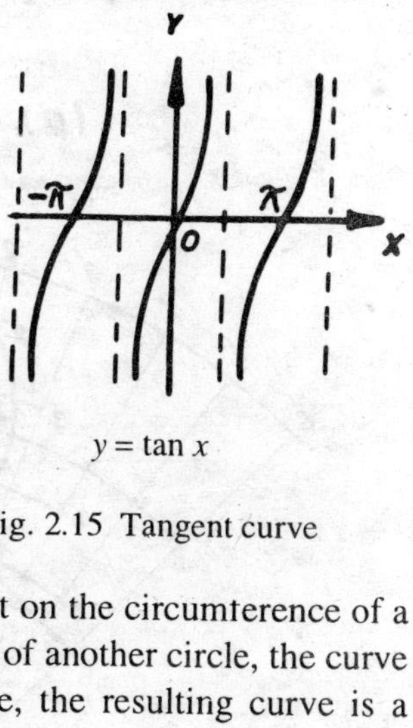

$y = \tan x$

Fig. 2.15 Tangent curve

Construction of cycloidal curves : A cycloid is the curve traced by a point on the circumference of a circle when it rolls on a straight line. When the circle is rolled on the outside of another circle, the curve generated by the point is called an epicycloid. When rolled on the inside, the resulting curve is a hypocycloid.

Cycloid : Referring to Fig 2.16(a), let the circle of radius $O - T$ roll on the line $T - 6'$. One half of the circle is then divided into any number of equal parts, say six, marked by the arcs $T - 1$, 2-3, ... 5-6. The arcs are next laid off on the line $T - 6'$, so that arc T-1 = line $T - 1'$, arc 1-2 = line $1' - 2'$, *etc.* Perpendiculars are then drawn on the points $1'$, $2'$, $3'$ *etc.* Horizontal lines are then drawn from points 1, 2, 3, *etc.*, of the semi-circle. These lines intersect the diameter at 1_1, 2_1, 0, 4_1, *etc.* They also cut the perpendiculars at

1″, 2″, *etc*. From 1″, the horizontal line $1'' - 1_1$ is cut in such a way that $1'' - a = 1 - 1_1$. Similarly, $2'' - b = 2 - 2_1$, *etc*., are laid off. Points, *a, b, c, etc*. are thus located. A smooth curve through these points is then drawn and the curve $T - a - b - c - d - e - f\,(6'')$ is the required cycloid curve.

Epicycloid : Referring to Fig. 2.16(b), the rolling circle is divided into six equal arcs as before. On the directing circle, arcs are laid off so that arc $T - 1'$ = arc $T - 1$, arc $1' - 2'$ = arc. 1-2, *etc*. Radial lines are drawn joining the centre *A* with 1′, 2′, 3′ *etc*. Arcs are next drawn through points 1, 2, 3, *etc*., which cut *T*-6 at 1_1, 2_1, *etc*., and also the radials at 1″, 2″, *etc*. Next, $1'' - a = 1 - 1_1$ is laid off, locating the point *a*. Proceeding as before, the epicycloid $T - a - b - c - d - e - f\,(6'')$ is drawn.

Hypoxcyloid : The hypocycloid, $T - a - b - c - d - e - f\,(6'')$ is drawn in a similar manner, Fig. 2.16(c).

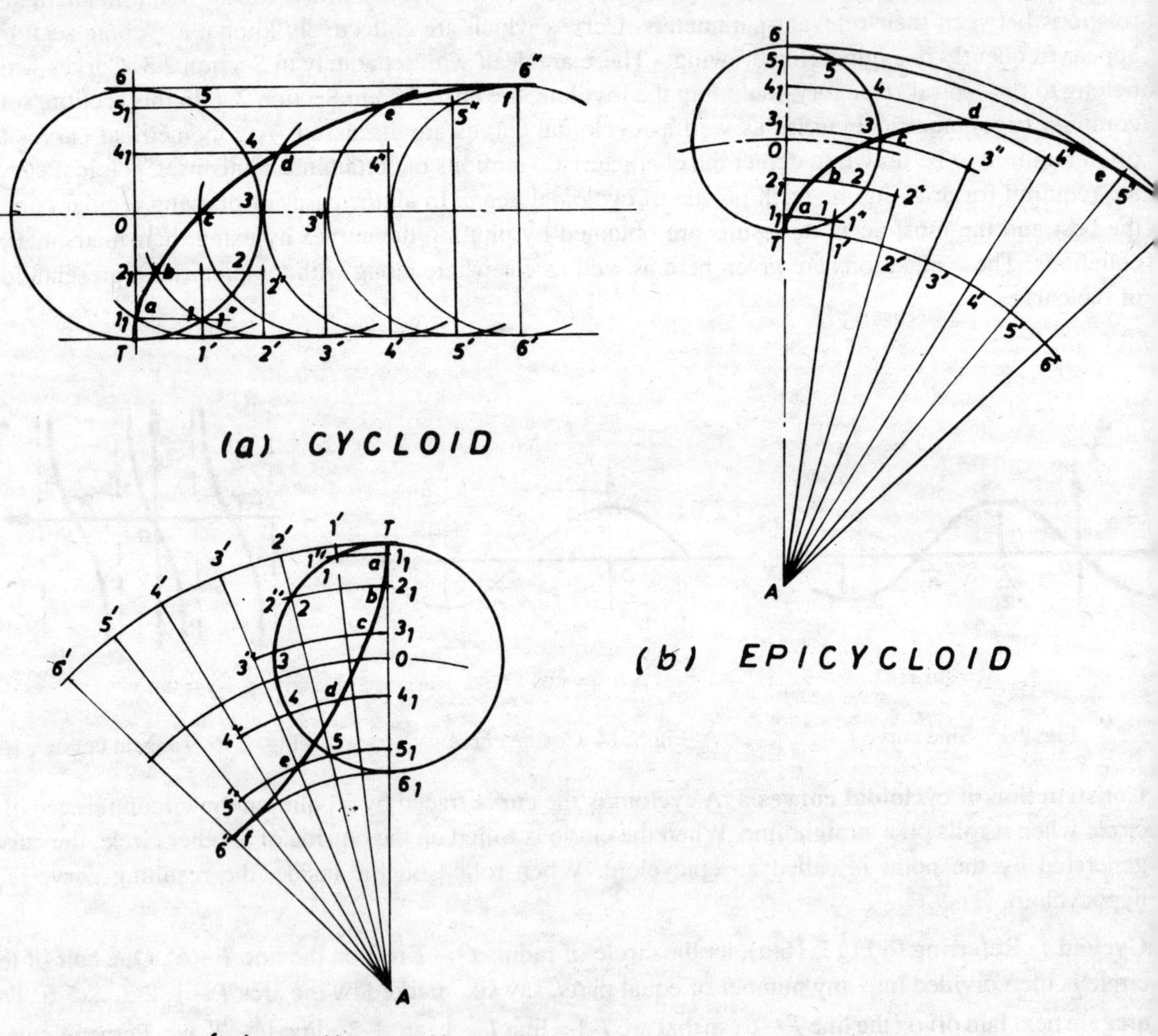

Fig. 2.16 Construction of cycloidal curves

2.3 CONIC SECTIONS AND THEIR CONSTRUCTION

When a right circular cone is intersected by a plane in various positions, seven different geometrical cases arise. These cases are shown in Fig. 2.17 in which the first column from the left shows the positions of the intersecting plane, the middle column shows the pictorial views of the resulting curves, and the column on the right shows these curves in their true forms. The cases are described below, from top to bottom :

1. When the plane just touches the apex of the cone, it produces a point.
2. When the plane just touches the slant surface of the cone, a line is produced.
3. When the plane cuts the cone, passing through the apex, the resulting figure is a triangle.
4. When the plane passes through the cone at right angles to the cone axis, it produces a circle.
5. When the plane passes in such a way that it makes an angle with the cone axis which is smaller than half the cone angle, the resulting curve is a hyperbola. In the figure, this angle is zero. That is, the plane is parallel to the axis.
6. When the plane cuts parallel to the slant surface element, a parabola ensues. Here, the cutting angle is equal to half the cone angle.
7. When the cutting plane is oblique to the axis so that the cutting angle is greater than half the cone angle but less than 90°, the resulting curve is an ellipse.

Out of the seven cases, the four types of curves, *viz.*, the circle, the hyperbola, the parabola and the ellipse are collectively known as the "conic sections". These curves are of great interest in mathematics, and two of these curves, namely, the circle and the ellipse, are to be found in majority of the industrial drawings.

As in the case of any curve, the conic sections can be most accurately drawn by using the mathematical equations relevant to the curves. Figure 2.18 and 2.19 show the parabola and the hyperbola respectively along with their mathematical expressions.

Since the ellipse is the conic section most commonly encountered in engineering drawings besides the circle, a practical method to construct the ellipse is given here. There are various methods of drawing an ellipse with varying degrees of accuracy. However, the method described here is good enough for all practical purposes. Construction of ellipse in isometric and dimetric drawings are given in Section 3.2.

Construction of ellipse : Referring to Fig. 2.20, *AB* and *CD* are the major and the minor axes of the ellipse to be constructed. Points *A* and *C* are joined. With *M* as centre and *MA* as radius, an arc is drawn which cuts the vertical central line at *E*. With *C* as centre and *CE* as radius, an arc is drawn which cuts *AC* at *F*. A perpendicular bisector of *AF* is drawn which cuts the two axes at M_2 and M_1. Arcs with radii r_1 and r_2 are drawn with M_1 and M_2 as centres respectively. *ADBC* is the required ellipse.

2.4 CONSTRUCTION OF COMMON SPIRALS

Spirals are curves which are quite prevalent in nature. Some of these spirals are also made use of in the engineering field. Common spirals are : Archimedian spiral, logarithmic spiral, hyperbolic spiral, involute spiral, helix, *etc*. There are methods for making accurate drawings of the different kinds of spirals, but the simplified and approximate methods are generally used for drawing spirals for engineering purposes. Some of these are discussed below :

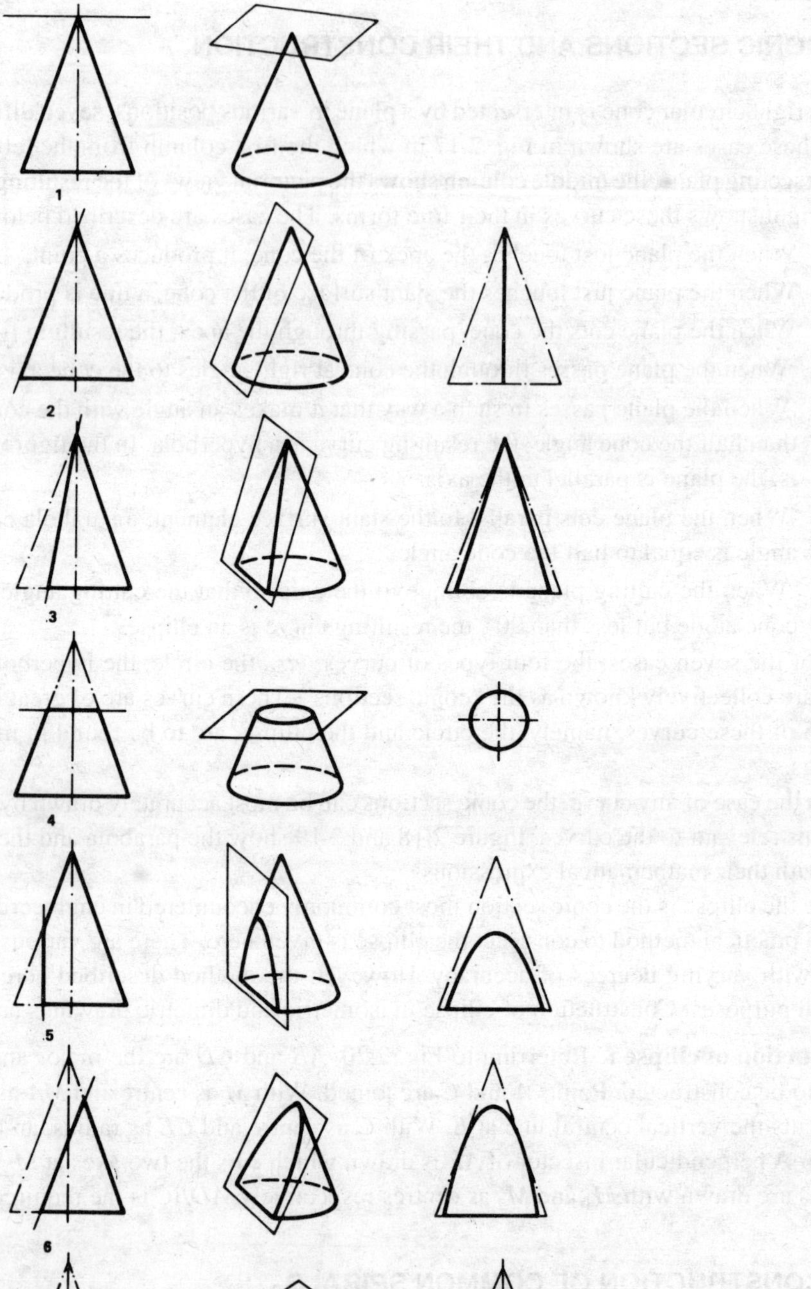

Fig. 2.17 Conic sections
Based on *Fachzeichnen Metalltechnik*, Fachstufe, 1982, Fig. No. 11.1 - 11.7, P. 11,
Verlag Europa - Lehrmittel, Wuppertal, Germany.

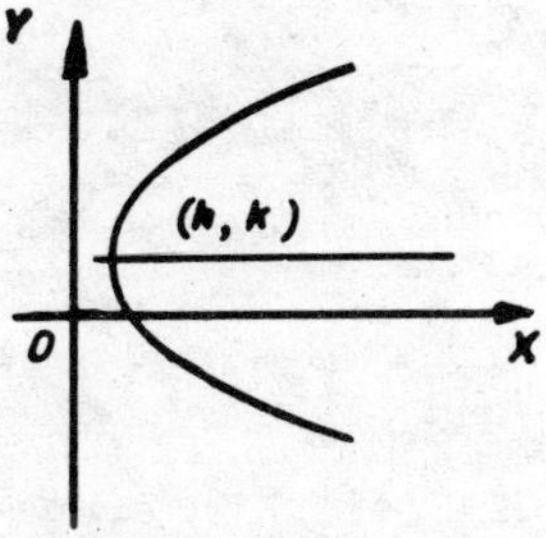

$(y - k)^2 = 4\,a\,(x - h)$
Vertex at (h, k), axis parallel to OX. (Here, a is positive).

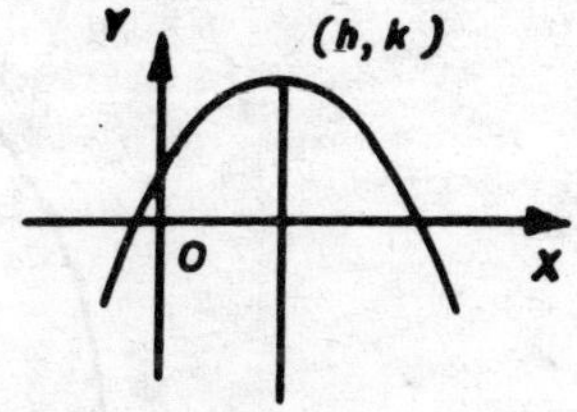

$(x - h)^2 = 4\,a\,(y - k)$
Vertex at (h, k), axis parallel to OY. (Here, a is positive).

Fig. 2.18 Parabola

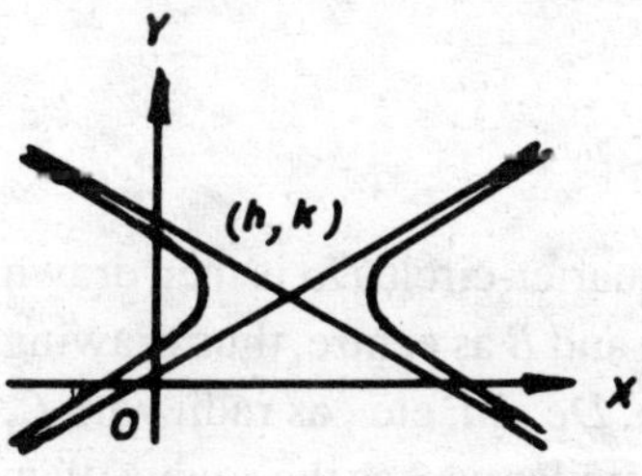

$$\frac{(x-h)^2}{a^2} - \frac{(y-k)^2}{b^2} = 1$$

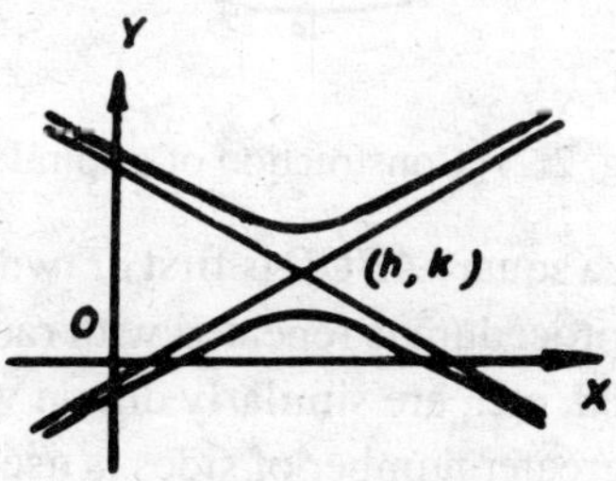

$$\frac{(y-k)^2}{a^2} - \frac{(x-h)^2}{b^2} = 1$$

Fig. 2.19 Hyperbola

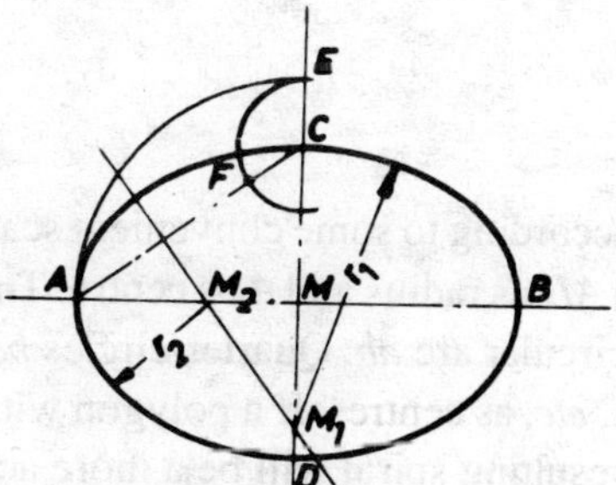

Fig. 2.20. Ellipse

An Archimedian spiral is a plane curve which is generated by a point which moves uniformly along a straight line while the line rotates about a fixed point with uniform angular velocity. In mathematical term, the polar equation of this spiral is expressed by the relation :

$$\rho = a\,\theta$$

where ρ is the radius vector and θ is the angle swept by this line. The mathematical expression for the logarithmic spiral is given by :

$$\rho = e^{a\,\theta} \text{ or } \log_e \rho = a\,\theta.$$

Spirals - as in the case of any curve following a mathematical relation - can be most accurately drawn by first assigning appropriate values to the parameters in the relevant mathematical equations, plotting the points or coordinates on (preferably graph) papers and then passing smooth curves through these points with the help of French curves. To draw a spiral approximately, the following method may be used (see Fig. 2.21).

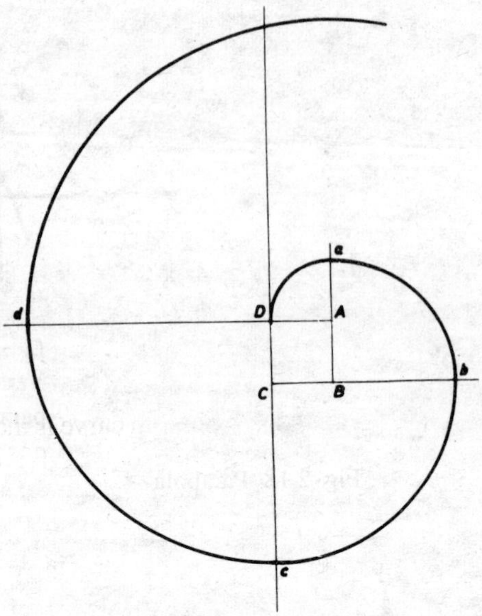

Fig. 2.21 Construction of a spiral

According to some convenient scale, a square *ABCD* is first drawn. The quarter-circle *Da* is then drawn with *AD* as radius and *A* as centre. The procedure is repeated with radius *Ba* and *B* as centre, thus drawing the circular arc *ab*. Quarter-circles *bc, cd*, etc., are similarly drawn with *Cb, Dc, Ad*, etc., as radii with *C, D, A, etc*, as centres. If a polygon with greater number of sides is used in a similar way as the square, then the resulting spiral will be a more accurate one.

The Archimedian and the logarithmic spirals are shown in Figs. 2.22 and 2.23.

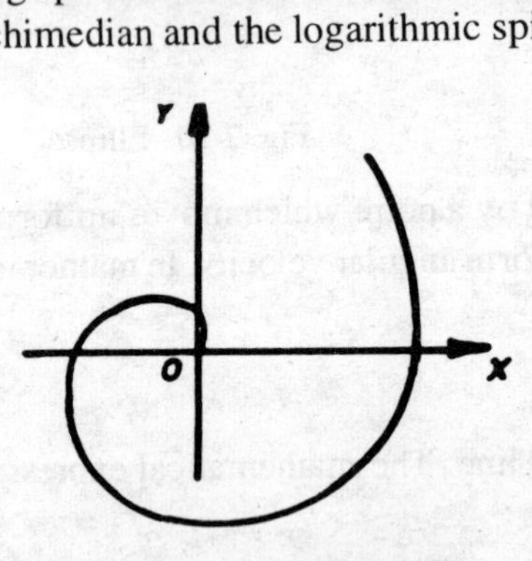

Fig. 2.22 Archimedian spiral

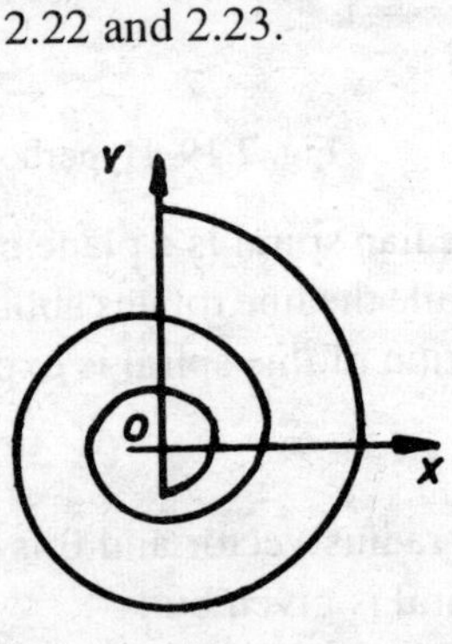

Fig. 2.23 Logarithmic spiral

A helix is a space curve which is generated by a point moving around and along the surface of a cylinder with uniform angular velocity about the axis and with uniform linear velocity in the direction of the axis. Instead of a cylinder, a helical curve can also be generated around a cone, but the term "helix" is generally meant in case of a cylindrical helix. A helix is a common curve much in use in the industry, *viz.*, screw threads, worms, helical gears, *etc.* If the cylinder containing the helix is developed as a plane surface, the helix becomes a straight line which is the hypotenuse of a right-angled triangle whose base is equal to

$\pi\, d$ where d is the diameter of the cylinder and the height of the triangle is the distance traversed by the point, measured parallel to the axis, when it makes a complete revolution around the cylinder. This height is called the "lead".

To draw a helix, two views of the cylinder are first drawn as shown in Fig. 2.24. The plan is a circle with d as the diameter. The elevation is a rectangle with sides d and p, where p is the lead (or pitch). A right-angled triangle is next drawn with base equal to $\pi\, d$ and the height equal to p, and is placed by the side of the rectangle as shown. The circumference of the plan circle is then divided into an equal number of parts, 12 in the case shown. The base of the triangle is also divided into the same number of parts. Both the circumference and the base are numbered correspondingly. Vertical lines are drawn from the circle and the base as shown. Horizontal lines from the hypotenuse are drawn to meet the vertical lines from the circle, both bearing the same number, *i.e.*, 1 with 1, 2 with 2 and so on. The points of intersection of these lines on the rectangle are the points on the required helix. A smooth curve is then drawn to complete the helix.

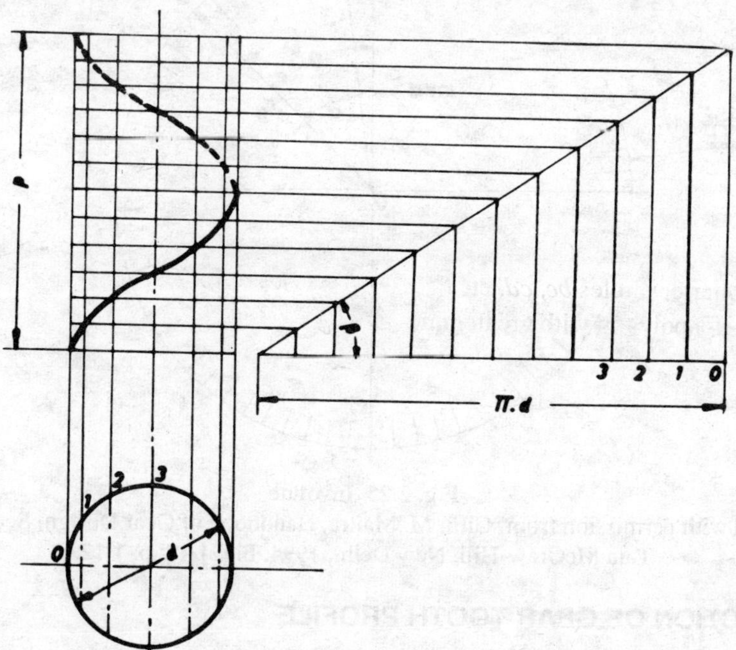

Fig. 2.24 Helix

An involute is a spiral curve which is traced by a point on a taut cord which is unwound from a circle. The involute curve is used as the profile of a gear teeth. The construction of an involute gear teeth is given in Section 2.5. Figure 2.25 shows an involute curve which can be drawn by using the parametric equations :

$$x = r_b\,(\sin\theta - \theta\cos\theta)$$

$$y = r_b\,(\cos\theta + \theta\sin\theta)$$

The parameters used in the equations are apparent from the figure. r_b is the radius of the base circle from which the involute is generated. R is the radius vector of the curve and P represents the instantaneous tracing point at the end of taut cord.

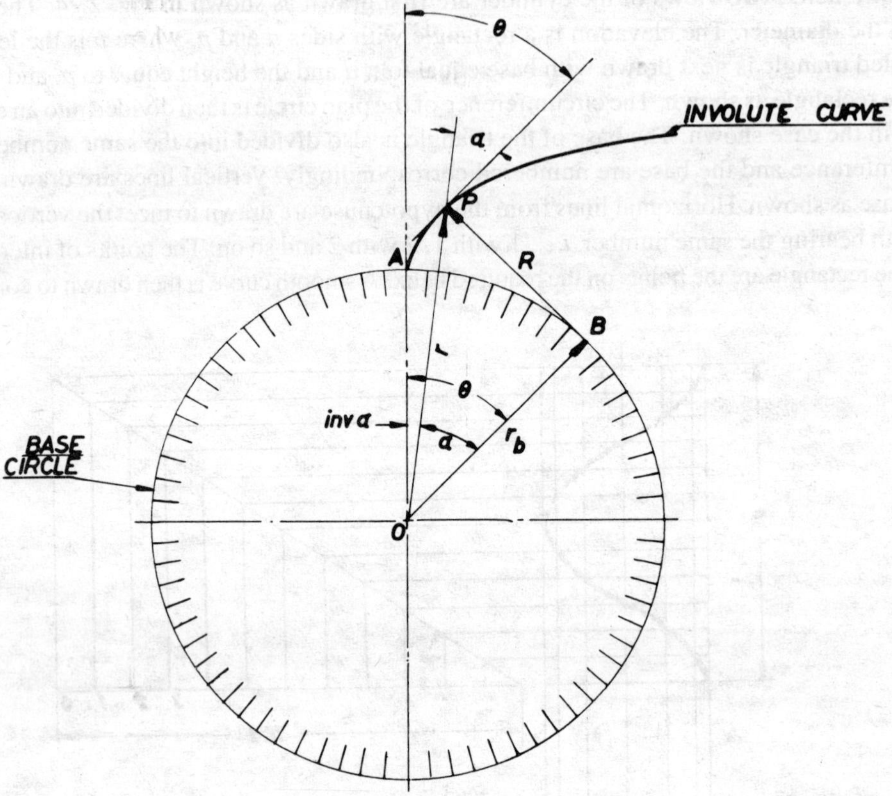

Fig. 2.25 Involute
Reproduced with permission from: Gitin M. Maitra, Handbook of Gear Design, Second Edition, Tata McGraw-Hill, New Delhi, 1994, Fig. 1.12, p. 1.12.

2.5 CONSTRUCTION OF GEAR TOOTH PROFILE

The involute gear tooth profile can be drawn by different methods with varying degree of accuracy. Three methods of construction will be discussed of which the first and the second are accurate enough for representing gear tooth in engineering drawings. The third method is the most accurate one and is used for making precise template of gear tooth profile for shop use.

1. Construction of gear tooth profile by drawing involute curves : The involute tooth profile of a standard tooth having 20° pressure angle and conforming to the Basic Rack (IS : 2535) can be drawn in the following manner : the basic parameters of the tooth are first calculated by using the formulae :

$$\text{Pitch circle diameter, } d = mz,$$

$$\text{Tip circle diameter, } d_a = mz + 2m,$$

$$\text{Root circle diameter, } d_f = mz - 2 \times 1.25\, m,$$

where m is the module in mm and z is the number of teeth of the gear.

The base circle diameter from which the involute is generated is given by,

$$d_b = d \cos 20^\circ$$

The vertical centre line is first drawn and the circles with diameters, d, d_a and d_f are laid out with O as centre. The centre line cuts the pitch circle at C, locating the pitch point. The line of action is then drawn which makes an angle of 20° with the horizontal passing through C. A perpendicular from O to the line of action meets at A through which a circle is passed with O as centre. This is the base circle. The line of action on both sides of A is divided into equal parts, locating the points E_2', E_1', E, E_1, E_2, E_3, E_4 etc. Here, points A and E coincide. With E as centre, arcs are drawn from points E_4, E_3, *etc*., onto the base circle giving points A_4, A_3 ....A_2' etc. From points A_4, A_3, etc., tangents are drawn on which points C_4, C_3 ....C_2' are marked off, making $A_4C_4 = E_4C$, $A_3C_3 = E_3C$, ... $A_1C_1 = E_1C$, *etc.* The required involute curve is obtained by joining these points which extends from the base circle to the tip circle. In Fig. 2.26(a), many points, *e.g.*, E_1, E_2 E_1', E_2', A_1, A_2, C_1, C_2', etc., have not been incorporated for the sake of clarity. From the base circle, the profile is extended upto the root circle as a radial. A suitable fillet is provided with radius equal to 0.38 to 0.45 m where the radial meets the root circle. The profile of the other side of the tooth can be similarly drawn. For normal, uncorrected gear tooth, the circular tooth thickness at the pitch circle is equal to half the circular pitch. That is,

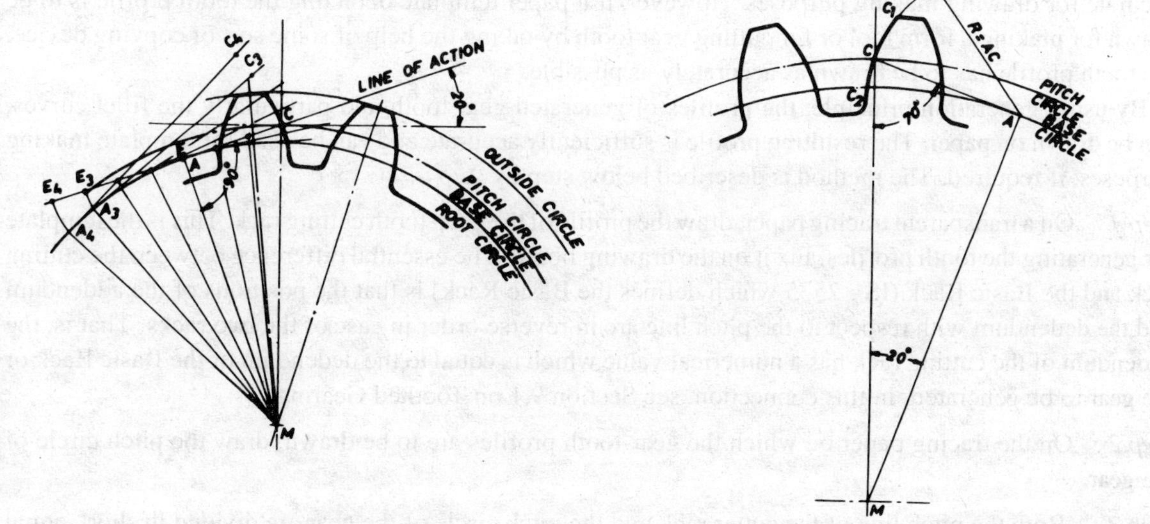

(a) (b)

Fig. 2.26 Construction of gear tooth profile

Reproduced with permission from: Gitin M. Maitra, Handbook of Gear Design, Second Edition, Tata McGraw-Hill, New Delhi, 1994, Fig. A.1, p. A.4.

$$\text{Circular teeth thickness} = \frac{1}{2} \times \text{circular pitch} = \frac{1}{2} . \pi . m$$

The other side of the tooth is the mirror image of the side drawn. The point on the other side which corresponds to the point C of this side is located at the end of the arc representing the circular tooth thickness.

2. Construction of gear tooth profile by circular arcs : Figure 2.26(b) shows a simplified method of drawing gear tooth profile by means of circular arcs. The method is approximate but it is adequate for representing the tooth profile on a drawing when accuracy is not required.

After drawing the pitch circle, tip circle and the root circle as before, the point A on the base circle is located at the intersection of CA and OA. From the pitch point C, the line CA is drawn, making an angle of 70° with the vertical centre line. The line OA is drawn, making an angle of 20° with the centre line. The base circle is passed through the point A.

An arc is then drawn with radius $R = AC$ and the centre at A. This arc, C_1CC_2, represents a reasonable approximation of the involute profile and is adequate for all practical purposes when only the graphic representation is warranted. The construction of the portion of the profile from the base circle to the root circle is the same as in the previous method.

In the same way, the other half of the tooth profile is drawn by first locating the corresponding point of C on the other side of the tooth profile.

3. Construction of gear tooth profile by generation method : The methods described so far for drawing gear tooth profile are approximate ones. The fillet curve portion of a gear tooth is a trochoid and this aspect is not taken care of in the above methods. The deviation of a tooth profile drawn by the conventional methods from the actual gear tooth profile is marginal and the methods are sufficiently accurate for drawing making purposes. However, if a paper template depicting the tooth profile is to be drawn for making a form tool or for cutting gear tooth by taking the help of some sort of copying device, the tooth profile has to be drawn as accurately as possible.

By using generation principle, the profiles of generated gear tooth and particularly the fillet curves, can be drawn on paper. The resulting profile is sufficiently accurate and can be used for template making purposes, if required. The method is described below step by step :

Step 1 : On a transparent tracing paper, draw the profile of the gear-tooth cutting rack. This is the template for generating the tooth profiles. Fix it on the drawing board. The essential difference between the cutting rack and the Basic Rack (IS : 2535 which defines the Basic Rack) is that the positions of the addendum and the dedendum with respect to the pitch line are in reverse order in case of the two racks. That is, the addendum of the cutting rack has a numerical value which is equal to the dedendum of the Basic Rack or the gear to be generated. In this connection, see Section 9.1 on Toothed Gearing.

Step 2 : On the tracing paper on which the gear-tooth profiles are to be drawn, draw the pitch circle of the gear.

Step 3 : Both the pitch line of the cutter rack and the pitch circle of the gear are divided in short, equal parts, say, 5 or 10 mm each, and numbered *a, b, c,* ... as well as a', b', c' ... and 1, 2, 3, ... as well as 1′, 2′, 3′, ... respectively on both sides of the vertical centre line. To facilitate correct alignment, draw line through these points parallel to the centre line in case of the rack and radial line in case of the gear to be drawn.

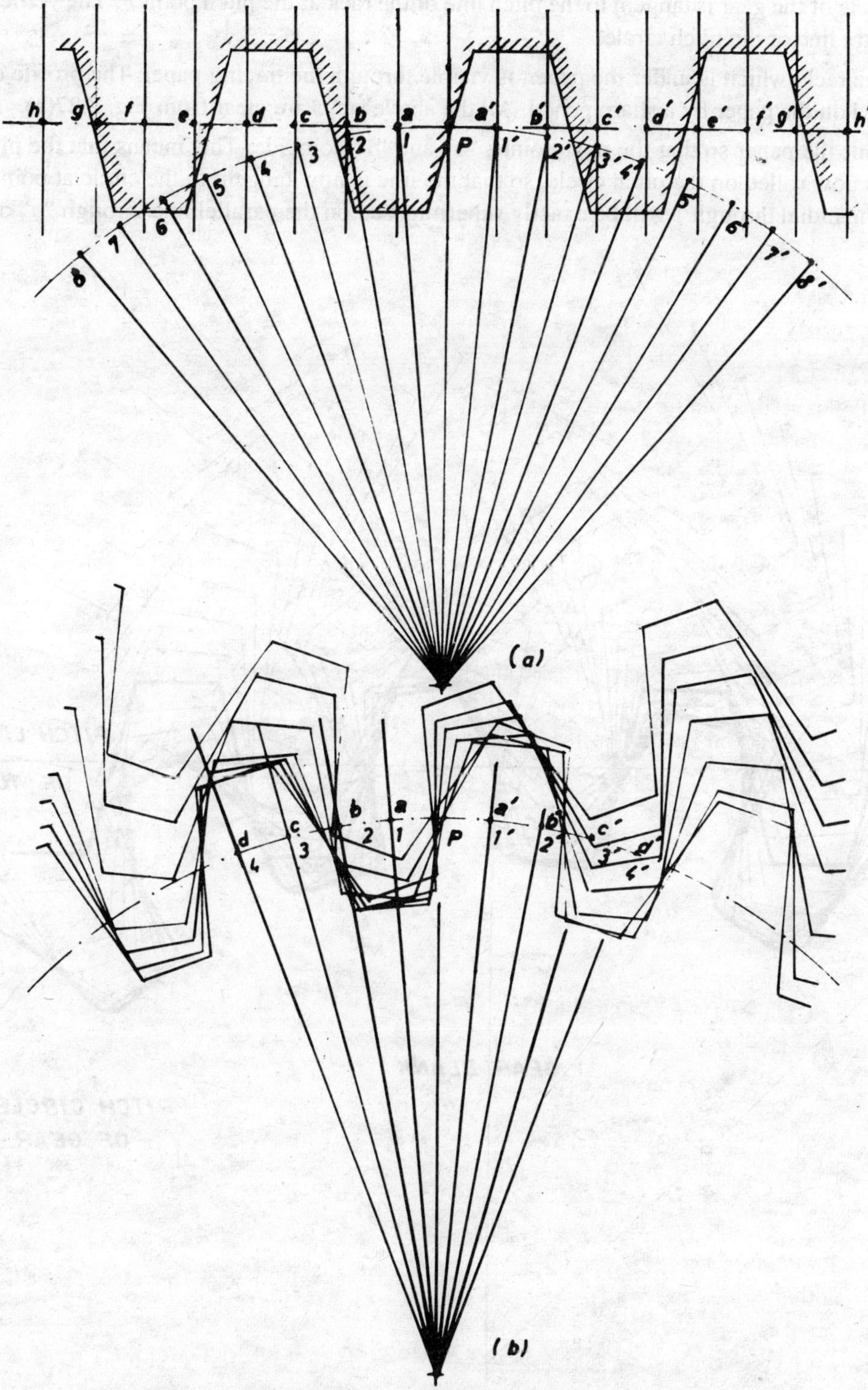

Fig. 2.27 Generation method of constructing the gear tooth profile
Reproduced with permission from: Gitin M. Maitra, Handbook of Gear Design, Second Edition, Tata McGraw-Hill, New Delhi, 1994, Fig. A.2, p. A.6.

Step 4 : Over the template, place the tracing paper meant for the gear tooth profile in such a manner that the pitch circle of the gear is tangent to the pitch line of the rack at the pitch point *P*. The vertical through *P* is the centre line of the pitch circle.

Step 5 : The rack, which is under the paper, is visible through the tracing paper. The profile of the rack is now traced on the paper by a sharp pencil. All the above steps are clear from Fig. 2.27(a).

Step 6 : Shift the paper so that the next points, "*a*" and "1" coincide. This means that the pitch line of the rack has now rolled on the pitch circle, so that the line is now tangent to the circle at point 1. In this condition, the radial through 1 will be exactly superimposed on the parallel line through "*a*" of the rack.

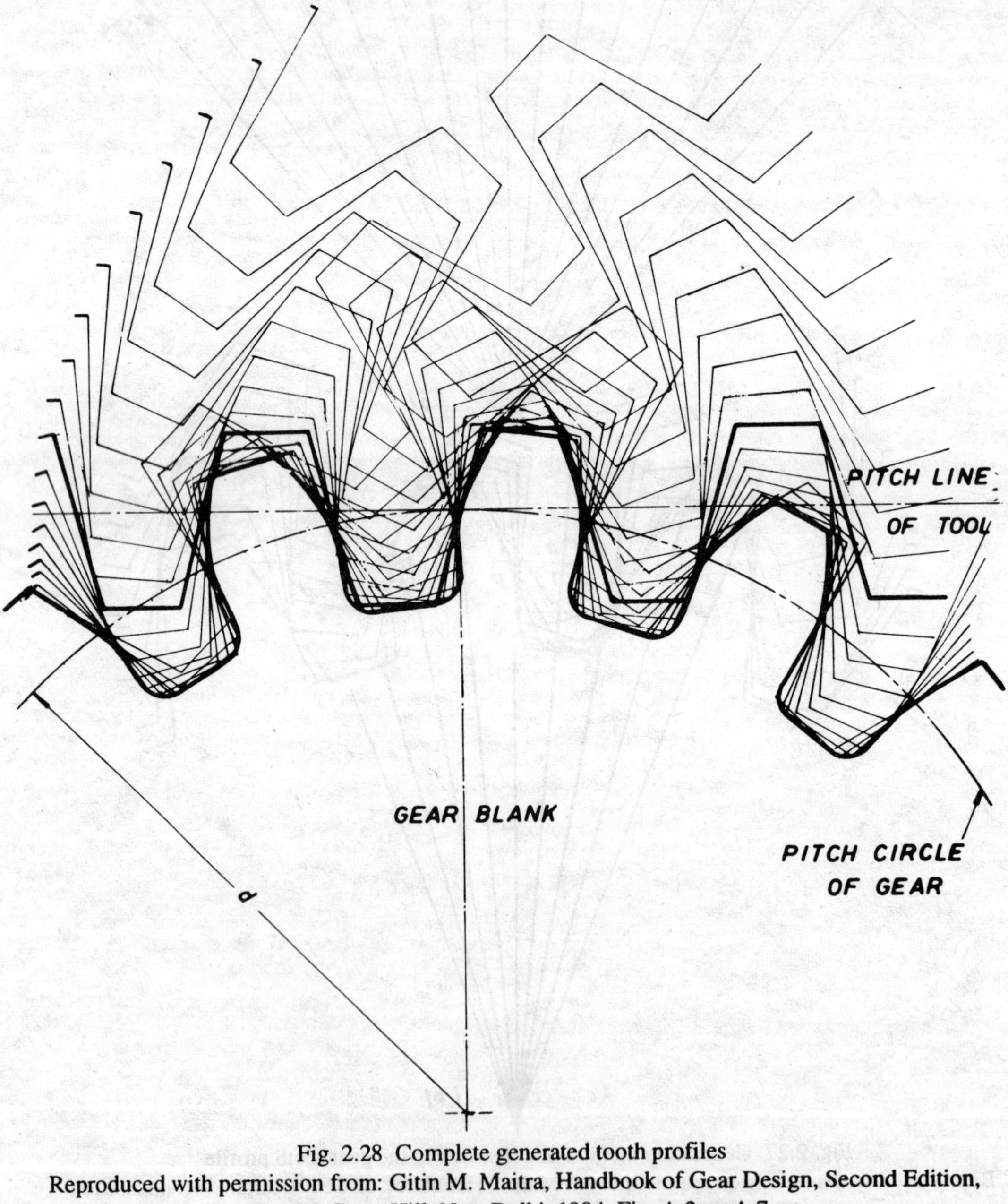

Fig. 2.28 Complete generated tooth profiles
Reproduced with permission from: Gitin M. Maitra, Handbook of Gear Design, Second Edition, Tata McGraw-Hill, New Delhi, 1994, Fig. A.3, p. A.7.

Draw the outline of the rack as before. Repeat the whole process at each successive positions, drawing the rack profiles and always making sure that the pitch line is tangent at these points. The process is carried out on both sides of the centre line.

Step 7 : The desired profile is the envelope of the rack profiles thus drawn. Pass a smooth curve to delineate this envelope. It can be seen from Fig. 2.27(b) that the tooth profiles are gradually taking shape. The complete tooth profiles are shown in Fig. 2.28.

2.6 CONSTRUCTION OF SPROCKET WHEEL

For power transmission, chain drives are often resorted to. This type of drive has been briefly described in Section 9.4. In this section, the geometrical construction of the sprocket wheel on which the chain runs and the relevant formulae thereof are discussed. The subject matter of this section is to be read along with that given in Section 9.4 where figures and other information pertaining to transmission chains are given.

The profiles of the tooth and the gap of the sprocket wheel should conform to the parameters of the corresponding chain. That is, the shape of the wheel is governed by the design and dimensions of the chain it is required to drive. As the efficiency of a chain drive largely depends on the quality and accuracy of the sprocket, it is imperative that the sprocket should be precisely manufactured and its calculated dimensions should be strictly adhered to. Figure 2.29 shows the shape of the wheel tooth and gap. For chain, see Fig. 9.17 in Section 9.4. The required parameters are calculated as per the formula given in IS : 2403.

The basic dimension of a chain drive is the pitch p. This is usually expressed in inch units or its metric equivalent. The pitch circle diameter of the sprocket is determined by the expression :

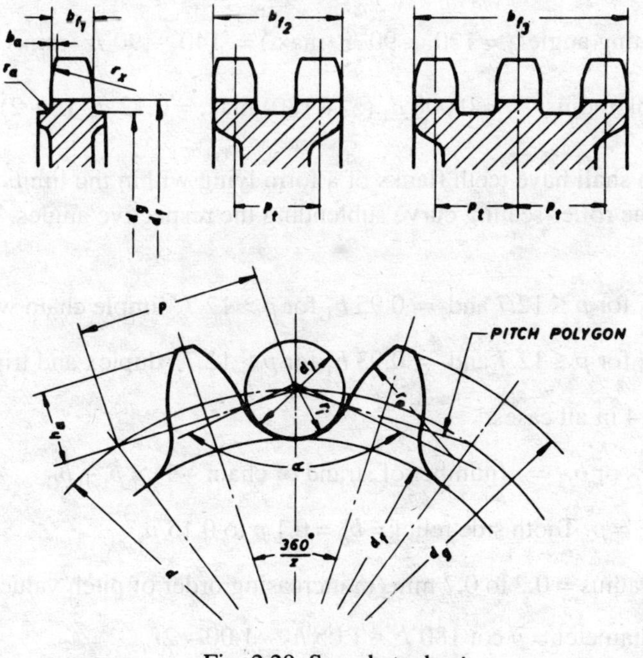

Fig. 2.29 Sprocket wheel

$$d = \frac{P}{\sin\frac{180^\circ}{z}} \approx \frac{p.z}{\pi},$$

where z is the number of tooth of the sprocket wheel.

The pitch p, roller diameter d_1 and other related data are given in Table 2.2. For the construction of sprocket tooth, an element of the pitch polygon is first laid out. This is the polygon which is inscribed in the pitch circle. Each side of this polygon equals the pitch as shown in Fig. 2.29. The relevant dimensions can be calculated from the following formulae :

d_f = Root diameter = $d - d_1$

Tolerance : When d_f is equal to or less than 127 mm, tolerance $= {}^{\;\;0}_{-0.25}$ mm

when d_f is above 127, upto 250 mm, tolerance $= {}^{\;\;0}_{-0.30}$ mm

When d_f is above 250 mm, tolerance = h_{11}

d_a = top diameter = $d + 1.25\,p - d_1$ (max.) = $d + p\,(1 - 1.6/z) - d_1$ (min)

h_a = height of tooth and pitch polygon $= 0.625\,p - 0.5\,d_1 + \dfrac{0.8\,p}{z}$ (max.)

$= 0.5\,(p - d_1)$ (min.)

r_1 = roller seating radius $= (0.505\,d_1 + 0.069\sqrt[3]{d_1})$ (max.)

$= 0.0505\,d_1$ (min.)

α = roller seating angle $= 120^\circ - 90^\circ\,z$ (max.) $= 140^\circ - 90^\circ/z$ (min.)

r_e = tooth flank radius $= 0.008\,d_1\,(z^2 + 180)$ max. $= 0.12\,d_1\,(z + 2)$ (min.)

The actual tooth gap form shall have teeth flanks of a form lying within the limits of the flank radii and blending smoothly with the roller seating curve subtending the respective angles. The tooth width b_{f1} is given as follows :

$b_{f1} = 0.93\,b_1$ for $p \le 12.7$ and $= 0.95\,b_1$ for $p > 12.7$, simple chain wheel

$= 0.91\,b_1$ for $p \le 12.7$ and $= 0.93\,b_1$ for $p > 12.7$, duplex and triplex chain wheel

The tolerance on b_{f1} is h 14 in all cases.

b_{f2} or b_{f3} = (number of strand of chain – 1) $\times p_t + b_{f1}$

Min. tooth side radius = $r_x = p$. Tooth side relief = $b_a = 0.1\,p$ to $0.15\,p$.

r_a = shroud radius = 0.2 to 0.7 mm (in increasing order of pitch values).

d_g = shroud diameter = $p \cot 180^\circ/z - 1.05\,h_2 - 1.00 - 2r_a$

2.7 CONSTRUCTION OF CAM PROFILE

Cams are machine elements used for imparting sliding or swinging movement to the mating components. They are extensively used in different classes of modern machineries and equipments to produce mechanical movements which are not readily obtained by other types of mechanisms. Typical applications where cams play a very important role are internal combustion engines, machine tools, computers and various instruments.

The cam usually executes rotary motion and the mating component, called the follower, moves in a pre-determined manner, depending on the designed contour of the cam. The motion of the follower may or may not be as per some law of motion. The cam body may be in the form of a plate with its edge having the desired curve, or it may be cylindrical in shape, having a groove cut along the cylindrical surface to direct the movement of the follower which is in contact with the groove. Although other forms are also in vogue, the above two types are generally used, and since the plate-type cam is most commonly used in general engineering practice, only this type will be discussed in this section. The curved working surface of a cam is laid out in accordance with the desired mechanical motion of the follower and as such, while laying out a plate cam, the problem lies in determining the shape or contour of the cam to effect the necessary resultant movement of the follower. In determining the desired contour of the cam and in studying the motions of the follower, it is useful to lay out a "displacement diagram". In constructing a displacement diagram, the successive angular cam positions are laid out as abscissas with the corresponding follower displacements as ordinates. See Fig. 2.30(c), (d), and (e). In effect, the displacement diagram shows the rise (or fall) of the follower with respect to the angular positions of the cam. The base line (abscissa) represents one full (360°) or one-half (180°) revolution of the cam shaft, and is divided in equal parts as per convenience.

Since cams can be designed to produce almost any variety of movement, from the simplest to the most complex, only those types which are commonly used will be discussed here.

Cam for producing uniform motion of follower : The determination of the profile of a cam to produce uniform motion of the follower can be best illustrated by means of an example. In case of uniform motion, the follower starts to ascend with a velocity which remains constant throughout till it reaches the top at the end of the stroke. It then reverses its direction, that is, it begins to descend in the same manner back to the starting point, while all the time the cam shaft rotates with uniform angular velocity in revolutions per minute (RPM) or radians per second.

Referring to Fig. 2.30(a), *C* is the centre of the cam circle and the follower is shown at its initial position. In actual practice, however, the follower is usually a cylindrical rod which is provided with a hardened steel roller at the end which comes in contact with the cam contour and rolls over it to reduce friction. In some cam design, the follower ends in a plate at right angles to the follower axis.

Fig. 2.30(a) shows only one-half since the other half is the mirror-image of the first half. On the extension of the diameter, a distance equal to the stroke length is laid off. The stroke length is also called the rise or the displacement. It is then divided into any equal parts to suit convenience. In this example, the stroke length 0-6 is divided into six equal parts. The semi-circle is also divided into six angular divisions. Radials $C-1'$, $C-2'$, ... $C-6'$ are then drawn. Arcs $1-1'$, $2-2'$, etc., are now drawn with centre at C as shown. A smooth curve is next drawn through the points of intersection of the radial

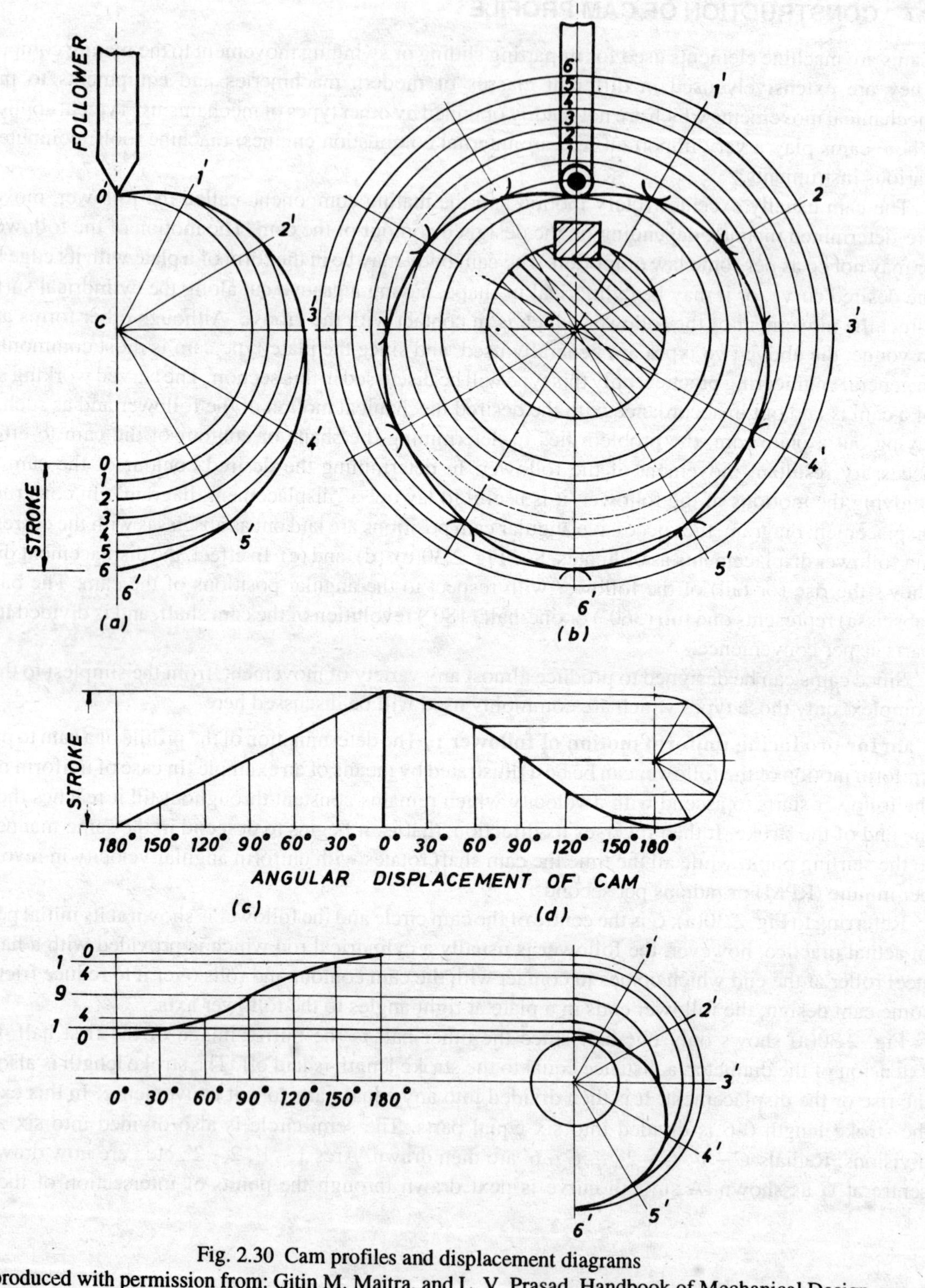

Fig. 2.30 Cam profiles and displacement diagrams
Reproduced with permission from: Gitin M. Maitra, and L. V. Prasad, Handbook of Mechanical Design, Second Edition, Tata McGraw-Hill, New Delhi, 1995, Fig. 9.40, p. 9.136.

$C - 1'$ and arc $1 - 1'$, $C - 2'$, and arc $2 - 2'$, etc. The resulting curve is one half of the desired contour of the cam. The other half can be similarly constructed.

An actual cam is shown in Fig. 2.30(b). C is the centre of the cam shaft. Divide the stroke 0 – 6 in six equal parts and draw radials $C - 1'$, $C - 2'$, *etc.* as before. Draw concentric circles through the points 1 to 6. Locate the points of intersection of the circle through 1 and radial $C - 1'$, etc., and describe arcs with roller radius, taking these points as centres as shown. A smooth curve is then drawn making it tangent to these series of arcs. The process is repeated for the other half as before. The follower is pushed up by the cam and the descent may be either by gravity or it can be actuated by a spring.

Some system may be so designed that the follower is allowed to have a rest period, generally at the end of the stroke or rise. This stationary period is usually known as "dwell". The contour of such a cam is constructed in a similar way. In this case, the rise is completed by a cam rotation of, say, 150° (and not 180° as in the previous cases), and the dwell portion of the follower is 60° (30° + 30°).

Cam for producing motion avoiding shocks : In a cam - follower system where the motion of the follower is uniform, the rate of the movement of the follower, that is, its velocity is constant throughout its rise or fall. The follower is, thus, forced to start and step abruptly at the beginning and the end of each stroke respectively. For a mechanism in which the components move with slow speeds, the uniform velocity system may not pose serious problems, but for a high speed mechanism, such a system is detrimental to the equipment as this results in the generation of undesirable forces and vibration which is not conducive to the operation of the equipment. To alleviate the bad effects of such a situation, the follower is made to move with accelerated and then decelerated motions during the course of motion or duration of stroke. Two systems are usually employed : the simple harmonic motion (S.H.M.) and the uniform acceleration-deceleration motion of the follower.

Plate cam producing simple harmonic motion : To obtain the cam contour which will produce a simple harmonic motion for the follower, a displacement diagram is drawn in the following manner from which the desired cam contour can be developed : referring to Fig. 2.30(d), a semi-circle, the diameter of which is equal to the stroke of the cam, is drawn and then divided into an equal number of divisions. The horizontal line represents the developed circumference of the cam circle, showing the angular displacement of the cam. It is divided into the same number of equal parts as the semi-circle. Intersection points of vertical and horizontal lines are noted. A smooth curve passing through these points represents the displacement diagram for half of the cam. For the sake of comparison, Fig. 2.30(e) shows the displacement diagram of a cam for uniform motion.

Plate cam producing uniformly accelerated-decelerated motion : For many high speed cams, the contour of the cam which produces the smoothest motion is a parabola. This contour, which is somewhat better than a curve giving a simple harmonic motion is designed to produce a uniformly accelerated-decelerated motion of the follower :

In this system, when the cam rotates a quarter of one revolution (*i.e.*, 90°) from the initial position, the follower rises and executes a uniformly accelerated motion during this period. For the next quarter, there is a reversal of this process. That is, the follower executes a uniformly decelerated motion till it reaches the end of its stroke. This process is repeated during the fall of the follower and one full revolution (360°) of the cam is completed.

In the above system, the distance through which the follower moves at the end of a given number of time units (or angular movement of the cam) varies as the square of the number of such units. For example, if the acceleration is 2 cm/sec^2, then the follower moves a distance of $(1/2) \times 2 \times (1)^2 = 1$ cm at the end of the first second, then a total distance of $(1/2) \times 2 \times (2)^2 = 4$ cm at the end of the next second, and so on. The same rule is followed in case of deceleration. A cam curve based on this law allows the follower to have a uniformly increasing velocity from the beginning upto the middle of the stroke and then to have a uniformly decreasing velocity till the follower reaches the end of the stroke. In case of the simple harmonic motion system, the acceleration and retardation are slightly irregular.

To lay a plate cam based on the above principle, the following procedure is adopted : on any convenient circle, equally spaced (say, 6) radial lines are drawn for one half of the circle. The stroke length is vertically laid out as shown in Fig. 2.30(e). One half of the stroke length is divided into 9 equal divisions. The divisions indicating the squares of 1, 2, and 3 that is, 1, 4, and 9 are marked off on the line representing the stroke. Circular arcs are struck off from the divisions 1, 4, and 9 on the stroke-line as shown. Radials are then drawn from the centre through points 1′, 2′, 3′ etc. The intersection points of arc 1 and radial 1′, arc 4 and radial 2′, arc 9 and radial 3′ are established. A smooth curve through these points gives the cam contour which produces a uniformly accelerated motion for the follower. The processs is reversed for the remaining half of the stroke-line to produce a uniformly retarded motion of the follower. The other half of the cam contour can be similarly drawn. The left half of Fig. 2.30(e) shows the relevant replacement diagram.

2.8 SEGMENTS OF CIRCLES

In the practical engineering field, particularly in the sheet metal industries where pieces made out of galvanised, tinned and other kinds of sheets are mass produced, it is often required to find out the relevant parameters of circular segments quickly. Detailed and accurate calculations are quite involved and entail complicated mathematical formulae.

To facilitate quick and easy determination of parameters, Table 2.1 can be used. Notations and figure are as per the illustration entitled "Segment" in Table 10.6, Section 10.

Table 2.1 Data of circular segments

$h : s$	$r : s$	$b : s$	$A : s^2$	ϕ	$h : s$	$r : s$	$b : s$	$A : s^2$	ϕ
0,005	25,003	1,0001	0,00313	2° 17′ 30″	0,255	0,61770	1,1653	0,17854	108° 5′ 10″
0,010	12,505	1,0003	0,00672	4° 34′ 59″	0,260	0,61077	1,1715	0,18237	109° 53′ 51″
0,015	8,3408	1,0006	0,01000	6° 52′ 24″	0,266	0,60420	1,1778	0,18623	111° 41′ 39″
0,020	6,2600	1,0011	0,01332	9° 9′ 45″	0,270	0,59796	1,1843	0,19010	113° 28′ 34″
0,025	5,0125	1,0017	0,01668	11° 26′ 58″	0,275	0,59205	1,1903	0,19399	115° 14′ 35″
0,030	4,1817	1,0024	0,02000	13° 44′ 4″	0,280	0,58643	1,1975	0,19790	116° 59′ 42″
0,035	3,5889	1,0033	0,02334	16° 1′ 0″	0,285	0,58110	1,2042	0,20183	118° 43′ 57″
0,040	3,1450	1,0043	0,02669	18° 17′ 44″	0,290	0,57603	1,2110	0,20578	120° 27′ 17″
0,045	2.8003	1,0054	0,03004	20° 34′ 15″	0,295	0,57123	1,2179	0,20975	122° 9′ 44″
0,050	2,5250	1,0067	0,0339	22° 50′ 32″	0,300	0,56667	1,2249	0,21374	123° 51′ 17″

$h : s$	$r : s$	$b : s$	$A : s^2$	ϕ	$h : s$	$r : s$	$b : s$	$A : s^2$	ϕ
0,055	2,3002	1,0080	0,03675	25° 6′ 33″	0,305	0,56234	1,2321	0,21775	125° 31′ 57″
0,060	2,1133	1,0096	0,04011	27° 22′ 16″	0,310	0,55823	1,2393	0,22178	127° 11′ 44″
0,065	1,9556	1,0112	0,04348	29° 37′ 39″	0,315	0,55433	1,2465	0,22583	128° 50′ 37″
0,070	1,8207	1,0130	0,04685	31° 52′ 42″	0,320	0,55063	1,2539	0,22990	130° 28′ 37″
0,075	1,7042	1,0149	0,05022	34° 7′ 23″	0,325	0,54712	1,2614	0,23400	132° 5′ 43″
0,080	1,6025	1,0170	0,05360	36° 21′ 40″	0,330	0,54379	1,2689	0,23812	133° 41′ 56″
0,085	1,5131	1,0192	0,05699	38° 35′ 32″	0,035	0,54063	1,2766	0,24226	135° 17′ 17″
0,090	1,4339	1,0215	0,06039	40° 48′ 57″	0,340	0,53765	1,2843	0,24642	136° 51′ 45″
0,095	1,3633	1,0239	0,06379	43° 1′ 55″	0,345	0,53482	1,2921	0,25061	138° 25′ 21″
0,100	1,3000	1,0265	0,06720	45° 14′ 23″	0,350	0,53214	1,3000	0,25481	139° 58′ 5″
0,105	1,2430	1,0291	0,07061	47° 26′ 21″	0,355	0,52961	1,3079	0,25904	141° 29′ 58″
0,110	1,1914	1,0320	0,07404	49° 37′ 17″	0,360	0,52538	1,8168	0,26336	143° 29′ 56″
0,115	1,1416	1,8546	0,07747	51° 49′ 40″	0,365	0,52497	1,3211	0,26758	144° 31′ 0″
0,120	1,1017	1,0380	0,08091	53° 58′ 58″	0,370	0,52284	1,3323	0,27188	146° 0′ 20″
0,125	1,0625	1,0412	0,08431	56° 8′ 42″	0,375	0,52083	1,3406	0,27620	147° 28′ 46″
0,130	1,0263	1,0445	0,08783	58° 17′ 49″	0,380	0,51895	1,3490	0,28058	148° 56′ 21″
0,135	0,99343	1,0479	0,09130	60° 26′ 18″	0,385	0,51718	1,3574	0,28493	150° 23′ 6″
0,140	0,96286	1,0515	0,09478	62° 34′ 8″	0,390	0,51551	1,3660	0,28932	151° 49′ 0″
0,145	0,93457	1,0552	0,09827	64° 41′ 19″	0,395	0,51396	1,3745	0,29375	153° 14′ 5″
0,150	0,90833	1,0590	0,10178	66° 47′ 49″	0,400	0,51250	1,3832	0,29820	154° 38′ 21″
0,155	0.88395	1,0629	0,10529	68° 53′ 37″	0,405	0,51114	1,3920	0,30267	156° 1′ 47″
0,160	0,86125	1,0669	0,10882	70° 58′ 43″	0,410	0,50988	1,4008	0,30717	157° 24′ 25″
0,165	0,84008	1,0711	0,11236	73° 3′ 5″	0,415	0,50870	1,4097	0,31170	158° 46′ 14″
0,170	0,82029	1,0754	0,11591	75° 6′ 44″	0,420	0,50762	1,4186	0,31625	160° 7′ 15″
0,175	0,80179	1,0798	0,11948	77° 9′ 36″	0,425	0,50662	1,4276	0,32082	161° 27′ 29″
0,180	0,78444	1,0843	0,12305	79° 11′ 44″	0,430	0,50570	1,4367	0,32543	163° 46′ 55″
0,185	0,76818	1,0889	0,12665	81° 13′ 4″	0,435	0,50486	1,4459	0,33005	164° 5′ 35″
0,190	0,75289	1,0936	0,13025	83° 13′ 38″	0,440	0,50409	1,4551	0,33471	165° 23′ 27″
0,195	0,73853	1,0985	0,13387	85° 13′ 23″	0,445	0,50340	1,4644	0,33939	166° 40′ 34″
0,200	0,72500	1,1035	0,13751	87° 12′ 20″	0,450	0,50278	1,4738	0,34410	167° 56′ 55″
0,205	0,71226	1,1085	0,14117	89° 10′ 28″	0,455	0,50223	1,4832	0,34884	169° 12′ 31″
0,210	0,70024	1,1137	0,14482	91° 7′ 46″	0,460	0,50174	1,4927	0,35360	170° 27′ 22″
0,215	0,68895	1,1190	0,14850	93° 4′ 15″	0,465	0,50132	1,5022	0,35839	171° 41′ 28″
0,220	0,67818	1,1244	0,15220	94° 59′ 53″	0,470	0,50096	1,5118	0,36321	172° 54′ 50″
0,225	0,66806	1,1300	0,15591	96° 54′ 39″	0,475	0,50066	1,5215	0,36805	174° 7′ 28″
0,230	0,65848	1,1356	0,15964	98° 48′ 35″	0,480	0,50042	1,5313	0,37293	175° 19′ 24″
0,235	0,64941	1,1413	0,16338	100° 41′ 38″	0,485	0,50023	1,5411	0,37783	176° 30′ 36″
0,240	0,64083	1,1471	0,16715	102° 33′ 50″	0,490	0,50016	1,5509	0,38276	177° 41′ 6″
0,245	0,63270	1,1531	0,17093	104° 25′ 9″	0,495	0,50003	1,5608	0,38771	178° 50′ 54″
0,250	0,62500	1,1591	0,17472	106° 15′ 37″	0,500	0,50000	1,5708	0,39270	180° 0′ 0″

Example : *Given height h = 8 cm and chord length s = 40 cm. To find the radius r, arc length b, area A and the angle ϕ of the circular segment.*

Solution: $h : s = 8 : 40 = 0.200$. From table, $r : s = 0.725$, $\therefore\ r = 40 \times 0.725 = 29$ cm. Similarly, $b = 40 \times 1.1035 = 44.14$ cm, $A = 40^2 \times 0.13751 = 220.016\ \text{cm}^2$, $\phi = 87°\ 12'\ 20''$.

In case where the height and the radius are known, the ratio $h : r$ can be found from the table thus :

$$h : s / r : s = \frac{h}{s} \cdot \frac{s}{r} = h : r$$

The ratio $h : r$ varies from 0.0001999 to 1 corresponding to the $h : s$ values of 0.005 to 0.500 (as given in the table) respectively. h and r being known, the ratio $h : r$ can be calculated and then compared to the $h : r$ values calculated from the data given in the table as shown. Segment parameters can then be obtained from the nearest $h : r$ value which matches with the calculated one.

Table 2.2 Transmission chain data (All dimensions in mm)

ISO Chain Number	*Pitch P*	*Roller Diameter d_1 Max*	*Width Between Inner Plates b_1 Min*	*Bearing Pin Body Diameter d_2 Max*	*Chain Path Depth h_1 Min*	*Inner Plate Depth h_2 Max*	*Outer Intermediate Plate Depth h_3 Max*	*Transverse Pitch pt*	*Width over Inner Link b_2 Max*	*Width Between Outer Plates b_3 Min*	*Width over Boaring Pins*		
											Simple b_4 Max	*Duplex b_5 Max*	*Triplex b_6 Max*
05B	8.00	5.00	3.00	2.31	7.37	7.11	7.11	5.64	4.77	4.90	8.6	14.3	19.9
06B	9.525	6.35	5.72	3.28	8.52	8.26	8.26	10.24	8.53	8.66	13.5	23.8	34.0
08B	12.70	8.51	7.75	4.45	12.07	11.81	10.92	13.92	11.30	11.43	17.0	31.0	44.9
10B	15.875	10.16	9.65	5.08	14.99	14.73	13.72	16.59	13.28	13.41	19.6	36.2	52.8
12B	19.05	12.07	11.63	5.72	16.39	16.13	16.13	19.46	15.62	15.75	22.7	42.2	61.7
16B	25.40	15.88	17.02	8.28	21.34	21.08	21.08	31.88	25.45	25.58	36.1	68.0	99.0
20B	31.75	19.05	19.56	10.19	26.68	26.42	26.42	36.45	29.01	29.14	43.2	79.7	116.1
24B	38.10	25.40	25.40	14.63	33.73	33.40	33.40	48.36	37.92	38.05	53.4	101.8	150.2
28B	44.45	27.94	30.99	15.90	37.46	37.08	37.08	59.56	46.58	46.71	65.1	124.7	184.3
32B	50.80	29.21	30.99	17.81	42.72	42.29	42.29	58.55	45.57	45.70	67.4	126.0	184.5
40B	63.50	39.37	38.10	22.89	53.49	52.96	52.96	72.29	55.75	55.88	82.6	154.9	227.2
48B	76.20	48.26	45.72	29.24	64.52	63.88	63.88	91.21	70.56	70.69	99.1	190.4	281.6
56B	88.90	53.98	53.34	34.32	78.64	77.85	77.85	106.60	81.33	81.46	114.6	221.2	—
64B	101.60	63.50	60.96	39.40	91.08	90.17	90.17	119.89	92.02	92.15	130.0	250.8	—
72B	114.30	72.39	68.58	44.48	104.67	103.63	103.63	136.27	103.81	103.94	147.4	283.7	—

CHAPTER 3

METHODS OF PROJECTION AND VIEWS

3.1 MULTIVIEW PROJECTION

Of all the various methods of representing a part in a drawing, the multiview projection method is the most commonly used one in engineering and manufacturing industries. As mentioned in Chapter 1, an engineering drawing can be correctly and meaningfully interpreted only by trained personnel. And since virtually all the engineering drawings are now made using the multiview projection method, a thorough grasp of the subject is imperative for engineers and draughtsmen.

In technical parlance, a view of an object is called a "projection" when it is thought to be implanted or "projected" onto a plane. This plane is called the "plane of projection". If the observer viewing the object is stationed at an infinite distance from the object, then the "lines of sight" emanating from the object will appear as parallel rays when they reach the observer. If a plane is placed in front or behind the object, the projection on this plane will be a true image of the object dimensionally. This is not so in the case of perspective view, where the projection is either smaller or bigger than the actual object, depending on whether the plane is placed in front or behind the object respectively. This difference between the multiview projection and the perspective projection is apparent in Fig. 1.5.

Angles of projections: In Section 1.8, reference has been made to the "angle of projection". In engineering nomenclature, this term refers to the quadrant in which the object is conceived to be situated in space. The four angles of projection are the four quadrants which are pictorially depicted in Fig. 3.1. The two frontal planes and the two horizontal planes meet at a line called the "ground line". In Fig. 3.2, a part is shown suspended in space in the first angle or quadrant and in the third angle or quadrant. When the direction of sight is horizontal, an image is conceived to be implanted on the vertical plane. This image is termed as the "front view" or "front elevation" or simply "elevation" of the object. Note that in the first angle, the image is projected behind the object, while in the third angle the plane of projection is in front of the object as viewed by the observer. When the direction of sight is vertical, the image is projected on the horizontal plane in both cases of the angles of projection. This image is known as the "top view" or "plan" of the object. In case of plan also, the placement of the view of the object vis-a-vis the plane of projection follows the same convention as in the case of elevation in the two angles of projection.

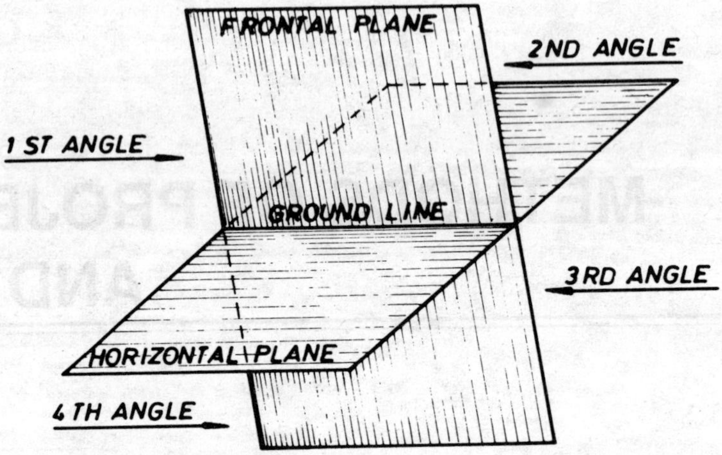

Fig. 3.1 Pictorial view of the four angles of projection

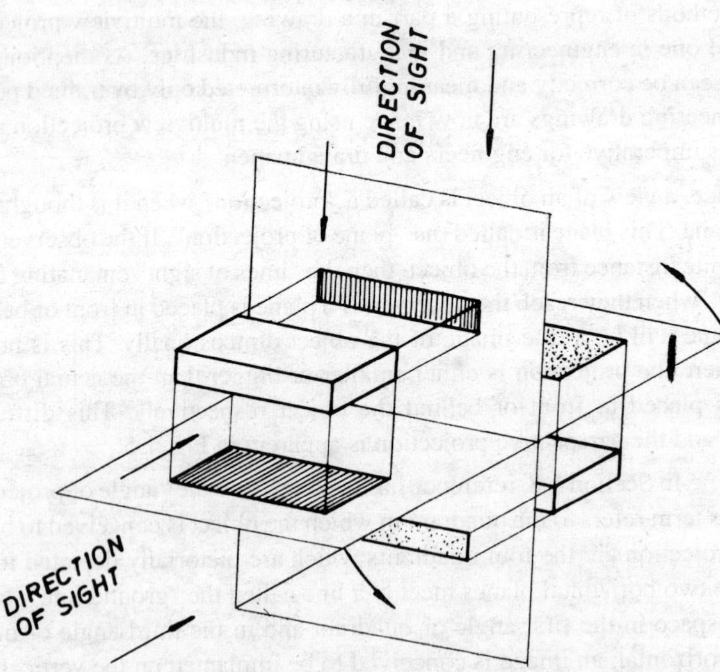

Fig. 3.2 Object in first and third angle of projection

To show the two views, the plan and the elevation, on a two dimensional paper, the convention is to rotate the horizontal plane anti-clockwise as indicated by the arrows. This makes the horizontal plane (carrying the plan) flush with the vertical plane. The result is shown in Fig. 3.3. Note that the placement of the front views and the top view are in reverse order in the first and the third angle of projection. This is as it should be as per the convention of rotation of the horizontal plane as stated above. In actual engineering drawings, the ground line is not shown. In Section 1.8, the adoption of the first angle projection method and the reasons thereof have been discussed. It is interesting to note here that the second angle and the fourth angle of projection are not used because in both the cases, the plan and the elevation will superimpose on each other, above and below the ground line as the case may be, making the resulting drawing indecipherable and failing in purpose for which an engineering drawing is made. The reader can easily make it out by studying Fig. 3.2.

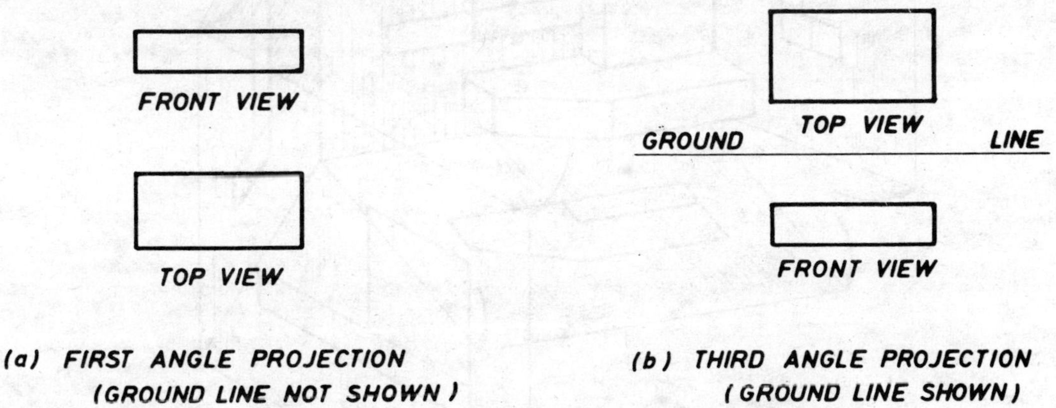

Fig. 3.3 First and third angle projection

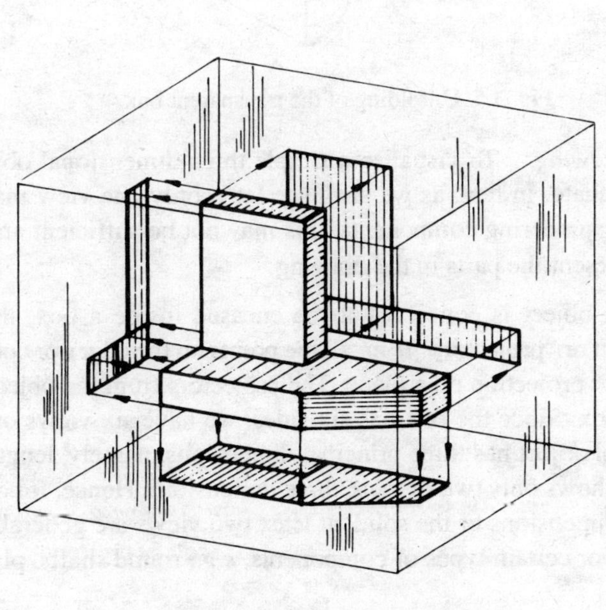

Fig. 3.4 Object in transparent box

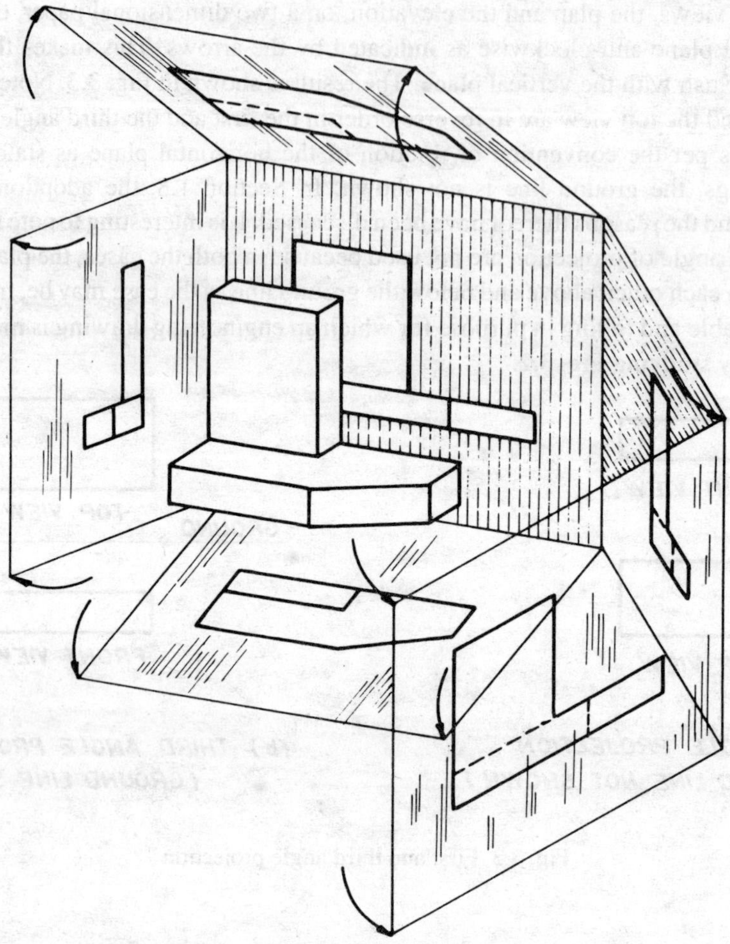

Fig. 3.5 Unfolding of the transparent box

Formation of a multiview drawing: To visualise a simple three-dimensional object, two views - plan and elevation - may be adequate. In fact, as we shall see later, only one view may serve the purposes. However, for complicated engineering components, this may not be sufficient and more views may be required to be drawn to represent the parts in the drawing.

Referring to Fig. 3.4, an object is conceived to be encased inside a box, the sides of which are transparent. The lines of sight or "projectors" from all the points on the edges or contours of the part form a two - dimensional image or projection on a plane. The projectors from the object collectively form an image on each side of the box. Since the box has six sides, we have six views of the object on the six sides. Any three dimensional object has three principal dimensions, namely, length, breadth and height. Any one of these six views shows only two of these three dimensions. Hence, for obtaining the complete information as regards the dimensions of the solid, at least two views are generally required for correct visualisation of the object. For certain types of components, *e.g.*, round shafts, plates, and other similar objects, only one view with the relevant written information is sufficient for the interpretation of the object, and other views are superfluous. However, in actual practice, engineering parts are generally complicated enough to warrant more views, sections, and other pictorial representations.

A drawing is executed as a flat sheet which is a two dimensional paper. Hence, the transparent box has to be unfolded so that all views lie on a plane. This unfolding process is shown in Fig. 3.5. When the box is completely unfolded and spread on a plane, it will look as shown in Fig. 3.6. In this figure, the placement of the views is as per the first angle projection method. Fig. 3.7 shows a comparison between the orientations of views in the first angle and the third angle projections. In both the angles of projection, the front view or the elevation is the principal view and the other views are arranged with reference to this view. All the views should be placed as shown in these two cases in accordance with the orientations of views depicted in Fig. 3.7. Only the back or rear view can be placed either at left or right to suit convenience in both the angles of projection.

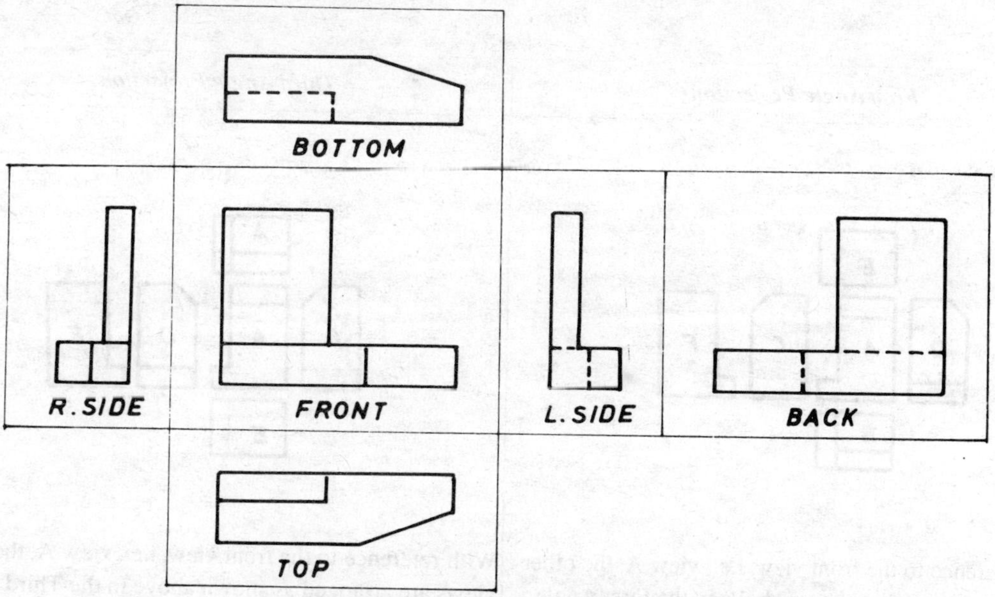

Fig. 3.6 Transparent box unfolded

In Fig. 3.6, the folding lines which denote the edges of the transparent box have been shown. In actual practice, these lines are omitted as in the case of ground line discussed before, and only the views are shown with the relevant dimensions, symbols and other data. Fig. 3.7 also shows the distinctive symbols to differentiate between the first and the third angle projection methods, and the relevant symbol is sometimes incorporated in the title block of the drawings of many engineering concerns to avoid confusion and to be precise as to the angle of projection followed by a particular firm. This practice, however, is no longer needed as all the manufacturing firms, design offices, consultants, and other engineering drawing making concerns are supposed to execute the drawings in the first angle projection only.

True length of a line : In a multiview projection system, a line appears in its true length in a view only when the line is parallel to the plane of projection containing that view. If the line is perpendicular the

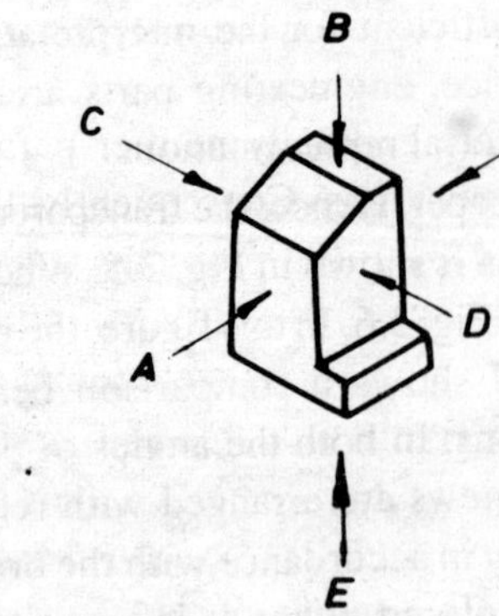

The following nomenclatures are used to designate the different views.

In direction A = Front view or elevation
In direction B = Top view
In direction C = Left side view
In direction D = Right side view
In direction E = Bottom view
In direction F = Back view

First Angle Projection	*Third Angle Projcetion*
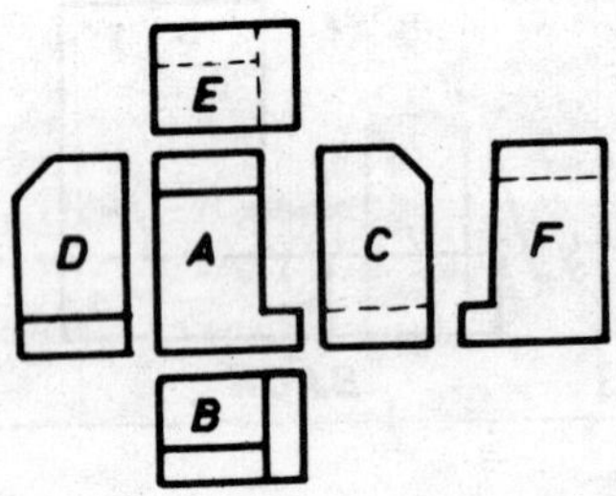	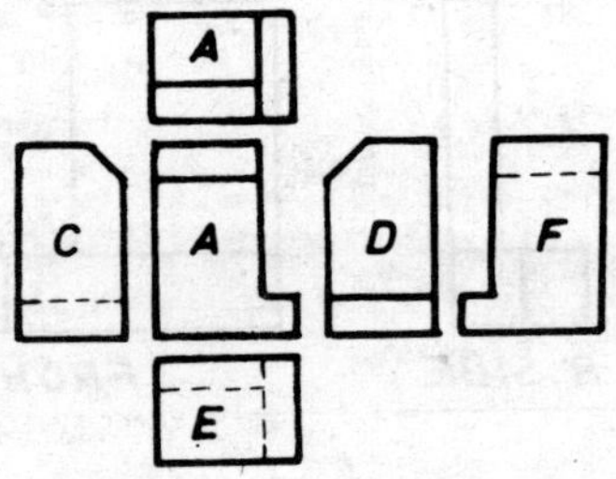
With reference to the front view, i.e., view A, the other views are arranged as shown above in the First Angle Projection	With reference to the front view, i.e., view A, the other views are arranged as shown above in the Third Angle Projection
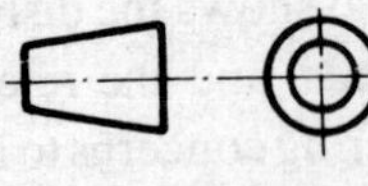	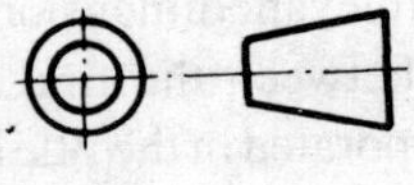
The Distinctive symbol of 1st Angle Projection.	The Distinctive symbol of 3rd Angle Projection.

Fig. 3.7 Orientation of views in first and third angle projections

plane, it appears as a point. The projection of the line appears to be shorter than its true length when it is inclined to the plane.

The lines defining the edges of objects in engineering drawings are generally neither parallel nor perpendicular to the planes of projection, but are oriented in inclined positions. Consequently, the lines appear as foreshortened ones in views and sections. In many cases concerning problems in practical engineering, especially those connected with development, it becomes often necessary to find the true length of a line where two or more views are given in a drawing. There are various methods to find out the true length from the views. In the method commonly used, the underlying principle is to revolve the line until it is parallel to a plane of projection.

In Fig. 3.8, the line *AB* appears to be shortened in all the major planes of projection, *viz*., planes containing elevation, side view and plan. In this pictorial representation, the elevation, side view and plan of the line *AB* are *A‴ B‴*, *A″ B″* and *A′ B′* respectively.

In Fig. 3.9, the line *AB* is revolved so that its true length appears in the elevation.

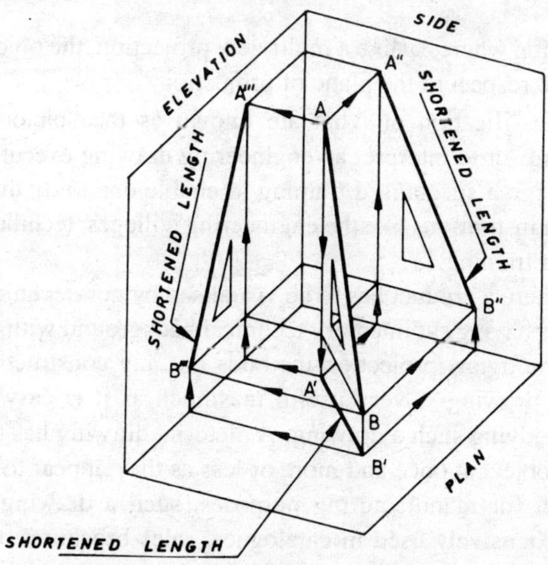

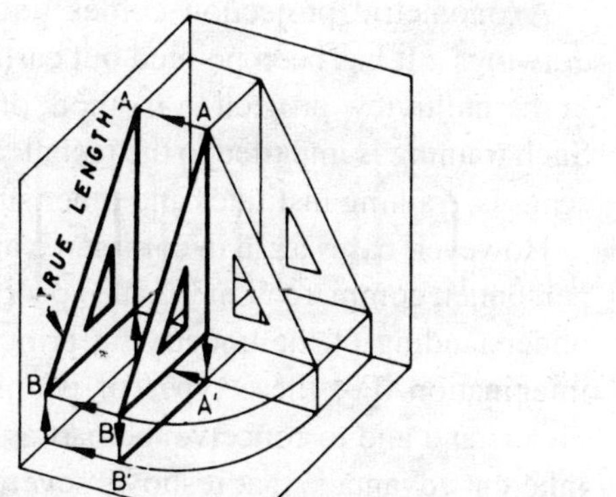

Fig. 3.8 Pictorial view of a line in space　　Fig. 3.9 Revolved pictorial view of the line
Based on Fachzeichnen Metalltechnik, Fachstufe, 1982, Fig. No. 43-44, P. 43-44, Verlag Europa-Lehrmittel, Wuppertal, Germany.

Referring to Fig. 3.10, in which the three views of the line *AB* appear in two-dimensional drawing, the true length of the line on the elevation can be found as follows : the projection of the line on the plan, *A′ B′*, is first rotated about *A′* until it becomes parallel to the horizontal projection axis and assumes the position *A′* B_1, thereby making it to lie parallel to the projection plane of the elevation. (The rotation is indicated by an arrow). A perpendicular is next drawn from B_1 which meets the extended line *B″ B‴* at B_2. The point B_2 is then joined with *A‴*. The line *A‴* B_2 is the required true length of the line *AB*. True lengths in plan and in side view can be similarly constructed as shown in Fig. 3.10.

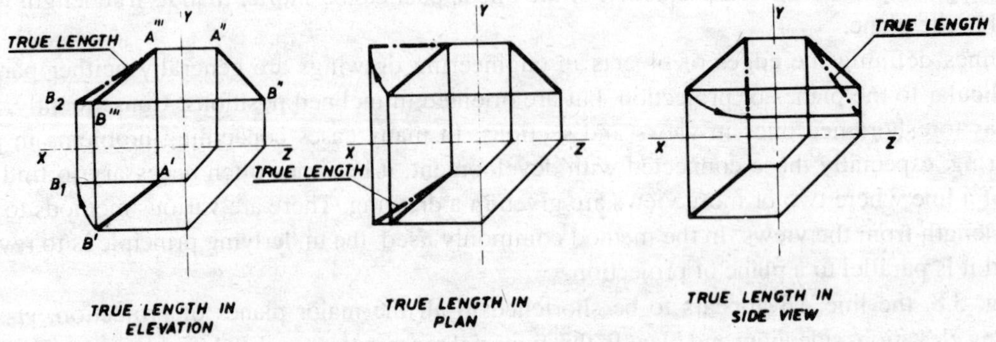

Fig. 3.10 Determination of the true length of a line in different views

3.2 AXONOMETRIC PROJECTION

An axonometric projection is an orthographic projection where, unlike a multiview projection, the object is viewed as if it is placed in an inclined position with respect to the plane of projection.

Axonometric projection comes under general classification of what are known as the "pictorial drawings". It has been pointed out earlier that to "read" or to interpret an engineering drawing executed in the multiview projection method, one has to undergo a specialised training to enable one to do that. Such training is imparted to the technical personnel in institutions like the engineering colleges, technical schools, training institutes and other similar places of training.

However, cases arise in commerce and industry where a product has to be visualised by non-technical personnel, commercial men, sales people and persons having rudimentary technical background with no understanding of the underlying principles of the multiview projection methods nor any constructive imagination. For these types of people, a pictorial drawing is very useful inasmuch as it is easy to understand and to conceive the part as a whole by studying such a drawing. A pictorial drawing has the inherent advantage that it shows several faces of an object at once and more or less as they appear to an observer in three-dimensional configuration though for manufacturing purposes, such a drawing is inadequate. These drawings in various forms are extensively used in catalogues, sales brochures and technical literatures.

In a pictorial drawing, several sides of the object are shown in a single view. The pictorial drawings in common usage are:

(i) Axonometric drawings,

(ii) Oblique drawings, and

(iii) Perspective drawings.

In this section, we will discuss the axonometric drawings only. Though the object may be placed at an infinite number of inclined positions with respect to the plane of projection, only three types of axonometric projections are generally considered. These are:

(i) Isometric projection,

(ii) Dimetric projection, and

(iii) Trimetric projection.

Isometric projection : In an isometric projection, the object is so placed that its principal edges make equal angles with the plane of projection. The edges are thus foreshortened equally. If we want to draw the isometric view of a cube, the cube is first rotated through an angle of 45° about an imaginary vertical axis. It is then tilted forward about an imaginary horizontal axis until the three frontal edges of the cube make equal angles with the frontal plane of projection. These edges are, therefore, shortened equally. The ratio of the length of an edge and the length of its projection is $1 : \sqrt{2/3}$ or 1 : 0.816 approximately. The projected length is thus around 80% of the true length. The top face of an isometric cube slopes approximately 35° 16′.

If a correct isometric projection is to be drawn, the foreshortening of the features from the actual case shall be considered. For this purpose, an isometric scale can be prepared which bear the ratio : isometric scale: true scale = 9 : 11 approximately.

A true isometric projection should be drawn according to this scale, but since pictorially there is no difference between the foreshortened isometric projection and a full-size isometric drawing, the usual practice followed in the drawing offices is to make true scale drawing, disregarding the isometric scale. This is much easier to execute and it is satisfactory enough for all practical purposes.

Dimetric projection : A dimetric projection is similar to an isometric projection, except that the object is so placed that two of its axes or edges make equal angles with the plane of projection while the third axis makes a different angle. Hence, the foreshortenings of the two equal-angle axes and of the third axis are in different ratios. A dimetric projection drawing is useful when one face is to be shown more prominently compared to other faces. These drawings are used mainly for the rectangular objects.

Trimetric projection : In this axonometric projection, no two axes make equal angles with the plane of projection. Therefore, the foreshortening ratios are different in all cases. When compared with the isometric and the dimetric, distortion is reduced in trimetric projection. Trimetric projection method is seldom used as it is slower to execute.

Properties in pictorial views : Comparative illustrations of the three common pictorial views, namely, the isometric, the dimetric and the oblique projections are shown in Fig. 3.11. With reference to this figure, the ratios of sides are as follows:

For isometric projection : $a : b : c = 1 : 1 : 1$

For dimetric projection : $a : b : c = 1 : 1 : 0.5$

For oblique projection : $a : b : c = 1 : 1 : 0.5$

In case of an isometric view, the horizontal edges of the object are drawn making an angle of 30° with the horizontal line of the drawing as shown.

In case of dimetric view, the horizontal edges are drawn at 7° and 42° approximately. In case of an oblique view, the usual angle is 45°. The view shown here is a cabinet projection which is the usual projection for an oblique view (see Section 3.3). In all the above-stated three types of pictorial projections, the vertical edge of the object is shown vertical in the drawing.

Ellipses in pictorial views : In any plane which is not parallel to the plane of projection, a circle which lies in that plane will appear as an ellipse. There are a number of ways to draw these ellipses. Some of these methods are described below:

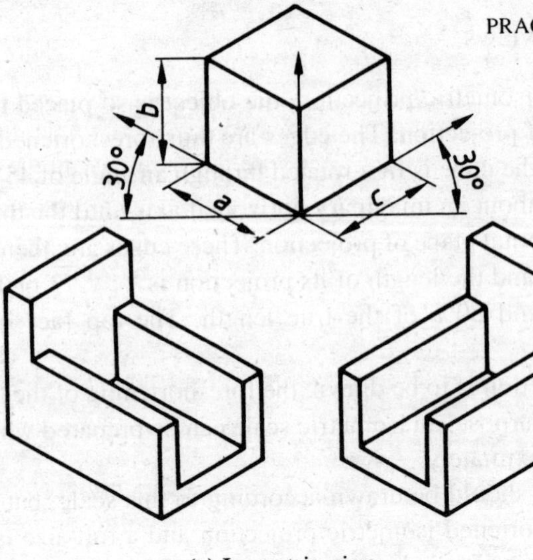

(a) Isometric view

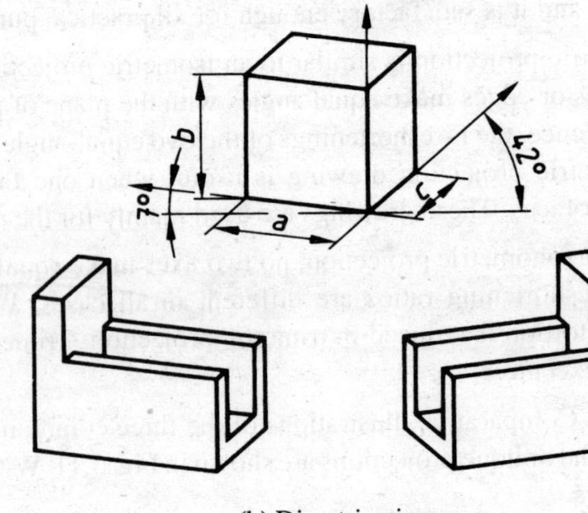

(b) Dimetric view

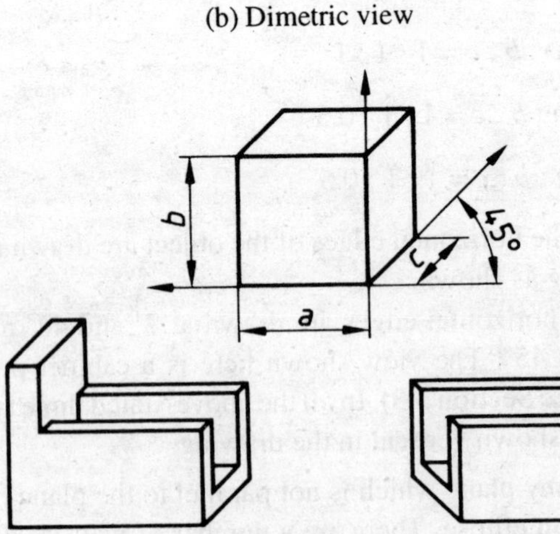

(c) Oblique view

Fig. 3.11 Pictorial views
Based on Fachzeichnen Metalltechnik, Grundstufe, 1984, Fig. No. 42, P. 42,
Verlag Eurepa - Lehrmittel, Wuppertal, Germany.

In an isometric cube, the ellipses on the three sizes can be drawn by following the relation:

$$\text{Major axis, } D \approx 1.22 \times a \text{ and minor axis, } d \approx D/1.7$$

The ellipse is then drawn by means of circular arcs with the radii having the following relation:

$$R \approx 1.06 \times a, \text{ and } r \approx 0.2 \times a$$

The ellipses drawn by using the above relations are approximate, but they are good enough for all practical drawing making purposes. The method is shown in Fig. 3.12. The centres of the circular arcs lie as shown in the figure which can be easily located as their positions are clear from the figure.

In Fig. 3.12 (right side), a simplified version of ellipse construction inside a rhombus has been shown. This method requires no calculation. The relations between the different parameters of the ellipse can be easily seen in the figure. E and F are the mid-points of AD and BC respectively.

Referring to Fig. 3.13, the ellipses in a dimetric projection are drawn as per the relation: Ellipse E_1 is drawn as a circle for simplification. The dimensions of the other ellipses are: For E_2 and E_3, diameter d : major axis $a = 1 : 1.06$, major axis a : minor axis $b = 3 : 1$

3.3 OBLIQUE PROJECTION

As mentioned in Section 1.8, an oblique projection is a kind of parallel projection. In case of an oblique

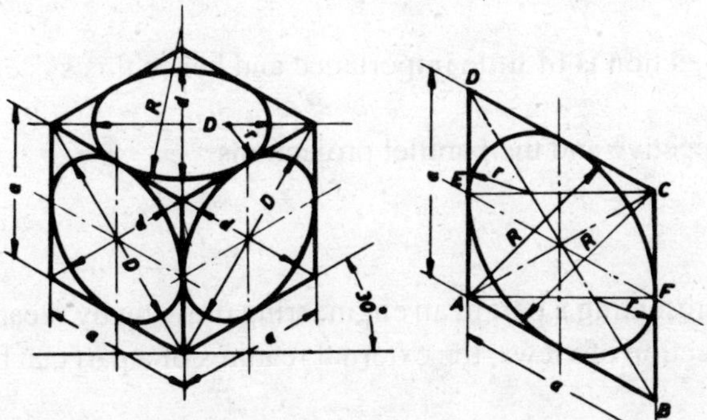

Fig. 3.12 Construction of ellipse in isometric projection

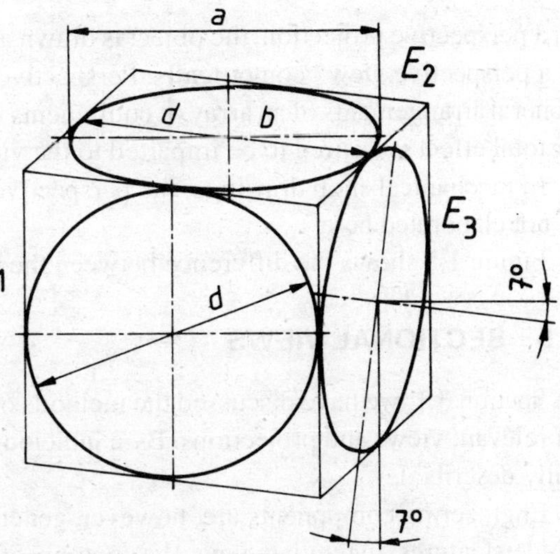

Fig. 3.13 Construction of ellipse in dimetric projection

projection, the distance from the observer to the plane of projection is infinite as in the case of an orthographic projection. But unlike the case with the orthographic projection, the direction of projectors are oblique to the plane of projection. The projectors are parallel to one another in each case of orthographic and oblique projections.

Comparative pictorial views of the orthographic (multiview) projection and the oblique projection are given in Fig. 1.6. Though both are considered to be pictorial views, an oblique projection is sometimes preferred to an axonometric (isometric, *etc.*) projection because it is easier to draw. In any type of axonometric projection, circles on the object practically always lie in surfaces which are inclined to the plane of projection. Consequently, these appear as eiiipses. In an oblique projection, the object can be so positioned that the faces containing the circles are parallel to the plane of projection. Consequently, the circles project as true, full-size circles which can obviously be easily drawn with instruments. For its comparative simplicity, the oblique drawings are much in use in technical literatures, sales brochures for machines and equipments, and company catalogues.

However, the oblique views have certain disadvantages. The human eye is accustomed to seeing objects in which the receding parallel lines appear as converging lines. In an oblique view, they do not converge, giving rise to an unnatural appearance. The distortion does not appear to be serious for objects of short depths. But objects which have comparatively large length should not be drawn in oblique view with the long dimension perpendicular to the projection plane. This distortion effect can be substantially reduced by decreasing the length of the receding lines.

Two types of oblique projections are generally in use : the cavalier projection and the cabinet projection. In cavalier projection, the receding lines are true lengths with the projectors making an angle of 45° with the plane of projection. In a cabinet projection, the receding lines are drawn to half their actual sizes.

3.4 PERSPECTIVE PROJECTION

In a perspective projection, the object is drawn as it approximately appears to a human eye. A photograph is a perspective view geometrically. Perspective projection is mainly used by the architects, designers of general arrangements of an array of equipments in a factory, and similar work where an overall impression or total effect is desired to be imparted to the viewer.

In mechanical shop drawings, the perspective projection is of little importance and hence this system is not elaborated here.

Figure 1.5 shows the difference between the perspective and the parallel projections.

3.5 SECTIONAL VIEWS

In Section 3.1, we have discussed the methods of representing a part in an engineering drawing by means of relevant views and projections. By a judicious selection of views, the external features of a part can be fully described.

Engineering components are, however, generally complicated in design and they contain a number of hidden features. In regular views, the contours of these hidden portions will come as dotted lines. A regular orthographic projection will thus contain so many full and dotted lines that the resulting view will be utterly confusing and virtually impossible to interpret correctly. Moreover, some full lines may coincide with the dotted lines in which case the full lines take the precedence over the dotted lines (as explained in Chapter 1) adding to further confusions.

To alleviate this difficulty, the common practice is to show the internal details in views which depict a condition as if some portion has been cut away to reveal the interior details. A view of the object in cutaway condition is then drawn. Such a view is called a "sectional view" or simply "section".

In this system, an imaginary cutting plane is assumed to slice the object in two halves. These halves are not necessarily symmetrical and the ensuing sectional view will depend upon the position of the cutting plane. The convention is to remove one of the cutaway portion and then make a drawing of the remaining portion, containing the interior configurations as it appears to the observer in sliced condition.

In a sectional view, the areas of these portions which have been in actual contact with the (imaginary) cutting plane are to be differentiated from those portions which are not cut by the plane. For this, the convention is to "hatch" the former portions, leaving the latter portions (which are visible) as unhatched. The convention as regards hatching lines is shown in Fig. 4.34, but nowadays the number of the variety of materials used in the engineering industries are so many and diverse in nature that only the hatching lines applicable to cast iron and steel are used to represent all materials irrespective of the materials actually specified for the part in the drawing. This section-lining is now considered as general-purpose symbol for hatching. The relevant materials are, of course, mentioned in the Bill of Materials correlating the relevant parts numbers, as that there is no scope of ambiguity as regards the part vis-a-vis its material. In some assembly drawings is section, the characteristic section linings as given in Fig. 4.34 are sometimes used to distinguish the different materials used in the equipments.

Section lines are drawn at 45° with the horizontal, unless some other angle is used for any particular reason (see Section 3). Depending on the size of the drawing, these lines should be around 2 to 3 mm apart. They should be spaced uniformly and should be of uniform thickness.

In a sectional view, all visible edges and contours behind the cutting plane should be shown. A section must not appear as disconnected. It should appear as a composite picture. Only in some special cases, lines representing visible features behind the cutting planes are emitted to avoid complexity or in cases where these are not necessary for making the part clear to read the drawing. Normally, a section should not contain hidden lines. Hidden lines, *i.e.*, dotted lines, should be entered in a sectional view only when they add to the clarity of the drawing for better understanding. In cases where the features come as hidden lines in a section which can otherwise be made clear in other views, the common practice is to avoid incorporating these lines, because these non-essential lines in a section only lead to confusion. However, these lines are sometimes given in a section if by so doing, it is possible to omit an extra view.

In conventional practices, certain items are not sectioned although the cutting plane passes through them. Such items are belts, rivets and other fasteners, pins, shafts and similar parts. Arms and spokes of pulleys and other similar objects are not sectioned. Portions of many other machine elements are not sectioned by convention. Teeth of gears and sprocket wheels, vanes of impellers, ribs, lugs, *etc.*, fall in this category.

A part is generally shown in full section. That is, the cutting plane completely slices the part, thereby revealing the internal details. Sometimes for symmetrical objects, a view is drawn in such a way as if one half of the object is in section while the other half is shown as a regular exterior view. Such a section is known as a half section. This practice saves space because in one picture, both the linear and the outer details are shown. It should be noted here that the common line demarcating the regular and the sectional views is a centre line (chain-dotted line) and not a solid line.

Illustrations concerning the different aspects of sectional and other views are given in Sections 3.7, 3.8, 4.2, 4.3, and 4.6.

3.6 AUXILIARY VIEWS

In Section 3.1, we have seen that views of parts in an engineering drawing are projected onto the main planes of projection which are mutually perpendicular to one another. If a feature of the part is parallel to a plane of projection, it will show its true shape dimensionally when it is projected on that particular plane. The lines of sight or projectors reaching the observer from the various points of the feature are perpendicular to that plane of projection. Such type of view is also called a "normal" view in the sense that in geometrical nomenclature, a "normal" is another name of the perpendicular. But any projection, where the viewing direction is perpendicular to the object feature and is made on a plane which is parallel to the object feature, can also be termed as a "normal" view. Such a plane of projection need not be any of the regular planes of projection as discussed in Section 3.1. A view of an object which is projected on a plane, which is non-regular but parallel to the object face, is called an "auxiliary" view. Such a plane is known as an "auxiliary" plane.

In making a practical engineering drawing of an object, it may so happen that a feature of that object is not parallel to the six regular planes of projection discussed hitherto. There can be one or more such inclined surfaces where it is necessary to show their true shape for clarity, especially if these features are irregular in outline. When projected on regular planes, such features appear distorted and a drawing-reader is liable to be confused. He may make wrong interpretation of such features, thus defeating the very purpose of a technical drawing.

The chief reason for incorporating an auxiliary view in a practical drawing is to show the true shape of an inclined feature in a localised manner. In an auxiliary view, the normal practice is to show only the feature in question, omitting other features which are foreshortened. An auxiliary view is thus a partial view or a localised view.

Sometimes it may so happen that it is necessary to show a view which is in contravention to the method of projection followed in the drawing, *e.g.*, a particular view shown in the third angle in a drawing which is otherwise drawn in the first angle. Such a view is known as a special view. The direction of that view is indicated by an arrow and identified by a letter (see Fig. 3.14).

Since the main purpose of an auxiliary view is to show a particular feature in its true shape which can not be shown otherwise, it is sometimes adequate to show the auxiliary view along with one of the six regular views to describe the part completely, thus rendering the other views as unnecessary or superfluous. Economy of space is thus achieved and man-hour is curtailed.

Some examples of auxiliary views are shown in Fig. 3.14.

3.7 REVOLVED VIEWS

In shop drawings, a practice called "revolved views" (or more correctly: "revolved section") is resorted to whereby the shapes in cross-section of bar, arm, spoke and other elongated features are incorporated in the respective views themselves. In this way, the drawing of a further view or section can be spared. A revolved section is made by assuming a cutting plane which is perpendicular to the axis of the arm, spoke and similar objects, and then revolving the plane containing the section through 90° with respect to the axis. The hatched section is then superimposed on the view of the feature as shown in Fig. 3.15. This practice is very common in illustrating the cross-sectional contour of pulleys, sheaves, fly-wheels,

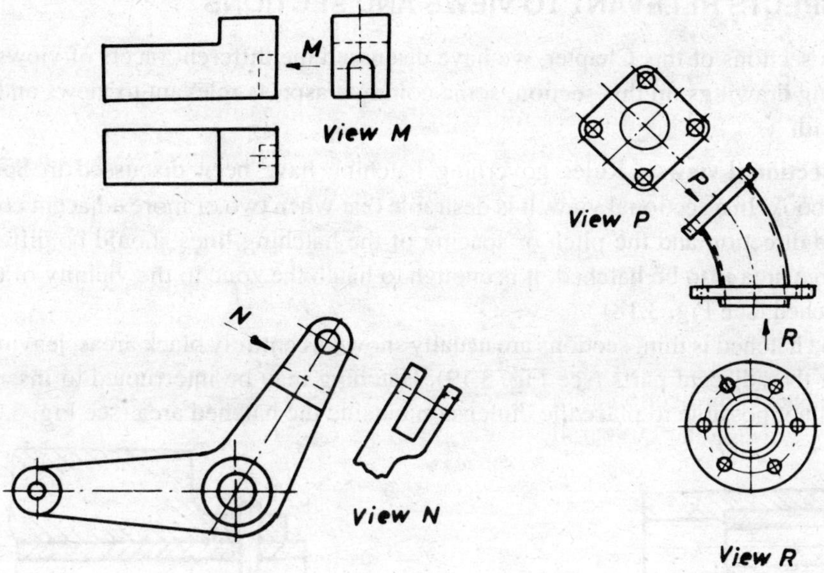

Fig. 3.14 Auxiliary views

structural sections and other similar parts. If desired, the visible lines of the part in the vicinity of the revolved section may be broken for better clarity.

A revolved section is sometimes shown away from the object. Such a section is shown separately and is placed in the neighbourhood of the object where the section is taken as shown in Fig. 3.16. This type

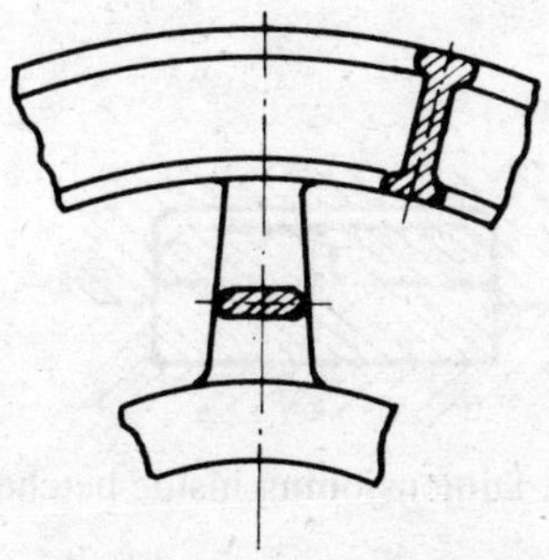

Fig. 3.15 Revolved view

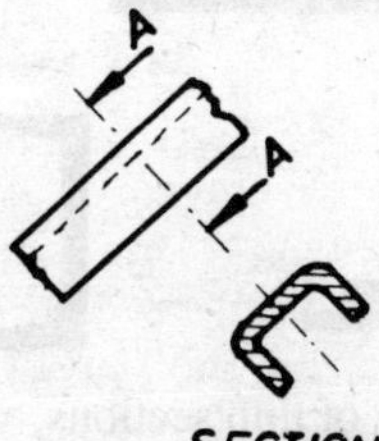

Fig.3.16 Removed section

of section is called a "removed section" and is extensively used in the drawings of parts like cross-hooks and other objects where the cross-sections of the elongated portions vary at frequent intervals. Sometimes these removed sections are drawn to a larger scale than the original scale to show small features in detail or to incorporate the relevant dimensions more clearly. Often, a removed section is a partial section. Though it is the common practice to place the removed sections in the vicinity of the object, sometimes such sections are carried over to a different sheet. In that case, cross-references in the form of notes should be given on the related sheets.

3.8 SOME ASPECTS RELEVANT TO VIEWS AND SECTIONS

In a few previous sections of this Chapter, we have discussed the different facets of views and sections used in engineering drawings. In this section, some common aspects relevant to views and sections will be briefly dealt with.

1. Hatching in sectional views: Rules governing hatching have been discussed in Section 3.5 and elsewhere in this book. In a sectional view, it is desirable that when two or more adjacent components are to be hatched, the direction and the pitch or spacing of the hatching lines should be different (see Fig. 3.17). When a large area is to be hatched, it is enough to hatch the zone in the vicinity of the contour of the area to be hatched (see Fig. 3.18)

If the parts to be hatched is thin, sections are usually shown as entirely black areas, leaving a thin white space in-between the adjacent parts (see Fig. 3.19). Hatching may be interrupted to insert the relevant dimension, if it is not possible to place the dimension outside the hatched area (see Fig. 3.20).

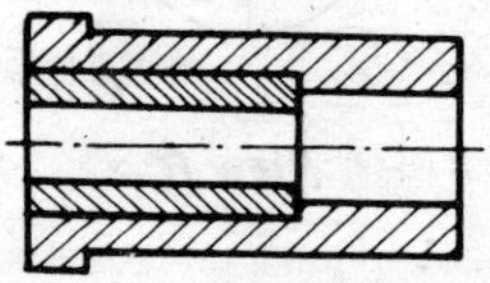

Fig. 3.17 Hatching of adjacent parts

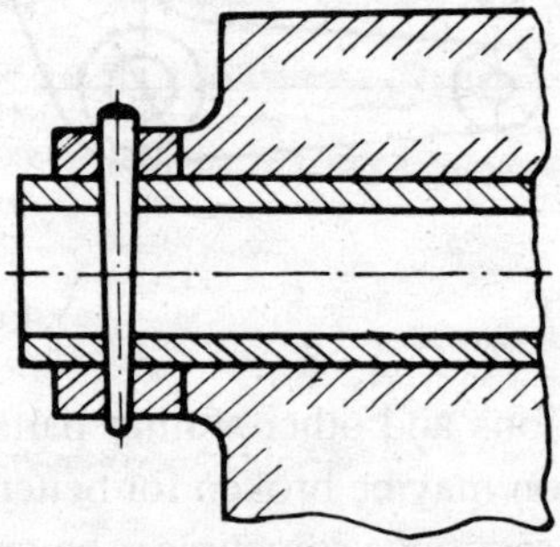

Fig. 3.18 Hatching of large area

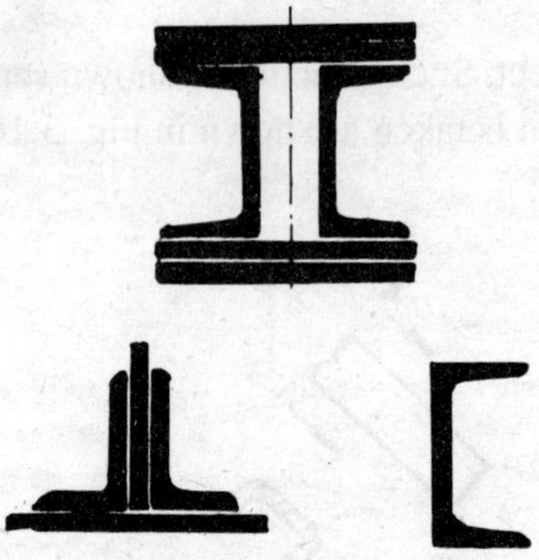

Fig. 3.19 Hatching of thin sections

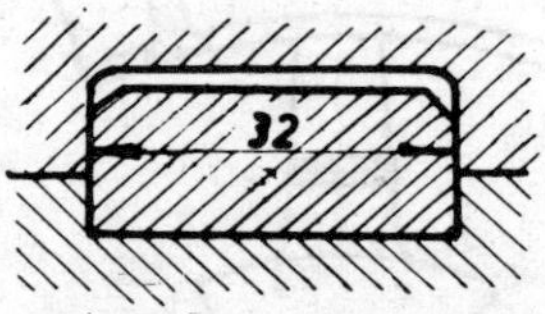

Fig. 3.20. Dimensioning inside hatched area

2. Half section and part section: In case of symmetrical parts, the drawing may be done partly as a view and partly as a section (see Fig. 3.21). If a complete or a half-section is not convenient, then a local or partial section may be drawn (see Fig. 3.22).

Fig. 3.21 Half section

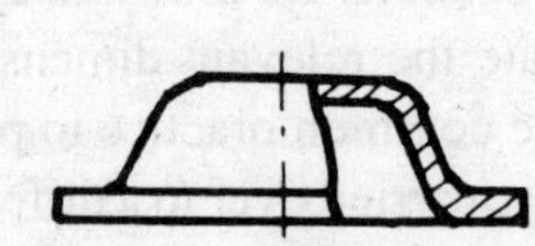

Fig. 3.22 Part section

3. Interrupted views: In case of symmetrical objects, it is enough to show only a fraction of the whole part to save time and space (see Fig. 3.23). Long objects can be shown as interrupted views when there is no scope for ambiguity (see Fig. 3.24). To show an interrupted view of a rolled section, the break-line should be a thin chain-dotted line (see Fig. 3.25).

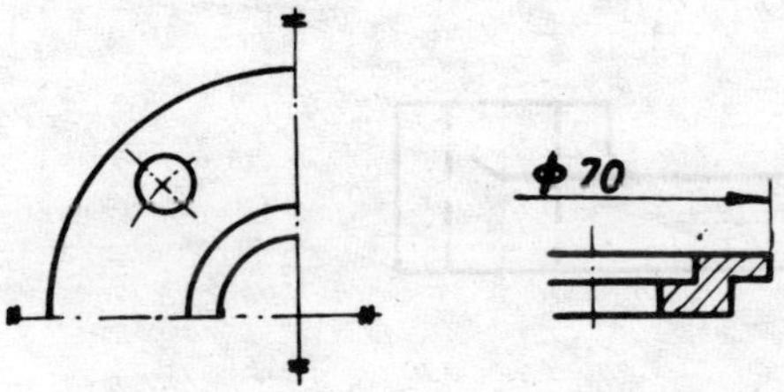

Fig. 3.23 Fractional views of symmetrical parts

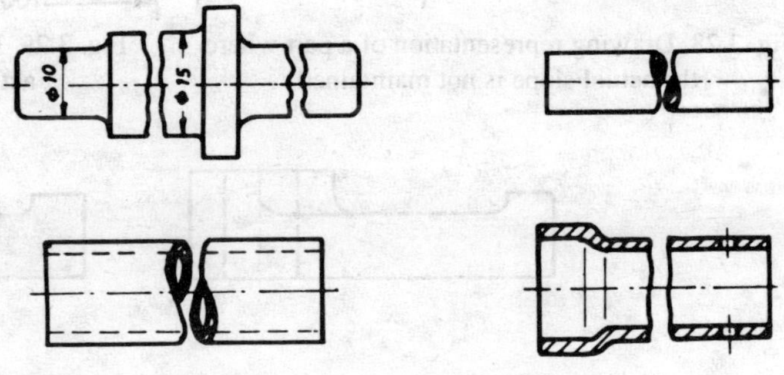

Fig. 3.24 Representation of long objects

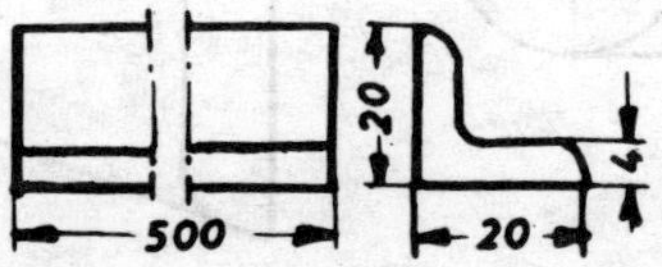

Fig. 3.25 Interrupted view of rolled section

4. Simplified views and sections: Views and sections are often represented in simplified forms. For example, to avoid an additional view, the bolt hole circle of a flange may be represented as shown in Fig. 3.26. Fig. 3.27 shows a part where a single view is sufficient to incorporate the required dimensions.

5. Objects with slope and taper: Interrupted parts with a slope or taper may be shown either with a line running from the big end to the small end disregarding the actual slope of the edge (see Fig. 3.28), or the true slope or taper may be adhered to and the interrupted parts may be represented as shown in Fig. 3.29. The value of the slope or taper may also be inserted as shown in Fig. 4.21 in Section 4.2 by triangular symbol which is the modern method of representing slopes and tapers.

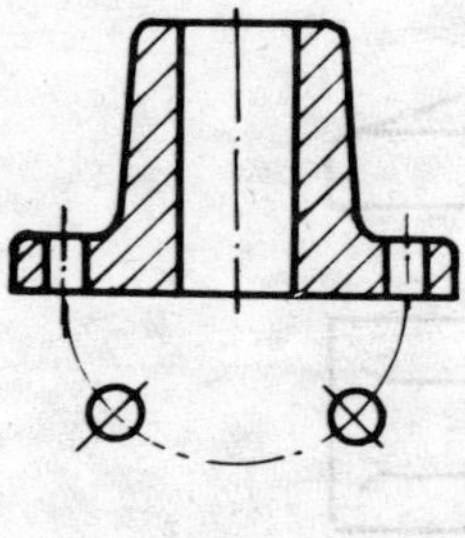

Fig. 3.26 Simplified view

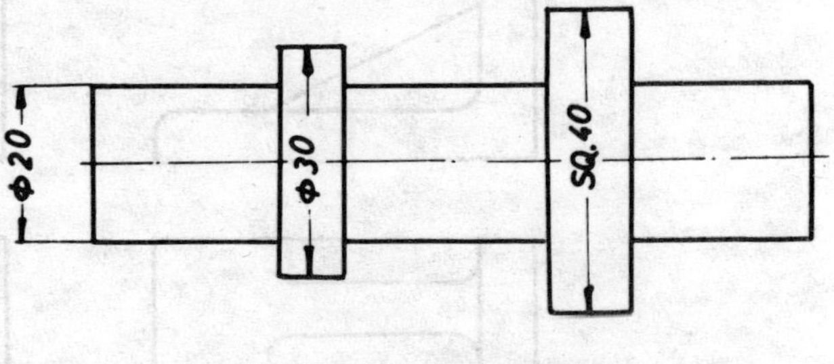

Fig. 3.27 Single view drawing of a part

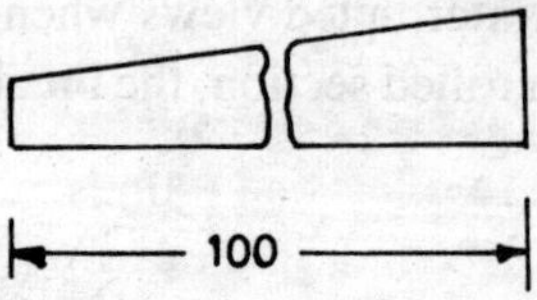

Fig. 3.28 Drawing representation of a part where the actual slope is not maintained

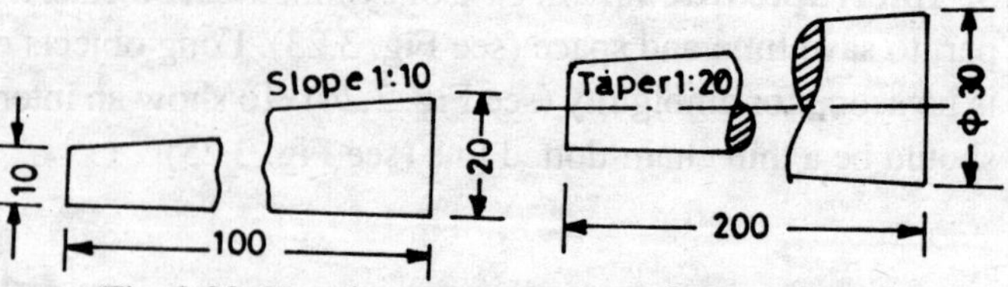

Fig. 3.29 Drawing representation of parts where the actual slope and taper are maintained

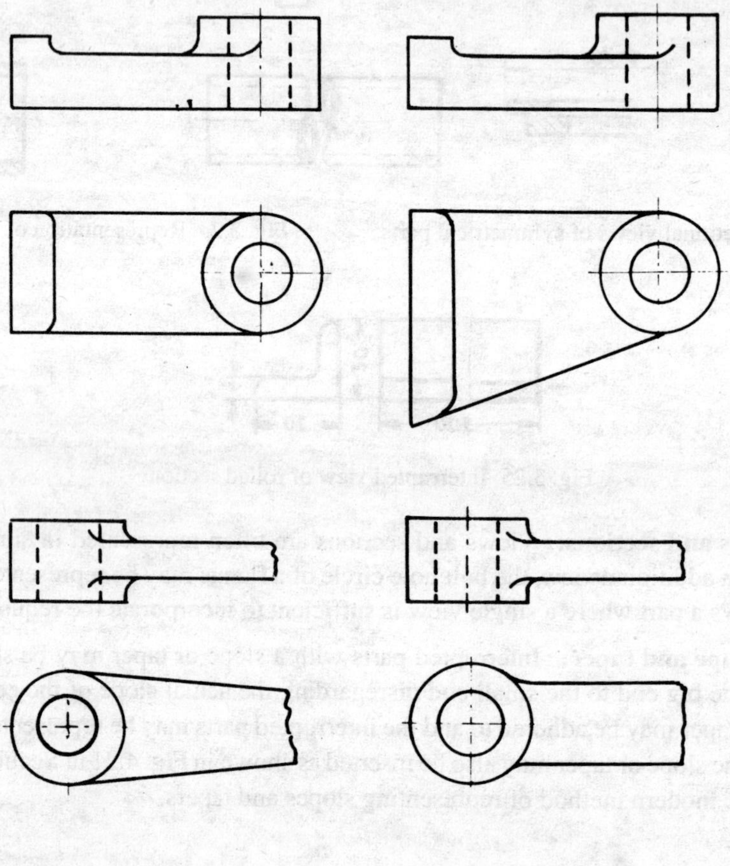

Fig. 3.30 Runouts in engineering drawings

6. Fillets and rounds in a drawing: In an engineering components, sharp edges and corners are generally avoided unless there is a specific reason to keep them. Sharp edges and corners are difficult to produce in case of cast or forged parts. Besides, they make the parts vulnerable to stress concentration and various types of failures. Drawings of parts are, therefore, provided with fillets and rounds.

A fillet is a rounded interior corner. A rounded exterior corner is called a round. In a drawing, fillets are represented by small curves when a plane surface meets a cylinder at a tangent. These curves are known as "run outs". A few typical examples of run outs are shown in Fig. 3.30. A run out has a radius equal to that of the fillet and its curved portion extends to around one-eighth of its circle. The curve ends at the projection of the point of tangency as shown in the figure.

CHAPTER 4

VARIOUS ASPECTS OF ENGINEERING DRAWING

4.1 TECHNICAL SKETCHING

As a preliminary step for making engineering drawings, free-hand sketches of parts are sometimes made. These are made by hand, that is, without the help of drawing instruments. However, some draughtsmen use compasses to draw arcs and circles.

A free-hand sketch may be made to serve as the initial picture which will be rendered into a full-fledged engineering drawing later on, or it can itself serve as a working drawing depending upon the exigency of situation, such as, an emergency in which the drawing of a part is required in a hurry to attend to a breakdown or to meet some similar contingency. It goes without saying that such a sketch must be a complete manufacturing drawing and should contain all the relevant dimensions and engineering data.

The draughtsman should endeavour to make the sketch in proportion to the real part as far as practicable. To facilitate this aspect, free-hand sketches in pencil should preferably be drawn on squared papers. Isometric papers are also available for making isometric sketches. For examples of free-hand sketches, see Figs. 4.1 to 4.4.

After making the sketch of the part, the relevant dimensions of the part should then be measured and entered in the sketch. If the sketch represents a new design-drawing based on the idea or the mental imagination of the designer, of course, no measuring instruments are required since there is no real part in such a case. If, however, the sketch is to be made from a part, instruments such as Vernier calipers, thread pitch gauge, radius gauge, steel scale, steel tape and other measuring gadgets are required. In this connection, the reader may refer to Section 10.4 dealing on spare part drawing.

As in the case of a drawing, the number of projections and views in a free-hand sketch should be limited to the barest minimum for adequate graphic representation of the part. The approximate scale, the required number of views to be drawn and other draughting considerations or constraints should be decided upon before starting to draw the sketch.

Very often the designers resort to free-hand sketches to record or clarify the nascent ideas about mechanical contrivances or equipments. The designers prepare these sketches prior to handling them over

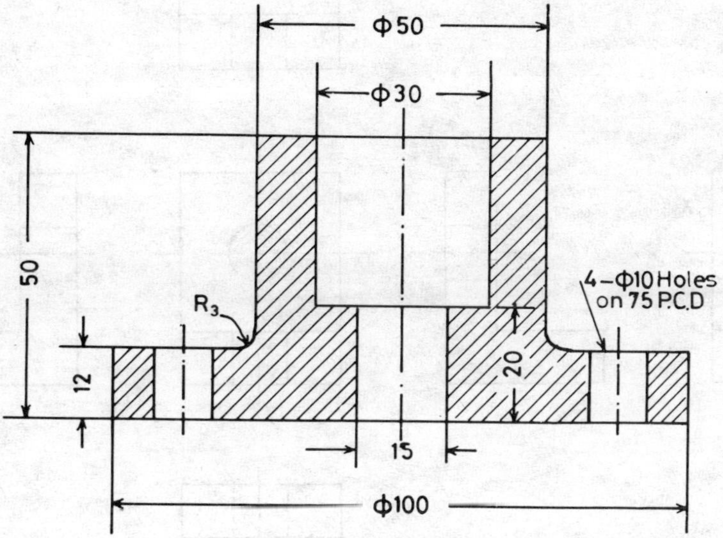

Fig. 4.1 Free-hand sketch of a flange

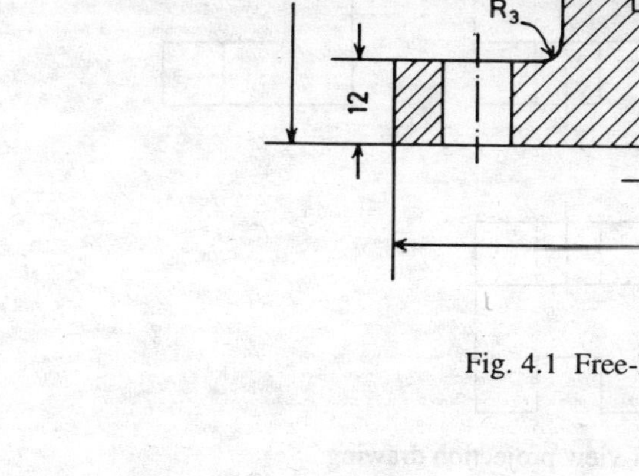

Fig. 4.2 Isometric sketch on plain paper

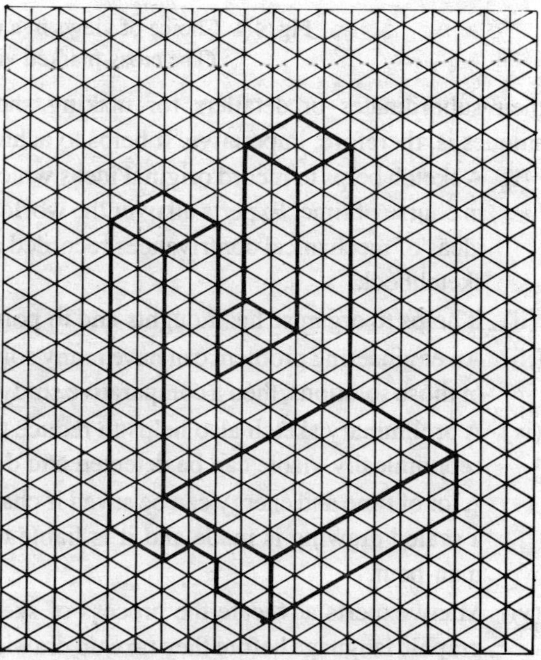

Fig. 4.3 Isometric sketch on isometric paper

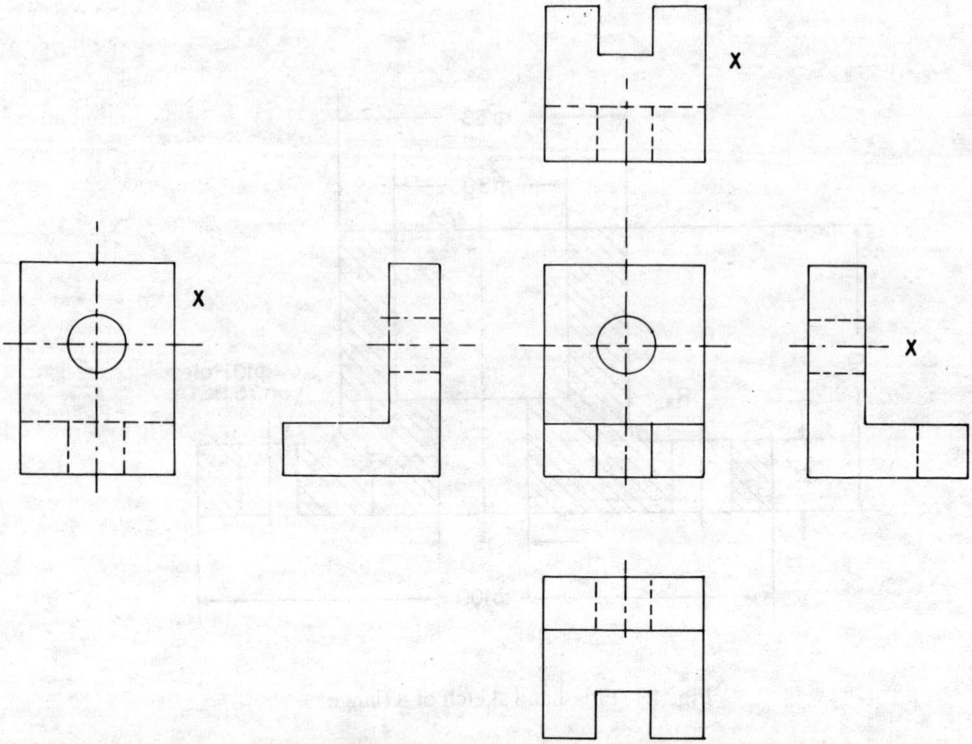

Fig. 4.4 Choice of views in a multi-view projection drawing.
(The cross-marked views are unnecessary)

to their subordinates or detailers for making shop drawings. A mastery over free-hand sketching techniques is, therefore, imperative. It helps to achieve the following objectives, among others:

(i) A sketch helps to record original ideas which would have otherwise escaped the designer's mind in course of time. Experts all over the world are still pondering and trying to decipher the original ideas of mechanical contrivances recorded as hand-sketches by Leonardo da Vinci in his famous sketchbook.

(ii) It helps to organise his thoughts and to point out the loop holes and inherent design defects.

(iii) An engineering problem can have many solutions. If several sketches are made representing each of those solutions, then a comparative study can be made and the best solution can be ascertained.

(iv) A preliminary sketch also helps to reduce the final cost since facets of the design which are not economically viable can be detected and discarded at the outset.

(v) A lot of time and energy can be saved if changes are made at the sketch making stage itself. Once a scaled drawing is made and it is liable to be altered, the process entails unnecessary loss of man-hour.

(vi) Many experts advocate the process of sketch making as a first step for beginners to learn the graphic language before they take up drawing making with instruments. Most technical drawings are made in orthographic projection and free-hand sketching is an excellent method for learning the fundamentals of this system quickly.

Different aspects of technical sketching : For sketching purpose, only paper, pencils and eraser are required. Normally, sketches are not made to scale. Parts are sketched by eye estimation and in their correct proportions as accurately as possible. Hand sketches can be made on any type of paper, but squared papers are preferable because free-hand straight lines can be drawn and the side of a square can be taken to represent some linear unit of convenience so that a more or less scaled and properly dimensioned picture can be made. Soft pencils, such as HB are preferable for sketching.

As indicated before, sketches may be made to express an idea on paper or they may be made from tangible objects. In both cases, sketches should be made as per some systematic order. In this connection, some points of relevance are discussed below:

1. Before the sketch of an object is made, a complete mental visualisation of the part is necessary. In the mind of the designer or the draughtsman, the mental image of the part must be very clear and definite. A clear mental picture leads to a clear actual picture.
2. Though orthographic multi-view projection method is generally followed while making technical sketches, sometimes pictorial methods are also employed. In a pictorial view, all dimensions can be entered in one view in most of the cases and notes as regards thickness, shape and other relevant information can be written by the side so that space and time can be saved while sketching. Such a sketch can be rendered into multi-view projection drawings in the final stage. The choice of representation (multi-view or pictorial) should therefore be decided before actual sketching.
3. Choice of views is one of the most important considerations in sketching. A shop drawing should ideally consist of only those views or sections which are necessary for a clear visualisation as to the shape or description of the object. Views and sections should be minimum in number and the draughtsman should make a judicious choice as to the selection of these, so that the essential contours and shapes are best represented. Preference should be given to those which contain the minimum number of hidden lines (Fig. 4.4). If the top, front and one of the side views are adequate to describe the part, it is unnecessary to include the rear, bottom and the other side views. In some cases, parts may have such shapes where only two views are sufficient (Fig. 4.6). Sometimes even one view is sufficient, *viz.*, the front view of a plate on which the plate thickness is written, a shaft and other round objects, *etc.* See Fig. 4.12 and 4.11.
4. In Fig. 4.5, the right ways of drawing hidden or dotted lines have been shown along with incorrect practices. As shown in Fig. 4.5(a), a gap should be provided between the visible and the dotted line. The joining points of dotted lines should be clearly shown as in Fig. 4.5(b). A visible line should preferably be drawn through the gap of a dotted line, Fig. 4.5(c). Parallel lines should be drawn in a staggered fashion, Fig. 4.5(d). Views should preferably be so chosen that they show the visible lines mostly. Only those hidden lines should be shown which are really necessary for clarity. Too many dotted lines unnecessarily confuse the reader of the drawing and are detrimental to proper understanding of the drawing.
5. Centre lines should be properly placed in a sketch. They indicate the axes of symmetrical parts or features. Also to represent bolt circles, pitch circles of gears, sprocket wheels and other transmission elements, neutral axes of VEE belts and pulleys, and in other features, centre lines are used.

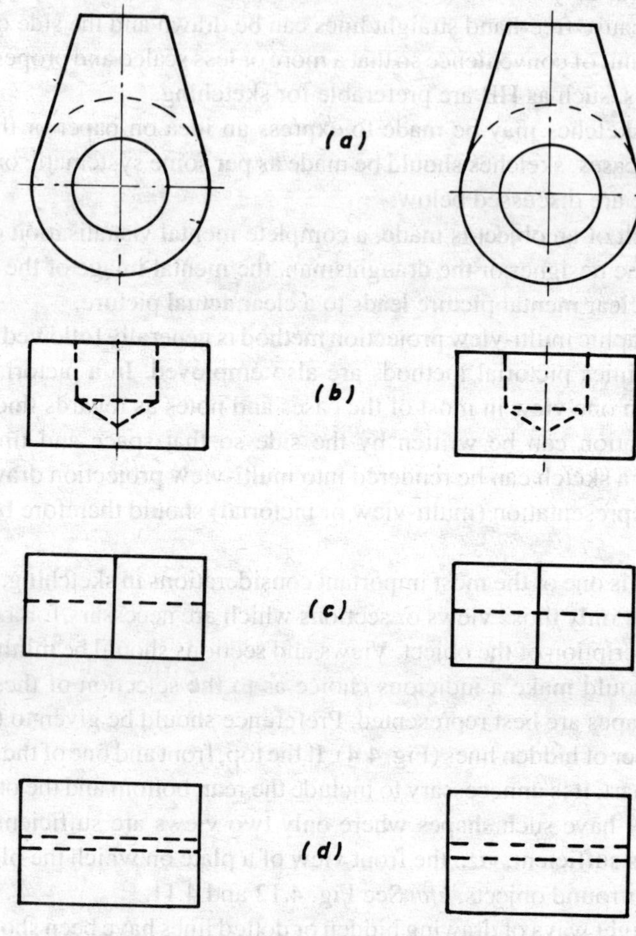

Fig. 4.5 Right ways of drawing hidden lines

Centre lines should not extend much outside the drawing of the part–around 3 to 5 mm is sufficient. These lines can be drawn as a chain-dotted line, *i.e.*, dash and dot alternating, or they may be drawn as alternating long and short dashes. Centre lines should be drawn as thin lines. Some examples of centre lines are shown in Fig. 4.6.

6. Sometimes it may so happen that visible (solid), hidden (dotted), and centre lines coincide in a drawing. In such cases, the order of precedence is : visible line, hidden line and then centre line (Fig. 4.7).

As regards lines used in technical drawings, the reader is advised to refer to Section 1.4.

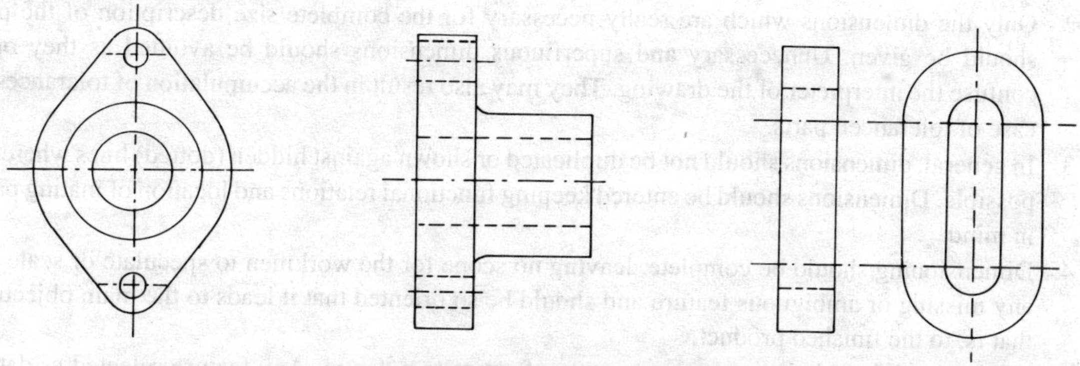

Fig. 4.6 Examples of centre lines

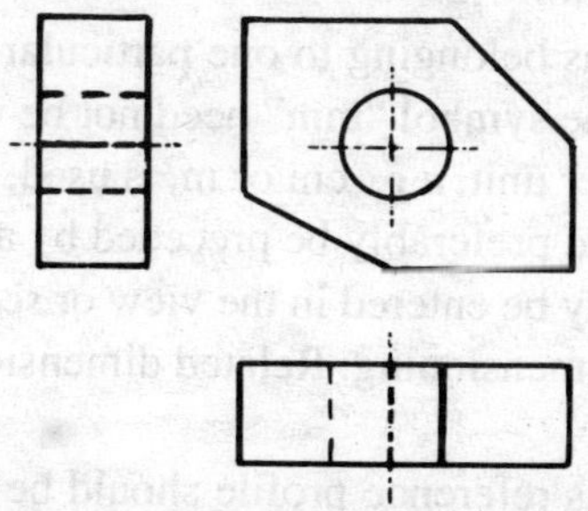

Fig. 4.7 Order of precedence of various types of lines

4.2 DIMENSIONING IN DRAWING

The orthographic and pictorial views describe the shape of an object. Such views and sections are to be supplemented by proper dimensions to arrive at the size of the object. In general, the shape, size and the relevant notes thereof supply the complete information in an engineering drawing for the manufacture or construction of the part.

Depending on the particular requirement, the dimensioning of parts on drawings are carried out as per the functions to which they are subjected or as per the finishing procedures adopted or as per the gauge and inspection requirements. The rules governing the dimensioning given in this book are in accordance with the codes laid down in IS: 696 and DIN 406. The rules apply to requirements which are frequently encountered in common shop practices. Engineering concerns which manufacture special products have their own codes for dimensioning commensurate with their peculiar needs. These are laid down in their work-norms and are followed in such concerns. This section deals with the cases of dimensioning engineering items used or manufactured in general mechanical shops. The salient points and guidelines as to the correct dimensioning procedures are summarised below with illustrations wherever necessary.

1. The necessity for the size description in the form of proper dimensioning stems from the fact that parts are now made at widely separated places and these are supposed to mate properly when assembled. Thus, the need for inter-changeability of items is the very basis of the modern dimensioning procedures.

2. Only the dimensions which are really necessary for the complete size description of the part should be given. Unnecessary and superfluous dimensions should be avoided as they only confuse the interpreter of the drawing. They may also result in the accumulation of tolerances in case of toleranced parts.
3. In general, dimensions should not be duplicated or shown against hidden (dotted) lines wherever possible. Dimensions should be entered keeping functional relations and location of mating parts in mind.
4. Dimensioning should be complete, leaving no scope for the workmen to speculate or scale out any missing or ambiguous feature and should be so oriented that it leads to the main objective, that is, to the finished product.
5. Dimensions should always be given with reference to a datum. Any feature selected as datum *e.g.*, lines, edges, *etc.*, should be so chosen as to be easily recognizable. Datum has been more elaborately dealt with in Section 7.2.
6. It is desirable that dimensions belonging to one particular drawing should be expressed in one unit only, *viz.*, millimetre. The symbol "mm" need not be written beside the numerical value of a dimension, but if some other unit, *e.g.*, cm or m, is used, the relevant symbol should be given. Numbers less than one should preferably be preceded by a zero, *viz.*, 0.87.
7. Dimensions should preferably be entered in the view or section where the shape of the part best represented at the place of dimensioning. Related dimensions should be given in the same view as far as practicable.
8. The profile which is to act as reference profile should be selected from design considerations. Such a feature, *e.g.*, a machined surface, edge of a part, *etc.*, once decided upon, should not be deviated from, and all dimensions should be given in accordance with that surface or edge.
9. Chordal and arcual dimensions should be identified by a bar and a bow respectively over the numerical values. These and the angular dimension should be entered as shown in Fig. 4.8.
10. The dimension lines, leader lines, arrow-ends, *etc.*, should be as shown in Fig. 4.9. When there is a lack of space, dots can be put instead of arrow-heads. Dimension lines should preferably be spaced 8 to 10 mm apart. To avoid the danger of cumulative tolerances, the intermediate dimension should not be entered when the overall dimension is given as in Fig. 4.10.
11. Dimensions can be entered in staggered fashion for dearth of space as shown in Fig. 4.11. For large symmetrical objects, it is not always necessary to show both the ends of the dimension lines. The symbol ϕ indicating diameter need not be used when the two arrow-heads at the ends of a diametral line terminate at the circumference of a circle. The word SPHERICAL should be written when such an object is shown in a single view.
12. Stepped dimensions should be entered with reference to a datum line or plane suitably chosen for the purpose. If a rectangular plane on a round object is machined for some purpose, it should be identified by entering diagonal crosses. The letters A/F indicating across flat, can be put to denote spanner size.
13. Co-ordinate system of dimensioning can be adopted in case of special geometrical shapes (see Fig. 4.12). In case the part consists of a plate for which one view is sufficient, the thickness of a plate can be written directly on the body of the drawing, *e.g.* THK

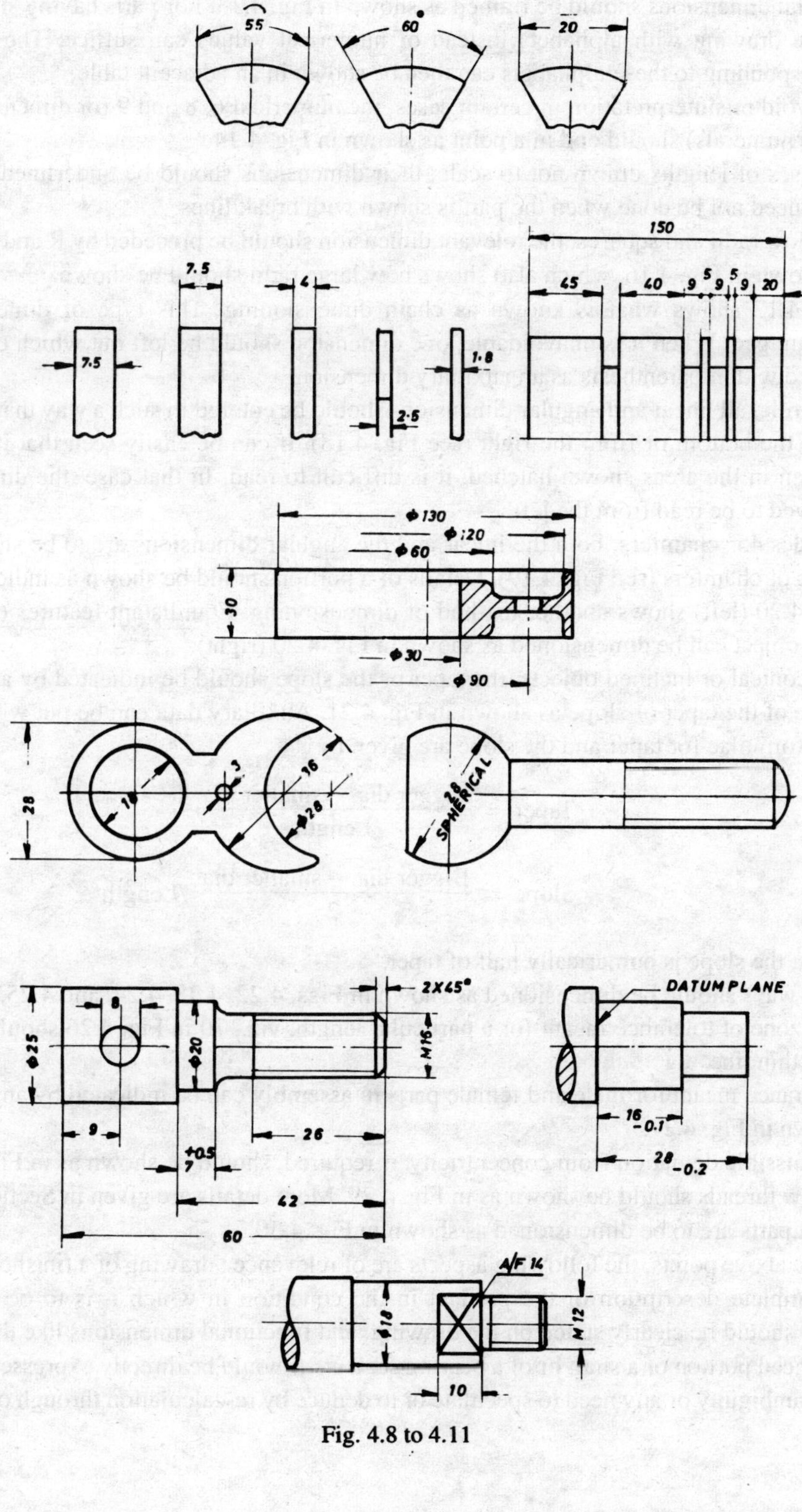
55
60°
20
150
7.5
4
45
40
9
5
9
5
9
20
7.5
1.8
2.5
Φ130
Φ120
Φ60
30
Φ30
Φ90
28
18
3
16
Φ28
28 SPHERICAL
2X45°
DATUMPLANE
8
Φ25
Φ20
M16
16 -0.1
9
26
+0.5 7
28 -0.2
42
60
A/F 14
Φ18
M12
10

Fig. 4.8 to 4.11

14. Special dimensions should be framed as shown in Fig. 4.13. For parts having similar features, a single drawing with alphabets instead of numerical values can suffice. The relevant values corresponding to these alphabets can then be shown in an adjacent table.
15. To avoid misinterpretation in certain cases, the numerical 6, 8 and 9 (or dimensions containing these numerals) should end in a point as shown in Fig. 4.14.
16. In cases of lengths drawn not to scale, their dimensions should be underlined (see Fig. 4.15). This need not be done when the part is shown with break lines.
17. To show radii and squares, the relevant dimension should be preceded by R and SQ respectively as shown in Fig. 4.16, which also shows how large radii should be shown.
18. Fig. 4.17 shows what is known as chain dimensioning. This type of dimensioning is not encouraged. When it is unavoidable, one dimension should be left out which can, however, be placed within parenthesis as an auxiliary dimension.
19. As a rule, all linear and angular dimensions should be entered in such a way that they are legible from the bottom or from the right (see Fig. 4.18). If can be easily seen that if a dimension is written in the areas shown hatched, it is difficult to read. In that case, the dimension may be allowed to be read from the left.
20. Besides 45° chamfers, both the linear and the angular dimensions are to be shown in all other cases of chamfers (see Fig. 4.19). Details of a portion should be shown as indicated.
21. Fig. 4.20 (left) shows another method of dimensioning. Equidistant features (holes, *etc.*) on a long object can be dimensioned as shown in Fig. 4.20 (right).
22. For conical or inclined objects, the taper or the slope should be indicated by a triangle and the value of the taper or slope as shown in Fig. 4.21. Auxiliary data can be put within parenthesis. The formulae for taper and the slope are given by :

$$\text{Taper} = \frac{\text{Bigger dia.} - \text{smaller dia.}}{\text{Length}}$$

$$\text{Slope} = \frac{\text{Bigger dia.} - \text{smaller dia.}}{2} /\text{Length}$$

Thus, the slope is numerically half of taper.

23. Key-ways should be dimensioned as shown in Figs. 4.22, 4.23, 4.24, and 4.25.
24. The zone of tolerance meant for a particular length, viz., 70 in Fig. 4.26 should be demarcated by a thin line.
25. Tolerance meant for male and female parts in assembly can be indicated by any of the methods shown in Fig. 4.27.
26. Permissible deviation from concentricity, if required, should be shown as in Fig. 4.28.
27. Screw threads should be shown as in Fig. 4.29. More details are given in Section 8.1.
28. Bent parts are to be dimensioned as shown in Fig. 4.30.

Besides the above points, the following aspects are of relevance : drawing of a finished product should contain a complete description of the product in the condition in which it is to be used, functional requirements should be clearly stated on the drawing, and functional dimensions like threaded length of a part, toleranced portion of a shaft or of a bearing seat, *etc.*, should be directly expressed on the drawing to avoid any ambiguity or any need to speculate or to deduce by re-calculation through other dimensions.

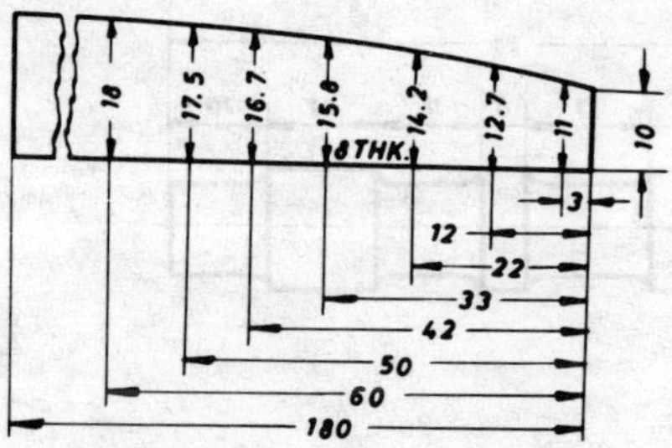

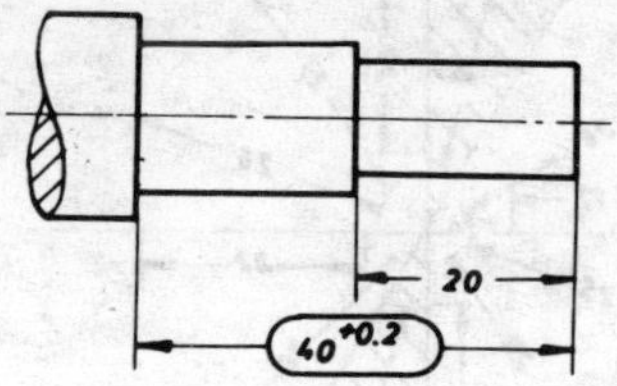

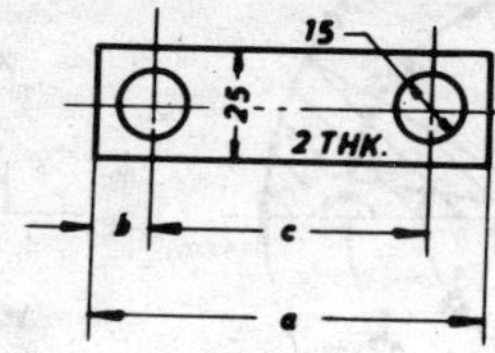

SL. NO.	a	b	c
1	100	15	70
2	140	15	110
3	220	20	180

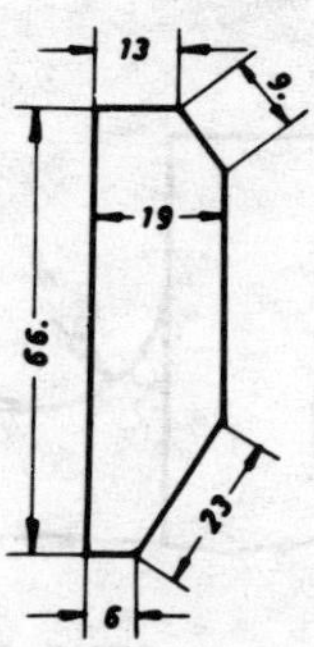

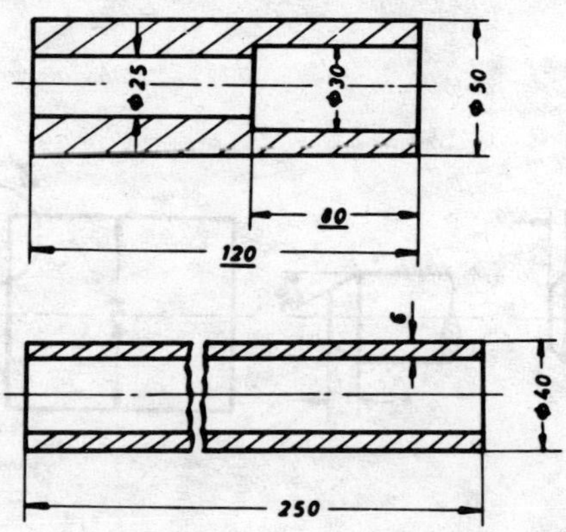

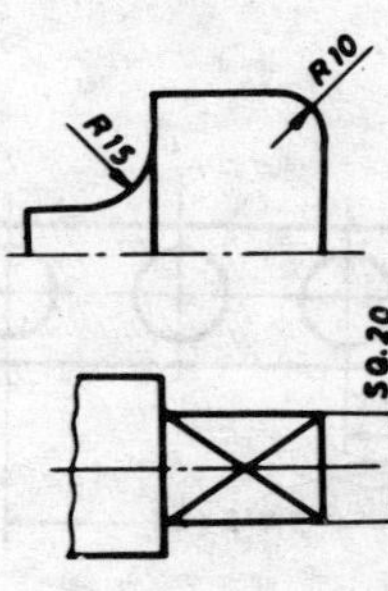

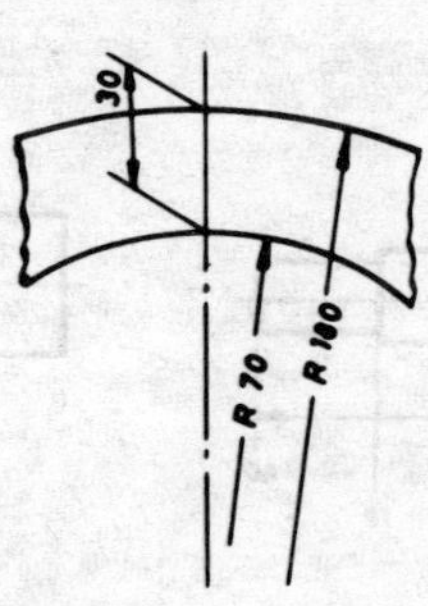

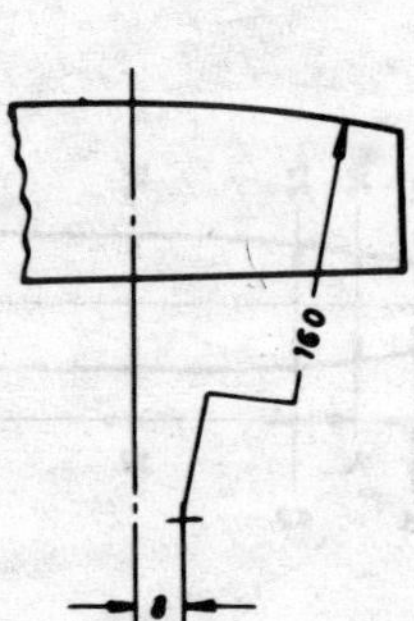

Fig. 4.12 to 4.16

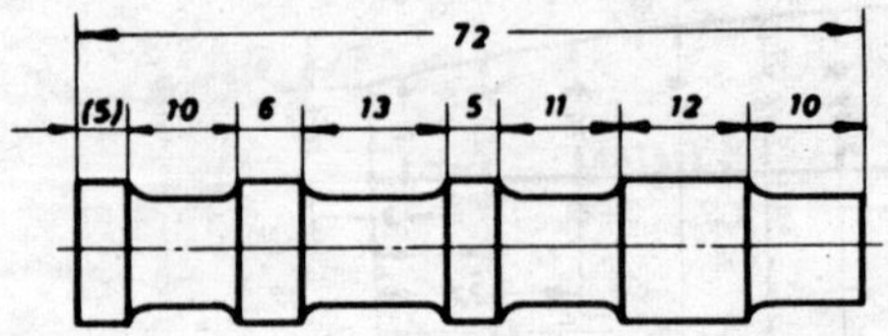

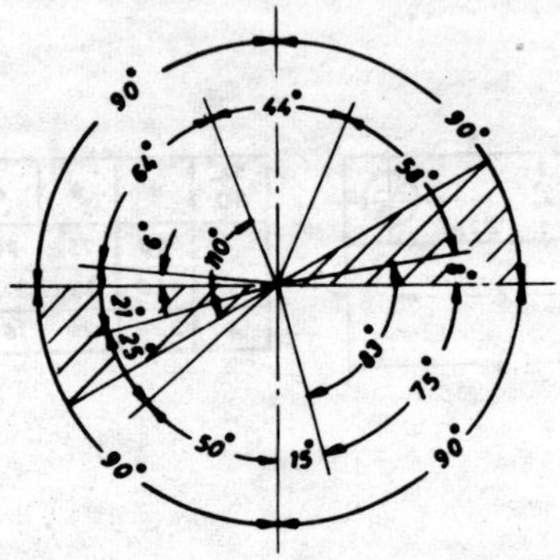

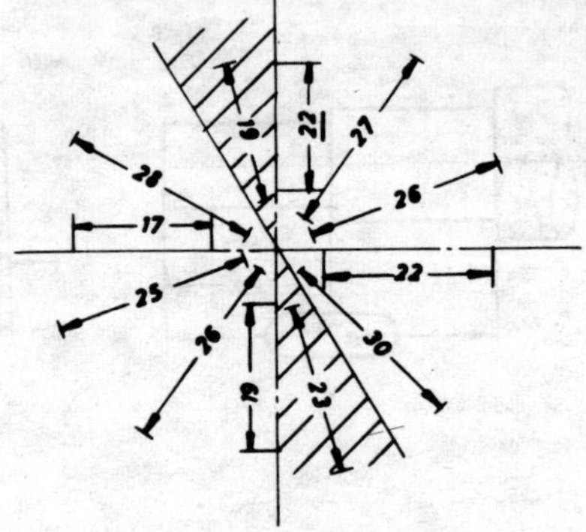

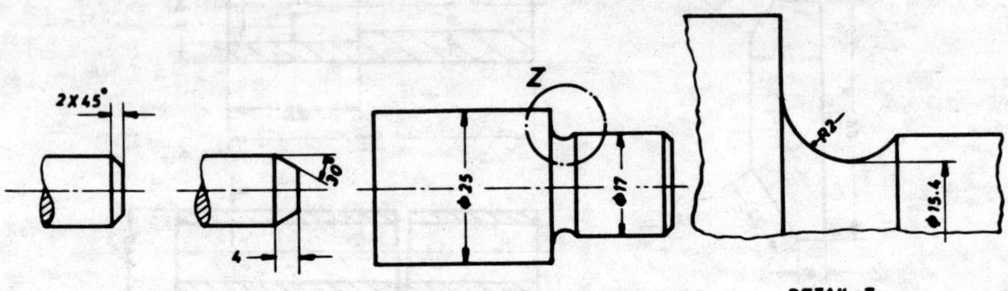

DETAIL-Z

5:1

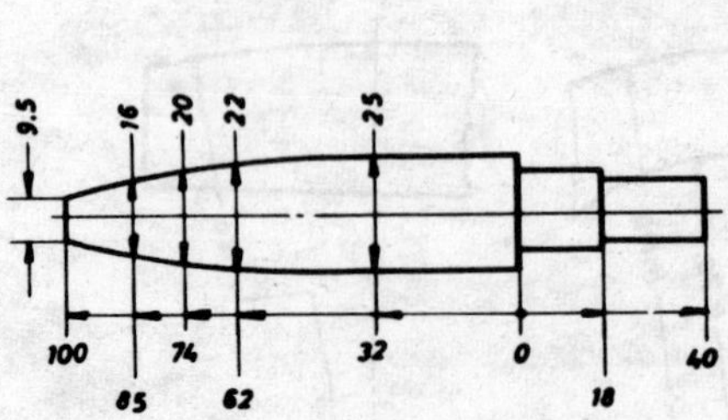

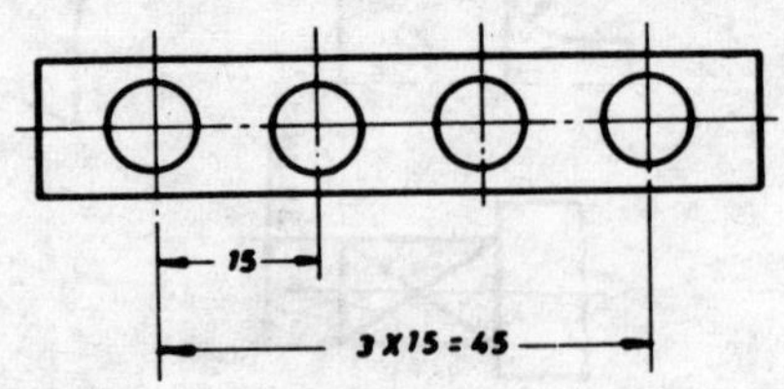

Fig. 4.17 to 4.20

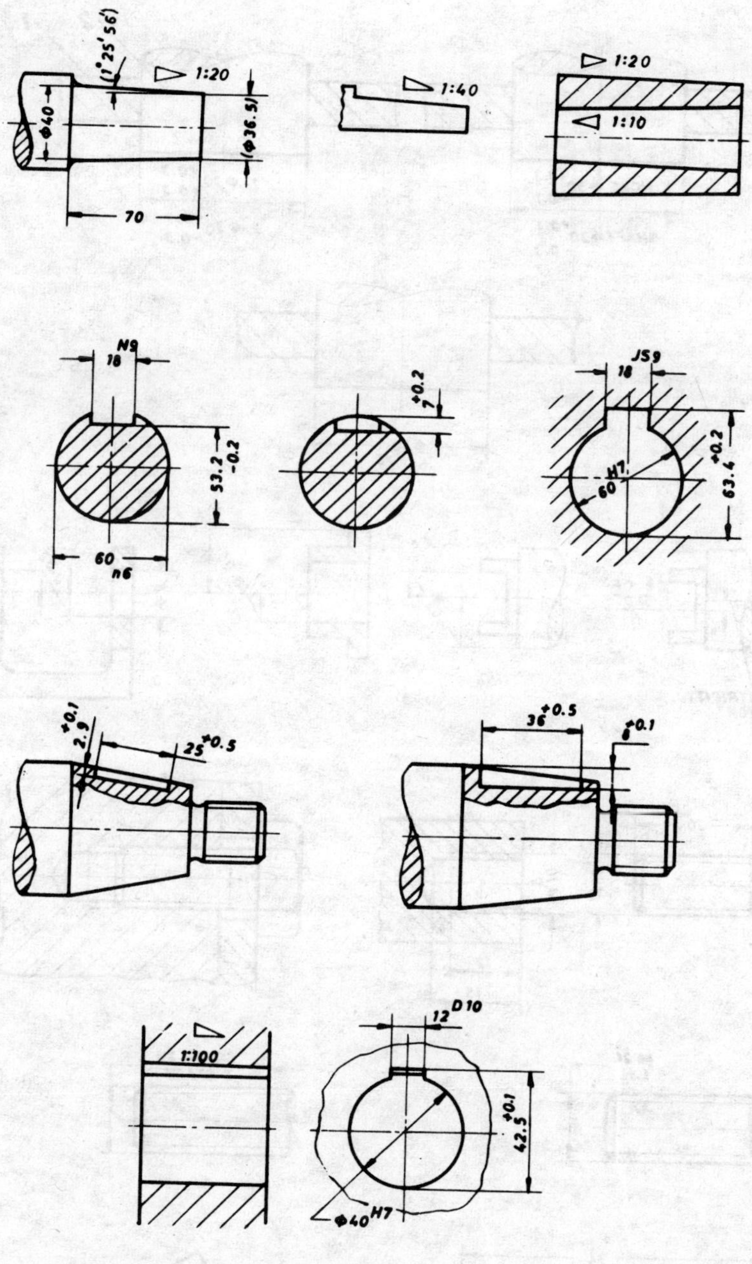

Fig. 4.21 to 4.26

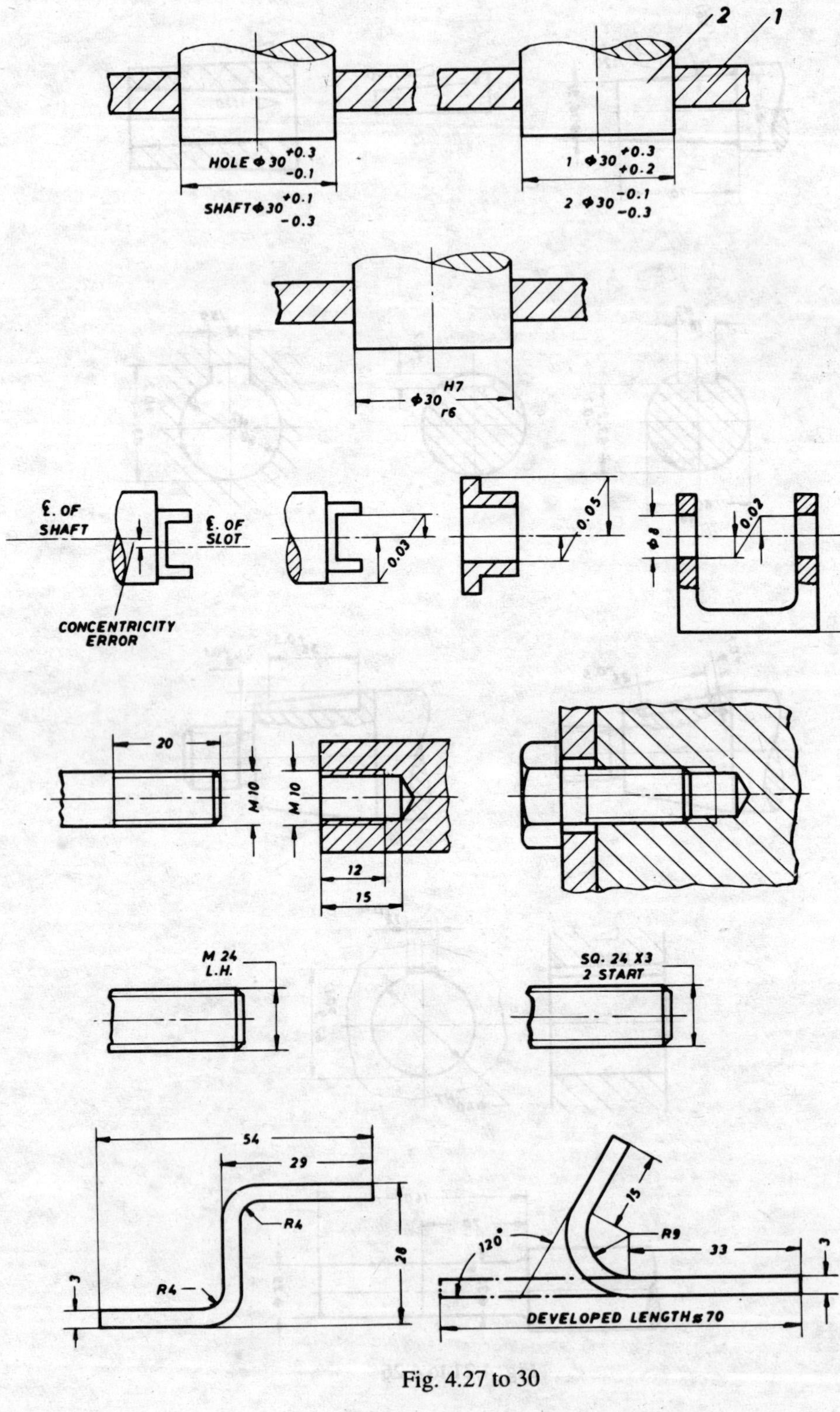
2
1
HOLE Φ30 +0.3 −0.1
SHAFT Φ30 +0.1 −0.3
1 Φ30 +0.3 +0.2
2 Φ30 −0.1 −0.3
Φ30 H7 r6
℄. OF SHAFT
℄. OF SLOT
CONCENTRICITY ERROR
0.03
0.05
0.02
Φ8
18
20
M 10
M 10
12
15
M 24 L.H.
SQ. 24 X3 2 START
54
29
R4
R4
28
3
120°
15
R9
33
3
DEVELOPED LENGTH≈70

Fig. 4.27 to 30

4.3 CORRECT AND INCORRECT REPRESENTATION IN DRAWINGS

In this section, the correct and the incorrect representations in engineering drawings have been discussed. The common incorrect representations which the engineering personnel working in a drawing office are liable to make have been pointed out and their correct versions incorporated.

Some of the representations shown under the heading "Incorrect" are not necessarily mistakes, but those corresponding representations shown in the "Correct" versions are better and therefore preferable.

INCORRECT	CORRECT
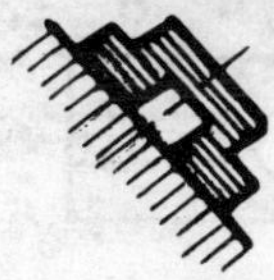	

Section lines are normally drawn at 45° to the horizontal. If the section is shown in an inclined position, then the hatching direction can be deviated from the usual practice.

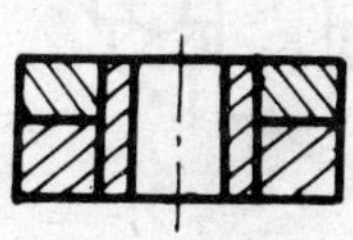

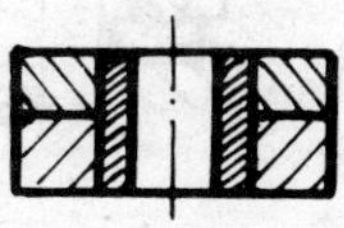

Parts in assembled condition should be hatched with different pitch and direction of section lines.

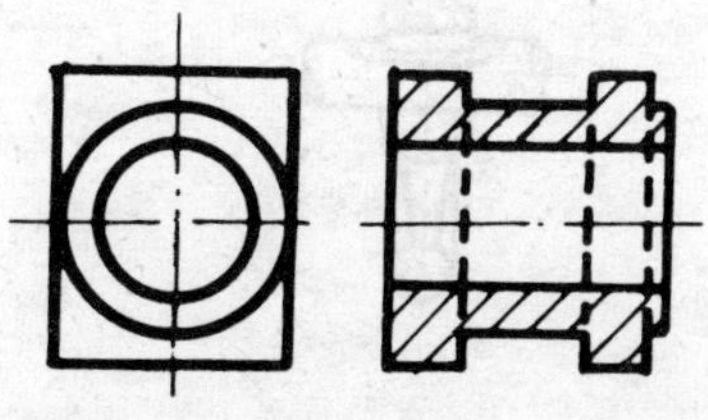

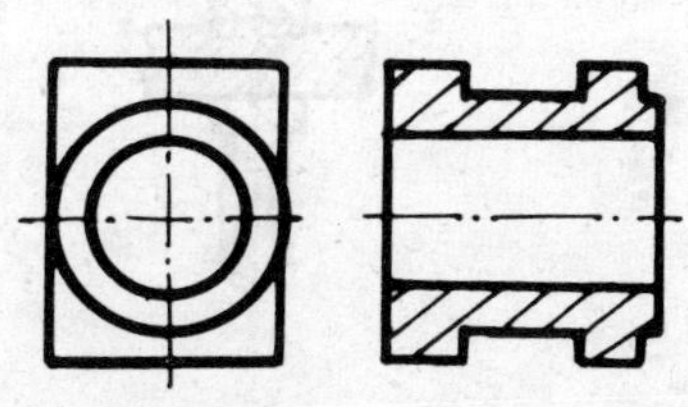

Superfluous dotted lines should be avoided. When these lines are essential for correct interpretation of the drawing, then only the dotted lines should be used in sectional views.

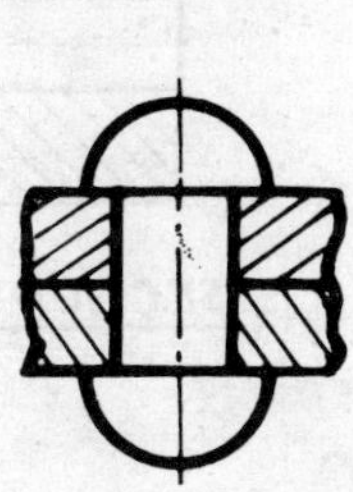

Rivets, shafts, ribs, pins, arms of pulleys, etc., should not be batched.

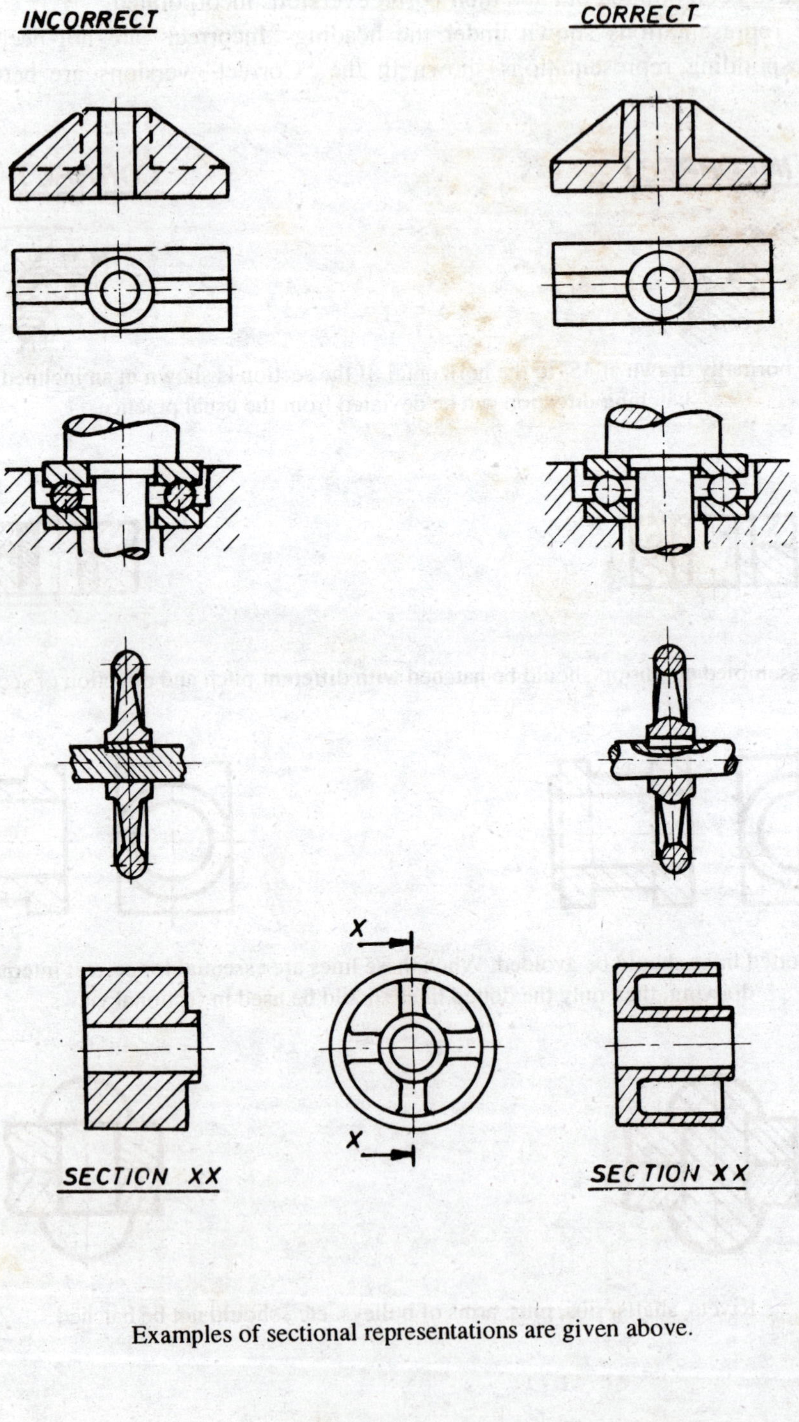

Examples of sectional representations are given above.

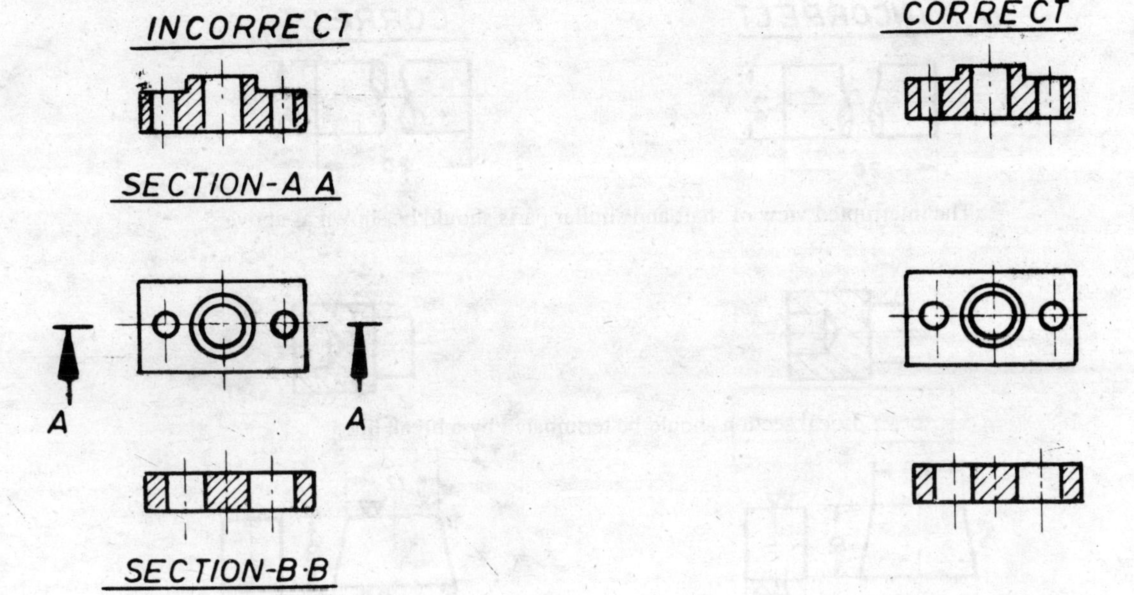

In simple cases, indication of cutting plane can be avoided.

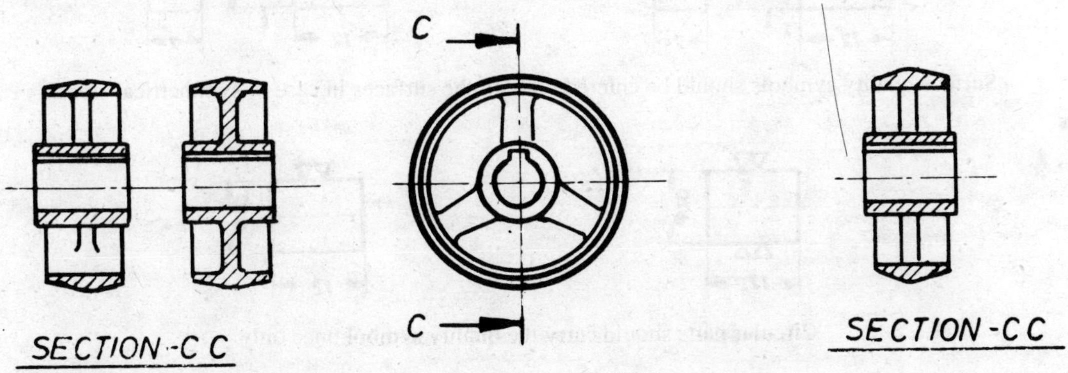

In case of parts with odd-number of arms, the conventional sectional view should be as shown.

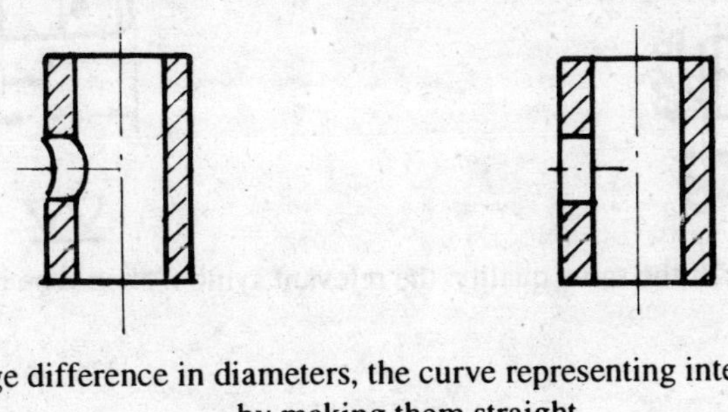

When there is a large difference in diameters, the curve representing intersection can be simplified by making them straight.

INCORRECT CORRECT

The interrupted view of shaft and similar parts should be shown as above.

Local section should be terminated by a break line.

It is preferable to incorporate surface quality symbols on the surfaces which carry the relevant dimensions.

Surface quality symbols should be entered on both the surfaces in case of symmetrical sides.

Circular parts should carry the quality symbol once only.

When all the surfaces are of the same quality, the relevant symbol should be indicated as shown.

CORRECT

INCORRECT

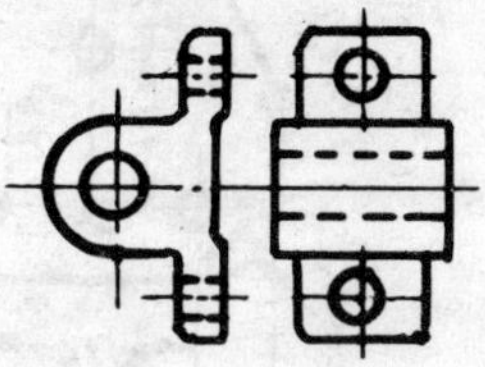

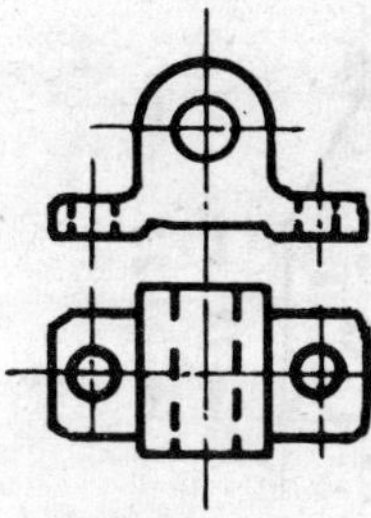

It is preferable to draw parts in their working position.

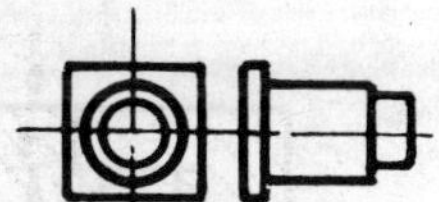

Cylindrical parts like pins, shafts, bolts, etc., should preferably be drawn in horizontal position. Thinner end should point towards right.

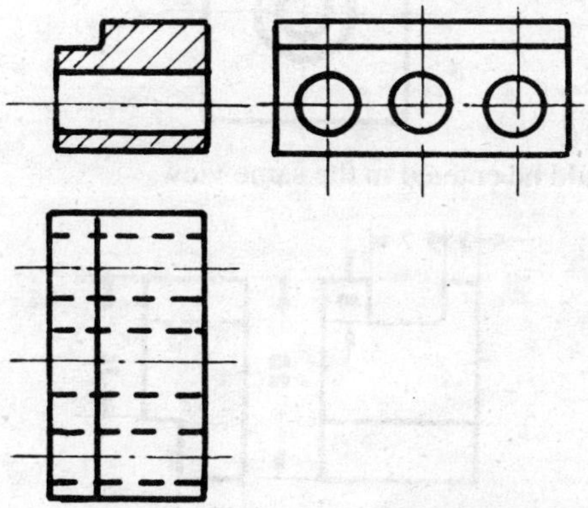

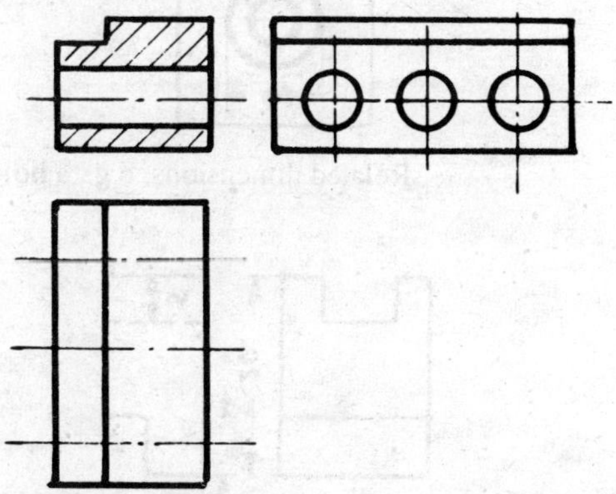

Unnecessary dotted lines in views should be avoided.

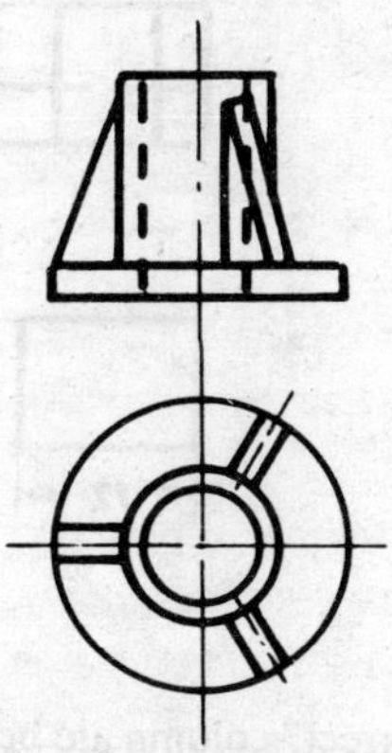

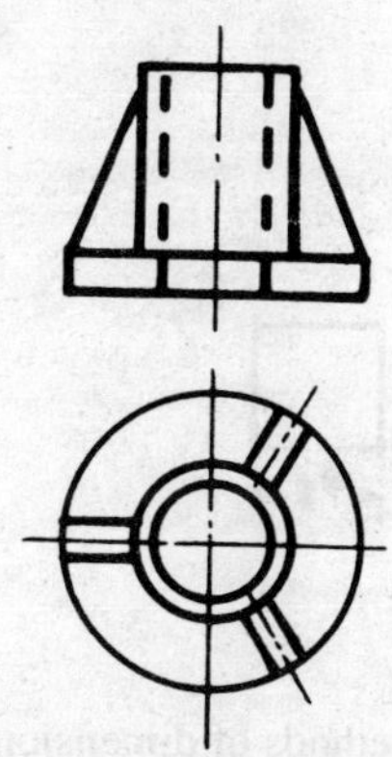

Ribs, etc., should be shown symmetrical as in the example above.

INCORRECT CORRECT

Symmetrical objects can be so dimensioned that only one side carries the dimensional values.

Related dimensions, e.g., a hole and its depth, etc., should be entered in the same view.

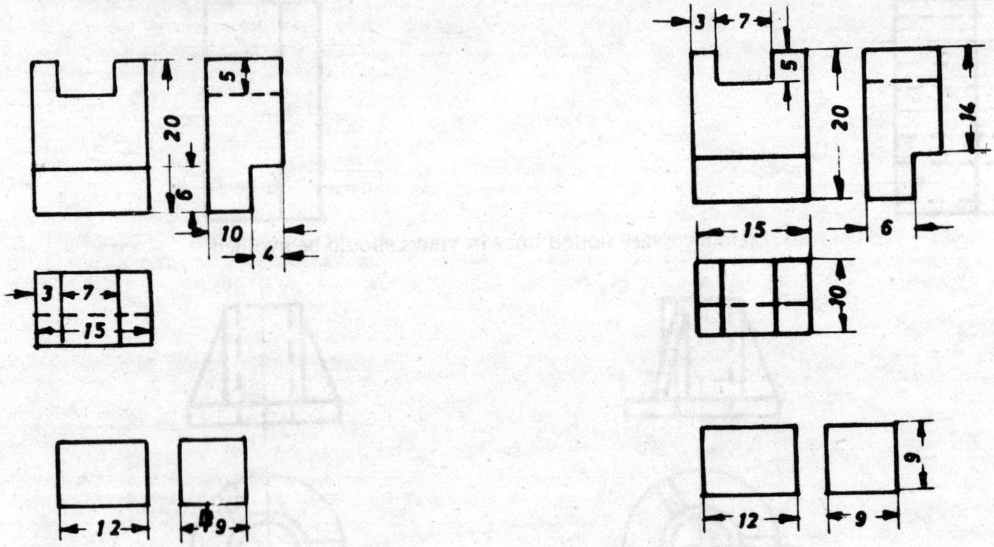

Methods of dimensioning shown under "Correct" column are better.

INCORRECT CORRECT

The centre lines are to be shown as indicated in the "Correct" column.

Unless there is a chance of confusion, symmetrical dimensions need not be shown.

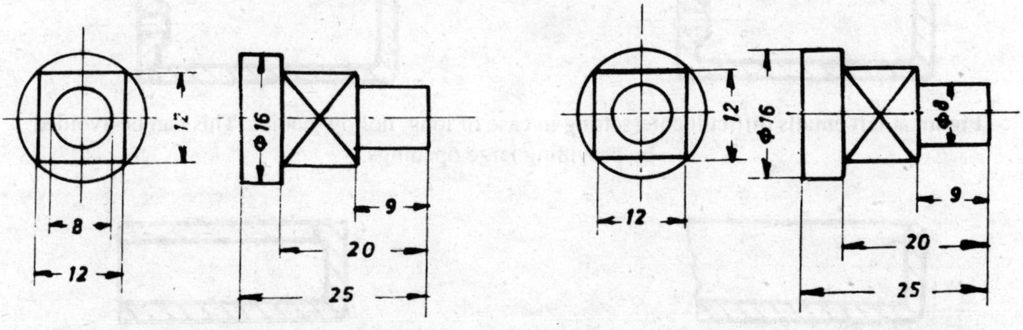

It is preferable that the lengths and the relevant diameters be shown in the same view.

Unless required for some special reason, dimensions are to be shown on the visible outlines and not on hidden areas.

INCORRECT **CORRECT**

It is desirable to have equal wall-thicknesses as far as possible. Unequal wall-thicknesses may result in faulty casting.

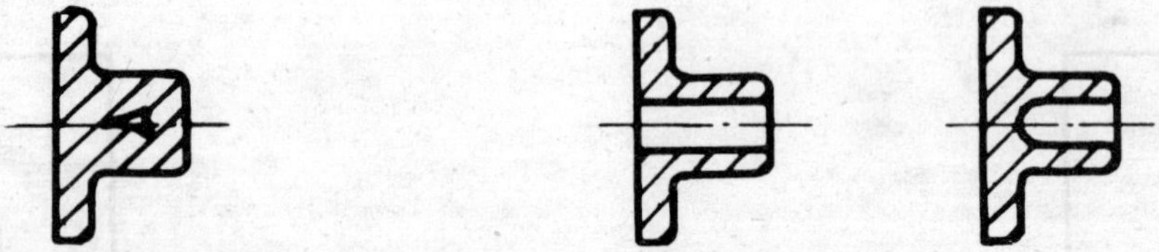

Hubs, flanges and similar configurations with cores usually help to avoid blow-holes.

Spot-faced seats for receiving bolt heads, nuts, etc., are cheaper than raised face.

Figure at left entails difficult core-setting in case of long, hollow bodies. This can be avoided by providing large openings.

Defects can arise when cooling. Edges and long ribs with projections alleviate such possibilities.

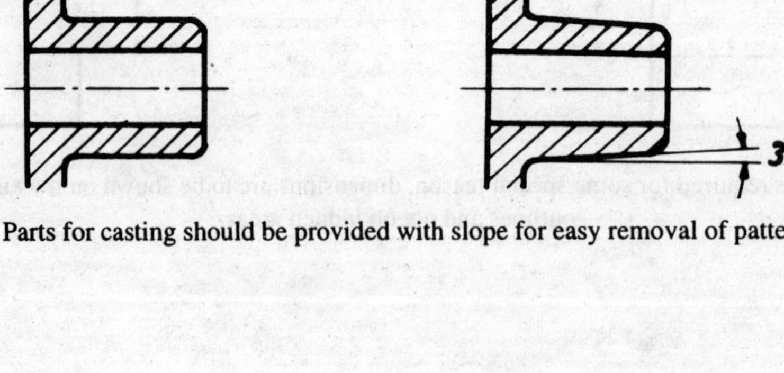

Parts for casting should be provided with slope for easy removal of pattern.

4.4 CONVENTIONAL REPRESENTATION OF COMMON ENGINEERING ITEMS AND MATERIALS

In an engineering drawing, it is not always necessary that a part should be drawn faithfully as per its dimensions to produce an exact representation of the shape or form of the part. Often it is sufficient to draw a part symbolically or in a simplified manner. To achieve this purpose, conventional representations of common engineering parts and features have been codified in Indian Standards and Standards of other countries as well as in some Company norms. These simplified representations save time, drawing space and cost.

Some of these representations are shown in Figs. 4.31, 4.32, 4.33. Besides these, conventional representations of other common machine elements and engineering features have been covered elsewhere in this book where these are discussed, *e.g.*, splines, chain wheels, threads, fasteners, pipe fittings, oil seals, bearings, *etc.* In Fig. 4.34, the conventional representation of different materials has been shown. However, in modern day engineering practice, the materials used in industries have become so diversified

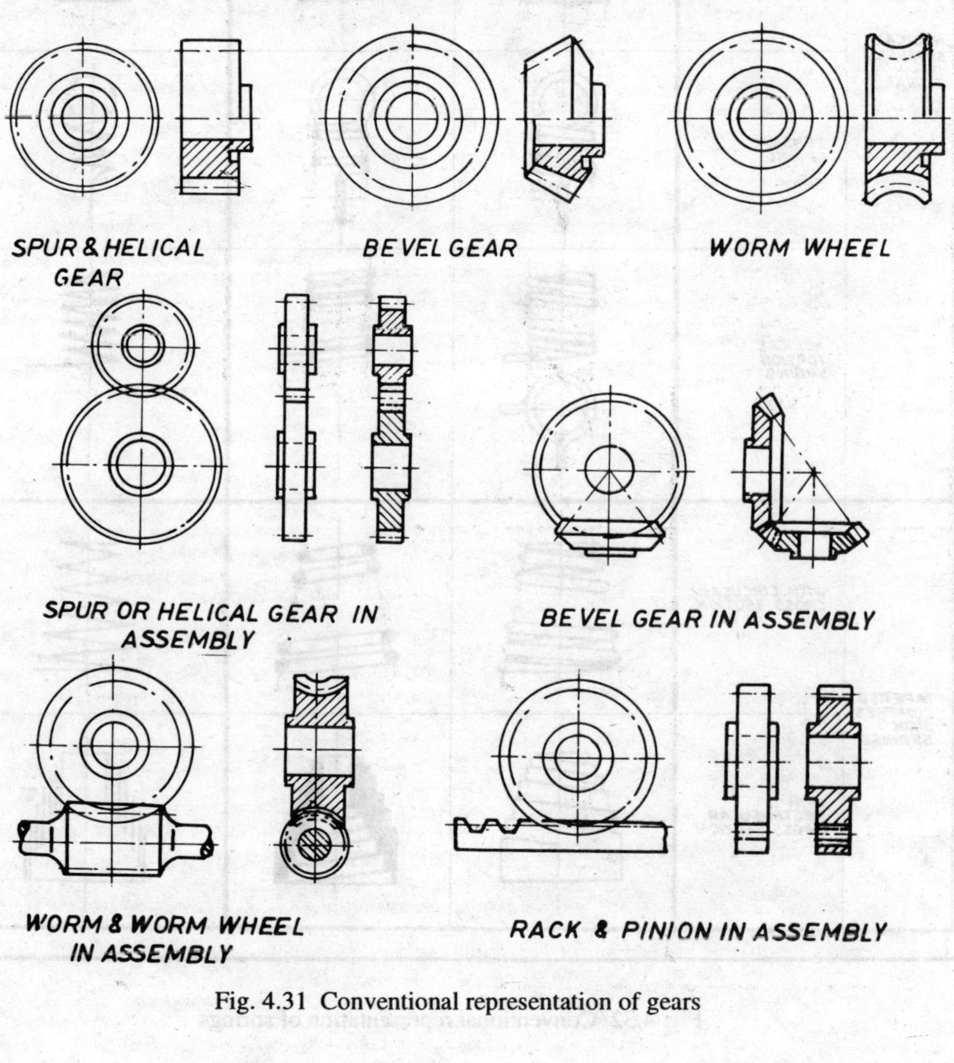

Fig. 4.31 Conventional representation of gears

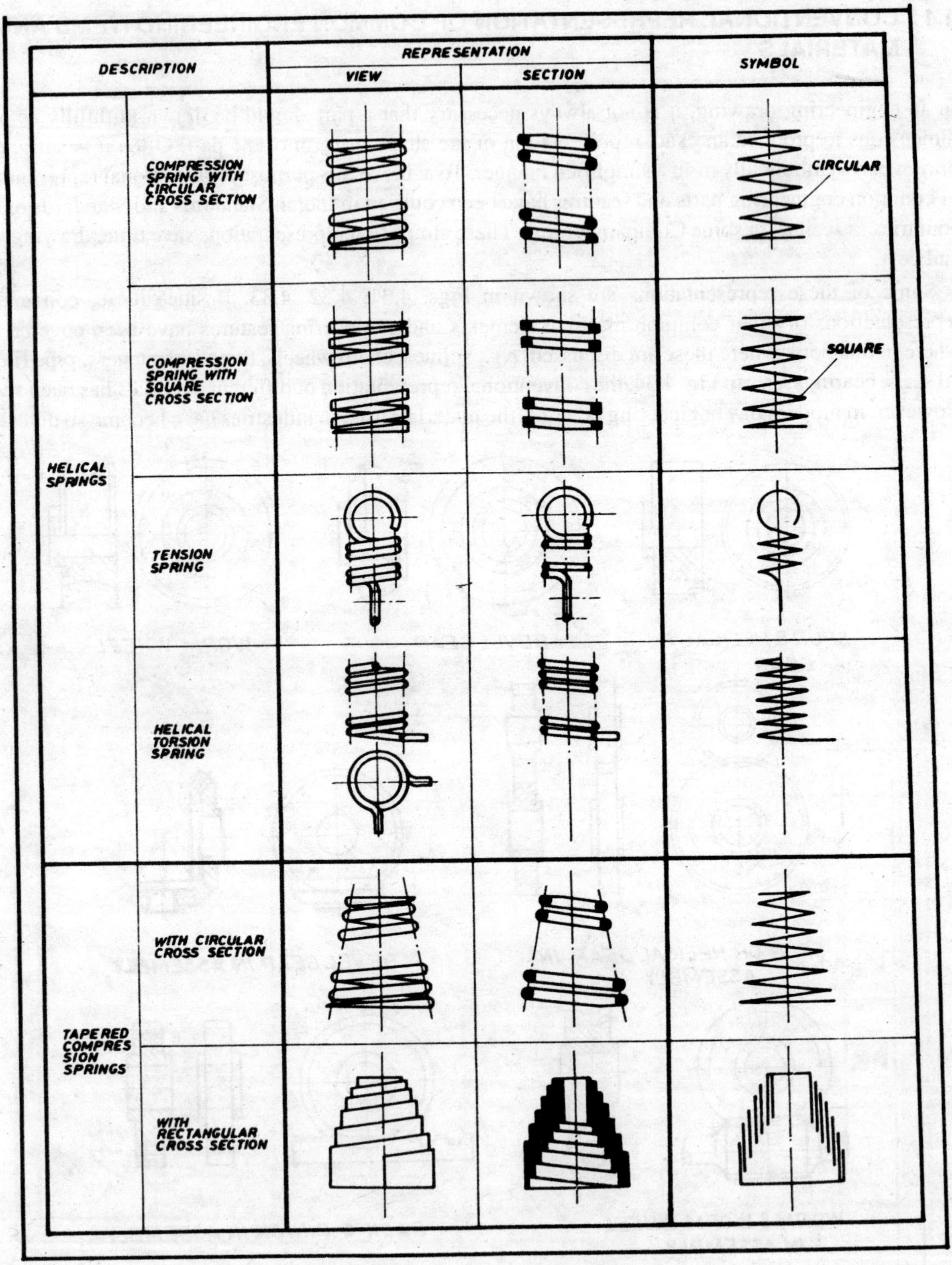

Fig. 4.32 Conventional representation of springs

DESCRIPTION		REPRESENTATION: VIEW, SECTION	SYMBOL
DISK SPRINGS	SINGLE SPRING	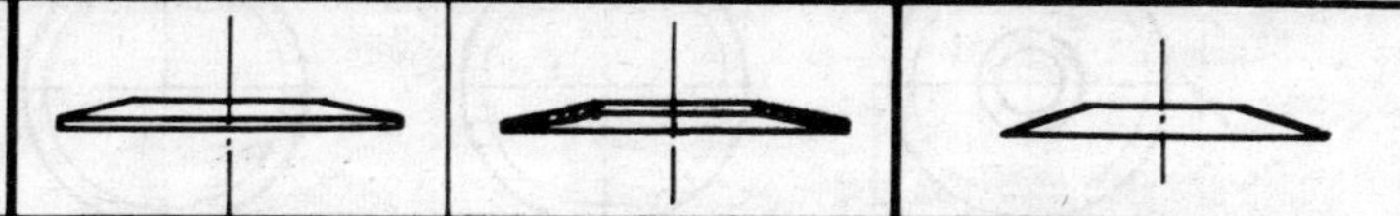	
	SPRING ASSEMBLY	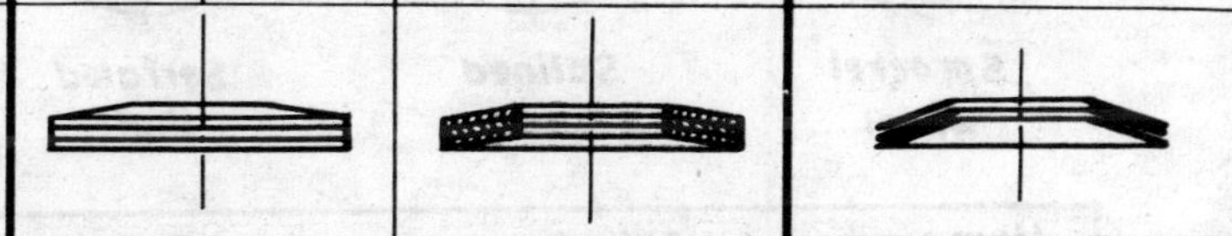	
	SPRING STACK	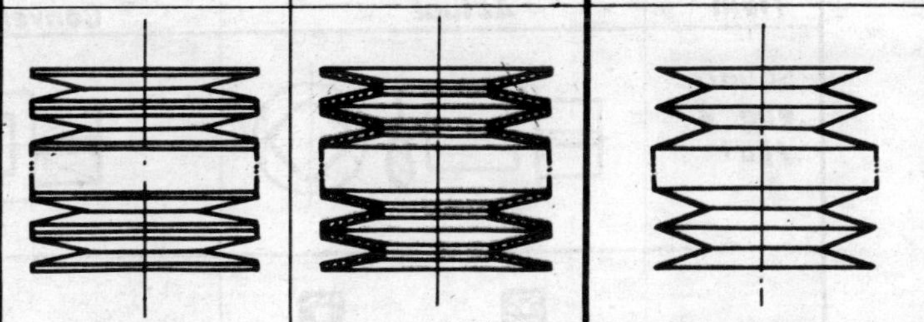	
SPIRAL POWER SPRINGS	SPIRAL POWER SPRING UNWOUND		
	SPIRAL POWER SPRING WITH BARREL WOUND		
LEAF SPRINGS	WITHOUT EYES		
	WITH EYES		
	WITHOUT EYES, WITH CENTRE BAND		
	WITH EYES, WITH CENTRE BAND		

4.3

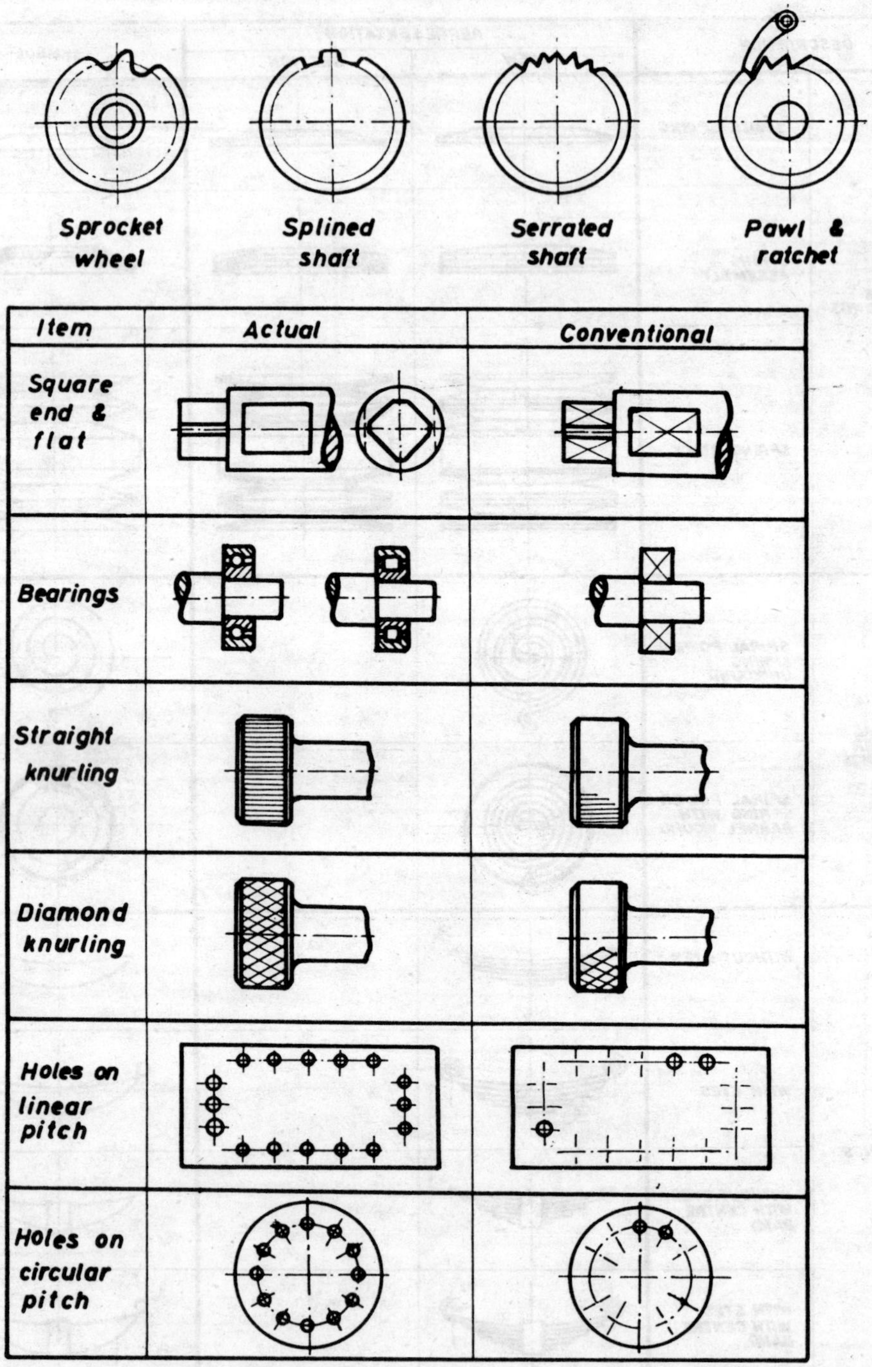

Fig. 4.33 Conventional representation of engineering items

TYPE	CONVENTION	MATERIAL
METALS	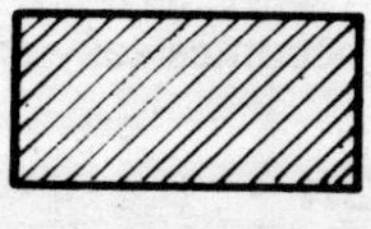	STEEL, CAST IRON, COPPER AND ITS ALLOYS, ALUMINIUM AND ITS ALLOYS, ETC.
		LEAD, ZINC, TIN, WHITE - METAL, ETC.
GLASS		GLASS
PACKING & INSULATING MATERIAL		PORCELAIN, STONEWARE, MARBLE, SLATE, ETC.
	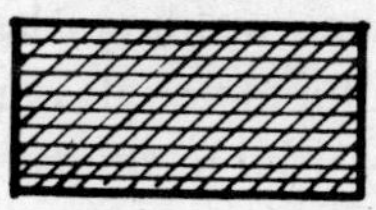	ASBESTOS, FIBRE, FELT, SYNTHETIC RESIN PRODUCTS, PAPER, CORK, LINOLEUM, RUBBER, LEATHER, WAX, INSULATING & FILLING — MATERIALS
LIQUIDS		WATER, OIL, PETROL, KEROSENE ETC.
WOOD		WOOD, PLYWOOD ETC.
CONCRETE	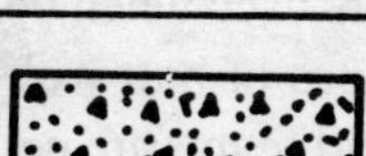	—

Fig. 4.34 Conventional representation of engineering materials

hat it is virtually impossible to assign a particular conventional or symbolic representation to each and very individual engineering material. Hence, irrespective of the material actually used, the conventional epresentation meant for cast iron and steel is used in all cases while showing the hatched portions in sectional drawings and other places, except in a few special cases like concrete, rubber, glass, *etc.* Such uniform representation for all items made of different materials in drawings should pose no problem as each part is assigned a reference number which is incorporated in the parts list which contains the material specification of each item, and hence correlation becomes easy and there is no scope for ambiguity.

4.5 WELDING AND ITS REPRESENTATION

Welding can be defined as the process of fusion or unification of two pieces of metals by the application of heat or by a combination of both heat and pressure. It can be performed with or without the application of pressure, and with or without the use of a filler metal. The metals will be completely joined if they are brought into such close contact that there is an actual union of the inter-atomic bonds between the two pieces. In case of iron which is the most common engineering base metal used, this fusion will take place only when the pieces to be joined are heated to a degree where they are very soft–almost near the melting point. If necessary, they may be forced together by the application of pressure. In general, steels with low carbon content are easier to weld.

The process of welding has now become highly developed and diversified. The great variety of welding methods can be classified into various categories as per different considerations. Thus, we have electric welding and gas welding, pressure welding processes and fusion welding processes, and so on. For fabrication and other industrial processes involving welding, electric arc welding is the most common method employed. Besides, gas welding, resistance welding, thermit welding, induction welding and forge welding are also in vogue. Almost all metallic materials can be welded, using different processes and conditions. In electric arc welding commonly used in shops, the electrode is made of various types of steel which is deposited at the joints as filler metals. Coatings containing fluxes are applied to the electrodes to provide shielding for the arc from the atmosphere.

Since welding is now extensively used in industrial applications, determination of welding data should not be left to the discretion of the shop people. A technical drawing itself should contain the relevant data, *e.g.*, edge preparation of parts to be joined, type of weld seam, size of weldments, electrode specifications and heat treatment (if necessary), *etc.* A system of welding symbols has been developed which has been laid down in IS : 813 as shown in Fig. 4.35. Note that the vertical side of the triangular welding symbol is always placed on the left, symbol is placed below the reference line (leg of the bent arrow) when welding is done on the side to which the arrow points, symbol is placed above when welding is done on the other side, and both upper and lower symbols are used when both sides are welded. Figure 4.36 shows the types of welded joints, Fig. 4.37 shows the types of butt welded joints, and Fig. 4.38 shows curves from which weld sizes can be read off when the weld seams are neither subjected to nor supposed to carry any load, that is, welding is used simply for the purpose of joining the parts only. For load carrying seams, elaborate calculations are needed. As per the Indian specifications, the size of the weld is given by the leg or one side of the fillet weld symbol. In German and some other continental practice, the throat indicates the weld size, so that. Throat of weld (a) = cos 45° × side of weld (h) , or $a = 0.707\ h$. Welding offers many advantages, *e.g.*, light weight constructions, cheaper end product, flexibility of design, facility to use standard structural sections or rolled shapes, etc. Apart from fabrication, welding is also extensively used for repair and maintenance of damaged or worn out components.

SKETCH OF WELD	REPRESENTATION	DESCRIPTION
	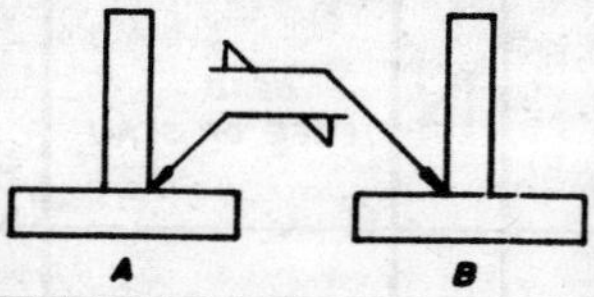	A. FILLET WELDS TO BE MADE ON THE ARROW SIDE OF THE JOINT. B. FILLET WELDS TO BE MADE ON THE OTHER SIDE OF THE JOINT.
	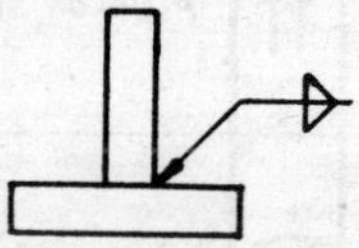	FILLET WELDS TO BE MADE ON EITHER SIDE OF THE JOINT.
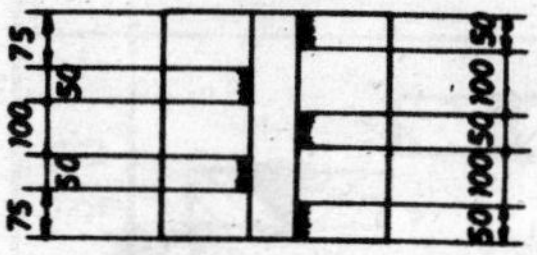	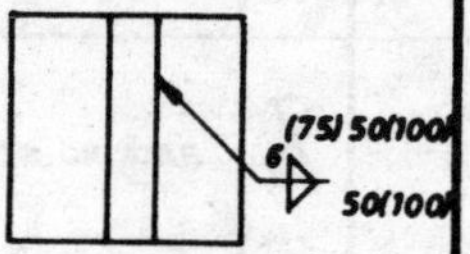	STAGGERED INTERMITTENT FILLET WELDS ON EITHER SIDE
	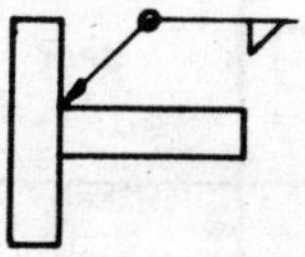	FILLET WELD TO BE MADE ALL ROUND.
	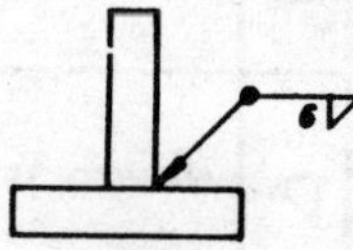	FILLET WELD TO BE MADE AT SITE & SIZE OF FILLET WELD IS TO BE 6mm.
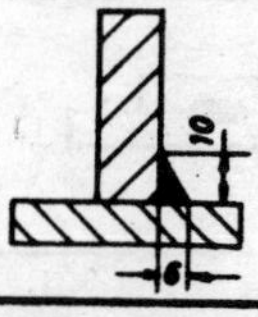	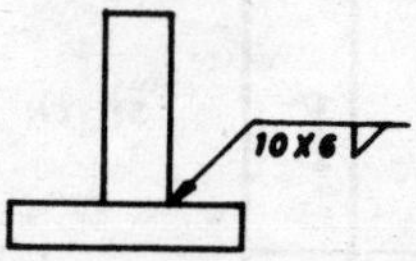	SIZE OF UN-EQUAL FILLET WELD.
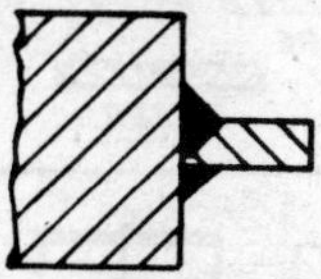	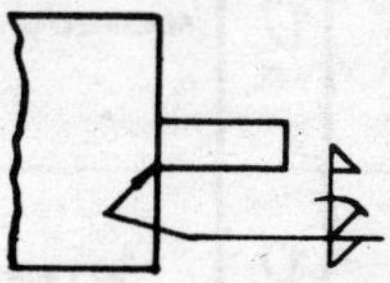	COMPOUND WELD (FILLET SUPER-IMPOSED ON PARTIALLY PENETRATED SINGLE-BEVEL BUTT) WITH FILLET WELD ON ARROW SIDE.
	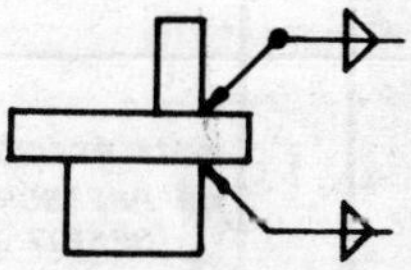	FILLET WELD IS TO BE MADE AT SHOP (BOTTOM PORTION) AND AT SITE (TOP PORTION)

Fig. 4.35 Welding symbols

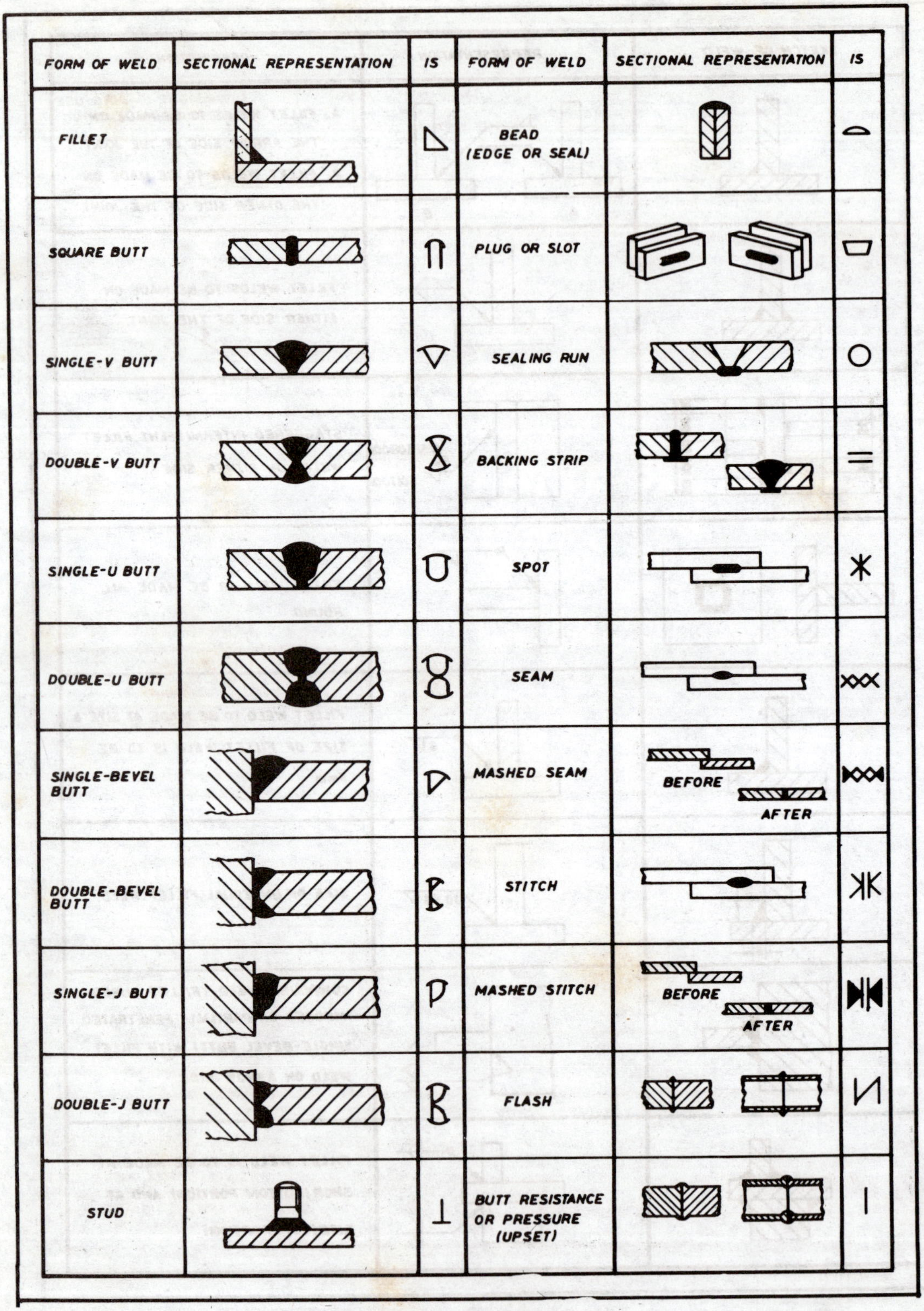

Fig. 4.36 Types of welded joints

S.NO.	TYPE	DETAILS OF WELDS	SIZE T	G	α	R	r	REMARKS
1	SQUARE BUTT	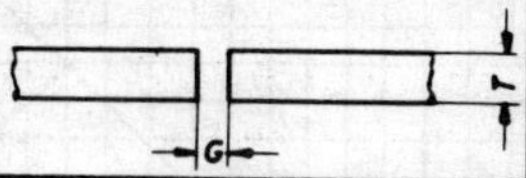	3.2 to 5 5 to 6	1.5 to 3 3				FOR LIGHT LOADING. USE BACKUP OR QUENCHING PLATE TO CONTROL WAR-PAGE OR DISTORTION.
2	SINGLE V BUTT	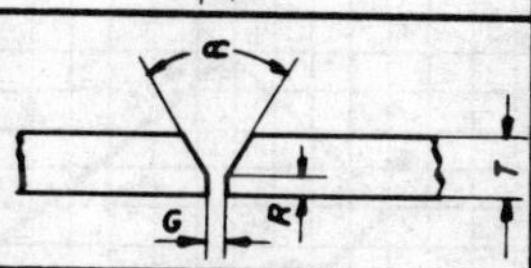	5 to 10 Over 10	1.5 to 5 3 to 6	60° 60°	0 to 1.5 0 to 3		FOR GENERAL PURPOSE. UNSUITABLE FOR IMPACT, FATIGUE OR TENSION AT ROOT WHEN ONE SIDE WELDED. PREFERABLY REINFORCED FROM BACK.
3	DOUBLE V BUTT A. EQUAL B. UNEQUAL	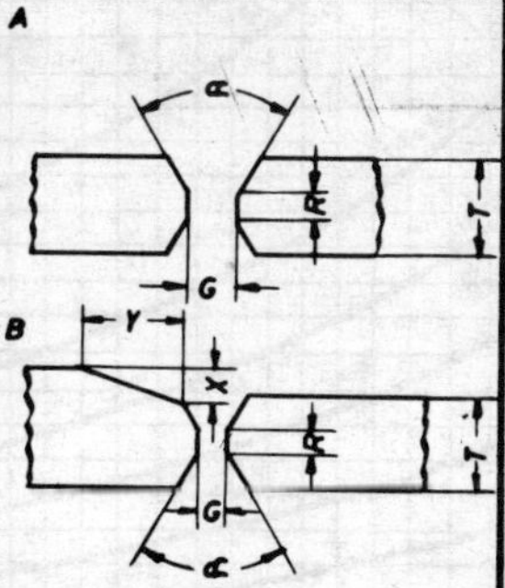	Over 12 Y = 5 X	3 to 6	60°	0 to 3		A: WHEN FULLY EXTENDED DEVELOPS FULL STRE-NGTH. ECONOMISES ELECTRODES. B: WITH PLATES OF UN-EQUAL THICKNESSES.
4	SINGLE U BUTT	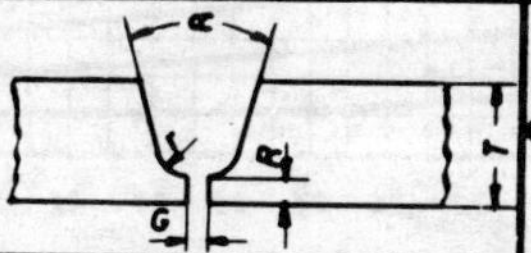	Over 12	0 to 3	10° to 20°	1.5 to 5	10 to 5	EXPENSIVE, FOR FATIGUE & IMPACT. WELD WITH REMOVABLE BACKING STRIP. SINGLE REINFOR-CING RUN FROM BACK PREFFERED.
5	DOUBLE U BUTT		Over 25	0 to 3	10° to 20°	1.5 to 3	10 to 5	EXPENSIVE. ECONOMICAL THAN DOUBLE V. DEVE-LOPS FULL STRENGTH. PERMITS USE OF LARGE ELECTRODES.
6	SINGLE J, TEE, BUTT AND CORNER	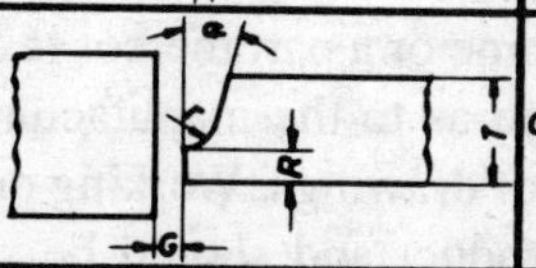	Over 12	0 to 3	10° to 20°	1.5 to 3	10 to 5	EXPENSIVE. FOR FATIGUE AND IMPACT LOADS WITH REMOVABLE BACKING STRIPS.
7	DOUBLE J, TEE, BUTT AND CORNER	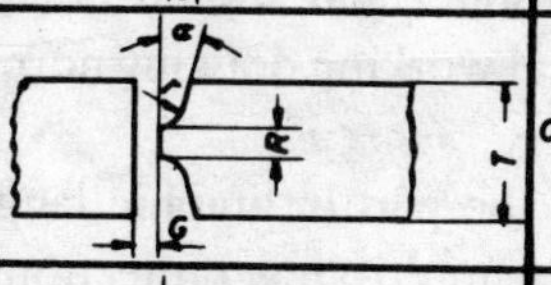	Over 25	0 to 3	10° to 20°	1.5 to 3	10 to 5	EXPENSIVE. CAN WITH-STAND SEVERE LOADS. ECONOMICAL THAN SINGLE J
8	SINGLE BEVEL BUTT, TEE & CORNER		5 to 10 Over 10	3 to 5 5 to 8	45° to 50° 45° to 50	0 to 1.5 0 to 3		UNDESIRABLE. WELD CAREFULLY FROM BOTH SIDES.
9	DOUBLE BEVEL BUTT TEE & CORNER	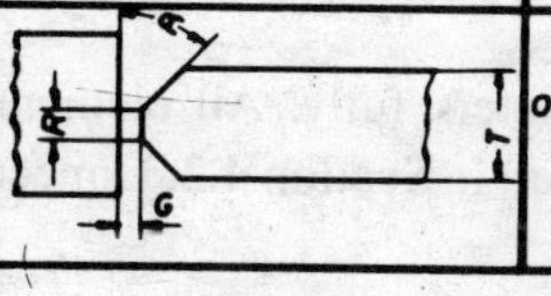	Over 12	5 to 8	45° to 50°	0 to 1.5		ECONOMICAL THAN SINGLE BEVEL. MORE CARE SHOULD BE TAKEN AT THE ROOT TO ENSURE WELD SOUNDNESS.

Fig. 4.37 Types of butt welded joints

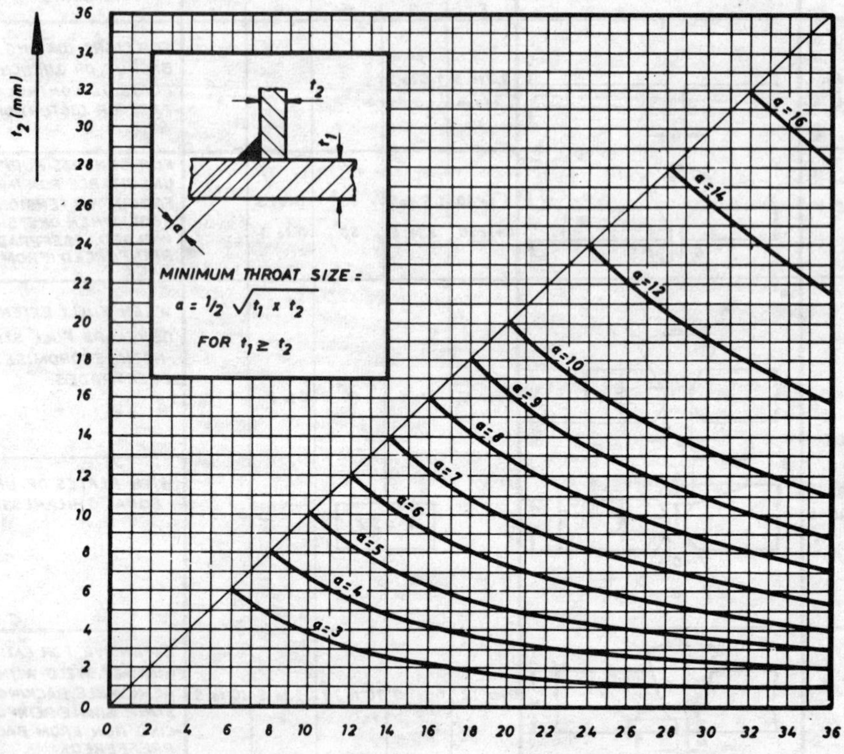

Fig. 4.38 Weld sizes for different thicknesses of plates

4.6 CHARACTERISTICS OF MANUFACTURING DRAWINGS

A drawing meant for production or construction of a machine or a part thereof is called a manufacturing or working drawing. For production, complete knowledge as to the manufacture of the equipment or component should be contained in such a drawing or a set of drawings. Working drawings should convey the relevant facts concerning the manufacture of the product and should be so explicit that further information and instructions are not necessary. A mechanical working drawing normally has the following characteristics.

1. The drawing is an engineering representation of the part in graphic language. It describes the shape of the part and the drawing is made as per the multi-view projection method in most of the cases in engineering industries, though other methods are also sometimes employed. The number of views and sections, their solution and orientation with respect to the drawing as a whole should be carefully determined. Unnecessary views and sections should be avoided as it only results in wastage of time and effort.
2. The drawing should include all the size parameters fully. All elements should be properly dimensioned as per the code of dimensioning given in Section 4.2. Unnecessary dimensions and

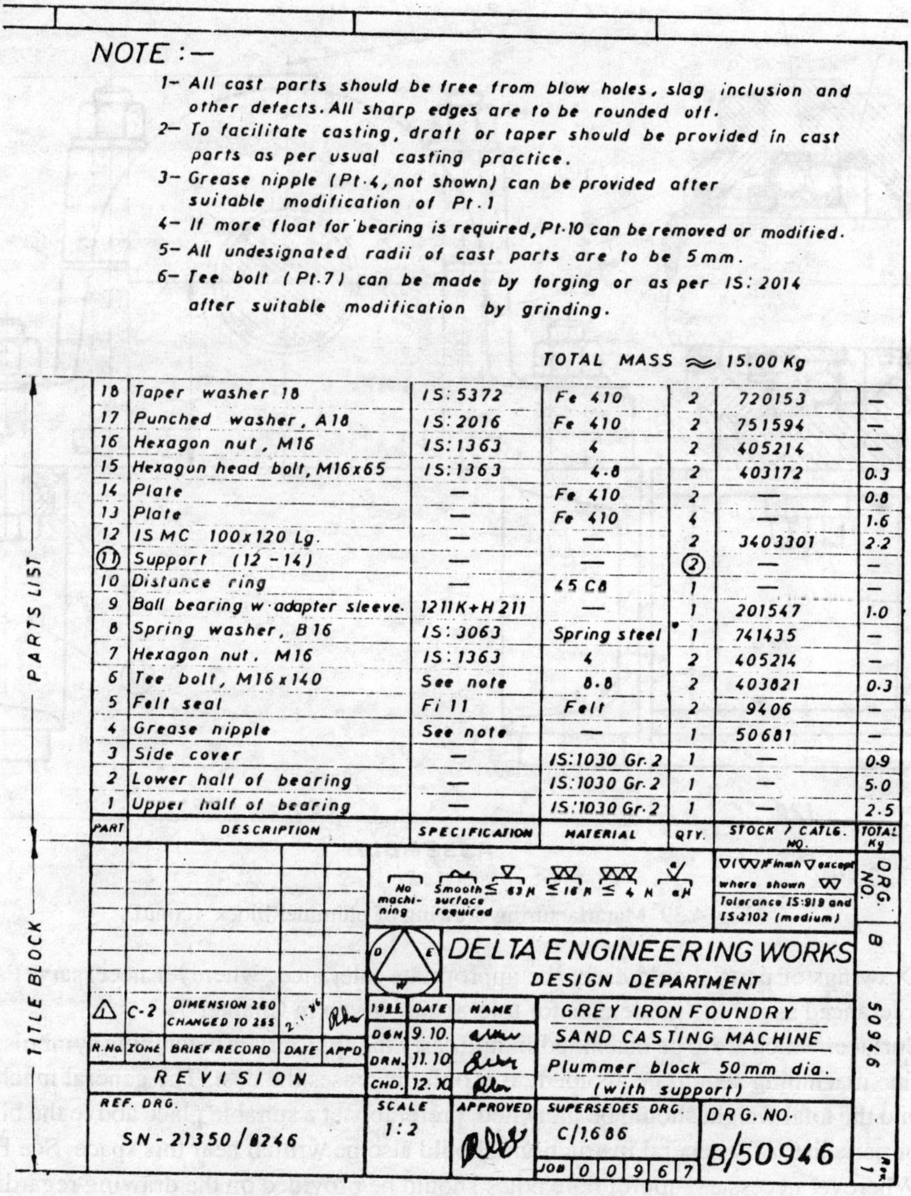

NOTE :—

1– All cast parts should be free from blow holes, slag inclusion and other defects. All sharp edges are to be rounded off.
2– To facilitate casting, draft or taper should be provided in cast parts as per usual casting practice.
3– Grease nipple (Pt·4, not shown) can be provided after suitable modification of Pt.1
4– If more float for bearing is required, Pt·10 can be removed or modified.
5– All undesignated radii of cast parts are to be 5 mm.
6– Tee bolt (Pt·7) can be made by forging or as per IS: 2014 after suitable modification by grinding.

TOTAL MASS ≈ 15.00 Kg

PARTS LIST

PART	DESCRIPTION	SPECIFICATION	MATERIAL	QTY.	STOCK / CATLG. NO.	TOTAL Kg
18	Taper washer 18	IS: 5372	Fe 410	2	720153	—
17	Punched washer, A18	IS: 2016	Fe 410	2	751594	—
16	Hexagon nut, M16	IS: 1363	4	2	405214	—
15	Hexagon head bolt, M16x65	IS: 1363	4.8	2	403172	0.3
14	Plate	—	Fe 410	2	—	0.8
13	Plate	—	Fe 410	4	—	1.6
12	IS MC 100x120 Lg.	—	—	2	3403301	2.2
(11)	Support (12 – 14)	—	—	(2)	—	—
10	Distance ring	—	45 C8	1	—	—
9	Ball bearing w·adapter sleeve.	1211K+H 211	—	1	201547	1.0
8	Spring washer, B 16	IS: 3063	Spring steel	1	741435	—
7	Hexagon nut, M16	IS: 1363	4	2	405214	—
6	Tee bolt, M16 x 140	See note	8.8	2	403821	0.3
5	Felt seal	Fi 11	Felt	2	9406	—
4	Grease nipple	See note	—	1	50681	—
3	Side cover	—	IS:1030 Gr. 2	1	—	0.9
2	Lower half of bearing	—	IS:1030 Gr. 2	1	—	5.0
1	Upper half of bearing	—	IS:1030 Gr. 2	1	—	2.5

TITLE BLOCK

No machining · Smooth surface required · ≦ 63 μ · ≦ 16 μ · ≦ 4 μ · a μ
∇(∇∇) Finish ∇ except where shown ∇∇
Tolerance IS:919 and IS:2102 (medium)

DELTA ENGINEERING WORKS
DESIGN DEPARTMENT

R.No	ZONE	BRIEF RECORD	DATE	APPD
1	C-2	DIMENSION 260 CHANGED TO 265	2·1·86	

REVISION

1986	DATE	NAME
DGN.	9.10.	
DRN.	11.10	
CHD.	12.10	

GREY IRON FOUNDRY
SAND CASTING MACHINE
Plummer block 50 mm dia. (with support)

REF. DRG. SN-21350/8246
SCALE 1:2
APPROVED
SUPERSEDES DRG. C/1686
JOB 00967
DRG. NO. B/50946
DRG. NO. B 50946
REV. 1

Fig. 4.39 Manufacturing drawing of plummer block

superfluous data should be avoided as this will only confuse the reader of the drawing. Elements drawn not to scale should carry underlined dimensions.

3. The materials used for all the components should be fully specified. As far as possible, common materials should be specified for reasons of availability and economy.

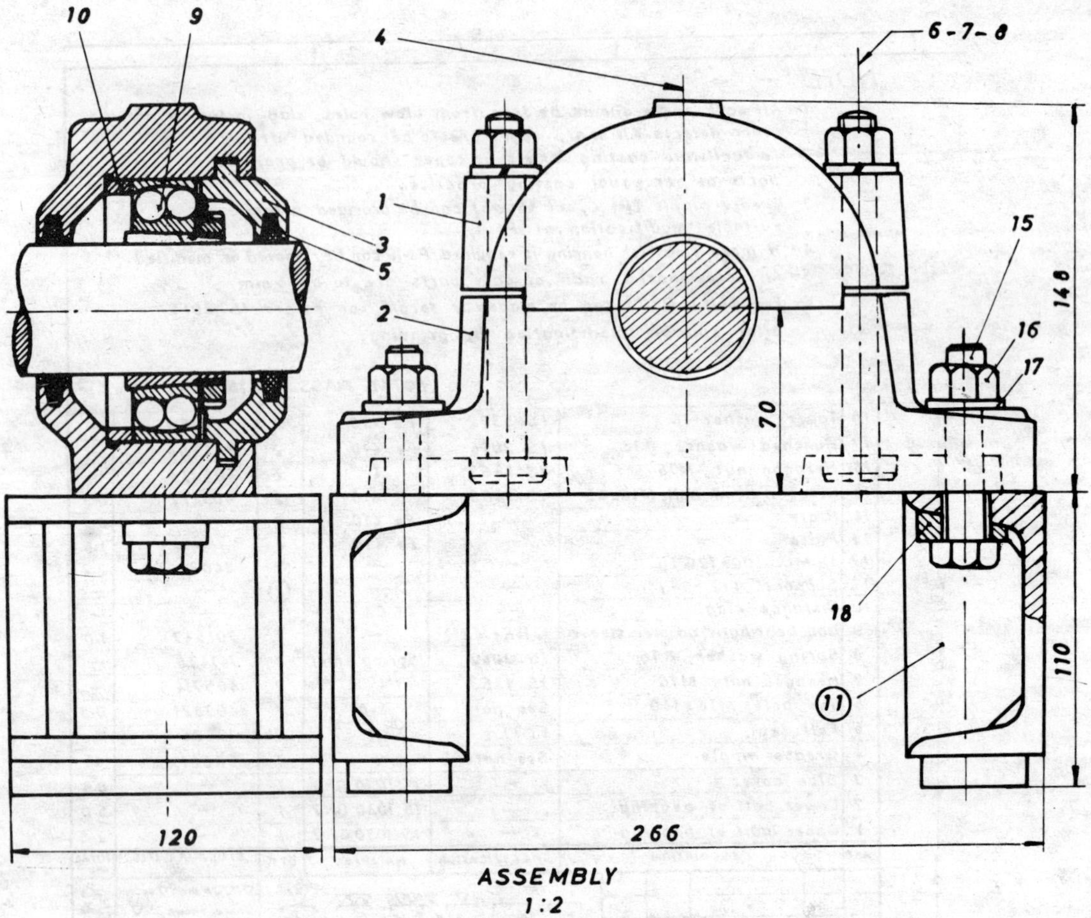

Fig. 4.39 Manufacturing drawing of plummer block (contd.)

4. Drawings of parts should carry the appropriate tolerances wherever necessary. Parts should be toleranced as per the guidelines for tolerancing given in Chapter 7.
5. Surfaces which are to be machined should carry the appropriate machining symbols. Unnecessary fine machining should be avoided, as it only increases the cost. The general machining symbol and the total weight should be indicated, preferably at a suitable place above the bill of materials or parts list. Any general instructions should also be written near this space. See Fig. 4.39.
6. Wherever necessary, appropriate notes should be provided on the drawing regarding heat-treatment, radiographic or ultrasonic testing, if any, and other relevant instructions as, for example, in case of casting drawings given below :
   (i) All cast parts should be free from blow-holes, slag inclusion and other defects.
   (ii) Minor casting flaws, *e.g.*, small sand and slag inclusions, small cold shuts, washes, *etc.*, should be removed by fettling or similar methods.
   (iii) Castings should be subjected to heat treatment as per specifications, if required.

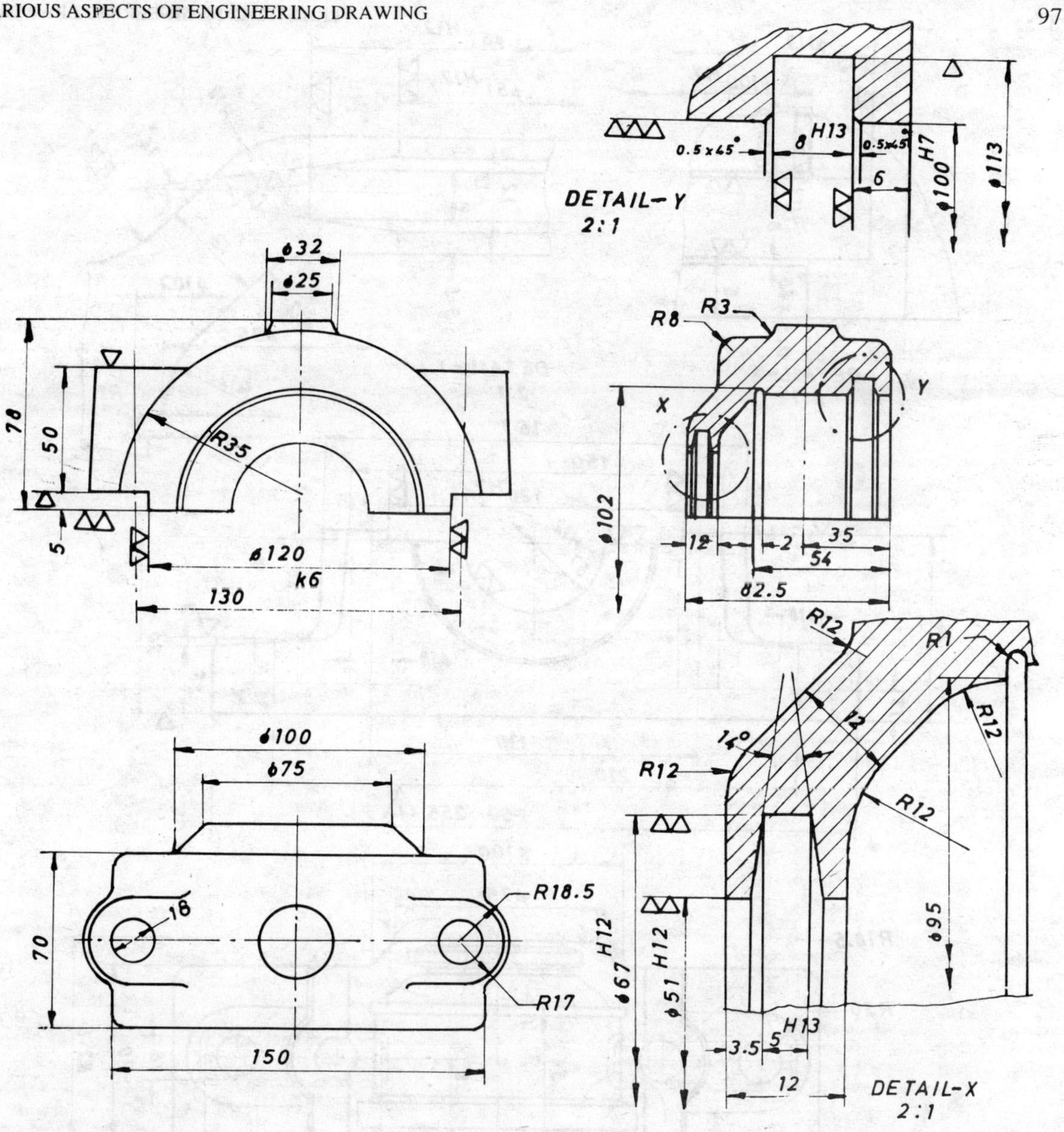

120	h6	0 / −0.022
ϕ100	H7	+0.035 / 0
ϕ51	H12	+0.300 / 0
ϕ67	H12	+0.300 / 0
5	H13	+0.180 / 0
8	H13	+0.220 / 0

1	Upper half of bearing	IS:1030	1	2.5
PART	DESCRIPTION	MAT.	QTY.	KG
Scale 1:2	DEW DESIGN DEPTT	B/50946/1		

Fig. 4.39 Manufacturing drawing of plummer block (contd.)

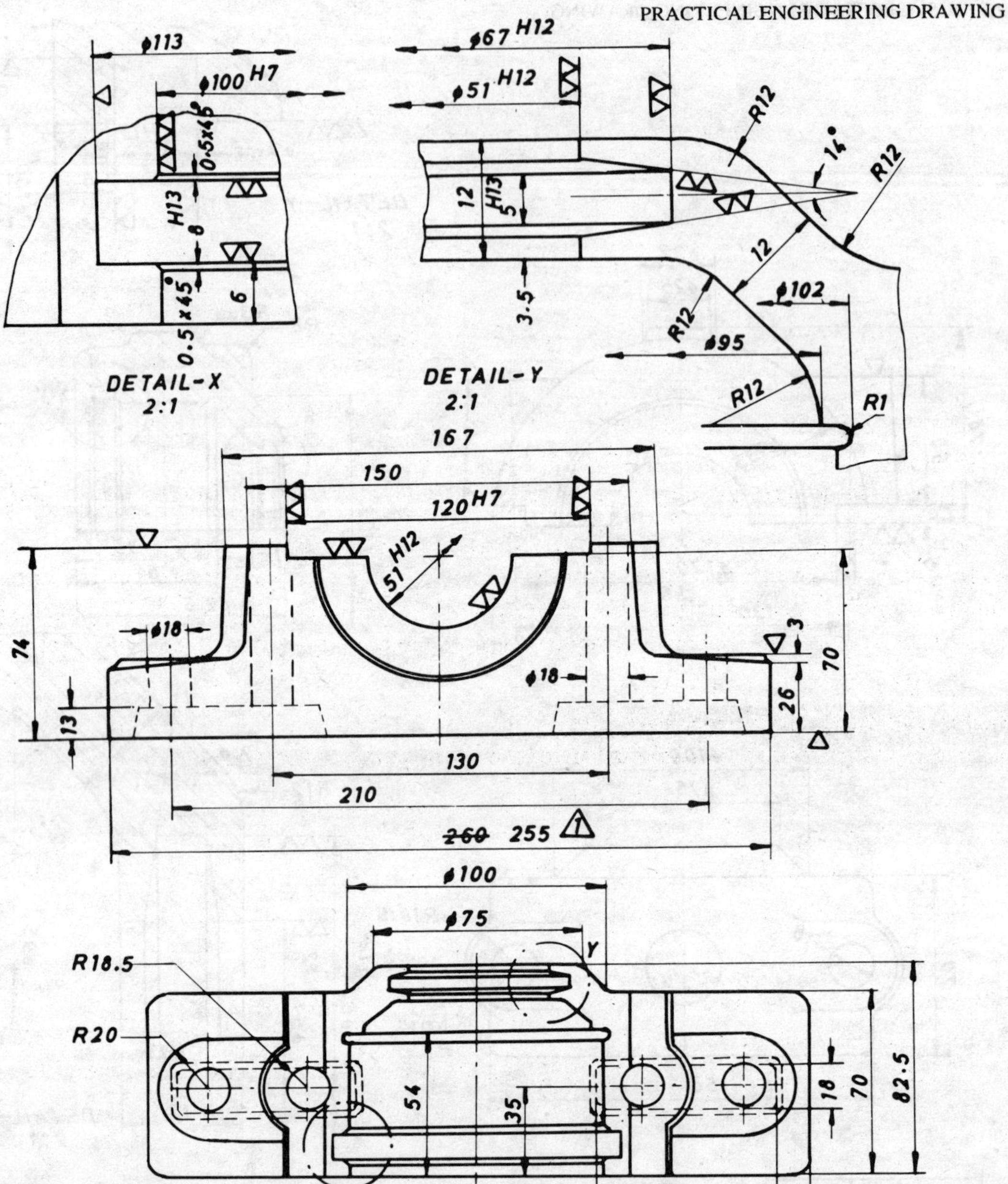

ϕ100	H7	+0.035 0
120	H7	+0.035 0
ϕ51	H12	+0.300 0
ϕ67	H12	+0.300 0
5	H13	+0.180 0
8	H13	+0.220 0

2	Lower half of bearing	IS:1030	1	5.0
PART	DESCRIPTION	MAT.	QTY.	KG
Scale 1:2	DEW DESIGN DEPTT.	B/50946/2		

Fig. 4.39 Manufacturing drawing of plummer block (contd.)

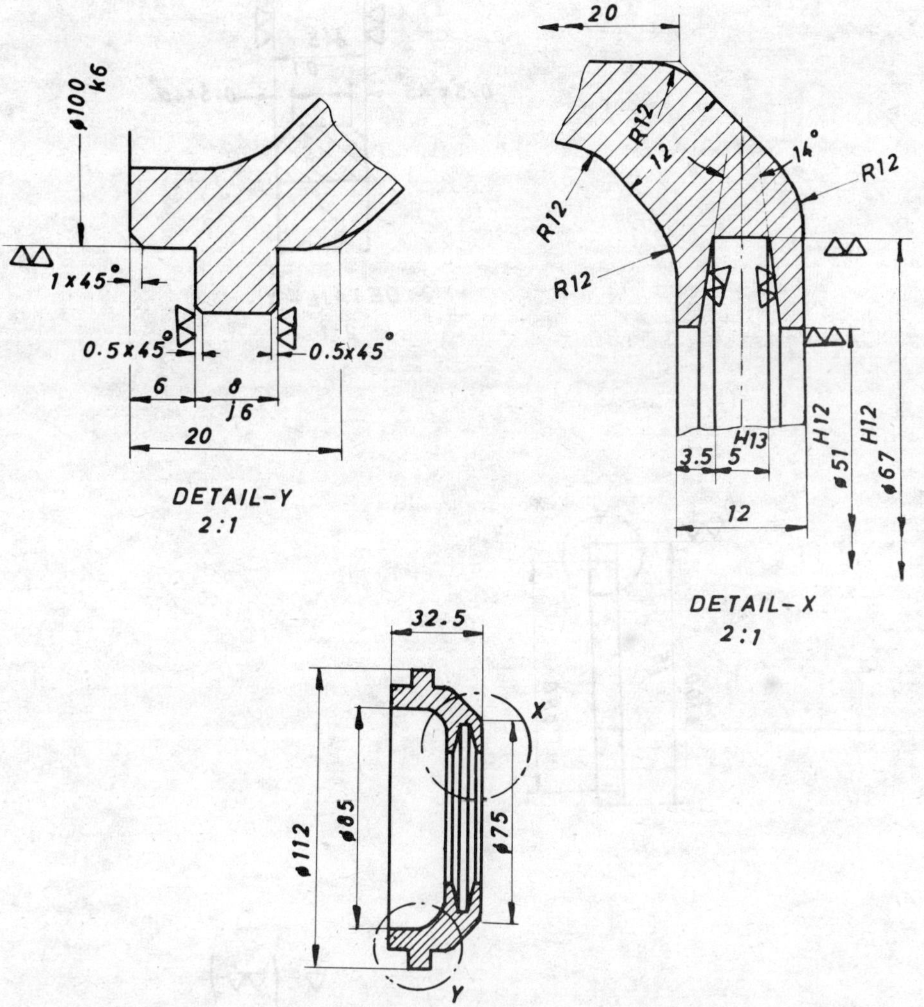

ø100	k6	+0.025 +0.003
ø67	H12	+0.300 0
ø51	H12	+0.300 0
5	H13	+0.180 0
8	j6	+0.007 -0.002

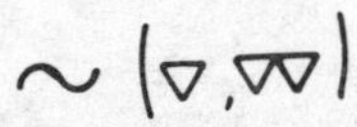

3	SIDE COVER	IS:1030	1	0.9
PART	DESCRIPTION	MAT.	QTY.	KG.
SCALE 1:2	DEW DESIGN DEPTT.	B/50946/3		

Fig. 4.39 Manufacturing drawing of plummer block (contd.)

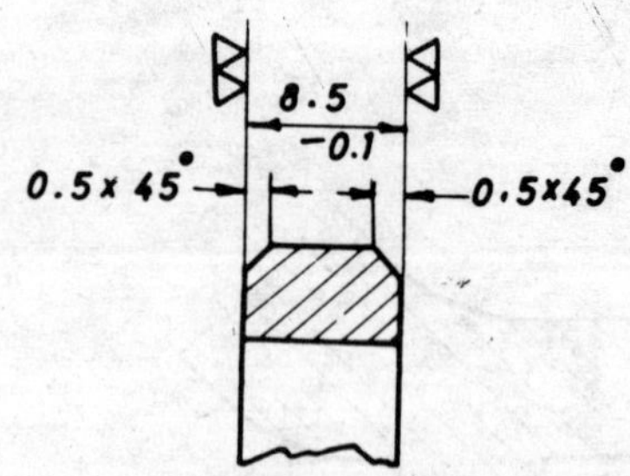

DETAIL– X
2:1

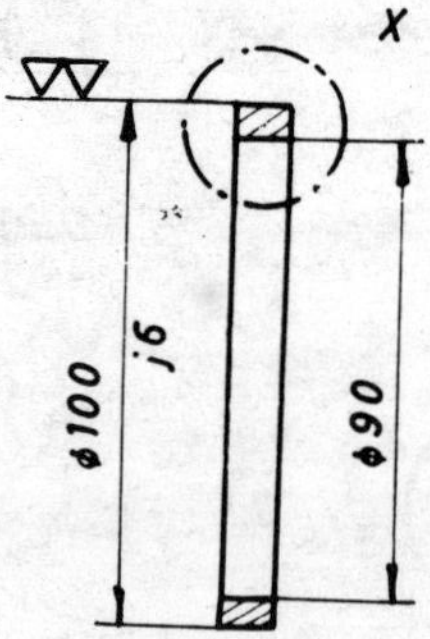

▽ (▽▽)

ϕ100	j6	+0.013 −0.009

10	DISTANCE RING		45 C8	1	—
PART	DESCRIPTION		MAT.	QTY.	KG.
Scale 1:2	DEW DESIGN DEPTT	B/50946/10			

Fig. 4.39 Manufacturing drawing of plummer block (contd.)

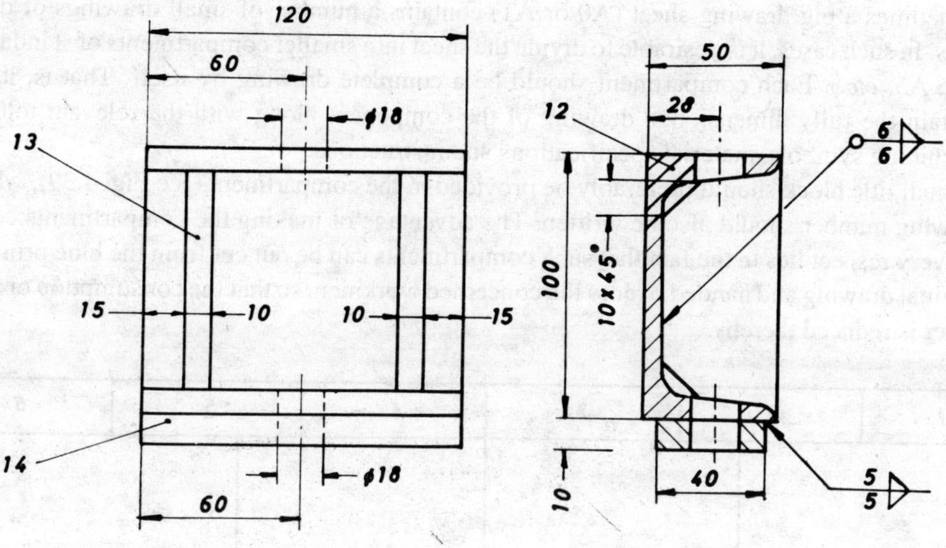

14	PLATE	Fe 410	2	0·8
13	PLATE	Fe 410	4	1·6
12	ISMC 100x120 Lg.	—	2	2·2
(11)	SUPPORT (12–14)	—	(2)	—
PART	DESCRIPTION	MAT.	QTY	KG.
Scale 1:2	DEW DESIGN DEPTT.	B/50946/11		

Fig. 4.39 Manufacturing drawing of plummer block (contd.)

(iv) Free measure tolerances should be as per IS: 4897 (steel castings), IS: 5519 (grey cast iron), IS: 8349 (malleable C.I.), *etc.*, as the case may be.

(v) If any testing is called for regarding the soundness of castings, *e.g.*, ultra-sonic test, X-ray or gamma ray test, *etc.*, this should be carried out as per the test codes.

(vi) All sharp edges and corners should be rounded off.

7. Sometimes a big drawing sheet (A0 or A1) contains a number of small drawings of different parts. In such cases, it is desirable to divide the sheet into smaller compartments of standard sizes (A4, A5, *etc.*). Each compartment should be a complete drawing by itself. That is, it should contain the fully dimensioned drawing of the component along with the relevant tolerances, machining symbols, material specifications and instructions.
   A small title block should preferably be provided in the compartment (see Fig. 4.39), where the drawing number should also be written. The advantage of making the compartments complete in every respect lies in the fact that such compartments can be out cut from the blue prints of the original drawing and handed over to the concerned workmen, so that the consumption of printing paper is reduced thereby.

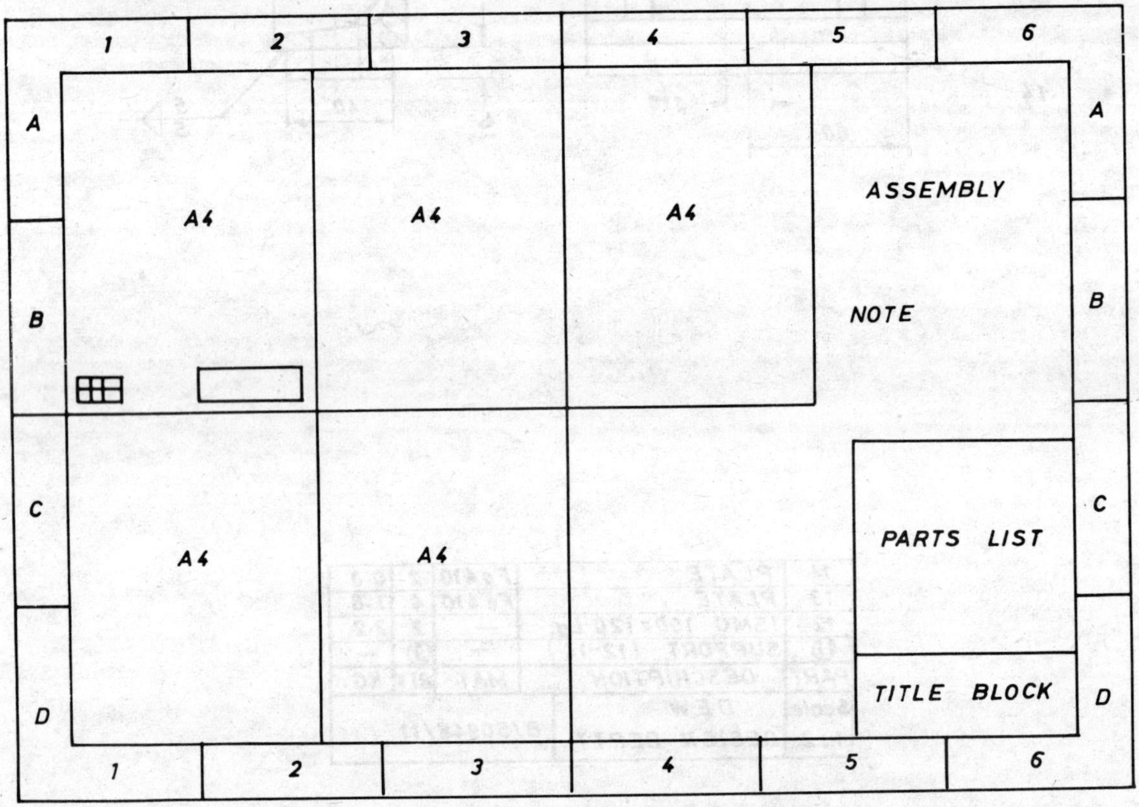

Fig. 4.40 Placement of compartments in a drawing

8. The main title block should be placed and filled-in as discussed in Section 1.7. Every manufacturing drawing should be numbered as per the practice followed by the firm. If the drawing is revised, the nature of revision should be clearly indicated in the space provided in the title block. Also, the revised dimension should be indicated in the body of the drawing by some symbol as shown in Fig. 4.39.

9. If a set of drawings are meant for the manufacture of the equipment, each drawing should carry the numbers of the other relevant drawings for correlation.
10. A manufacturing drawing should also show the components in the assembled condition. The assembly drawing and the parts list should be as per the discussion in Section 4.7.
11. Scales, types of lines and other features should be as per the guidelines given in Chapter 1.
12. Parts which are standardised components as per the national standard of the country, *e.g.*, the IS, or as per the Company Norms should be used wherever possible, though deviations can be allowed from the point of functional considerations. Manufacturing drawings of standard items are not supposed to be drawn and these items are only to be properly specified in the drawing, unless some change in the shape or size is required for a particular purpose. In that case, the modified drawing of the standard part is to be used.
13. If the drawing consists of several compartments, it is desirable to enter the numerical values of the tolerances used in each compartment in a tabular form. This table should preferably be placed at the bottom left-hand corner (see Fig. 4.39). Determination of the tolerance values should not be left to the shop people. If these values are given in the drawing in the Design Department itself, it eliminates the chances of mistake in the shop and also it saves time.
14. Under no circumstances should the shop people scale out any particular dimension not given in the drawing. Though this is universally understood in the technical world, some firms prefer to specifically mention it somewhere in the manufacturing drawing by means of printed instruction, *viz.*, " DO NOT SCALE. IF IN DOUBT, ASK ".
15. Besides long, cylindrical parts like shafts, bolts, and some other similar items which are generally shown in horizontal position irrespective of their actual placing in the equipment, parts should preferably be shown in the drawing in the same orientation as they appear in the working condition in the machine.
16. In some cases, parts which have identical shapes but different dimensional values are shown as tabular drawings. In such a case, only one drawing is made in which numerical values of dimensions are replaced by reference letters. In an adjacent table, the numerical values are given corresponding to the sizes and dimensions of the parts (see Fig. 4.13).

A typical manufacturing drawing has been illustrated in Fig. 4.41. It has been stated before that in common engineering practice, the basis of all working drawings is generally the orthographic projection method, or more precisely, the multi-view projection method. In a working drawing, it is customary to draw the front view of the face of the object as the main view which shows the largest dimension of the object in its operational position. Once this view is drawn, other necessary views should follow. For cylindrical parts, only one frontal view is normally sufficient (see Fig. 4.41). Details, such as, undercut, centre hole and other special features can be shown by taking local sections as in Fig. 4.39. The view should be so chosen that it contains as few dotted lines as possible. For symmetrical parts, namely, right-handed and left-handed halves of an equipment, it is not necessary to make manufacturing drawings of both the halves. One-half may be drawn and the other half may be designated as the "mirror image" of the half which has been drawn.

Sometimes an auxiliary view or an appropriate note is sufficient to avoid making another space-consuming view. Selection of views and sections, therefore, rests entirely on the discretion of the engineer or the draughtsman. Besides, since in commercial draughting, accuracy and speed are the two main

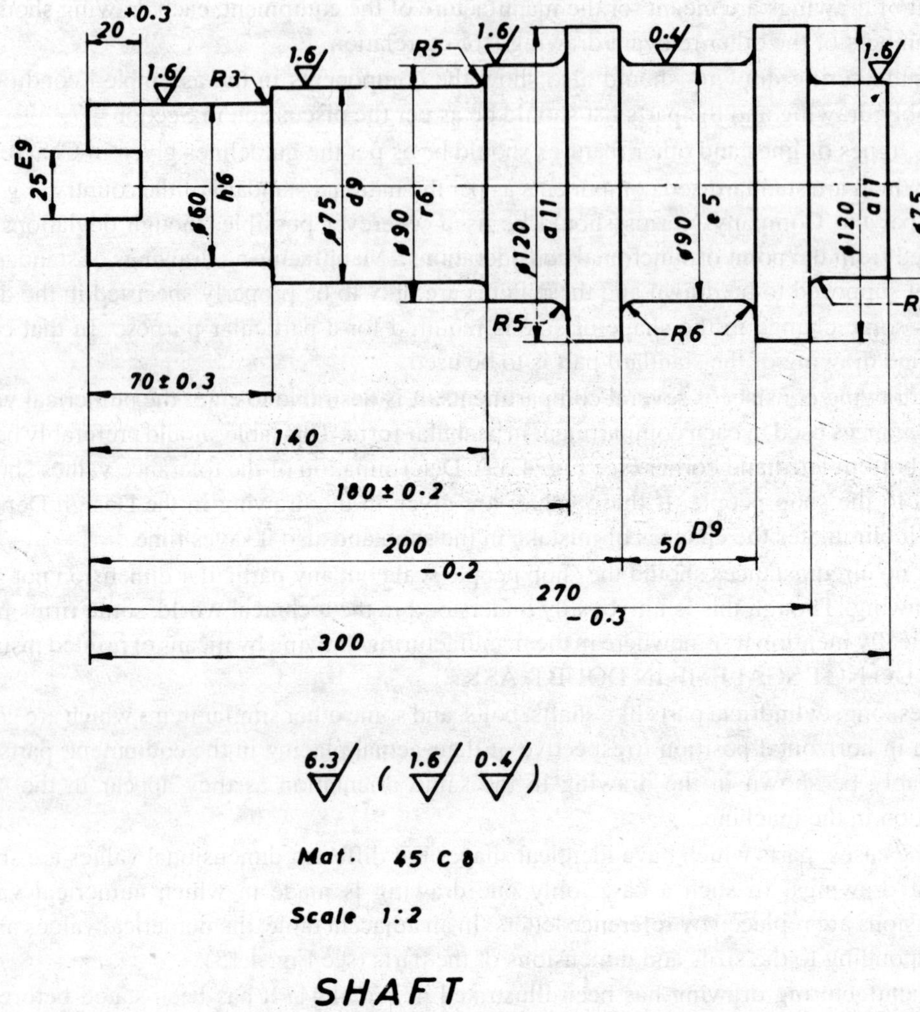

Fig. 4.41 Manufacturing drawing of a shaft

criteria, drawings should be made keeping these yardsticks in mind for overall efficiency of the man who makes the drawings.

In short, a manufacturing drawing should be optimum in every respect, yet it should contain all the relevant information, dimensions and data, so that there is no confusion when the shopman reads the drawing to make the part. If, however, the shopman has any doubt as regards the interpretation of the drawing, he should always consult the drawing office people for further action to manufacture the part.

In Fig. 4.39, manufacturing drawings of different components of a plummer block along with the title block and the assembly have been shown. A possible order of placement of the individual compartments containing the parts of the plummer block and the orientation of the assembly drawing with respect to the compartments and the parts list have been shown in Fig. 4.40.

4.7 ASSEMBLY DRAWING AND THEIR REQUIREMENTS

In general, an assembly drawing serves the purpose of showing the individual components comprising a unit in assembled condition. It shows how all the parts are fitted in a machine or put together in a structure so as to convey an overall picture of the unit. The main features and characteristics of an assembly drawing are summarised below:

1. An assembly drawing gives an idea as to what the construction of a unit is supposed to look like in a final working condition. The individual components comprising the unit are shown in their respective functional positions.
2. It shows the reader of the drawing the correct relative positions of each part in the assembled condition.
3. Since a single scale is used in an assembly drawing, the parts of which the unit is consisted of are shown in a particular scale so that the relative size of one part vis-a-vis that of another part can be promptly visualised.
4. The working relation between two or more components, if any, also becomes apparent by studying the assembly drawing of the unit.
5. The methods by which the individual parts are joined in an assembly, *e.g.*, by threaded fasteners, welding, riveting and other means, are also revealed in an assembly drawing.
6. An assembly drawing normally does not carry any dimension. Dimensions of individual parts are given in the detail drawings of these parts. However, overall dimensions are sometimes given to convey the idea as to the space requirements of the unit. Principal dimensions are also indicated which are needed for assembling the parts. In gear box assembly drawings, it is imperative to show the centre distances with tolerances between gears and pinion shafts. Dimensions which are required to be maintained from a particular datum line or reference plane must also be indicated. In short, all the relevant dimensions connected with the assembly of the unit should be shown.
7. Each part and allied items are assigned numbers for identification purposes in the detail drawing of these parts. These numerals, or sometimes alphabets, are called item references and are inserted in the assembly drawing as well as in the bill of materials for correlation with the detail drawings. Obviously, the reference number of a particular item remains the same in the detail drawing, in the assembly drawing and in the bill of materials.
8. An assembly drawing may be drawn in a separate sheet or it may be drawn in the same sheet where detail drawings of the parts are also drawn. In the latter case, the assembly drawing is usually put at the top right-hand portion of the drawing sheet. The placement of the assembly drawing should preferably be so oriented that the drawing represents the unit as it actually is in its working condition (see Fig. 4.39).
9. In case of complicated assemblies, drawings of sub-unit may be required to be shown to avoid any confusion. Such drawings are called sub-assemblies.
10. Normally, one view or section is adequate to represent all the components in an assembly drawing. More views or sections, however, may be needed sometimes to show some hidden parts or for some other reason. However, unless there is a scope for confusion, dotted lines to show the hidden features should be avoided as far as practicable.

11. In assigning numbers or item references of components, some sort of sequential order should be followed. In this connection, the following points are of relevance :
   (i) Numbering can be done according to the possible order in which the part has been assembled or it can be done as per the importance of the constituent parts or according to some other logical sequence.
   (ii) It is preferred that the item references are arranged in vertical columns and horizontal rows (see Fig. 4.39). They can also be written in a star-like fashion in some cases.
   (iii) Identical items are assigned one item reference only and when there is no risk of ambiguity, the numer standing for such an item need only be shown once against such an item in the assembly drawing.
   (iv) The same type and height of lettering should be used for writing all item references. These numerals should always be placed well outside the body of the assembly drawing, so that there is no scope for confusion.
   (v) The component and its item reference number in the assembly drawing should be connected by a leader line. At the component end, the leader line may terminate in an arrow or a dot or a short line at an angle. All three types of endings have been shown in Fig. 1.3, but only one type should be used in a drawing for uniformity.
   (vi) Leader lines should preferably be short, they should not intersect each another and should be drawn at an angle.
   (vii) If the item reference of a sub-assembly is shown as an encircled number in the bill of materials (Fig. 4.39), and if this complete sub-assembly is incorporated in the main assembly drawing, the sub-assembly need carry only that number. That is, the item references of individual parts comprising the sub-assembly need not be shown again.
   (viii) Reference numbers of related items can be shown against the same leader line (see Fig. 4.39).
12. Any special instruction as to the assembly of a component or a sub-assembly should be written alongside the assembly drawing to facilitate correct assembly (or dismantling) of the unit.

The nature of an assembly drawing will depend upon its specific use. The term "assembly drawing" may include preliminary layouts, general assemblies, working drawing assemblies, layout of piping systems, installation assemblies, erection drawings and other types besides the final and complete assembly drawings of machine units or structures discussed hitherto.

In a preliminary layout drawing, the designer puts forth his ideas and conceptions which he has in mind about the shape and size of the components of the unit as they are visualized in the assembled condition. In such a layout, the designer considers many aspects, namely, previous design of similar nature and improvements thereof, strength factors, space restrictions, clearance for moving parts, availability or otherwise of standard parts, selection of materials, overall cost and many other related factors.

General assembly drawings are those which are most commonly used in engineering industries. These have been discussed in detail in the beginning of this section. Such an assembly drawing can be made by tracing the parts from the detail drawings and then fitting them in proper positions. It can also be freshly drawn from the dimensions of the parts given in detail drawings.

In working drawing assemblies, the details and the assembly features are combined, so that these are available in a single drawing. That is, a view or a section contains the drawings of the components of the unit which are complete in every respects as regards dimensions, machining, tolerances and other relevant data, and at the same time the whole view or section shows how these components are put together in assembly. This sort of practice is sometimes followed in simple machine units, such as, valve assemblies, jigs and fixtures and other small machine parts.

Assembly drawings showing installation or erection specify how to install or erect an equipment or a structure. These are essentially outline drawings showing the outlines of machines and their relationship *vis-a-vis* exterior surroundings.

In an engineering drawing, a general assembly may be preceded by drawing a "check assembly". Many a time, changes are made in detail drawings when circumstances demand such changes. A check assembly is drawn accurately to scale so that any incorrectness or pitfalls which might have crept in are revealed in assembled condition. A correct general assembly can then follow the check assembly.

Besides the above types, assembly drawings are also made for incorporation in company catalogues and manuals. Such drawings convey a general idea of the equipment and are made for illustrative purposes only, for the benefit of the customers. Exterior shapes are shown and only principal dimensions are given. Sometimes when the picture is supposed to represent a number of sizes of the equipment, the main dimensions are entered in the picture in alphabetical form and the numerical values of these alphabets are given in an accompanying table.

Bill of materials : A bill of materials or parts list shown the description, item reference and other relevant data of the items or components used in an assembly or sub-assembly of a unit and is generally presented in a tabular form. The term "bill of materials" should normally be used in the case of structural and architectural drawings while the "parts list" is meant for machines drawings, but the two terms are now taken to be synonymous and no clear distinction is made now-a-days.

In most cases, the bill of materials is given in the drawing sheet itself and is usually placed above the title block (see Fig. 4.39). If the list contains a large number of parts, it is given on a separate sheet. The information which should be given in a parts list may vary from firm to firm, depending on the peculiarity of the product of the firm and other considerations. However, the following informations are generally included in a parts list :

(i) Item reference number;
(ii) Description of the part;
(iii) Specification of the part;
(iv) Material specification;
(v) Quantity required, and
(vi) Weight of the part.

Besides, a column may be provided for stock numbers and catalogue numbers of the parts for easy identification of those parts which are stock items in the stores of the manufacturing concern. Some parts' list also contains a column for pattern numbers of cast items, a column to indicate the raw stock dimensions, such as, rounds, forged pieces, structural sections from which the finished products are machined, and a remarks column.

There is no hard and fast rule as to the numbering of parts, but the general practice is to list the parts in the order of their importance or size. Thus for example in a gear box assembly, the cast or fabricated top and bottom housings are allotted the initial item reference numbers while standard parts such as bearings, threaded fasteners, circlips, seals and other items are written towards the end of the list.

If the parts list is placed on top of the title block, the item references should be numbered in numerically ascending order from bottom upwards, so that if necessary few items can be added afterwards. When written in separate sheets, the reverse order is usually followed.

On a parts list, a sub-assembly is sometimes shown as an encircled number, as shown in Fig. 4.39. The individual numbers comprising the sub-assembly receive the next numbers in numerically ascending order. The numbers of these components are not circled, and when there is a lack of space, these are put within parentheses after the description of the sub-assembly to indicate the numbers consisting the sub-assembly, as shown in Fig. 4.39. Otherwise if space permits, full description can be given, viz.,

"Support, consisting of parts 12 to 14".

The specification column should be used to specify the part only which it qualifies, and should not be used for specification of the material of the part which should be re-accommodated in the materials column.

Bought out items : In this age of specialization, there is hardly any manufacturing concern which produces all the components necessary for the end product of the concern. Hence, a number of items which are necessary for the final product are procured from outside agencies. Such items, which are not manufactured inside the factory but are purchased from other sources which specialise in the manufacture of those items, are called "bought out" items.

The advantages of such a practice are obvious. The bought out items are generally cheaper than they would be had the company itself undertaken to manufacture them, because these are mass produced by specialised concerns. The system assures standardised, quality products which also turn out to be cheaper. Such items may range from fasteners and accessories, such as, bolts, nuts, screws, washers and other similar items to anti-friction bearings, pumps, electric motors, oil seals, motors, gauges, pneumatic and hydraulic fittings, and a host of other components required for the final assembly of the main products. Besides, if such items are procured from outside, the company can concentrate and specialise in its main product, thus enhancing its quality.

Details of bought out items should be given in the bill of materials. If the space is insufficient in a parts list to fully specify a bought out item, the specification and other relevant data should be indicated in the drawing sheet somewhere in the vicinity of the parts list.

The drawing office should keep up-to-date company catalogues and trade journals of various, relevant engineering products for easy accessibility whenever the designers or the draughtsmen require them during their work.

4.8 CHECKING OF WORKING DRAWINGS

A working or manufacturing drawing meant for use in the shops should be thoroughly checked before it is approved by the competent authority and released for shop use. Such checking is vital because a faulty drawing may lead to a lot of wastage of money, effort and time. An inaccurate drawing is useless for practical purposes although it may look pleasing to the eye or may apparently seem to be all right when given a cursory examination. Even a supposedly insignificant aspect as an arrow-head may completely change the shape of the part if that arrow-head is misplaced. It is, therefore, imperative that the drawing (or the print thereof) is perfect before it leaves the drawing office.

A manufacturing drawing is normally checked by an experienced person. Obviously, it should not be checked by the person who has draughted it. A man belonging to the supervisory rank, *e.g.*, the chief draughtsman or the engineer incharge should be entrusted with the job. Sometimes checkers are given this task or the draughtsmen themselves may check one another's work, as is done in small drawing offices. In any case, the person who makes the drawing and the one who checks it are both responsible for the accuracy of the drawing. The drawing should always bear the signatures of the above persons. In some cases, double checking is done by a final checker to eliminate remaining errors, if any.

Checking should be done in a systematic and orderly manner. Although the primary purpose of checking is to locate any error that might have inadvertently crept in, a competent checker or a checker with innovative ideas may even suggest changes in the design or point out some intrinsic defect. This may lead to the redesign of the part before it is too late. Checking can be done on the tracing itself, but this practice should be discouraged as a lot of erasing is to be done afterwards by the draughtsman which may damage the tracing. A blue print or an ammonia print should be taken from the tracing and all corrections or alterations should be indicated on the print. Normally, a red pencil should be used to point out the errors or the desired alterations and some other colour, preferably yellow or green, should be used to tick the right dimensions, arrows, symbols and other salient features. After completion of checking, the draughtsman incorporates the corrections on the body of the tracing as per the check print.

For methodical checking a checking list should be followed. The following guidelines are given to facilitate the checking procedures :

1. The drawing sheet should be of correct and optimum size. The checker should ensure that the drawing is neither unduly crowded rendering it difficult to interpret nor should the sheet used be too large, leaving a lot of unnecessary empty spaces.
2. The views and sections drawn should be just adequate to serve the purpose, and those that are not required should be avoided. Proper placement of the views and sections on the drawing is very important to satisfy the rules governing the technical drawing procedures. Besides, it facilitates easy and correct "reading" of the drawing by the user, and also it is aesthetically pleasing.
3. Proper scale should be used. It is better to avoid a large number of scales in a single drawing sheet.
4. All dimensions should be checked as to their values and correct location. All lines, radii and curves should be checked to ensure that they have been drawn as per scale. Lines which are not drawn as per scale should be underlined. Arrowheads and leader lines should get particular attention. Sharp corners should be replaced by rounds and fillets. Dimensions should be entered on the drawing strictly as per the instructions given in Section 4.2 dealing on dimensioning in drawing.
5. The checker should see to it that dimensions shown on the drawing are adequate for processing the part in the shops and that the shop people are not left to arrive at a certain dimension by mathematical calculations. Under no circumstances should the shop people measure off an undimensioned line by putting a scale or a rule on the drawing. If in doubt, the shopman should always ask and consult the proper man. Superfluous dimensions, if any, should be scored off as they tend to confuse the reader of drawing and also may make the drawing vulnerable to misinterpretation.

6. Tolerances should be strictly checked. The zone and quality of tolerances should be so chosen that they meet the requirements of the fit necessary for the purpose. The quality should be neither too fine as it increases the cost of production unnecessarily nor should it be too coarse which may defeat the purpose (see Fig. 7.14 on cost vs. quality in Section 7.4). Normally, fits and tolerances are specified for standard items, such as, anti-friction bearings, couplings, keys, splines, gears, sprockets and other transmission elements as well as for various other machine elements and parts. The checker has only to ensure that the draughtsman has entered the specified standardised tolerances correctly.
7. Surface finish symbols should be shown as per the requirements to which the surfaces are subjected. As in the case of tolerance, an unnecessarily finer finish will only entail greater cost. Finish marks should always be placed on the lines to which dimensions are entered.
8. As far as possible, standard parts should be used for reasons of economy, availability and inventory control. If the firm has a stock control department, it is advisable to quote the stock or catalogue number of the standard part or bought out item in the bill of materials, so that the shop can procure the material easily.
9. The checker should ensure that the material of which a part is to be made, is judiciously chosen. This way, the material cost can be brought down. Costly material should not be used where a cheaper material can serve the purpose. If a part is to be made of steel, it is to be ensured whether that particular steel is available as rolled (or forged) stock in the market or whether the steel is castable. Where alloy steels, special bronzes and other similar materials are envisaged to be used, it is advisable to ascertain beforehand whether these materials are obtainable in the sizes required for the purpose.
10. The bill of material should be thoroughly checked. This includes, the requisite quantity of items, proper designation of parts, specification of standard items, proper choice of materials for ferrous, non-ferrous and non-metallic parts, catalogue number (if any) and the mass of individual and total items.
11. The title block should clearly indicate the department, unit and description of part or assembly. In case of an old drawing which is to be revised, then the revision carried out should be entered in the revision column with date and signature, and the drawing number should be changed accordingly, *e.g.*, by writing R1 or R2, as the case may be, after the original drawing number. Also, the revised dimension should be entered on the body of the drawing in the proper manner. In this connection, see Sections 1.7, 4.6 and 4.7.
12. If the firm has manufacturing facilities such as foundry, machine shop, heat-treatment furnaces and other process equipments, it is advisable that the part be so designed that it can be made within the shop complex of the firm itself. The cost is thus substantially reduced, because off-loading the part to other concern will naturally cost more. Use of costly tools should be avoided unless it is really necessary.
13. The checker is to see that proper notes and instructions are written in the drawing in case of cast parts as regards cleanliness, blow holes, proper draft and radii, cracks and other defects or features. Both in case of cast pieces and welded assemblies, proper heat-treatment and other relevant instructions, if any, should be written on the drawing, *e.g.*, stress relieving, normalising,

ultrasonic testing, gamma ray or X-ray examination, *etc.* However, these are costly test procedures and as such, they should be recommended with prudence.

14. Proper and adequate greasing points should be ensured, if the part needs lubrication.
15. Requisite clearance is very important. The checker should make sure that the part does not foul with adjacent parts in assembled condition. Sufficient room for operating the spanners and allied tools should be ensured in case of nuts and bolt heads.
16. Attention should be given to the following necessities : the seats of bolt heads, nuts, washers and similar objects should be spot faced. Bosses should be provided wherever required. Ground surfaces should be indicated by marking or writing and proper groove should be incorporated to provide clearance for the wheel while grinding. If needed, R.H. and L.H. parts should be clearly indicated.

Besides the above points, the checker should also consider the value engineering aspects of the parts or assemblies as a whole. Looking from a larger perspective, it is quite possible that while checking, he comes out with ideas from which a better and a cheaper design may ensure. For an example of check-print, see Fig. 4.42.

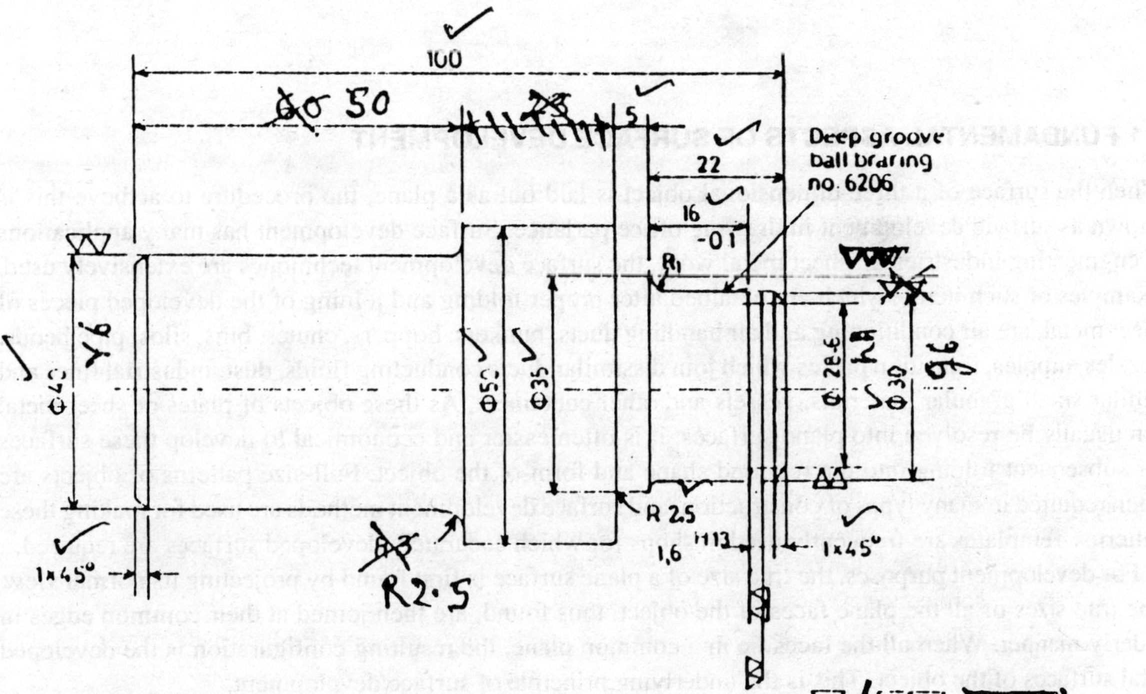

Fig. 4.42 A sample check print

CHAPTER 5

SURFACE DEVELOPMENT AND INTERSECTION

5.1 FUNDAMENTAL ASPECTS OF SURFACE DEVELOPMENT

When the surface of a three-dimensional object is laid out as a plane, the procedure to achieve this is known as surface development in drawing office parlance. Surface development has many applications in engineering industries. In sheet metal work, the surface development techniques are extensively used. Examples of such items, which are obtained after proper folding and joining of the developed pieces of sheet metal, are air conditioning and air handling ducts, bunkers, hoppers, chutes, bins, silos, pipe bends, nozzles, nipples, transition pieces which join dissimilar ducts conducting fluids, dust, industrial fines and similar small granular materials, vessels and other containers. As these objects of plates or sheet metal can usually be resolved into plane surfaces, it is often easier and economical to develop these surfaces for subsequent folding into the required shape and form of the object. Full-size patterns of objects are often required in many types of construction, and surface development methods are used for making these patterns. Templates are frequently used in shops for which accurately developed surfaces are required.

For development purposes, the true size of a plane surface is first found by projecting its normal view. The true sizes of all the plane faces of the object, thus found, are then joined at their common edges in orderly manner. When all the faces lie in a common plane, the resulting configuration is the developed total surfaces of the object. This is the underlying principle of surface development.

A plane surface can be easily developed. A single-curved surface of an item is generated by revolving an element called generatrix about an axis. A common example is a cylinder the surface of which is generated by revolving a line, that is, an element parallel to the axis, about the axis of the cylinder. The numerical values of such areas of revolution can be found by using the rule of Pappus (see Mensuration in Section 10.5). A warped surface cannot be developed. Examples of warped surfaces are saddle-shaped bodies, screw threads, aircraft wings, *etc*. A double-curved surface can be generated by revolving a curved line about an axis. Common examples are sphere, ellipsoid, paraboloid, torus, etc.

Figure 5.1 shows the surfaces of some common items obtained by unfoldment to illustrate the principles of surface development. A surface is developable only when the object can be unfolded or unrolled to lie

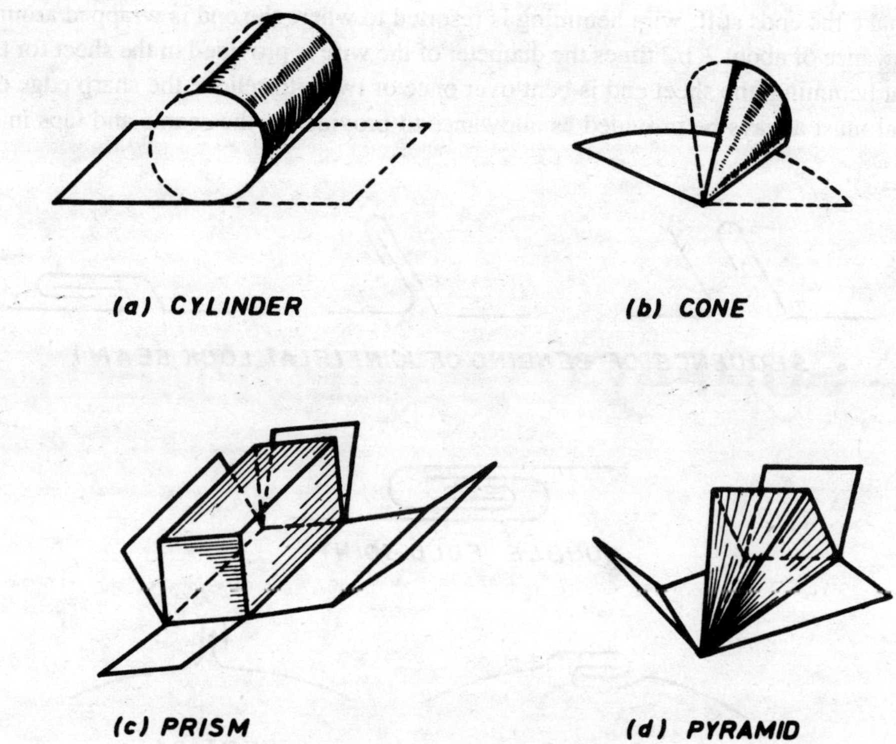

Fig. 5.1 Principles of surface development

on a plane. Plain objects like cubes, tetrahedrons, polyhedrons, prisms, pyramids, *etc.*, and single-curved surfaces like cylinders, cones, *etc.*, are developable. Warped surfaces and double-curved surfaces can be developed only approximately by dividing the entire surface into tiny units and then developing these units by methods used for regular, developable features. Example of such a unit is the lune of a sphere.

In case of small objects, surfaces are developed on plain papers in the drawing office which are then supplied to the workshop and pasted on sheets or plates. The developed surfaces of sheets or plates are then obtained by cutting along the edges of the pasted paper by means of shears, gas-cutting equipments or some other method. In case of large objects, the developed surfaces on papers can be used as templates to guide the cutting device by means of a copying machine. The stylus of the copying machine traces the edges of the paper and this actuates and guides the cutting device, *e.g.*, oxy-acetylene flame, which cuts the plate or sheet to produce the actual developed surface required in true size.

A plate or a sheet has a certain thickness and this must be considered while making development. When a plate is bent, the outer surface becomes stretched while the inner surface becomes crowded or compressed in the vicinity of the bent portions. The neutral surface, which is taken to be situated at a distance of around one-third the thickness of the plate measured from the inner surface, remains unaffected due to bending. Determination of the developed lengths of bent plates has been covered in Section 5.8.

Different types of joints are used for seaming sheet metal pieces and for connecting one piece to another. Some of these joints commonly used are shown in Fig. 5.2. For finishing raw edges of sheet ends, hemming is used. To make the ends stiff, wire hemming is resorted to where the end is wrapped around the wire. An extra allowance of about 2 1/2 times the diameter of the wire is provided in the sheet for the purpose. In case of flat hemming, the sheet end is bent over once or twice to relieve the sharp edge of the sheet. Extra material must always be provided as allowance to provide for the seams and laps in sheet metal work.

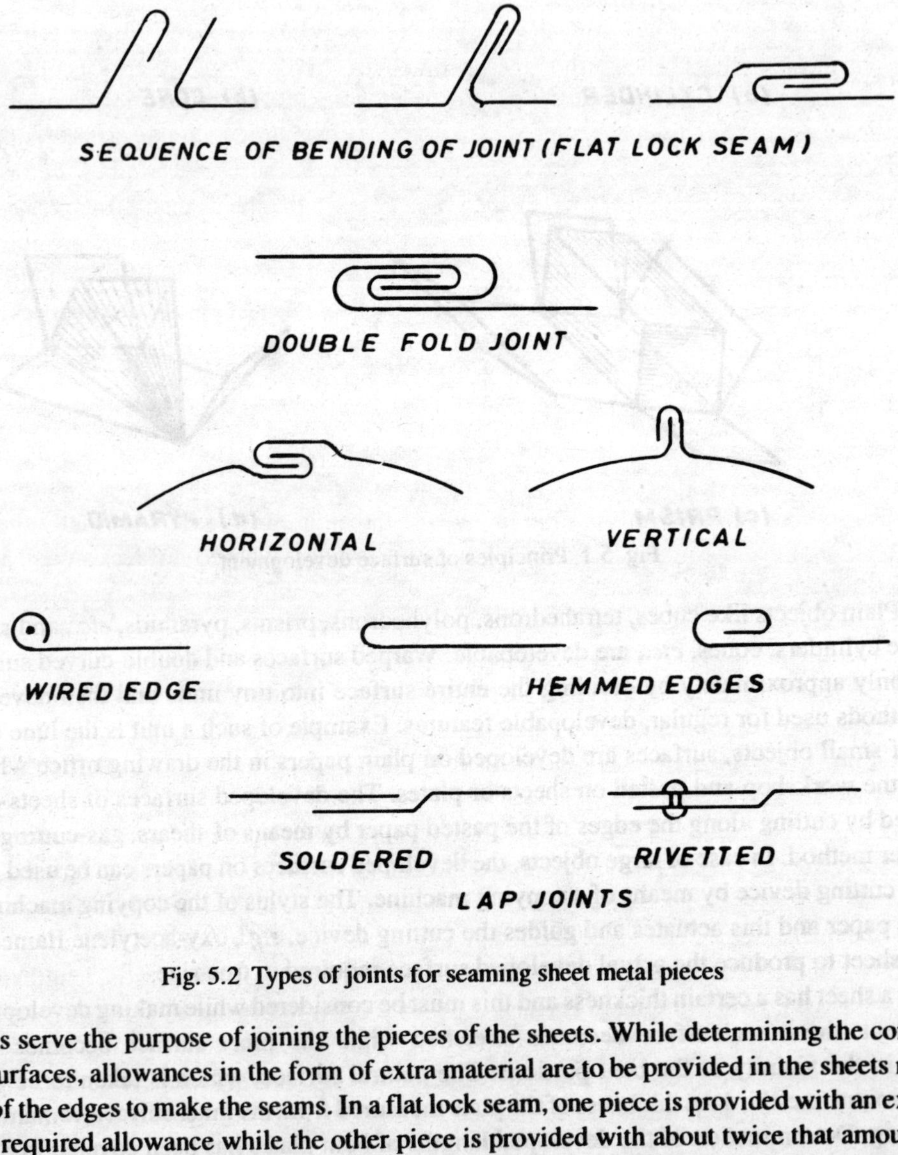

Fig. 5.2 Types of joints for seaming sheet metal pieces

The seams serve the purpose of joining the pieces of the sheets. While determining the contour of the developed surfaces, allowances in the form of extra material are to be provided in the sheets required for the folding of the edges to make the seams. In a flat lock seam, one piece is provided with an extra amount equal to the required allowance while the other piece is provided with about twice that amount so that it can bend and loop over the extra material of the first piece (see the sequence of bending in Fig. 5.2). The edge preparation of sheets for seam is done by special machines if quality work is envisaged. The width of seam cannot be specified as it depends on the thickness of the sheet. It is preferable to select small width which should serve the purpose.

5.2 SURFACE DEVELOPMENT BY TRIANGULATION

By the method of triangulation, surface development of bodies with complex shape can be achieved. The method is useful in case of bodies of which the surfaces cannot be resolved easily into fundamental geometrical surface elements. The method is very much used to develop the surfaces of transition pieces, oblique cones, *etc*. In this method, the surface of the object is resolved into a sufficient number of triangles. Triangular areas can be easily placed side by side on the plane on which the surface is developed. The method, however, entails a certain amount of approximation because curved lines are taken as straight lines, *viz*., the bases of the triangles. It is, therefore, necessary to take as many triangles as possible into consideration, so that the error is held to a minimum. The points on the surface are joined in zig-zag fashion. The whole developed surface is thereby divided into a number of narrow triangles of very short bases lying side by side. The true lengths of the longer sides of the triangles are geometrically determined on a separate figure. The final developed surface then consists of these triangles when they are placed side by side in their true sizes.

In short, the triangulation method presents a procedure for developing surfaces approximately which are non-developable by resolving such surfaces into narrow sections of developable surfaces, *viz*., strips of plane triangles with very short bases. It is a simple method in which the surfaces are merely divided into triangles as stated above, then the true lengths of the sides are found, and finally the triangles are constructed one by one and joined together at their common sides. Warped surfaces as well as developable surfaces like oblique cones can be developed much more easily and with reasonable accuracy by triangulation method. The following example will illustrate the method.

In Fig. 5.3 a part is shown which is used for connecting two pipes of unequal diameters. By using the method of triangulation, the developed surface of this object is found as follows : The top view and the elevation of the part are first drawn. The two circles of the plan are divided into equal parts - in this case, into 12 parts, and numbered alternatively as 0, 1, 2, ... 12, 13. Next from the two circles on the plan, vertical lines are drawn on the elevation corresponding to the points on the plan circles as shown in the figure. Zig-zag lines, *viz*., 0 - 1 - 2 - 3 - 4 - 5 are then drawn on both plan and elevation. This process demarcates the triangular strips to be developed. The next step is to find the true lengths of the longer sides of the triangles. To achieve this, a "true length diagram" is to be drawn. This is done in the following manner: By the side of the elevation, horizontal and vertical axes are drawn. On the vertical axis, a length *XY* is cut off making it equal to the height of the object. Along the horizontal axis, lengths 1 - 2, ... 6 - 7, ... 12 - 13, *etc*., are drawn making them equal to the plan lengths measured off from the plan. True lengths of these lines are found by joining the extremities of the plan lengths with the point *X*, as indicated by the line T.L. 1 - 2 in case of the line 1 - 2. Similarly, the true lengths of other lines are found. This constitutes a true length diagram. To construct the developed surface, the line 0 - 1 is first drawn. Now, arcs with radii '*t*' and T.L. 1 - 2 are drawn with centres at 0 and 1, respectively. These two arcs intersect at the point 2. Point 3 is found by drawing area from 2 with T.L. 2 - 3 as radius and from 1 with "T" as radius. The other points are found in a similar manner. Smooth curves are then drawn through these points of intersection. The bounded area shown in the required development of half the surface, the other half being the mirror-image of the first half. Here, the chordal lengths, "t''" and "T", approximate the circular lengths of the arcs.

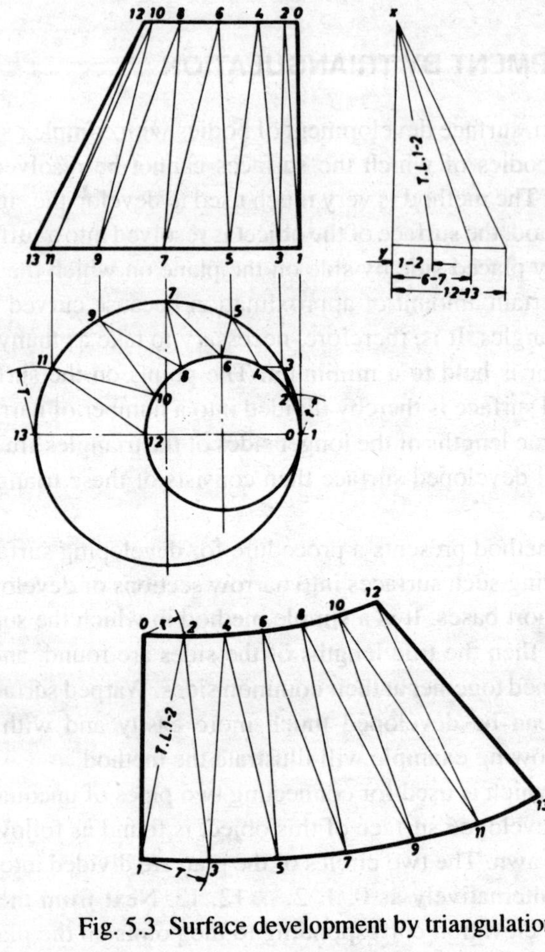

Fig. 5.3 Surface development by triangulation

5.3 SURFACE DEVELOPMENT OF CONES

In general, a cone is an object which has a triangular shape in longitudinal sections. From base, a cone tapers off to an apex. In case of a truncated cone, the top portion ends in a plane which may be parallel or inclined to the plane of the base. The base is generally of circular shape, but it may be of elliptical cross-section also. Conical objects with various shapes and configurations are common items in industries. Hence, the development techniques of different types of conical objects are very important. The surfaces of most of the conical objects can be developed by common methods, but for certain special cases the method of triangulation is employed. Some of these developments are given below :

Right circular cone : To develop a right circular cone with base diameter D, vertical height H and slant height L, a circular sector is laid out with L as radius and included angle α which is given by :

$$\alpha = \frac{D}{L} \times 180^\circ$$

The circular length of the arc of this sector is obviously equal to the circumference of the base circle and is equal to $\pi. D$. Slant height is given by :

$$L = \sqrt{(D/2)^2 + H^2}$$

The surface development of a parallel truncated cone is done in a similar manner.
Referring to Fig. 5.4,

$$L = \sqrt{(D/2)^2 + H^2}, H = \frac{D.h.}{D - d},$$

$$l = \sqrt{(d/2)^2 + x^2}, x = \frac{d.h}{D - d}$$

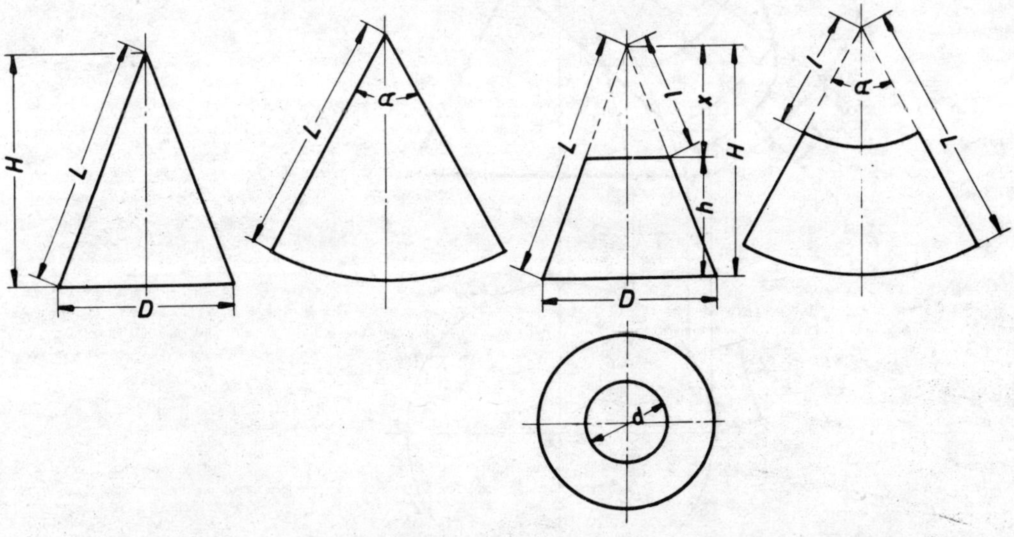

Fig. 5.4 Surface development of right circular cone

Inclined truncated cone : Referring to Fig. 5.5, the surface development of an inclined truncated cone is executed as follows : The plan and elevation of the cone are drawn first. The next step is to develop the surface of the complete, that is, untruncated cone, as discussed in the previous case. The plan is then divided into 12 equal parts. Vertical lines are drawn from the points 1, 2, 3, *etc*., locating the corresponding points on the elevation. These points are joined with the apex A. The developed circular arc is also divided into 12 equal parts as shown. Next, horizontal lines from the points of intersection of the oblique lines are drawn to meet the edge of the complete cone. With centre at A, these points are rotated to meet the corresponding radial lines from A to the circular arc. A smooth curve is then passed through these points of intersection. The bounded area is the required development of the inclined truncated cone.

Oblique cone : A right circular cone is a cone of revolution. An oblique cone, on the other hand, has elements of different lengths.

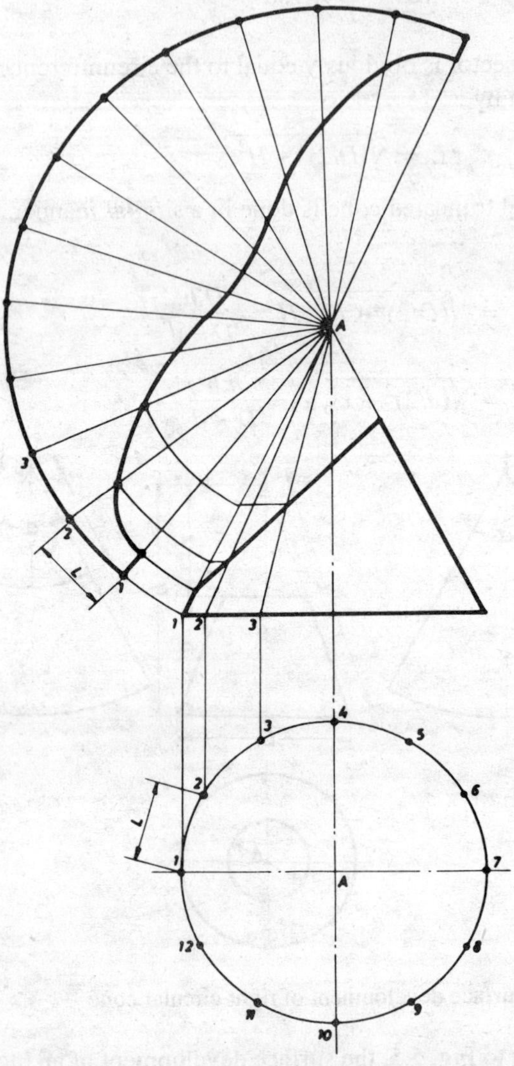

Fig. 5.5 Surface development of inclined truncated cone

Referring to Fig. 5.6, a "true length diagram" is to be drawn first to determine the true lengths of the generator at different points. The true length of a generator or element is the hypotenuse of the right-angled triangle with the projection of the element on the horizontal plane (plan length) as its base and the height of the cone as its other side. Plan and elevation are drawn as well as the vertical and horizontal axes. Height *H* is cut off on the vertical axis. The base circle is divided into 12 equal parts. On the plan, the points 1, 2, 3, etc., are joined with the apex *A*. Thus, the plan lengths of the generator at these points are located (1 - *A*, 2 - *A*, 3 - *A*, etc.). These are laid down on the horizontal axis of the true length diagram as

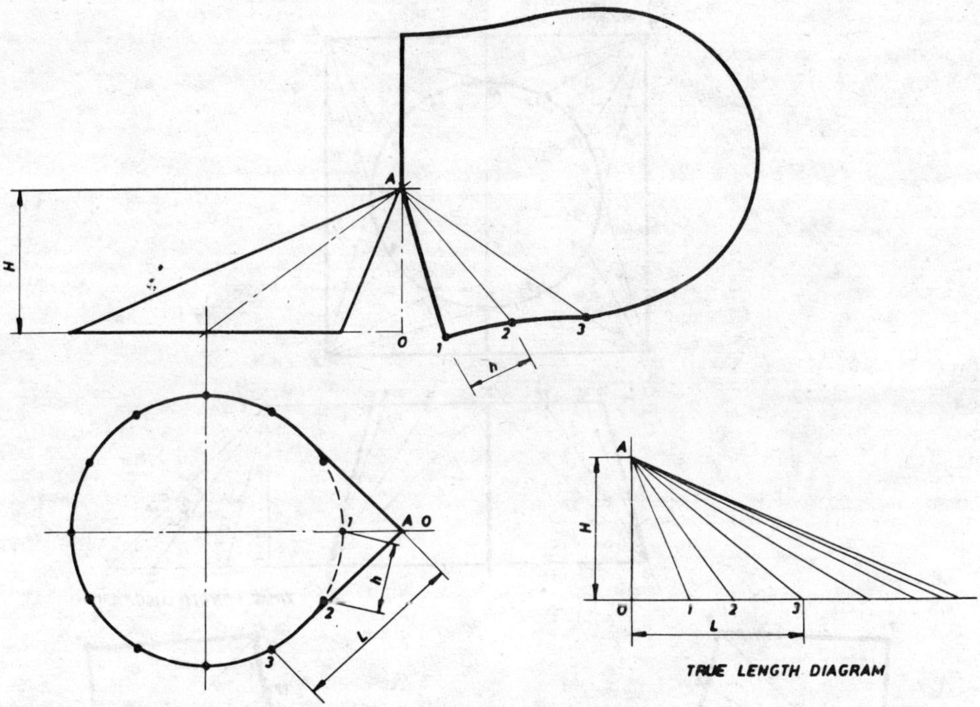

Fig. 5.6 Surface development of oblique cone

shown (01, 02, 03, etc.). These points are then connected with *A* and the lines *A*1, *A*2, etc., are the true lengths of the generators at these points.

Now, to develop the surface of the oblique cone, the true length of the shortest side *A* 1 is drawn conveniently from the apex *A* of the elevation. Arcs with radii *A* 2 (taken from the true length diagram) and chordal length "*h*" of the base circle (which is nearly equal to the circular length) are drawn with points *A* and 1 as centres. This locates point 2. Other points are similarly found corresponding to the true lengths *A*3, *A*4, etc., and the chordal length *h*. By passing a smooth curve, we get the required development.

5.4 DEVELOPMENT OF TRANSITION PIECES

Transition pieces, also known as transformers, are those objects which connect dissimilar pieces to maintain continuity. In industrial applications, such transition pieces are very common. Most of these are simple connectors, but there are many cases, especially in chemical plants, these pieces are very complicated and hence their development is a series of complex steps. Transition from circular to square or rectangular section is very common in piping, ducting and other industrial applications like hoppers, bunkers, bins, *etc*. One such example is illustrated in Fig. 5.7. The surface can be developed following the procedures laid down in Section 5.2 and 5.3.

An examination of the object will reveal the fact that the body consists of 4 triangles on 4 sides (*viz*., 3 - 6 - 7, 7 - 10 - 11, 11 - 14 - 15, and 15 - 2 - 3), intercepted by 4 curved surfaces (6 - 7 - 10, *etc*.), each

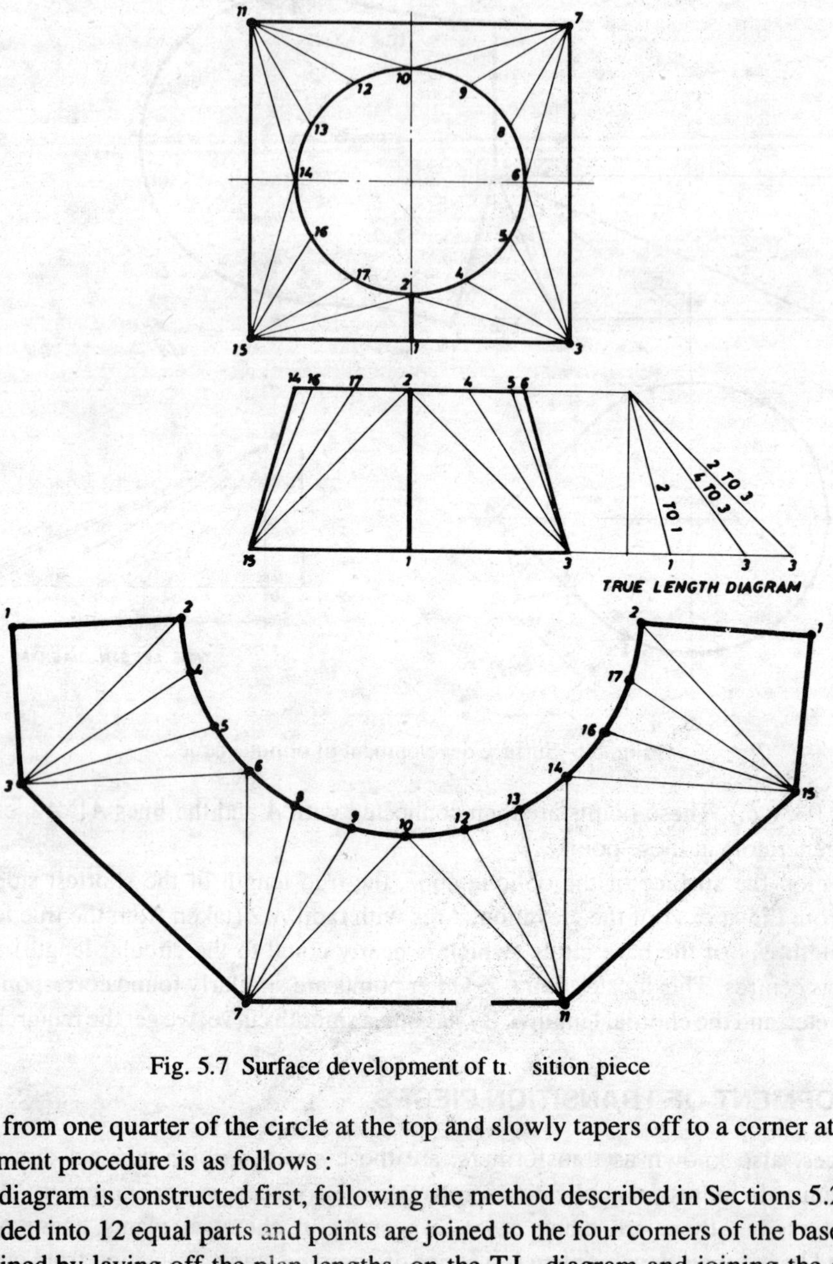

Fig. 5.7 Surface development of tı sition piece

of which begins from one quarter of the circle at the top and slowly tapers off to a corner at the base. The surface development procedure is as follows :

A true length diagram is constructed first, following the method described in Sections 5.2 and 5.3. The top circle is divided into 12 equal parts and points are joined to the four corners of the base square. True lengths are obtained by laying off the plan lengths on the T.L. diagram and joining the extremities of these lengths with the top point of the diagram corresponding to the height of the part.

To develop the surface, T.L. 2 - 1, which is the line of separation, is drawn first. Line 1 - 15, which is T.L. in the plan, is drawn at right-angles to T.L. 2 - 1. Arcs are drawn from points 2 and 15 with radii equal to chord $\overline{2-17}$ (taken from the plan circle) which approximate the circular arc $\widehat{2-17}$, and T.L. 15 - 17

(= T.L. 4 - 3). These arcs intersect at 17. Similarly, the points 16 and 14 are located. Intersection of the line 15 - 11 and the T.L. 14 - 11 (= T.L. 15 - 14 = T.L. 2 - 3) locates the point 11. We now have the developed surfaces of a half triangle, a curved surface and a full triangle. In a similar manner, other points are located.

We get the required development of the complete surface of the transformer by passing a smooth curve joining 2 with 2 and drawing the relevant straight lines as shown in the figure.

5.5 SURFACE DEVELOPMENT OF COMMON OBJECTS

Surface development of some common objects are discussed in this section.

Truncated hollow cylinder : Referring to Fig. 5.8, the plan and elevation of the truncated cylinder are drawn first. The plan circle is divided into 12 equal parts. A horizontal line equal in length to the developed length of the base is drawn. The numerical value of this developed length is π. *d*, where the diameter of the cylinder is "*d*".

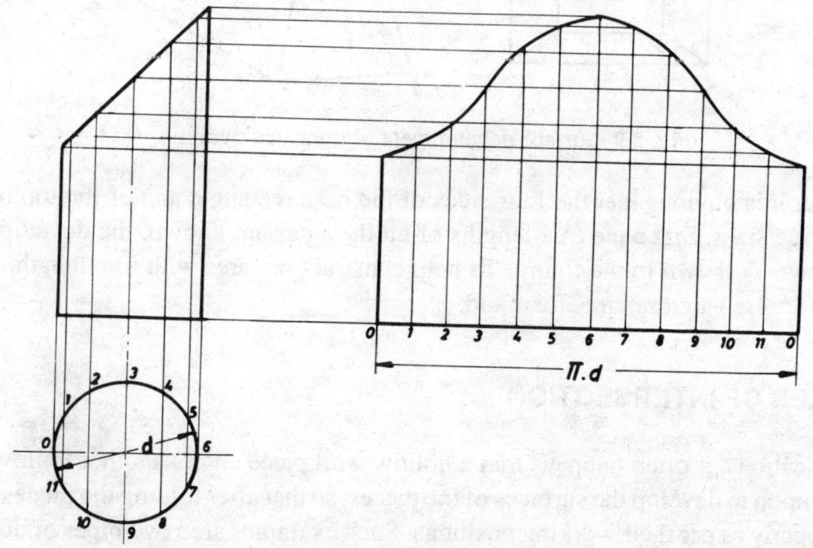

Fig. 5.8 Surface development of truncated hollow cylinder

The developed base line is also divided into 12 equal parts. Vertical lines are then drawn from the plan view to the elevation, cutting the inclined plane of the truncated cylinder as shown. From these points of intersection, horizontal lines are drawn to meet the vertical lines drawn from the developed length of the base circle. Care should be taken to ensure that the lines from the base circle with certain number intersect the corresponding lines from the developed base line with the correct numbers, matching the sets.

A smooth curve is then passed through these meeting points, and the area bounded as shown in the figure is the required developed surface.

Truncated pyramid : Referring to Fig. 5.9, the plan, elevation and side views of the truncated pyramid are first drawn. Taking 0 as centre in the plan view, arcs are drawn with radii 0 2 and 0 2′ upto the horizontal axis of the plan. From there, verticals are drawn to meet the extended base line and top line of the elevation as shown. The distance 0 2 - 0 2′ = 22′ is the true length of all the four slant sides of the pyramid.

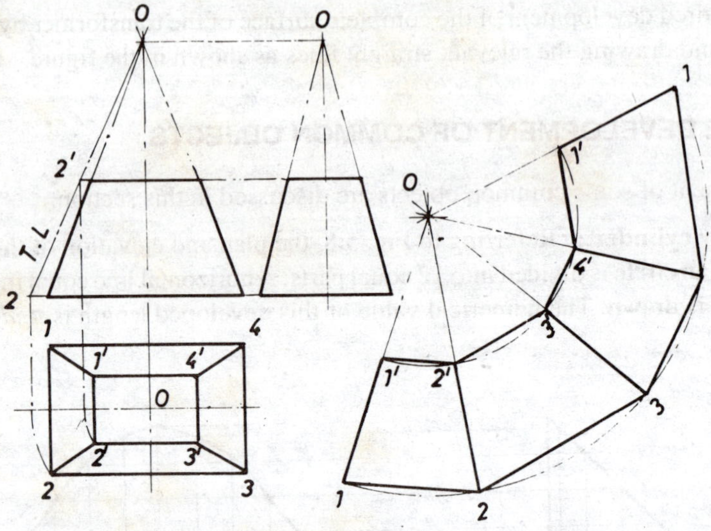

Fig. 5.9 Surface development of truncated pyramid

From the figure, it is obvious that the four sides of the base rectangle and of the top rectangle in the plan view are in true sizes. Since the true lengths of all the sides are known, the developed surface can now be easily drawn as shown in the figure. To help construction, arcs with true length radii 0 1 (= 02, *etc.*) and 0 1′ (= 0 2′, *etc.*) are drawn beforehand.

5.6 PRINCIPLES OF INTERSECTION

In industrial applications, it often happens that a hollow workpiece enters another hollow piece and the designer is called upon to develop the surfaces of the pieces, so that after folding the pieces to their correct shape, they fit properly as per their working positions. Such examples are : two pipes or ducts penetrating each other, either at right angles or at any other angle depending on the operational point of view, an inclined conical chute entering a pipe or vice versa, and other numerous configurations. After the pieces are brought together to their correct positions, the common contour is joined by welding, soldering or some other means.

In cases where the bodies consist of plane surfaces, *e.g.*, prisms, pyramids, cubes, *etc.*, the contours comprise straight lines. In case of bodies with curved surfaces, *e.g.* cylinders, cones, *etc.*, the contours are curvilinear. In Section 5.7, example of application of intersection is given with its surface development procedures.

Besides developing the surfaces, the drawing maker has to determine the correct shapes of the common contour when the developed surfaces of the two work-pieces are spread over a plane. A contour curve is determined point by point and a smooth curve is passed through these points to ascertain the final shape of the curve.

This correct determination of the contour curve in two-dimensional plane is very important because the two pieces are supposed to fit perfectly at the common contour after folding the pieces to the three-dimensional form. Otherwise, any faulty shape of the curve will result in non-uniform fitting which in turn will entail costly and unnecessary modifications. Many methods are in vogue to determine the shapes of contour curves and surface development.

5.7 INTERSECTION OF COMMON OBJECTS

Intersection of two hollow cylinders is very common in industrial applications like pipe-line complexes, ducting, *etc.* In Fig. 5.10 two pipes inter-penetrating each other have been shown. The development of the surfaces of the two individual can be attained by taking projection lines and joining corresponding points. The procedure for surface development of two intersecting pipes of equal diameter is as follows:

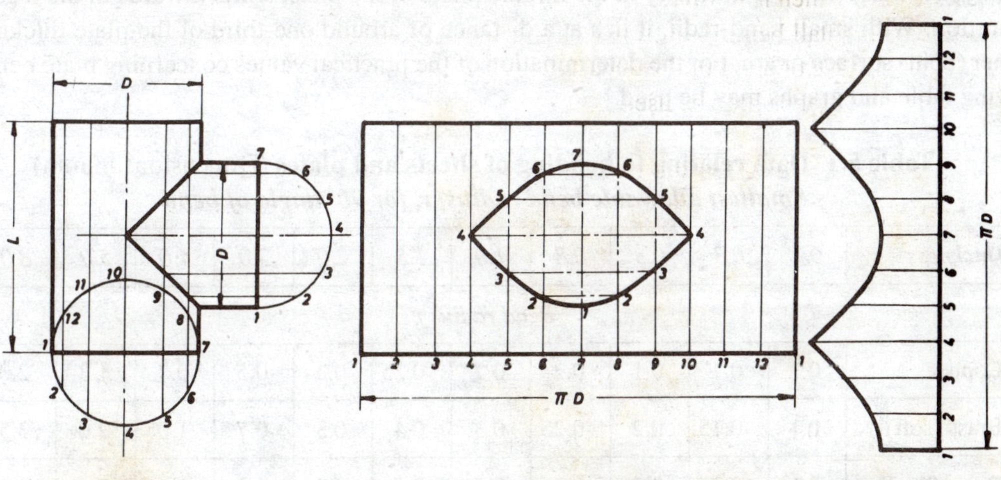

Fig. 5.10 Intersection of pipes

The elevation of the combined piece is drawn first on which the plan circle of the vertical pipe is shown superimposed. This base circle is divided into equal parts (12 in this case). The semi-circle in the elevation, corresponding to the horizontal pipe, is also divided equally in the same number of parts. The developed circumferences of the two pipes, which is equal to $\pi . D$ where D is the diameter of each pipe, are laid down separately as shown. Vertical lines are now drawn through the divisions on the plan circle. These lines meet the corresponding horizontal lines drawn through the semi-circle of the elevation. Since the pipe diameters are equal, the two extremities of the opening on the vertical pipe will be separated by half the developed circumference, that is, $\pi . D/2$ (points 4 and 4 in the figure). These two points are located on the central line of the developed pipe surface. Now, horizontal lines from the elevation are extended to meet the corresponding vertical lines on the developed surface (7 to 7, 6 to 6, *etc.*). Passing a smooth curve through the points of intersection demarcates the contour of the opening of the vertical pipe.

To determine the contour of the horizontal pipe which enters the vertical pipe, the developed circumference is divided into the same number of equal parts. Distances are measured off from the horizontal pipe in the elevation, and these are carried over and drawn as vertical lines from the corresponding points on the developed circumference. A smooth curve is now drawn to complete the development. The two pipes will now match when the developed surfaces are folded back to circular cross-section and the corresponding contours are fitted to each other.

5.8 BENDING OF SHEETS AND PLATES

As pointed out earlier in Section 5.1, a plate becomes stretched on the outer surface and compressed on the inner surface when the plate is bent. It goes without saying that a developed surface is theoretically perfect only when the thickness is zero. Since in practice, the plate has a certain thickness, a minimum bend-radius is to be maintained while bending. This minimum bend-radius is determined corresponding to the plate thickness as well as to the material of the plate.

The neutral axis, which is normally in the middle plane of the plate, shifts inwards in the region of the bent portion. With small bend-radii, it lies at a distance of around one-third of the plate thickness from the inner (bent) surface or arc. For the determination of the practical values concerning plate bending, the following table and graphs may be used.

Table 5.1 Data relating to bending of sheets and plates (Dimensions in mm)
Smallest allowable bend radius, r, for 90° angle of bend

	Thickness	*0.1*	*0.3*	*0.5*	*0.8*	*1.0*	*1.5*	*2.0*	*3.0*	*4.0*	*5.0*	*8.0*	*10.0*
							Bend radius r						
MATERIAL	Copper	0.1	0.1	0.1	0.2	0.2	0.25	0.3	0.5	0.7	1.3	2.0	2.5
	Brass (soft)	0.1	0.15	0.2	0.25	0.3	0.4	0.5	0.7	1.0	2.0	3.5	5.5
	Brass (hard)	0.1	0.2	0.3	0.4	0.5	0.6	0.8	1.2	1.5	2.5	4.0	6.0
	Steel (soft)	0.1	0.2	0.3	0.4	0.5	0.6	0.9	1.4	2.0	3.2	4.5	6.0
	Steel (medium hard)	0.1	0.2	0.3	0.4	0.5	0.6	1.0	1.5	2.5	4.5	7.0	10.0
	Al Cu Mg	0.2	0.4	0.6	1.2	2.0	3.0	4.0	6.0	8.0	12.0	14.0	16.0
	mg Al	0.5	1.0	2.0	4.0	6.0	9.0	12.0	18.0	24.0	35.0	50.0	60.0

Determination of developed length with 90° angle of bend

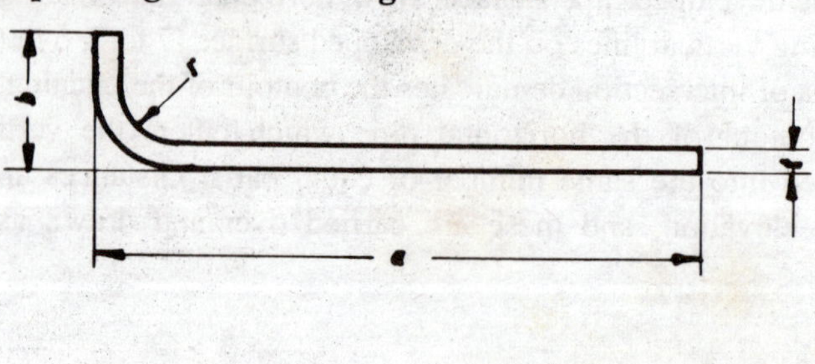

Developed length

$L = a + b - x$

Thickness t		0.5	0.8	1.0	1.5	2.0	2.5	3.0	3.5	4.0	4.5	5.0	6.0
						x							
RADIUS	0.1	0.7	1.2	1.4	2.1	2.8	3.5	4.3	5.1	5.8	6.5	7.4	8.6
	0.5	0.9	1.4	1.6	2.3	3.0	3.7	4.5	5.2	5.9	6.7	7.5	8.7
	1.0	1.2	1.6	1.8	2.5	3.2	3.9	4.7	5.4	6.1	6.9	7.7	8.9
	1.5	1.4	1.8	2.0	2.7	3.4	4.1	4.8	5.5	6.3	7.1	7.9	9.1
	2.0	1.7	2.0	2.3	2.9	3.6	4.3	5.0	5.7	6.5	7.3	8.1	9.3
	2.5	1.9	2.3	2.5	3.1	3.8	4.5	5.2	5.8	6.6	7.5	8.3	9.5
	3.0	2.1	2.5	2.7	3.3	4.0	4.7	5.3	6.0	6.8	7.7	8.5	9.8
	3.5	2.3	2.7	2.9	3.5	4.2	4.9	5.5	6.2	7.0	7.8	8.7	10.0
	4.0	2.6	3.0	3.2	3.7	4.4	5.0	5.7	6.3	7.2	8.0	8.9	10.3
	4.5	2.8	3.2	3.4	4.0	4.6	5.2	5.8	6.5	7.4	8.2	9.1	10.5
	5.0	3.0	3.4	3.6	4.2	4.8	5.4	6.0	6.7	7.5	8.4	9.3	10.7

Example : *Given, a = 20, b = 10, r = 1.5, t = 1*

To find L : From table, x = 2. Therefore. L = 20 + 10 – 2 = 28 mm

Dimensions in mm

Determination of developed length with any angle of bend

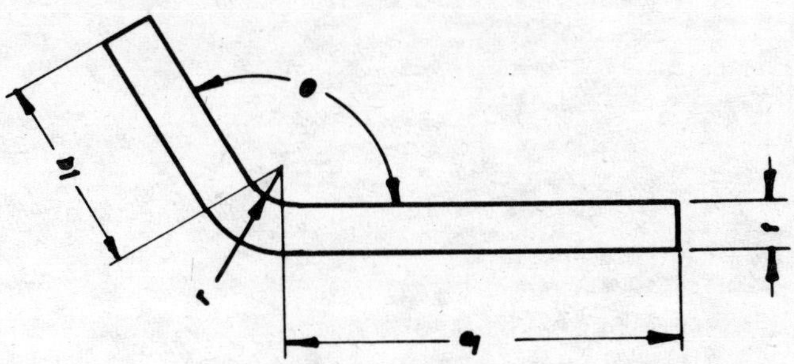

Developed length $$L = a_1 + \frac{\pi\,(180^\circ - \theta)}{180^\circ} \cdot \left(r + \frac{t}{2} \cdot y \right) + b_1$$

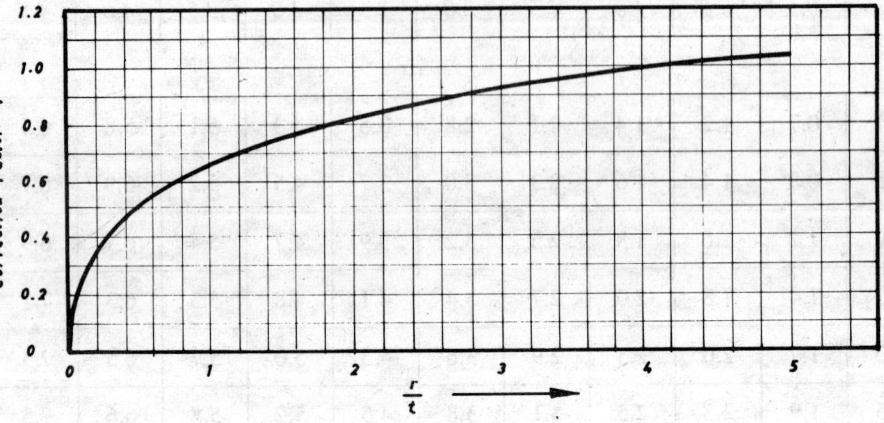

Example : *Given* $a_1 = 15$, $b_1 = 10$, $\theta = 120^\circ$, $r = 6$, $t = 5$,

To find L.

From graph, $y = 0.7$ for $\frac{r}{t} = \frac{6}{5} = 1.2$

Therefore $$L = 15 + \frac{3.1416\,(180^\circ - 120^\circ)}{180^\circ} \left(6 + \frac{5}{2} \times 0.7 \right) + 10$$

$= 33.1$ mm.

CHAPTER 6

STRUCTURAL, PIPING AND OTHER DRAWINGS

6.1 ASPECTS OF STRUCTURAL DRAWINGS

Structural drawings fall mainly in the domain of structural and architectural engineering. Since this book is primarily meant for mechanical draughting, only those aspects relevant to the work of a mechanical draughtsman will be discussed in this section. In many instances a mechanical draughtsman is often confronted with the task of drawing simple structures, such as, support for bearings, base of gear boxes and equipments, structural foundations of machines, and similar constructions. Standard symbols used in structural drawings have been given in Section 6.2 and common structural accessories are covered in sections 6.3. The standard sections as per the Structural Engineers Handbook by the ISI are given in Appendix D.

Basically, structural drawings concern the preparations of working drawings of buildings, towers, bridges and similar structures. The fundamental principles underlying the drawings of mechanical equipments and structural constructions are the same in both types of work. However, certain methods of representation are different in case of structural draughting. Besides, a structural draughtsman should also possess some knowledge about the design procedures involved in structural design work. Also, he should be able to select the proper materials, choose the proper sections from the standards on steel sections and should be thoroughly familiar with the various methods of fastening member within a structure, such as, bolting, rivetting and welding. Details of bolts, rivets and weldments are given in the relevant sections of this book.

Making of structural design drawing is primarily done by the design engineer. Such a drawing shows the overall dimensions of the structure, locations of structural members, such as columns, angles, beams and other shapes within a structural construction. The draughtsman is mainly concerned with the making of detailed shop drawings for the execution and fabrication of the construction. Such a drawing showns the details of all parts of the entire structure to show exactly the way the parts are to be made, fabricated and fitted. Connection details should strictly conform to the codes laid down for the purpose and only approved symbols should be incorporated in a structural drawing. Steel sections are available in standard shapes, such as, round and square bars, standard I - beams and wide flange beams, equal and unequal angles, bulb angles, toes, constructional pipes and various other shapes. It is desirable that the draughtsman should have the requisite knowledge of force and stress analysis involved, and he should be able to pick the most economical sections and entire construction is economically viable and unnecessary cost of material and fabrication is avoided. While a mechanical draughtsman needs not go into the detail design procedures in structural construction, nevertheless, he should have sufficient knowledge for making simple s ctural drawings of supports, frames, bases and allied structural constructions needed for mechanic equipments.

6.2 STANDARD SYMBOLS USED IN STRUCTURAL DRAWINGS

In IS:10702 - 1983 which is identical with ISO 5261-1981, guide-lines for technical drawings for structural metal work have been laid down. Standard symbols for representation of holes, bolts and rivets are given in Table 6.1. Figure 6.1 (a) and (b) show the method of dimensioning a structural construction. Symbols for bars and profile sections should be as per Tables 6.2 and 6.3. A typical example of the drawing of a structural construction is shown in Fig. 6.2.

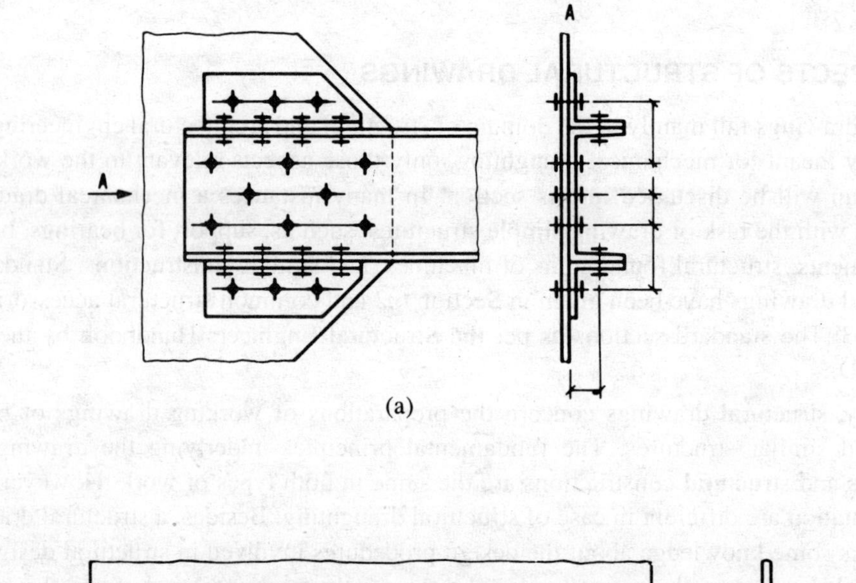

Fig. 6.1 Method of dimensioning a structural construction

Table 6.1 Standard symbols for hole, bolt and rivet in structural drawings

Hole	*Symbol for hole*			
	without countersinking	*countersunk on near side*	*countersunk on far side*	*countersunk on both sides*
drilled in the workshop				
drilled on site				

Bolt or rivet	*Symbol for bolt or rivet to fit in hole*			*Symbol for rivet to fit in hole countersunk on both sides*
	without countersinking	*countersunk on near side*	*countersunk on far side*	
fitted in the workshop				
fitted on site				
fitted on site and hole drilled on site				

Note : To distinguish bolts from rivets, the designation of bolts shall always begin with a prefix showing the type of screw thread. (Example : the designation for a bolt with metric screw thread is M 12×50, whilst that for a rivet is φ 12×50.)

Table 6.2 Symbols for bars

Description	*Designation*		*Signification of dimensions*
	Symbol	*Dimensions*	
Circular solid section	Ø	*d*	
Tube		*d*×*t*	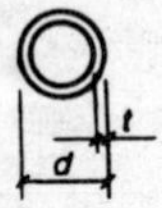
Square solid section	□	*b*	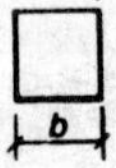
Square hollow section		*b*×*t*	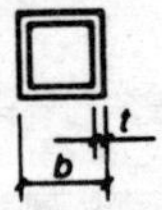
Rectangular solid section	▬	*b*×*h*	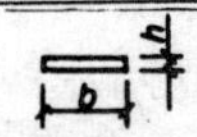
Rectangular hollow section		*b*×*h*×*t*	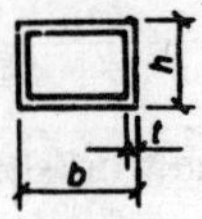
Hexagonal solid section	⬡	*s*	
Hexagonal hollow section		*s*×*t*	
Triangular solid section	△	*b*	
Semi-circular solid section	◠	*b*×*h*	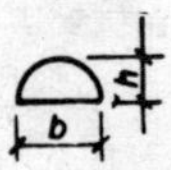

Table 6.3 Symbols for profile sections

Description	*Designation*		*Signification of dimensions*
	Symbol	*Dimensions*	
Angle section	L		If there are no International Standards or other relevant standards, profile section dimensions shall be indicated by means of their particular characteristics preceded by the symbol (Example : L 80 × 60 × 7 − 500)
T-section	T		
I-beam section	I		
Channel section	[		
Z-section			
Rail section			
Bulb angle section			
Bulb flat			

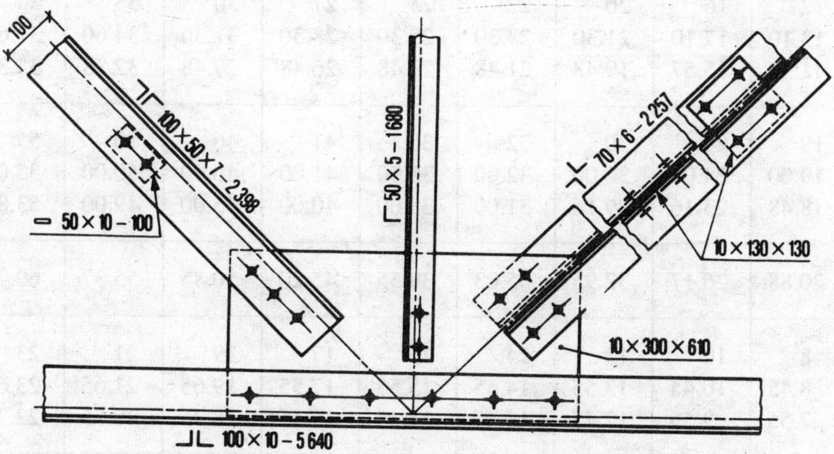

Fig. 6.2 Drawing of a structural construction

6.3 STRUCTURAL ACCESSORIES

Structural constructions consist of rolled structural sections (see Appendix-D) and structural accessories. Some of these common accessories are given in this section. Heavy washers used for the structurals are given in Table 6.6. Tables 6.7 and 6.8 show the taper washers meant for beams and channels. These are used for proper fitting of the bolts on the flanges of those structurals. Some common forms of foundation bolts are given in Table 6.9. The unthreaded portion or shank which is embedded in the concrete can be chosen from the different types shown. The dimensions given in Table 6.10 are to be maintained.

Table 6.4 shows the hexagon bolts used for steel structures and Table 6.5 shows the preferred length-diameter combinations along with the clamping lengths.

Table 6.4 Dimensions for hexagon bolts for steel structures

All dimensions in millimetres

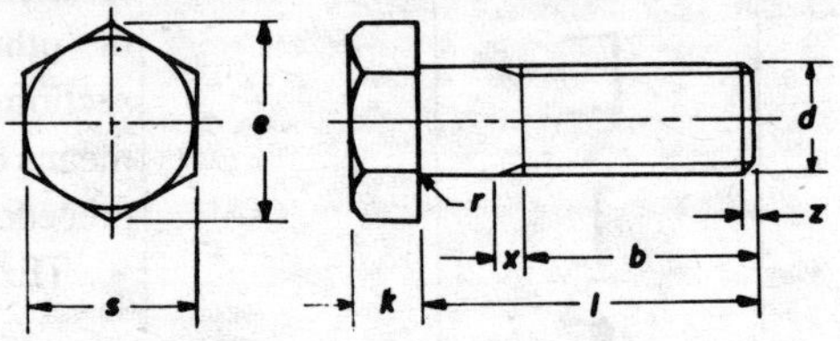

x according to IS:1369-1961 'Dimensions for screw threads run-outs and undercuts'.
z according to IS:1368-1967 'Dimensions for ends of bolts and screws (*first revision*)'.

Size		*M12*	*M16*	*M20*	*(M22)*	*M24*	*(M27)*	*M30*	*(M33)*	*M36)*	*(M39)*
d	*Nom*	12	16	20	22	24	27	30	33	36	39
	Max	13.10	17.10	21.30	23.30	25.30	28.30	31.30	34.60	37.60	40.60
	Min	11.57	15.57	19.48	21.48	23.48	26.48	29.48	32.38	35.38	38.38
s < 18 h 14 > 19 h 15	*Nom*	19	24	30	32	36	41	46	50	55	60
	Max	19.00	24.00	30.00	32.00	36.00	41.00	46.00	50.00	55.00	60.00
	Min	18.48	23.16	29.16	31.00	35.00	40.00	45.00	49.00	53.80	58.80
e	*Min*	20.88	26.17	32.95	35.03	39.55	45.20	50.85	55.37	60.79	68.44
k = 16	*Nom*	8	10	13	14	15	17	19	21	23	25
	Max	8.45	10.45	13.55	14.55	15.55	17.55	19.65	21.65	23.65	25.35
	Min	7.55	9.55	12.45	13.45	14.45	16.45	18.35	20.35	22.35	24.35
r	*Max*	1	1	1	1	1	1	1	1	1	1
		20	23	26	28	30	33	35	38	40	43

Note : Sizes shown in brackets are of second preference.

Table 6.5 Preferred length-diameter combinations and clamping lengths for hexagon bolts for steel strucutres

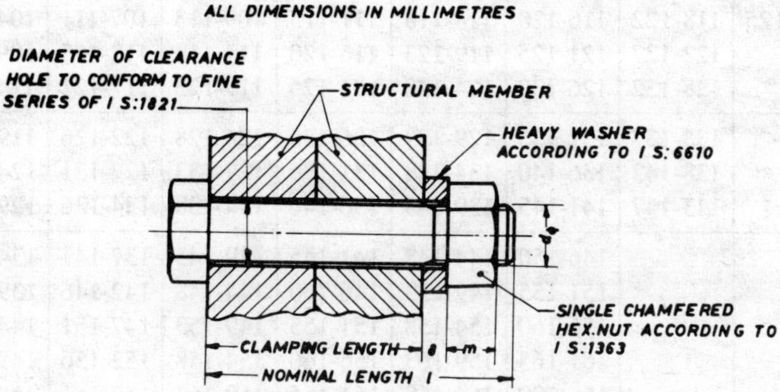

Size	*M12*	*M16*	*M20*	*(M22)*	*M24*	*(M27)*	*M30*	*(M33)*	*M36*	*(M39)*
m	*10*	*13*	*16*	*18*	*19*	*22*	*24*	*26*	*29*	*31*
Nominal Length l					*Clamping length*					
30	5-9									
35	10-14	6-10								
40	15-19	11-15	8-12	6-10						
45	20-24	16-20	13-17	11-15	9-13					
50	25-29	21-25	18-22	16-20	14-18					
55	30-34	26-30	23-27	21-25	19-23					
60	35-39	31-35	28-32	26-30	24-28	21-25				
65	40-44	36-40	33-37	31-35	29-33	26-30				
70	45-49	41-45	38-42	36-40	34-38	31-35				
75	50-54	46-50	43-47	41-45	39-43	36-40				
80	55-59	51-55	48-52	46-50	44-48	41-45	39-43			
85	60-64	56-60	53-57	51-55	49-53	46-50	44-48			
90	65-69	61-65	58-62	56-60	54-58	51-55	49-53			
95	70-74	66-70	63-67	61-65	59-63	56-60	54-58			
100	75-79	71-75	68-72	66-67	64-68	61-65	59-63	57-61	54-58	
105	80-84	76-80	73-77	71-75	69-73	66-70	64-68	62-66	59-63	
110	85-89	81-85	78-82	76-80	74-78	71-75	69-73	67-71	64-68	62-66
115	90-94	86-90	83-87	81-85	79-83	76-80	74-78	72-76	69-73	67-71
120	95-99	91-95	88-92	86-90	84-88	81-85	79-83	77-81	74-78	72-76
125		96-100	93-97	91-95	89-93	86-90	84-88	82-86	79-83	77-81
130		101-105	98-102	96-100	94-98	91-95	89-93	87-91	84-88	82-86

135		106-110	103-107	101-105	99-103	96-100	94-98	92-96	89-93	87-91
140		111-115	108-112	106-110	104-108	101-105	99-103	97-101	94-98	92.96
145		116-120	113-117	111-115	109-113	106-110	104-108	102-106	99-103	97-101
150		121-125	118-122	116-120	114-118	111-115	109-113	107-111	104-108	102-106
155			123-127	121-125	119-123	116-120	114-118	112-115	109-113	107-111
160			128-132	126-130	124-128	121-125	119-123	117-121	114-118	112-116
165			133-137	131-135	129-133	126-130	124-128	122-126	119-123	117-121
170			138-142	136-140	134-138	131-135	129-133	127-131	124-128	122-126
175			143-147	141-145	139-143	136-140	134-138	134-136	129-133	127-131
180				146-150	144-148	141-145	139-143	137-141	134-138	132-136
185				151-155	149-153	146-150	144-148	142-146	139-143	137-141
190				156-160	154-158	151-155	149-153	147-151	144-148	142-146
195				161-165	159-163	156-160	154-158	152-156	149-153	147-151
200				166-170	164-168	161-165	159-163	157-161	154-158	152-156

Note : Nominal lengths against which clamping lengths are indicated, are preferred.

Table 6.6 Heavy washers for steel structures

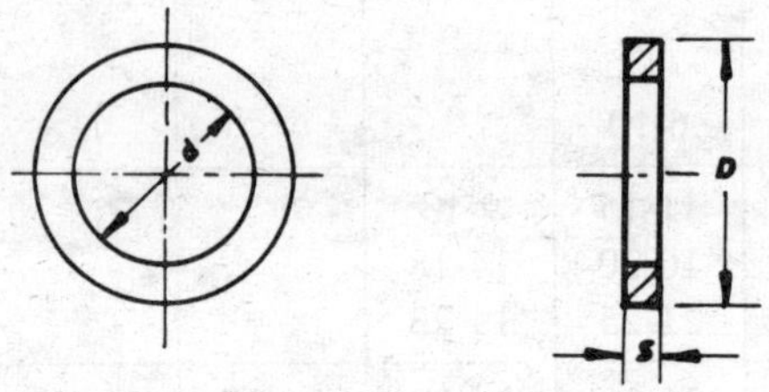

Nominal Size, d		D		s	Suitable for Bolt Size
Basic	Tolerance (H14)	Basic	Tolerance (h16)		
14	+ 0.43	24	– 1.30	8	M12
18	+ 0.43	30	– 1.30	8	M16
22	+ 0.52	37	– 1.60	8	M20
24	+ 0.52	39	– 1.60	8	M22
26	+ 0.52	44	– 1.60	8	M24
30	+ 0.52	50	– 1.60	8	M27
33	+ 0.62	56	– 1.90	8	M30
36	+ 0.62	60	– 1.90	8	M33
39	+ 0.62	66	– 1.90	8	M36
42	+ 0.62	72	– 1.90	8	M39

Table 6.7 Taper washers for I - beams

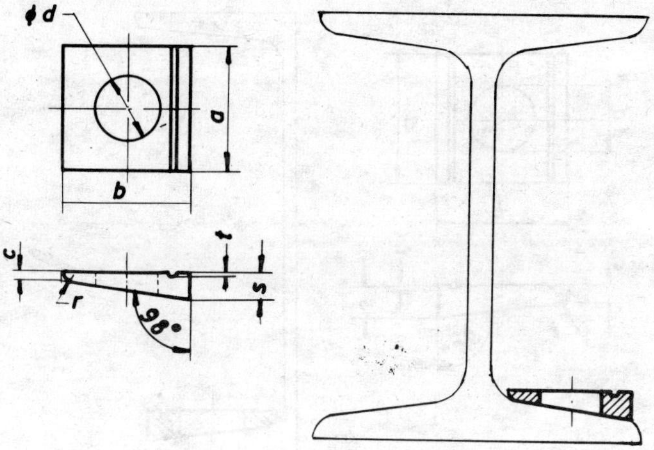

All dimensions in millimetres

Nominal Size d	*a*	*b*	*c*	*r*	*s*	*t Approx*	*For Bolt Size*
11	22	22	1.5	2.4	4.6	0.5	M10
14	26	30	2	2.4	6.2	0.7	M12
18	32	36	2.5	2.4	7.6	0.8	M16
22	40	44	3	3.2	9.2	0.9	M20
(24)	44	50	3	3.2	10.0	1	(M22)
26	56	56	3	4.0	10.8	1	M24
(30)	58	56	3	4.0	10.8	1	(M27)
33	62	62	3	4.0	11.7	1.2	M30
(36)	68	68	3	4.0	12.6	1.2	(M33)
39	75	75	3	4.8	13.6	1.2	M36
(42)	80	80	3	4.8	14.2	1.2	(M39)

Note : Sizes shown within brackets are of second preference.

Table 6.8 Taper washers for channels

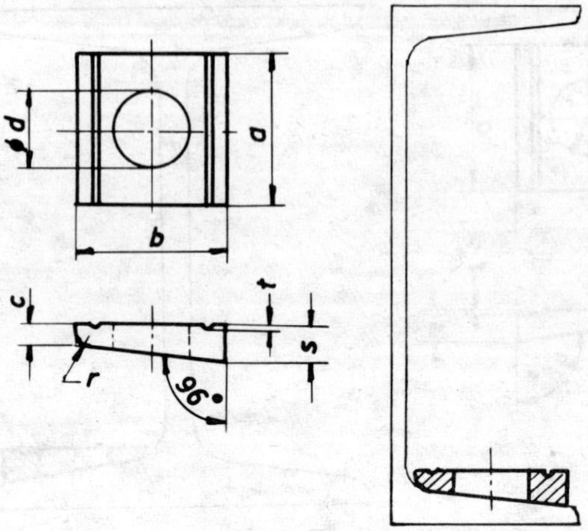

All dimension in millimetres

Nominal Size *d*	*a*	*b*	*c*	*r*	*s*	*t* *Approx*	*For Bolt Size*
11	22	22	2	2.4	4.3	0.5	M10
14	26	30	2.5	2.4	5.7	0.7	M12
18	32	36	3	2.4	5.8	0.8	M16
22	40	44	3.5	3.2	8.1	0.9	M20
(24)	44	50	4	3.2	9.2	1	(M22)
26	56	56	4	4.0	9.9	1	M24
(30)	58	56	4	4.0	9.9	1	(M27)
33	62	62	4	4.0	10.5	1.2	M30
(36)	68	68	4	4.0	11.2	1.2	(M33)
39	75	75	4	4.8	11.9	1.2	M36
(42)	80	80	4	4.8	12.4	1.2	(M39)

Note : Sizes shown within brackets are of second preference.

Table 6.9 Foundation bolts

ALL DIMENSIONS IN MM

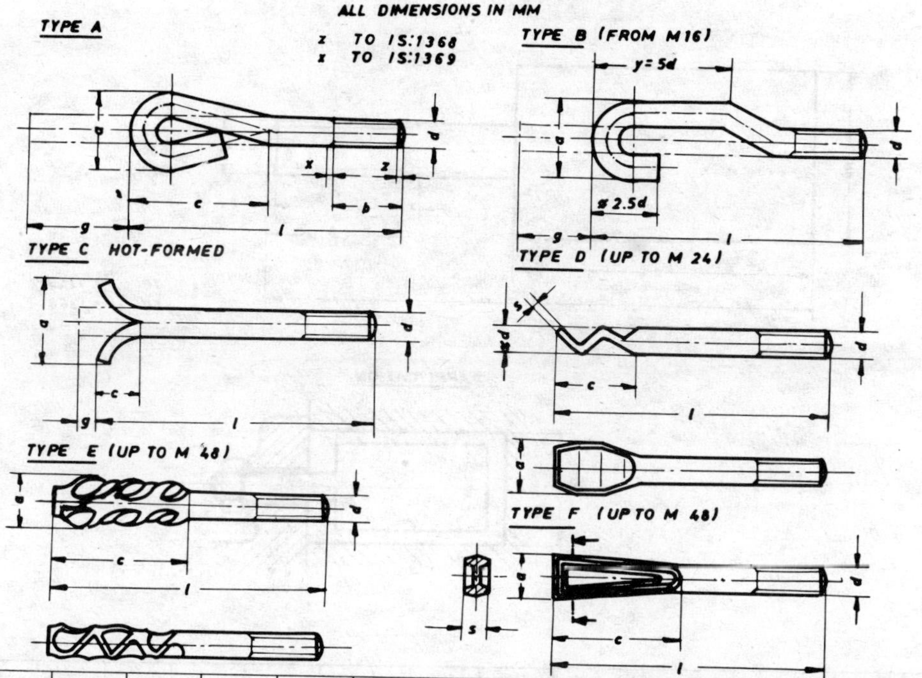

d		M 8	M 10	M 12	M 16	M 20	M 24	M 30	M 36	M 42	M 48	M 56	M 64	M72 × δ
b		20	25	30	40	50	60	75	90	105	120	140	160	180
a ≈	A, B, C	24	30	36	48	50	75	95	115	135	155	180	200	240
	D	16	20	24	32	40	48	—	—	—	—	—	—	—
	E	15	20	24	32	40	48	60	72	85	90	—	—	—
	F	14	16	20	25	30	35	45	55	65	75	—	—	—
c ≈	A	45	55	65	85	105	125	155	190	220	250	290	335	370
	C	12	15	18	24	30	36	45	54	63	72	84	96	110
	D	24	30	36	48	50	72	—	—	—	—	—	—	—
	E	45	55	70	90	100	135	150	180	250	250	—	—	—
	F	50	50	55	85	95	120	130	190	200	220	—	—	—
g ≈	A	30	38	45	60	75	90	115	135	155	180	210	235	260
	B	—	—	—	45	55	70	90	110	125	140	165	185	215
	C	5	7	8	11	14	18	24	30	34	40	45	50	60
s ≈	D	3	3.5	4	5	6	8	—	—	—	—	—	—	—
	F	6	8	10	14	18	22	25	30	36	42	—	—	—

Table 6.10 Dimensions for foundation bolt shanks

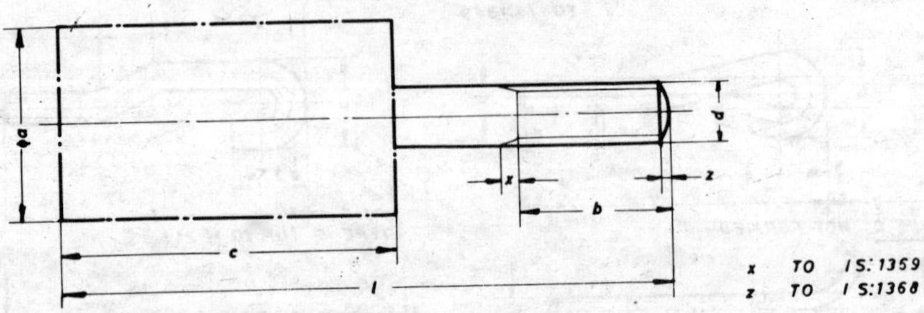

APPLICATION

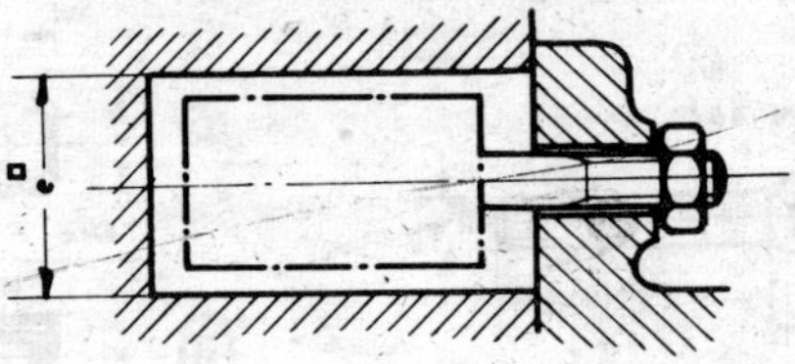

d	M8	M10	M12	M16	M20	M24	M30	M36	M42	M48	M56	M64	M72 × 6
b	20	25	30	40	50	60	75	90	105	120	140	160	180
a Maximum	25	32	40	55	65	80	100	120	140	160	185	210	250
c Maximum	45	55	70	90	110	140	160	190	260	260	290	340	370
e See under application	35	45	55	70	85	100	120	140	160	180	210	250	290
Length *l*													
80													
100													
125													
160													
200													
250													

320													
400													
500													
630													
800													
1000													
1250													
1600													
2000													
2500													
3200													

6.4 ASPECTS OF PIPING DRAWING

Pipes are primarily used for conveying liquids and gases. They are also used as members in a structural construction. Details about the dimensions of standard pipes are given in Section 6.6. In a pipe line complex transporting fluid materials, many accessories and fittings are provided. These are various kinds of valves, vessels, seal pots, flanges, clamps, fittings such as elbows, tees, plugs, bends, *etc.* Drawing them true to shape will only make the picture cumbersome. Hence in a pipe line, standardised symbols are used to represent those components. Some of these symbols commonly used are shown in Section 6.5. Dimensional details of common pipe fittings are given in Section 6.7.

Pipes are made of various materials such as, aluminium, copper, lead, concrete, plastic and other materials. A mechanical draughtsman, however, is primarily concerned with pipes made of steel and cast iron. Steel pipes for general purposes are mainly ERW (electric resistance welding) types and spirally welded types. Seamless pipes are used for more stringent operating conditions.

Pipe line drawings can be made as single-line drawings or double-line drawings. As mentioned earlier, various components used in a pipe line drawing are generally represented by approved symbols. Pipe line drawings can be made as multi-view drawings following the rules of orthographic projection. However, most of the workmen who actually lay out the line at site prefer some sort of pictorial view of the pipe line, so that the erector can easily visualise the pipe line complex in three-dimensional space. The pictorial view can be drawn on isometric paper (see Fig. 6.3). It is to be noted here that the dimensions incorporated in a pipe drawing line are mainly locational dimensions, given with respect to the centre lines and referred to some datum feature. Usually, in a large pipe line complex, such as, boiler house, power plant, *etc.*, where the magnitude of lengths are critical for space restrictions and other constraints, it is desirable to draw the lines, components, fittings, *etc.*, true to scale and shape. The pipes (which are usually flanged in such applications) and the fittings (valves, bends, *etc.*) should be true to actual layout in the drawing, so that there is no scope for fouling against equipments like pumps, vessels and other items. In open air plant layout drawings and other similar cases where the shape is not important, the diagrammatic representation of pipe line should suffice in most of the cases. In a pipe line conveying liquids and gases, the pipes should be painted according to the approved colour code to identify the material transported.

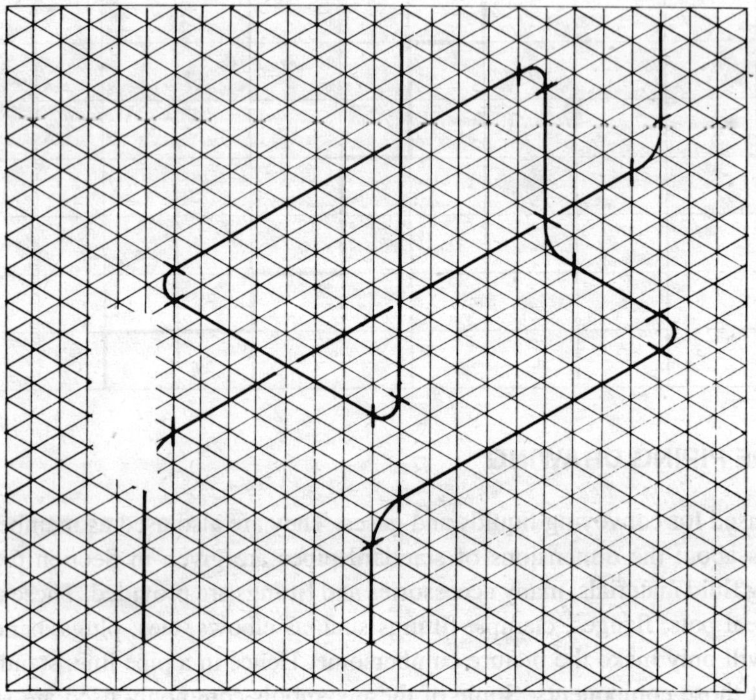

Fig. 6.3 Pictorial view pipe-line system

In laying underground pipes, elaborate protective measures are to be prepared against leakage, rusting and other factors detrimental to the pipe line. Approved codes are there for laying pipes in such conditions. In such a drawing, the relevant codes and standards pertaining to the soil conditions and surroundings are to be incorporated.

6.5 STANDARD SYMBOLS USED IN PIPING DRAWINGS

In Section 6.4, we have discussed the different methods of drawings to represent pipe-line systems. Though multi-view system of drawing pipe-lines is common, the actual erectors of pipe-line complex prefer the pictorial view of the complex. Such a pictorial view is shown in Fig. 6.3. This is simplified view. In actual site drawings, used by the erectors, the diagram is supplemented by insertion of the relevant pipe fittings at the appropriate places represented by the approved symbols.

The graphical symbols which are generally used in pipe-line work have been shown in Table 6.11.

Table 6.11 Symbols for pipe-line systems

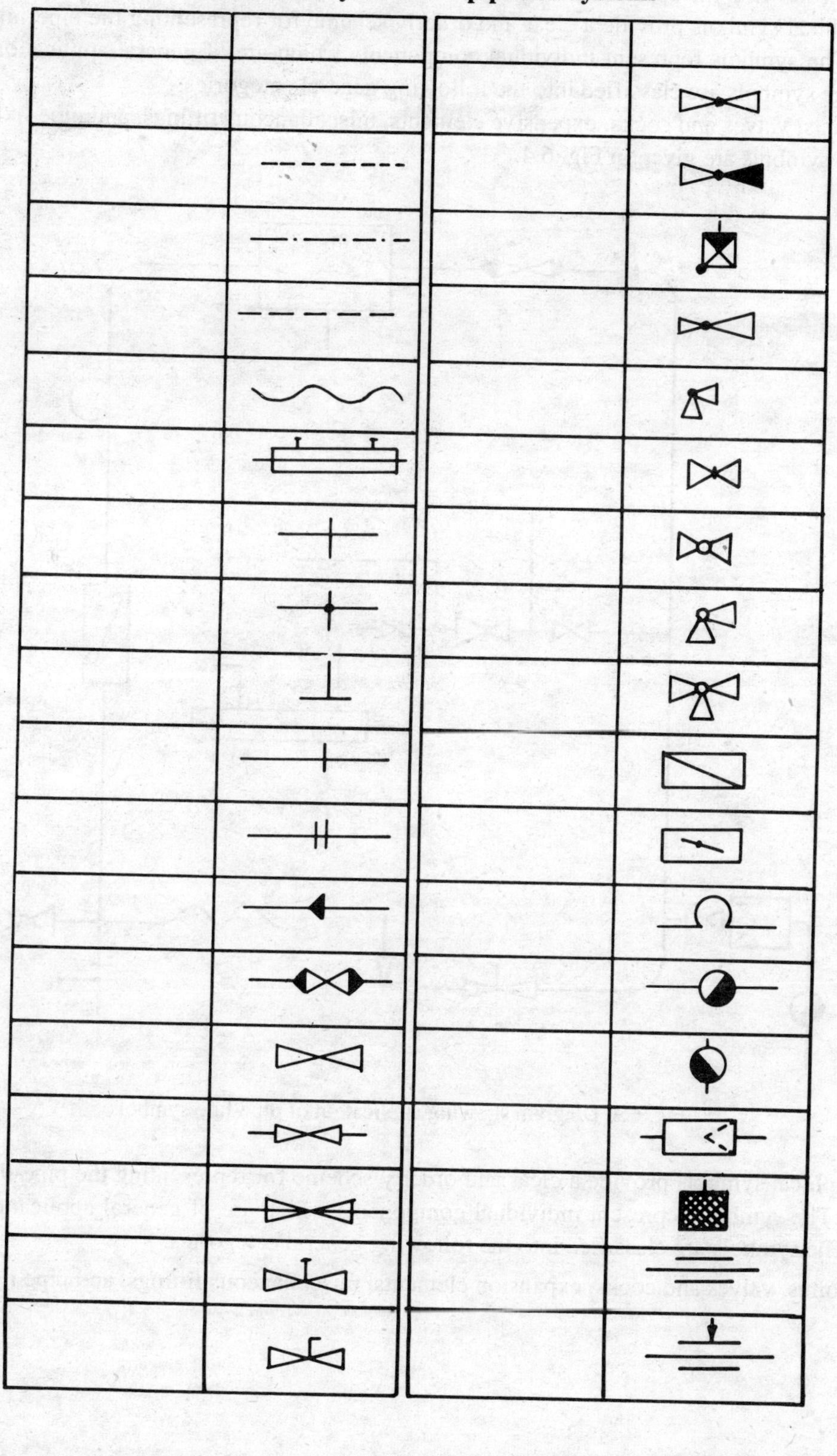

A diagram showing the application of pipe-line symbols is given in Fig. 6.4.

The graphical symbols provide a clear and orderly scheme for representing the pipe-line and control diagrams. The symbols represent individual components which are of general application in pipe-line systems. The symbols are classified into the following general categories:

Pipes, joints, valves and cocks, expensive elements, miscellaneous fittings, and pipe fixings.

Pipe-line symbols are given in Fig. 6.4.

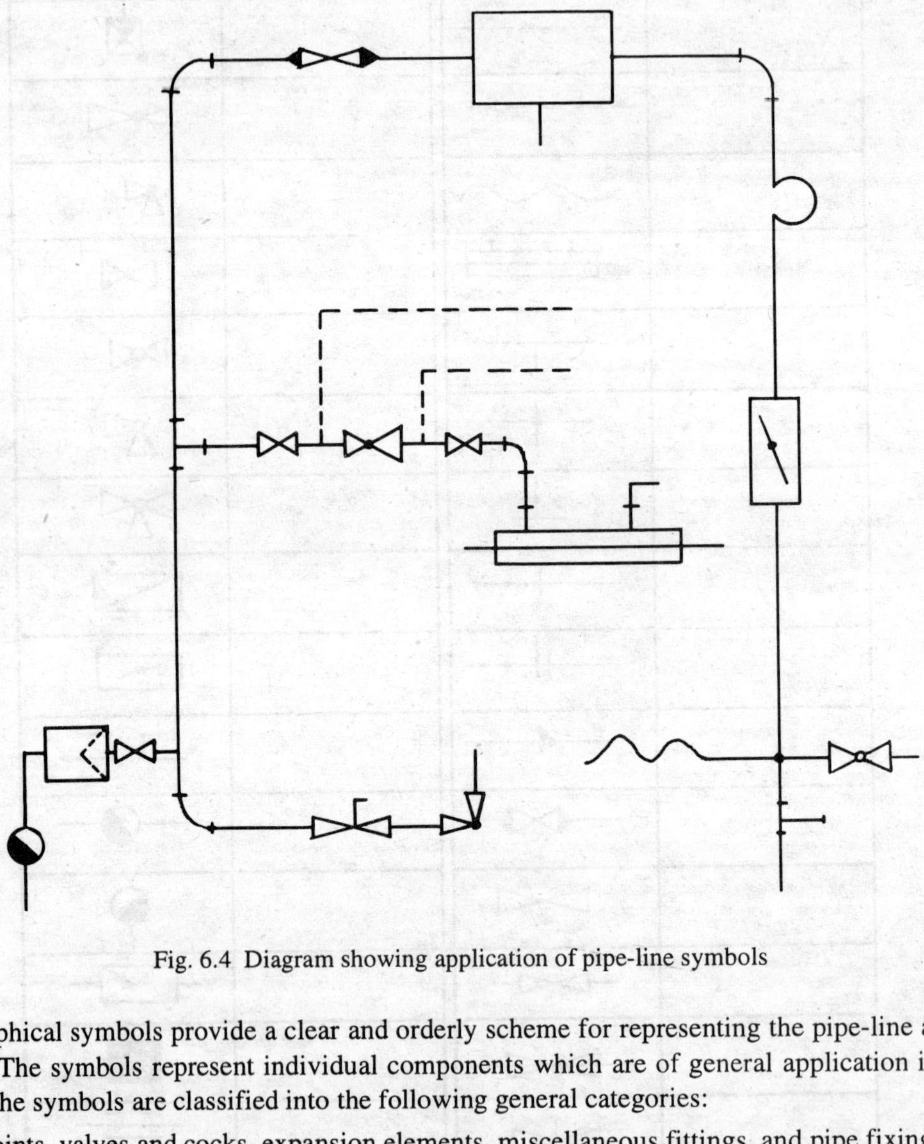

Fig. 6.4 Diagram showing application of pipe-line symbols

The graphical symbols provide a clear and orderly scheme for representing the pipe-line and control diagrams. The symbols represent individual components which are of general application in pipe-line systems. The symbols are classified into the following general categories:

Pipes, joints, valves and cocks, expansion elements, miscellaneous fittings, and pipe fixings.

6.6 DIMENSIONS OF STANDARD PIPES

Pipes belonging to commercial varieties come in different categories. As per the manufacturing process, steel pipes are generally divided into seamless and welded types. Pipes are designated either by the nominal bore or size (NB) or by the outside diameter and thickness of pipe. The nominal pressure (NP) is also to be mentioned while ordering a pipe. The designation NB approximately corresponds to the inside diameter or bore of the pipe, but it does not indicate the exact dimension of the inside bore. Thus, depending on the thickness of the pipes, a set of pipes grouped under the same NB size can have slight variations in the numerical values of their inside diameters. The concept of NB is also applied to pipe fittings such as, connecting pieces, flanges, bends, valves, compensators and other allied items.

Table 6.12 shows the main dimensions of mild steel pipes which are commonly available commercially. Pipes are supplied as plain-ended or they may be threaded at the ends or socketed. Pipe threads conform to IS : 554. These pipes can be butt-welded or seamless.

Table 6.12 Mild steel pipes

Nominal Bore	*Light (welded only)*					*Medium (welded or seamless)*					*Heavy (welded or seamless)*				
	Outside diameter		*Thickness*	*Nominal mass kg/m*		*Outside diameter*		*Thickness*	*Nominal mass kg/m*		*Outside diameter*		*Thickness*	*Nominal mass kg/m*	
NB	*Max.*	*Min*		*Plain ended*	*Socketed*	*Max.*	*Min*		*Plain ended*	*Socketed*	*Max.*	*Min.*		*Plain ended*	*Socketed*
6	10.1	9.7	1.8	0.378	0.381	10.6	9.8	2.0	0.410	0.413	10.6	9.8	2.65	0.493	0.496
8	13.6	13.2	1.8	0.522	0.526	14.0	13.2	2.35	0.639	0.543	14.0	13.2	2.9	0.762	0.766
10	17.1	16.7	1.8	0.688	0.694	17.5	16.7	2.35	0.850	0.856	17.5	15.7	2.9	1.03	1.04
15	21.4	21.0	2.0	0.96	0.97	21.8	21.0	2.65	1.21	1.22	21.8	21.0	3.25	1.44	1.45
20	26.9	26.4	2.35	1.41	1.42	27.3	26.5	2.65	1.57	1.58	27.3	26.5	3.25	1.89	1.90
25	33.8	33.2	2.65	2.01	2.03	34.2	33.3	3.25	2.42	2.44	34.2	33.3	4.0	2.95	2.97
32	42.5	41.9	2.65	2.57	2.60	42.9	42.0	3.25	3.11	3.14	42.9	42.0	4.05	3.81	3.84
40	48.4	47.8	2.9	3.27	3.31	48.8	47.9	3.25	3.59	3.63	48.8	47.9	4.05	4.41	4.45
50	60.2	59.6	2.9	4.14	4.21	60.8	59.7	3.65	5.07	5.14	60.8	59.7	4.5	6.17	6.24
65	76.0	75.2	3.25	5.80	5.92	76.6	75.3	3.65	6.49	6.61	76.6	75.3	4.5	7.92	8.04

80	88.7	87.9	3.25	6.81	6.98	89.5	88.0	4.05	8.43	8.60	89.5	88.0	4.85	10.10	10.27
90	101.2	100.3	3.65	8.76	8.98	102.1	100.4	4.05	9.70	9.92	102.1	100.4	4.85	11.90	12.12
100	113.9	113.0	3.65	9.90	10.2	115.0	113.1	4.5	12.10	12.30	115.0	113.1	5.4	14.50	14.80
125	–	–	–	–	–	140.8	138.5	4.85	16.60	17.00	140.8	138.5	5.4	17.90	18.30
150	–	–	–	–	–	166.5	153.9	4.85	19.70	20.20	166.5	163.9	5.4	21.20	21.70

Table 6.13 shows the data for pipes which are made by the electric resistance welding (ERW) process. As these pipes are still sometimes designated in the FPS units to conform to the API Specifications used in the petroleum industries, the data in Table 6.13 are given in both FPS and metric units.

Table 6.13 ERW Pipes

Outside diameter		*Wall thickness*		*Mass*	
Inches	*mm*	*Inches*	*mm*	*lb/ft*	*kg/m*
8 5/8	219.075	0.188	4.78	16.90	25.15
8 5/8	219.075	0.219	5.56	19.64	29.23
8 5/8	219.075	0.250	6.35	22.36	33.28
8 5/8	219.075	0.277	7.04	24.70	36.76
8 5/8	219.075	0.312	7.92	27.74	41.28
8 5/8	219.075	0.322	8.18	28.55	42.49
8 5/8	219.075	0.344	8.74	30.40	45.24
8 5/8	219.075	0.375	9.52	33.04	49.17
10 3/4	273.050	0.188	4.78	21.15	31.48
10 3/4	273.050	0.250	6.35	28.04	41.73
10 3/4	273.050	0.279	7.09	31.20	46.43
10 3/4	273.050	0.307	7.80	34.24	50.96
10 3/4	273.050	0.344	8.74	38.20	56.85
10 3/4	273.050	0.365	9.27	40.48	60.24
12 3/4	323.850	0.219	5.56	29.28	43.57
12 3/4	323.850	0.250	6.35	33.38	49.68
12 3/4	323.850	0.281	7.14	37.45	55.73
12 3/4	323.850	0.312	7.92	41.51	61.78
12 3/4	323.850	0.330	8.38	43.77	65.14

12 3/4	323.850	0.344	8.74	45.55	67.79
12 3/4	323.850	0.375	9.52	49.56	73.76
14	355.600	0.250	6.35	36.71	54.63
14	355.600	0.281	7.14	41.21	61.33
14	355.600	0.312	7.92	45.68	67.98
14	355.600	0.344	8.74	50.14	74.62
14	355.600	0.375	9.52	54.57	81.21
16	406.400	0.250	6.35	42.05	62.58
16	406.400	0.281	7.14	47.22	70.29
16	406.400	0.312	7.92	52.36	77.92
16	406.400	0.344	8.74	57.48	85.54
16	406.400	0.375	9.52	62.58	93.13
18	457.200	0.281	7.14	53.22	79.20
18	457.200	0.312	7.92	59.03	87.85
18	457.200	0.344	8.74	64.82	96.47
18	457.200	0.375	9.52	70.59	105.05

Pipes can also be used for structural purposes as constructional members. Data regarding such pipes are given in Table 6.14.

Table 6.14 Pipes for structural purposes

Nominal Bore	*Out side Diameter*	*Grade*	*Thickness*	*Mass kg/m*	*Area of Cross Section cm^2*	*Moment of Inertia cm^4*	*Modulus of Section cm^3*	*Radius of Gyration cm*
15	21.3	Light	2.0	0.962	1.21	0.57	0.54	0.69
		Medium	2.6	1.21	1.53	0.68	0.64	0.67
		Heavy	3.2	1.44	1.82	0.77	0.72	0.65
20	26.9	Light	2.3	1.41	1.78	1.36	1.01	0.87
		Medium	2.6	1.57	1.98	1.48	1.10	0.86
		Heavy	3.2	1.89	2.38	1.70	1.27	0.85
25	33.7	Light	2.6	2.01	2.54	3.09	1.84	1.10
		Medium	3.2	2.42	3.07	3.60	2.14	1.08
		Heavy	4.0	2.95	3.73	4.19	2.49	1.06
32	42.4	Light	2.6	2.57	3.25	6.46	3.05	1.41
		Medium	3.2	3.11	3.94	7.62	3.59	1.39

		Heavy	4.0	3.81	4.83	8.99	4.24	1.36
40	48.3	Light	2.9	3.27	4.14	10.70	4.43	1.61
		Medium	3.2	3.59	4.53	11.59	4.80	1.60
		Heavy	4.0	4.41	5.57	13.77	5.70	1.57
50	60.3	Light1	2.9	4.14	5.23	21.59	7.16	2.03
		Light2	3.2	4.54	5.74	23.47	7.78	2.02
		Medium	3.6	5.07	6.41	25.87	8.58	2.01
		Heavy	4.5	5.17	7.89	30.90	10.2	1.98
65	76.1	Light	3.2	6.80	7.33	48.78	12.8	2.58
		Medium	3.6	6.49	8.20	54.01	14.2	2.57
		Heavy	4.5	6.92	10.1	65.12	17.1	2.54
80	88.9	Light	3.2	6.81	8.62	79.21	17.8	3.03
		Medium	4.0	6.43	10.7	96.34	21.7	3.00
		Heavy	5.0	10.3	13.2	116.4	26.2	2.97
90	101.6	Light	3.6	8.76	11.1	133.2	26.2	3.47
		Medium	4.0	9.70	12.3	146.3	28.8	3.45
		Heavy	5.0	11.9	15.2	177.5	34.9	3.42
100	114.3	Light	3.6	9.9	12.5	192.0	33.6	3.92
		Medium	4.5	12.1	15.5	234.3	41.0	3.89
		Heavy	5.4	14.5	18.5	274.5	48.0	3.85
110	127.0	Light	4.5	12.2	17.3	325.3	51.2	4.33
		Medium	5.0	15.0	19.2	357.1	56.2	4.32
		Heavy	5.4	16.2	20.6	382.0	60.2	4.30
125	139.7	Light	4.5	14.9	19.1	437.2	62.6	4.78
		Medium	5.0	16.6	21.2	480.5	68.8	4.77
		Heavy	5.4	17.9	22.8	514.5	73.7	4.75
135	152.4	Light	4.5	16.4	20.9	572.2	75.1	5.23
		Medium	5.0	18.2	23.2	629.5	82.6	5.21
		Heavy	5.4	19.5	24.9	674.5	88.5	5.20
150	165.1	Light	4.5	17.8	22.7	732.6	88.7	5.68
		Medium	5.0	19.7	25.1	806.6	97.7	5.66
		Heavy	5.4	21.2	27.1	864.7	105.0	5.65
150	168.3	Light	4.5	18.1	23.2	777.2	92.4	5.79
		Medium	5.0	20.1	25.7	855.8	102	5.78
		Heavy1	5.4	21.7	27.6	917.7	109	5.76
		Heavy2	6.3	25.3	32.1	1053	125	5.73

175	193.7	Light	5.0	23.3	29.6	1320	136	6.67
		Medium	5.4	25.0	31.9	1417	146	6.66
		Heavy	5.9	27.3	34.8	1536	159	6.64
200	219.1	Light	5.0	26.4	33.6	1928	176	7.57
		Medium	5.6	29.4	37.6	2142	195	7.55
		Heavy	5.9	31.0	39.5	2247	205	7.54
225	244.5	Heavy	5.9	34.2	44.2	3149	258	8.44

6.7 PIPE FITTINGS AND ACCESSORIES

Pipe fittings include valves, cocks, traps, seal pots, bends, pipe clamps, tees, elbows, and various other items used in a pipe line. Most of these are bought out items and detailed drawings of them are generally not required to be made. Some items are, however, locally made in a workshop and for these items, the draughtsman attached to the drawing office of the factory is often required to make manufacturing drawings. Data regarding some of these common parts are given here.

Pipe bends: Pipe bends are required for changing the flow of direction of the medium which is conveyed through a pipe line. Such bends are either made beforehand in the workshop and then welded to the pipelines or bolted to the pipes of flanged types, or they may be made at site to suit site conditions. Bends made by bending pipe lengths are generally of the following constructional types (see Fig. 6.5).

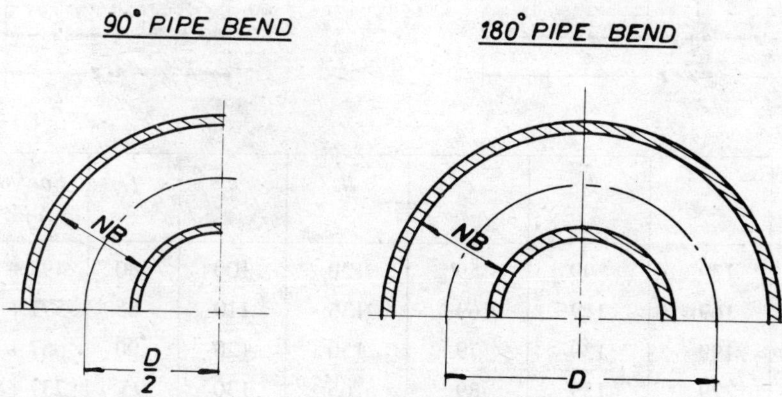

Fig. 6.5 Pipe bends made by bending pipe lengths

$$D3 \text{ type: } D = 3 \times \text{NB}$$

$$D5 \text{ type: } D = 5 \times \text{NB}$$

where D is the diameter of the bend, and NB is the nominal bore of the pipe.

Table 6.15 Fabricated pipe bends

Dimensions in mm

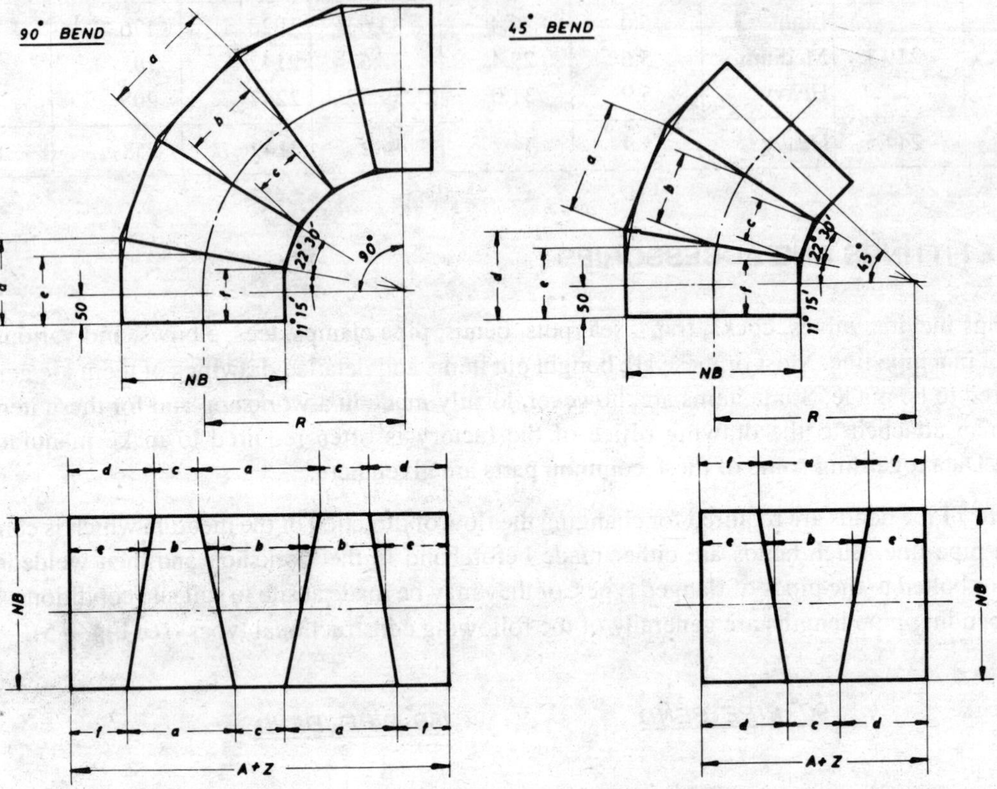

Nominal bore (N B)	*R*	*a*	*b*	*c*	*d*	*e*	*f*	*For 90° A+Z*	*For 45° A+ Z*
200	250	139	99	59	120	100	80	497 + 13	299 + 13
250	300	169	119	69	135	110	85	577 + 13	339 + 13
300	350	199	139	79	150	120	90	657 + 13	379 + 13
350	400	229	159	89	165	130	95	737 + 13	419 + 13
400	450	259	179	99	180	140	100	817 + 13	459 + 13
500	550	319	219	119	210	160	110	977 + 13	539 + 13
600	650	379	259	139	239	180	120	1137 + 13	619 + 13
700	750	439	299	159	269	200	130	1297 + 13	699 + 13
800	850	497	338	179	299	219	140	1453 + 13	777 + 13
900	950	557	378	199	328	239	149	1611 + 13	855 + 13

1000	1050	617	418	219	358	259	159	1771 + 13	935 + 13
1100	1150	678	458	239	389	279	170	1932 + 13	1016 + 13
1200	1250	736	497	259	418	299	179	2089 + 13	1095 + 13
1300	1350	789	538	279	449	319	190	2252 + 13	1176 + 13

Table 6.15 shows bends which are fabricated by welding together pipe pieces which are precisely gas-cut, cleaned and smoothened.

Table 6.16 Pipe clamp (plate type)

Dimensions in mm

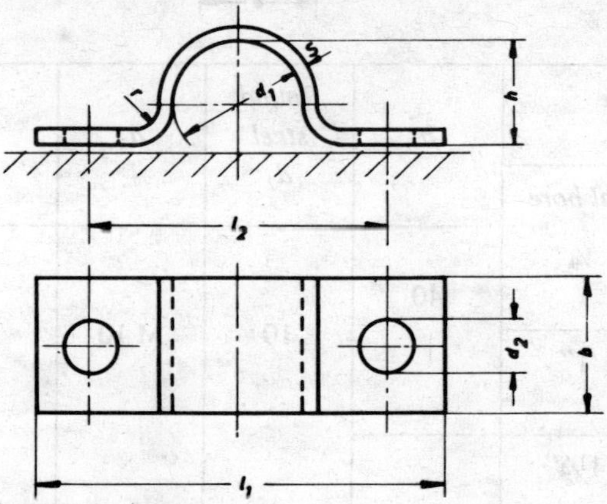

d_1	b	s	d_2	For bolts	h	l_1	l_2	r	Stretched length ≈	Mass kg/piece ≈
15.5	20	2	7	M 6	12	56	40	4	68	0.020
19	20	2	7	M 6	15	40	44	4	78	0.023
23	25	3	11.5	M 10	19	82	56	4	103	0.056
28.5	25	3	11.5	M 10	24	90	64	4	120	0.066
36	30	5	11.5	M 10	32	106	80	5	145	0.162
46	30	5	14	M 12	41	122	90	5	176	0.194
52	40	5	14	M 12	47	138	106	5	200	0.302
58	40	5	14	M 12	52	142	110	5	216	0.326
61	40	5	14	M 12	57	142	110	6	223	0.335
77	40	5	18	M 16	72	176	136	6	278	0.416
91	40	8	18	M 16	85	198	158	6	314	0.756
109	40	8	18	M 16	104	220	180	6	369	0.894

Table 6.17 Pipe clamp (steel strap type)

Dimensions in mm

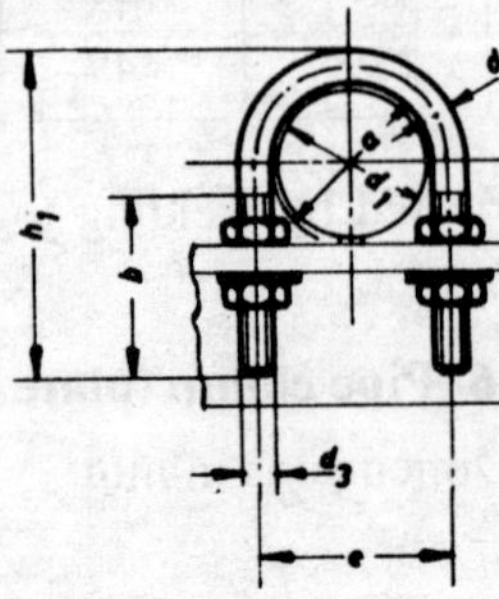

d	*For use of pipe with* d_1	*Nominal bore*		*b*	*Round steel* d_2	d_3	*e*	h_b	*Mass kg/100 pieces*
30	25 26.9	20	¾″	40	10	M 10	40	70	11.5
38	30 33.7	25	1″				48	76	12.6
46	38 42.4	32	1¼″	50			56	85	14.0
52	44.5 48.3	40	1½″				62	92	15.0
64	57 60.3	50	2″		12	M12	76	109	24.6
82	76.1	65	2½″				94	125	28.7
94	88.9	80	3″				106	138	31.7
120	108 114.3	100	4″	60	16	M15	136	171	69.5
128	133 139.7	125	-				164	191	77.6
176	159 168.3	150	-				192	217	98.1
202	(191) 193.7	(175)	-				218	249	101.1

228	216 219.1	200	-				248	283	170.5
282	267 273	250	-		20	M20	302	334	212.5
332	318 323.9	300	-	70			352	385	245.0
378	355.6 368	350	-				402	435	276.5
428	406.4 419	400	-		24	M24	452	487	309.7
530	508 521	500	-				554	589	374.3

6.8 DRAWING DATA FOR BELT CONVEYOR SYSTEM

The belt conveyor system is the most common method of transportation used in the industries for conveying bulk materials. Nowadays, belt conveyors are utilised for material handling purposes in case of practically all kinds of pulverised, granular and lumpy materials. The system has the advantages of simplicity of construction, low frictional losses, high efficiency, low power requirements and economical maintenance.

Conveyors may be horizontal, inclined or a combination of both. Two main types are generally used: flat and troughed. In maximum industrial plants, troughed belt conveyors are generally used for handling bulk materials as they have greater carrying capacity than flat belt systems. Also, troughed types are more suitable for handling material that would slide or roll off the flat belts.

The essential parts of a belt conveyor are: an endless belt to carry the material, drive arrangement for the belt, supporting rollers or idlers, take-up devices to maintain belt tension, loading and unloading arrangements, and arrangements for cleaning and protection of the belt.

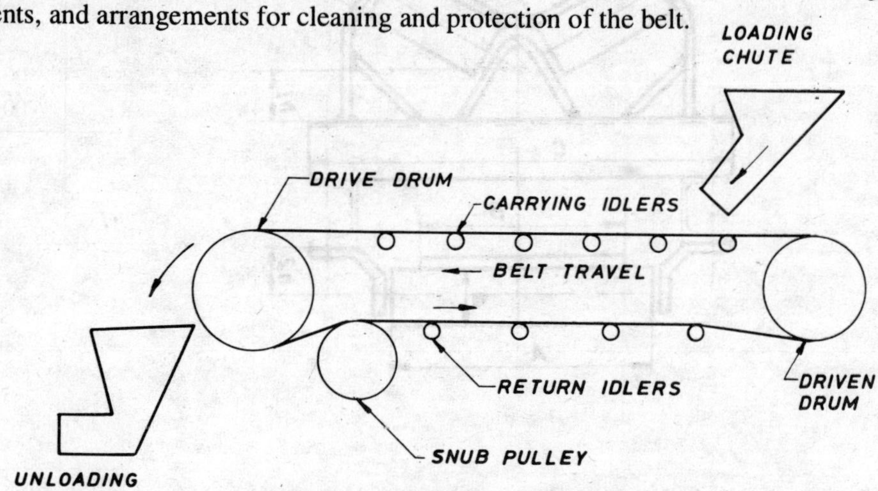

Fig. 6.6 A simple belt conveyor system

Figure 6.6 shows the schematic diagram of a simple belt conveyor system. The conveyor carries material on the upper run of the belt and discharges the load over the head or drive pulley. The empty belt returns below. The drive pulley or drum is powered by a motor through a suitable reduction unit. The driven pulley or drum acts as the take-up or tail pulley. A snub pulley is used to increase the arc of contact at the drive side, thereby proportionately increasing the belt traction. The conveyor belt runs on rollers called idler rollers or simply idlers. Running freely on anti-friction bearings, these rollers do not need power to run them. In most conveyor installations, three-roll troughing idlers are provided for the carrying side. The return side is equipped with a single roll idler. In a carrying idler, the side rollers are mostly fitted in inclined manner at an angle of 20° and 30°. The rollers are usually made of steel. Cast iron is used for heavy duty rollers.

After the overall dimensions and data of a belt conveyor system have been fixed by the designer and entered in a general arrangement drawing, the draughtsman is often asked to work out the detailed drawings. The idlers, drums, belts, etc., have been covered by different Indian Standards and other publications, such as, the company catalogues of the relevant manufacturers. To facilitate the making of such drawings, data regarding the idlers, drive and driven drums commonly in use are given in Tables 6.18 to 6.20. All dimensions are in mm.

For the same belt size and length of drum (*L*), the drum can have different outside diameters (*D*). This depends on the number of plies and thickness of the belts used, the thicker belts requiring bigger drums to avoid damage of the belts.

Table 6.18 Different types of idlers in a conveyor system

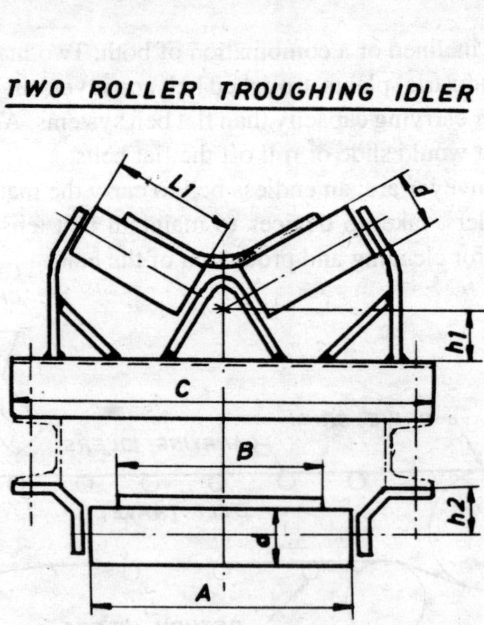

THREE ROLLER TROUGHING IDLER

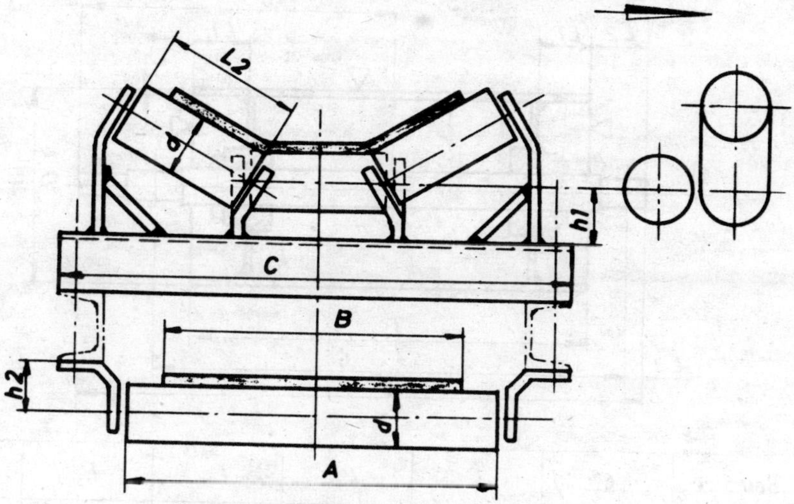

STRAIGHT CARRYING IDLER

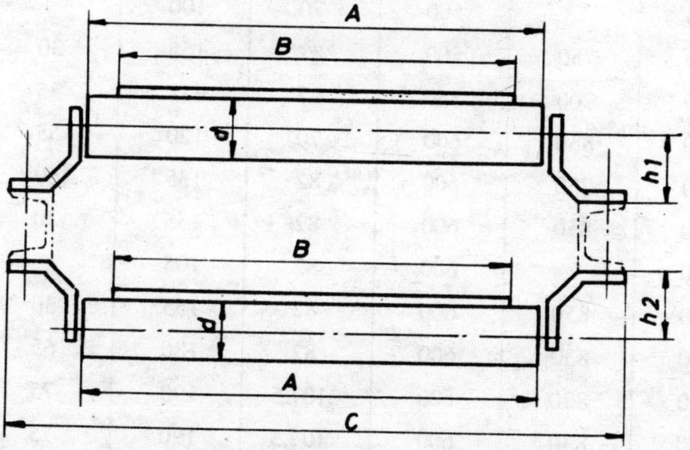

Belt width B	Dia. of roller d	Length of roller A	L_1	L_2	h_1	h_2	C
400	88.9	500	250	–	85	75	700
500	88.9	600	300	200	85	75	800
650	114.3	750	–	250	85	75	950
800	114.3	950	–	315	85	75	1150
1000	139.7	1150	–	380	100	90	1350
1200	139.7	1400	–	465	100	90	1600
1400	139.7	1600	–	530	100	90	1800

Table 6.19 Conveyor belt drive drums

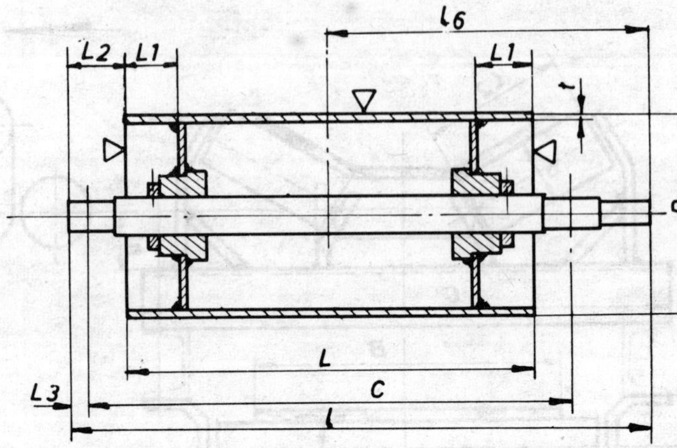

Drum Size	*Belt Size*	*C*	*L*	L_1	L_2	L_3	*t*	*D*
DR01	350		450	82	130		10	422
DR02	400		500	70	100		8	315
DR03	400	750	500	82	155	30	8	318
DR04	450	700	550	97	118	35	12	500
DR05	500	830	600	70	120	55	10	315
DR06	500	750	600	82	135	60	10	315
DR07	500	850	600	82	155	30	8	318
DR08	500		600	82	105		10	415
DR09	500	850	600	82	155	30	12	419
DR10	500	830	600	82	180	65	10	400
DR11	500	830	600	107.5	190	75	12	500
DR12	500	830	600	107.5	190	75	14	630
DR13	650		750	82	105		8	415
DR14	650	1000	750	82	155	30	10	419
DR15	650	980	750	82	180	65	10	400
DR16	650	1000	750	97	160	35	10	419
DR17	650	980	750	97	185	67.5	10	400
DR18	650	1000	750	97	190	65	15	400
DR19	650	980	750	97	185	70	12	500
DR20	650	980	750	107.5	190	75	12	500

DR21	650	950	750	107.5	197.5	77.5	14	630
DR22	650	980	750	107.5	190	75	14	630
DR23	650	950	750	125	185	85	14	630
DR24	650	980	750	125	205	90	14	630
DR25	650	980	750	125	205	90	14	630
DR26	650	980	750	125	205	90	14	630
DR27	650	950	750	145	192.5	92.5	14	630
DR28	650	1000	750	145	170	45	12	1000
DR29	650	1000	750	145	175	50	12	1000
DR30	800	1060	950	82	95	40	10	340
DR31	800	1200	950	97	160	35	12	521
DR32	800	1180	950	97	185	70	12	500
DR33	800	1180	950	125	205	90	12	500

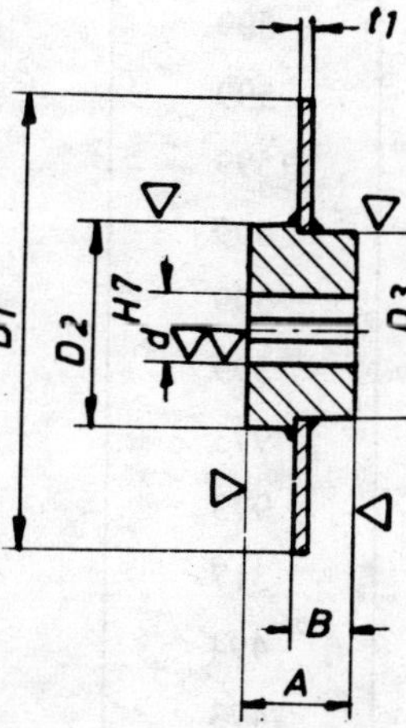

A	B	t_1	D_1	D_2	D_3	d
110	60	10	399	130	125	70
100	55	10	296	110	105	60
110	60	8	299	130	125	70
120	65	10	473	140	135	80
100	55	10	296	110	105	60
110	60	10	292	130	125	70
110	60	8	299	130	125	70

110	60	12	392	130	125	70
110	60	20	392	130	125	70
110	60	10	377	130	125	70
140	75	10	473	160	155	90
140	75	10	599	160	155	90
110	60	12	396	130	125	70
110	60	10	396	130	125	70
110	60	10	377	130	125	70
120	65	10	396	140	135	80
120	65	10	377	140	135	80
120	65	10	365	140	135	80
120	65	10	473	140	135	80
140	75	10	473	160	155	90
140	75	10	599	160	155	90
140	75	10	599	160	155	90
150	80	10	599	180	175	100
150	80	10	599	180	175	100
150	80	10	599	180	175	100
150	80	10	599	180	175	100
170	90	10	599	190	185	110
170	90	10	973	190	185	110
170	90	10	973	190	185	110
110	60	10	317	130	125	70
120	65	10	494	140	135	80
120	65	10	473	140	135	80
150	80	10	473	180	175	100

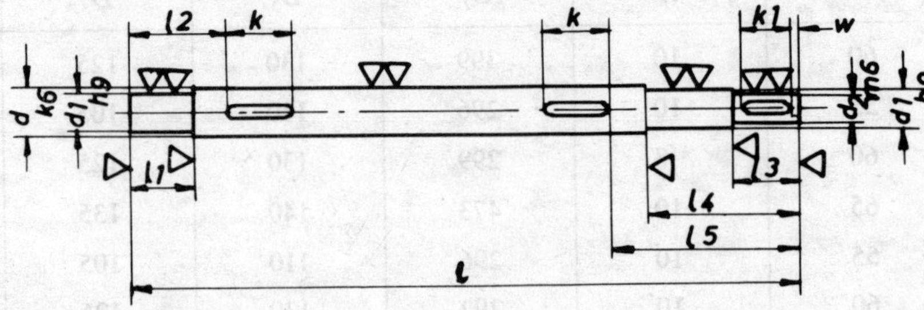

d_1	d_2	l	l_1	l_2	l_3	l_4	l_5	l_6	w	Key $b \times h \times k$	Key $b_1 \times h_1 \times k_1$
60	60	835	90	165	95	215	290	480	5	20 × 12 × 100	18 × 11 × 80
50	50	840	100	130	95	240	270	490	5	18 × 11 × 100	14 × 9 × 80
60	50	935	90	190	95	215	315	530	5	20 × 12 × 100	14 × 9 × 80
70	70	940	110	155	95	280	325	555	5	22 × 14 × 120	20 × 12 × 100
50	50	1060	170	200	95	290	320	590	5	18 × 11 × 100	14 × 9 × 80
60	50	960	135	170	95	225	260	525	5	20 × 12 × 100	14 × 9 × 80
60	50	1035	190	190	95	215	315	580	5	20 × 12 × 100	14 × 9 × 80
60	60	980	105	145	115	275	315	575	5	20 × 12 × 100	18 × 11 × 100
60	50	1035	90	190	95	215	315	580	5	20 × 12 × 100	14 × 9 × 80
60	60	1090	180	220	115	310	350	610	5	20 × 12 × 100	18 × 11 × 100
80	80	1140	190	240	135	350	400	650	5	25 × 14 × 120	22 × 14 × 130
80	80	1185	190	240	175	395	445	695	5	25 × 14 × 120	22 × 14 × 140
60	60	1130	105	145	115	275	315	650	5	20 × 12 × 100	18 × 11 × 100
60	50	1185	90	190	95	215	315	655	5	20 × 12 × 100	14 × 9 × 80
60	60	1240	180	220	115	310	350	685	5	20 × 12 × 100	18 × 11 × 100
70	60	1215	100	200	115	245	345	680	5	22 × 14 × 120	18 × 11 × 100
70	60	1230	185	230	95	295	340	672.5	5	22 × 14 × 120	18 × 11 × 90
70	70	1240	130	235	115	240	345	675	5	22 × 14 × 120	20 × 12 × 100
70	70	1250	185	230	115	315	360	690	5	22 × 14 × 120	20 × 12 × 100
80	80	1290	190	240	135	350	400	725	5	25 × 14 × 120	22 × 14 × 130
80	70	1247.5	160	225	135	302.5	367.5	695	5	25 × 14 × 120	20 × 12 × 120
80	80	1290	190	240	135	350	400	725	5	25 × 14 × 120	22 × 14 × 130
90	80	1295	175	245	140	350	420	735	5	28 × 16 × 140	22 × 14 × 140
90	80	1310	205	265	140	355	415	730	5	28 × 16 × 140	22 × 14 × 130
90	80	1350	205	265	180	395	455	770	5	28 × 16 × 140	22 × 14 × 140
90	90	1350	205	265	175	395	455	770	5	28 × 16 × 140	25 × 14 × 160
100	90	1307.5	190	262.5	180	362.5	435	740	5	28 × 16 × 160	25 × 14 × 160
100	90	1165	135	240	78	210	315	620	5	28 × 16 × 160	25 × 14 × 70
100	90	1315	145	245	180	360	440	765	5	28 × 16 × 160	25 × 14 × 160
55	55	1255	95	130	90	210	245	685	5	20 × 12 × 100	16 × 10 × 80
70	60	1415	100	200	115	245	345	780	5	22 × 14 × 120	18 × 11 × 100
70	70	1450	185	235	115	315	365	790	5	22 × 14 × 120	20 × 12 × 110
90	80	1460	205	265	100	305	365	780	5	28 × 16 × 140	22 × 14 × 90

Table 6.20 Conveyor belt driven drums

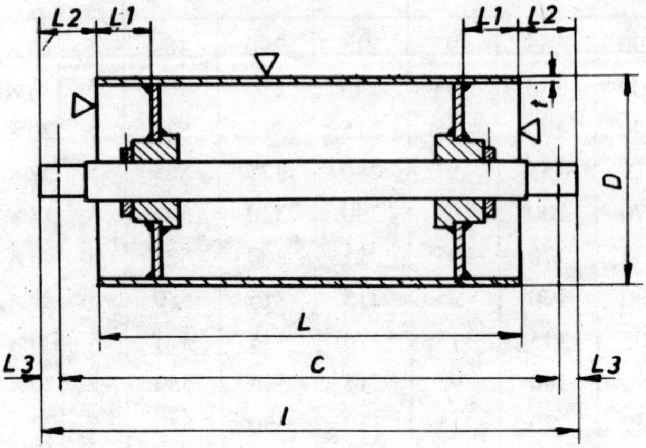

Drum Size	Belt Size	C	L	L_1	L_2	L_3	t	D
DN01	400	635	500	71.5	97.5	27.5	8	315
DN02	500		600	60	95		8	200
DN03	500	740	600	60	95	25	8	315
DN04	500	830	600	71.5	145	30	10	315
DN05	500	740	600	104.5	145	75	8	315
DN06	500	740	600	85	100	30	14	315
DN07	500	740	600	85	100	30	10	419
DN08	650	980	750	71.5	145	30	10	315
DN09	650	895	750	71.5	97.5	25	10	419
DN10	650	890	750	85	100	30	10	419
DN11	650	980	750	85	150	35	10	400
DN12	650	885	750	104.5	102.5	35	12	419
DN13	650	1050	750	104.5	215	65	16	400
DN14	650	1017	750	71.5	150	16.5	12	500
DN15	650	850	750	85	82.5	32.5	12	500
DN16	650	980	750	104.5	155	40	12	500
DN17	650	1017	750	71.5	150	16.5	14	630
DN18	650	880	750	120.5	107.5	42.5	12	800
DN19	800	1217	950	71.5	150	17.5	10	415
DN20	800	1060	950	85	95	40	10	400

DN21	800	1180	950	104.5	155	40	10	400
DN22	800	1050	950	85	82.5	32.5	12	500
	800	1090	950	85	100	30	12	521
DN24	800	1085	950	104.5	102.5	35	14	630
DN25	800	1180	950	135	170	55	12	800
DN26	1000	1380	1150	85	150	35	10	400

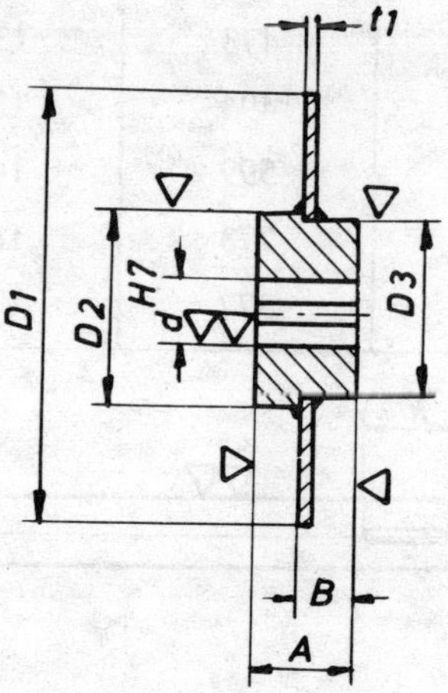

A	B	t_1	D_1	D_2	D_3	d
100	55	8	296	110	105	60
80	45	10	181	90	85	50
80	45	8	296	90	85	50
100	55	10	292	110	105	60
120	65	8	[illegible]6	140	135	80
110	60	10	[illegible]4	130	125	70
110	60	10	396	130	125	70
100	55	10	297	110	105	60
100	55	10	396	110	105	60
110	60	10	396	130	125	70
110	60	10	377	130	125	70
120	65	10	392	140	135	80
120	65	10	365	140	135	80
100	55	10	473	110	105	60

110	60	10	473	130	125	70
120	65	10	473	140	135	80
100	55	10	599	110	105	60
140	75	10	773	160	155	90
100	55	10	392	110	105	60
110	60	10	377	130	125	70
120	65	10	377	140	135	80
110	60	10	473	130	125	70
110	60	10	494	130	125	70
120	65	10	599	140	135	80
150	80	10	773	180	175	110
110	60	10	377	130	125	70

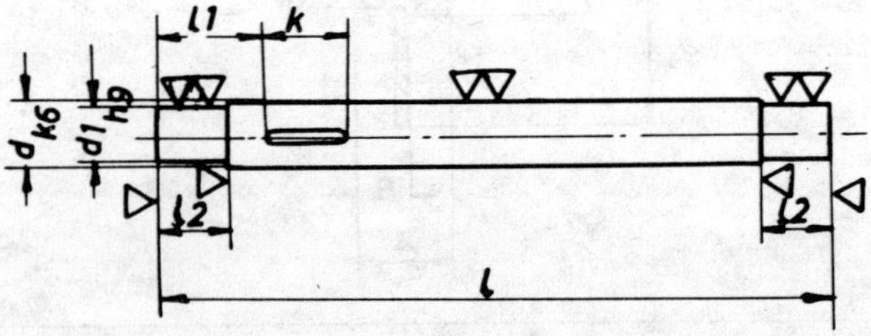

d_1	l	l_1	l_2	Key $b \times h \times k$
50	695	127	77	18 × 11 × 90
40	790	120	70	14 × 9 × 70
40	790	120	70	14 × 9 × 70
50	890	175	140	18 × 11 × 90
70	890	195	160	22 × 14 × 120
60	800	140	90	20 × 12 × 100
60	800	140	90	20 × 12 × 100
50	1040	175	140	18 × 11 × 90
50	945	127	77	18 × 11 × 90
60	950	140	90	20 × 12 × 100
60	1050	190	145	20 × 12 × 100
70	955	150	100	22 × 14 × 120

70	1180	265	125	$22 \times 14 \times 120$
50	1050	180	77	$18 \times 11 \times 90$
60	915	125	90	$20 \times 12 \times 100$
70	1060	205	150	$22 \times 14 \times 120$
50	1050	180	77	$18 \times 11 \times 90$
80	965	175	120	$25 \times 14 \times 120$
50	1250	180	77	$18 \times 11 \times 90$
60	1140	140	90	$20 \times 12 \times 100$
70	1260	205	150	$22 \times 14 \times 120$
60	1115	125	90	$20 \times 12 \times 100$
60	1150	140	90	$20 \times 12 \times 100$
70	1155	150	100	$22 \times 14 \times 120$
90	1290	240	170	$28 \times 16 \times 140$
60	1450	190	145	$20 \times 12 \times 100$

CHAPTER 7

FITS AND TOLERANCES

7.1 TERMINOLOGY

In these days of high technology, modern requirements warrant a very precise description of the work-piece. Interchangeability in workshop technology has become imperative from the standpoints of manufacture, functional requirements, cost and inspection. To achieve this objective, systems of fits and tolerances have been evolved which are extensively used in common engineering practice. The need for tolerancing a part stems from the fact that dimensional perfection can never be achieved virtually. Besides, from an economic point of view, this perfection is not even desirable, as manufacturing costs increase rapidly as the tolerances become finer (see Fig. 7.14). To attain manufacturing control, tolerancing of parts is necessary. The main factors which determine the considerations for assignment of correct tolerances for parts are functional requirements, ease of assembly of fitting parts, operational conditions, and the requisite capability of the manufacturing method employed to make the components as per the specified tolerances at a reasonable cost. In mass production, a smooth flow of products through the assembly line must be ensured. This condition can be attained only if all the mating parts fit together with minimum effort which in turn can be ensured only if the parts strictly adhere to the specified fits and tolerances. In this chapter, the general aspects of fits and tolerances which are of relevance for drawing purposes will be broadly discussed. The common terminology and nomenclatures associated with fits and tolerances are given in this section.

Fit : When two parts are intended to be mated so as to act together in assembled condition, the degree of tightness or looseness between the parts is termed as the fit of the parts. In other words, a fit is a generalised term used to designate the relation which may ensue as a result of the difference between the sizes of the mating parts caused by the application of a specified combination of tolerances assigned to the parts.

Nominal size : It is the designation or dimension of a part used for general identification purposes.

Tolerance : The tolerance is defined as the difference between the highest and the lowest permissible limits of size of a part.

Limits of size : It is the maximum and the minimum sizes which are prescribed or permissible for a specific dimension.

Allowance : An allowance is defined as the intentional difference between the maximum material limits of mating parts. In case of a clearance fit, allowance is the minimum clearance between the female and the male part. In interference fit, the maximum interference is the measure of the allowance. In both cases, the parts are at the maximum material condition. That is, the internal part is at its lowest limit and the external part is at its highest limit. A minimum clearance is known as a positive allowance and a maximum interference is known as a negative allowance.

Basic size : By basic size is meant that size from which the limits of size are arrived at by the application of allowances and tolerances.

Design size : The design size is the one from which the limits of size are derived by the application of tolerances. It is the same as the basic size when there is no allowance.

Actual size : This is the measured, actual size of a part.

Dimension : The dimension is defined as a geometrical characteristic of a part, *e.g.*, length, diameter, etc. When a numerical value is assigned to a dimension, it is referred to as the size of that dimension. In other words, the size signifies the magnitude of a dimension. However, this subtle difference is often overlooked in general engineering practice and both the terms, size and dimension, are used synonymously.

Deviation : The deviation is the algebraic difference between a size (*viz*., actual size, maximum size, *etc*.) and the corresponding base size. The difference between the maximum size and the basic size is known as the upper deviation, and that between the minimum size and the basic size is known as the lower deviation.

Unilateral tolerance : When the variation in size of a part is permitted only in one direction from the design size, the part is said to have a unilateral tolerance. The tolerance plan which follows this arrangement is known as the unilateral tolerance system.

Bilateral tolerance : When the size variation is permitted in both directions from the design size, a bilateral tolerance ensues. The plan which follows this arrangement constitutes the bilateral tolerance system.

Clearance fit : When a clearance always exists between the mating parts in assembly, it results in a clearance fit. This is achieved by assigning suitable limits of sizes.

Interference fit : When the limits of sizes of mating parts are so prescribed that there is always an interference between the parts in assembled condition, the resulting fit is termed as an interference fit.

Transition fit : This kind of fit results when the limits of sizes are chosen in such a way that they overlap, resulting in either a clearance fit or an interference fit in assembly, depending upon the actual sizes of the parts prior to assembly.

Hole : This term is used by convention to designate all internal features of a part, including those which are not cylindrical.

Shaft : All external features of a part, including those which are not cylindrical, are designated as shaft by convention.

Basic hole system : In a system of fits, if the design size of the hole is the basic size and the allowance is applied to the shaft, such an arrangement is known as the basic hole system.

Basic shaft system : This is the reverse of the basic hole system. Here, the design size of the shaft is the basic size and the allowance is applied to the hole.

Maximum material limit : This is the limit of a size which provides the maximum amount of material for the part. In case of an internal dimension (*e.g.*, a hole diameter), this is the minimum limit of size and in case of an external dimension (e.g., a shaft diameter), it is the maximum limit of size.

Minimum material limit : This is the limit of a size which provides the minimum amount of material for the part, and is the reverse case of the maximum material limit.

In Fig. 7.1, some of the terms discussed in the foregoing paragraph have been illustrated.

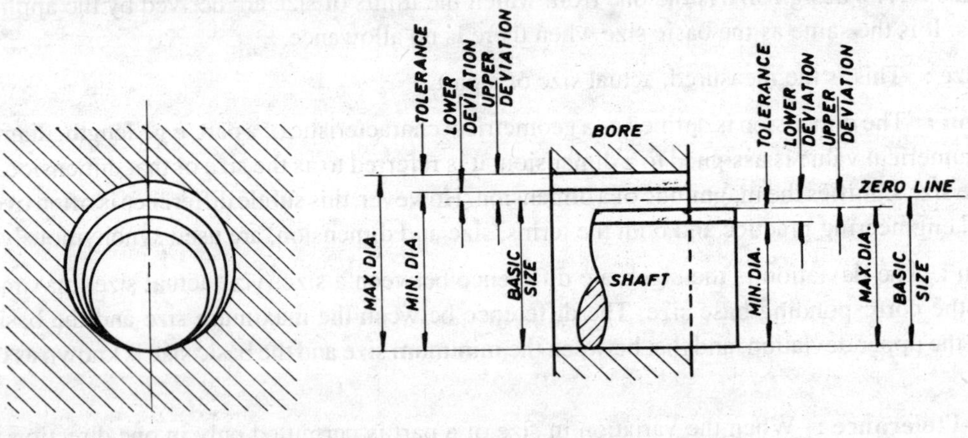

Fig. 7.1 Parameters pertaining to tolerance system

7.2 PRINCIPLES OF TOLERANCE SYSTEM

Basic concept and scope: We have seen in Section 7.1 that as it is virtually impossible to produce parts to exact sizes, it is imperative to introduce some kind of system to limit the permissible maximum and minimum sizes, so that the finished products whose final dimensions lie within this range are acceptable to serve the purpose for which they are meant. Such a system is termed as the "tolerance" system. The fundamental tenets which govern the tolerance system are discussed in Section 7.4 in which the various aspects concerning the tolerancing of engineering components are dealt with. In this section, we shall discuss the basic conceptual ideas on which the tolerance system is based.

It will be relevant here to recapitulate the following aspects which are discussed in Sections 7.4 and 7.6: An engineering workpiece is assigned dimensional tolerances when the functionability, operational parameters and interchangeability warrant it. A dimensional tolerance on a part sets close limits in the relevant nominal dimension of the part to be finished. This calls for special workmanship, added effort, proper machining and tooling facilities, inspecting gadgets and time. Besides, the single most important factor is cost, and finer the tolerance greater is the production cost which rises sharply as the quality becomes finer and finer. The number of rejects also increases steeply. These aspects have been elaborately dealt with in Section 7.4 and elsewhere. It is, therefore, prudent to avoid unnecessary tolerancing, and the requisite dimensional tolerances should be assigned only when it is necessary to serve the particular purpose for which the part is meant. Nevertheless, the determination of the finished dimensions of a part cannot be left to the workers, and it stands to reason that some sort of binding conditions on dimensional parameters should be followed, even if no tolerances are specified on the dimensions of a part, as otherwise there is a possibility that the finished dimensions may vary widely from the nominal dimensions in the absence of dimensional constraints. To this end, a systematised method for restricting the dimensions within reasonable limits has been evolved for dimensions which are not toleranced in the drawing. This aspect is treated in Section 7.6 on general tolerances.

Symbolic representation of tolerances and fits: The modern tolerance systems are based on the recommendations laid down in ISO/R 286 and other publications on the subject. According to these universally accepted standards, a tolerance is designated by a letter (in some cases, two letters) symbol and a numeral. The letter symbol signifies the position of the "zone" of tolerance with respect to the zero line which represents the basic size. The numeral stands for the value of this tolerance and is known as the "grade" or "quality" of tolerance. The actual numerical values of a tolerance grade depends upon the value of the basic size. In Fig. 7.2, the positions of tolerance zones corresponding to different letter symbols have been illustrated in relation to the basic size.

In modern engineering drawing practice, a toleranced size is thus designated by its basic value along with a letter subscript (representing zone) and a numeral (representing quality). For example,

$$60^{H7} \quad \text{or} \quad 60_{g6}$$

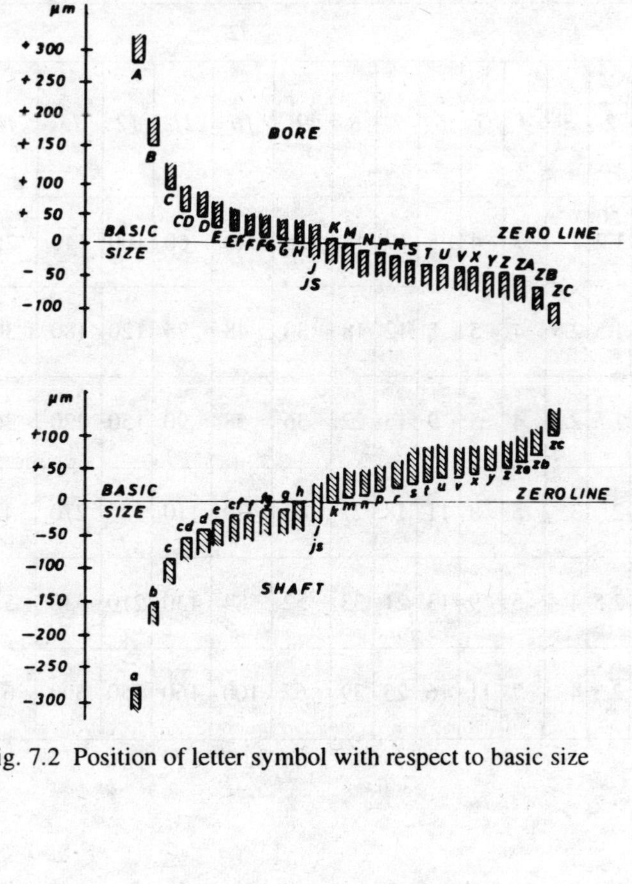

Fig. 7.2 Position of letter symbol with respect to basic size

represents a toleranced dimension in which 60 mm is the basic size, *H* (or *g*) is the zone and the quality is 7 (or 6). The dimension

$$60^{H7}_{g6}$$

represents a fit (in this case, a sliding fit) in which the common size of the hole and the shaft is 60 mm, the tolerance of the hole is *H* 7 and that of the shaft is *g* 6. Note that the hole tolerance and the shaft tolerance are written a little above and a little below the basic size respectively.

In assigning the zone of tolerance, capital letters are used for holes and small letters for shafts. A study of Fig. 7.2 will reveal the fact that the range of letter symbols span from *A* to *ZC* for holes and from *a* to *zc* for shafts. It can also be seen that these letters symbols represent the sense of tolerance according to their remoteness or closeness from the basic size, both in positive and in negative directions. Thus

$$60_{g6} = 60^{-0.010}_{-0.029} \quad \text{(See Appendix–A)}$$

means that the final closest toleranced value, viz., 59.99 mm is at least 0.01 mm away from the basic size, in this case in the negative direction.

To avoid confusion, the letters *I, L, O, Q, W*, and *i, l, o, q, w*, are not used.

All measured values correspond to the standard reference temperature of 293K (20° C).

Table 7.1 ISO standard tolerances (for sizes 1 to 500 mm)

Ranges of nominal dimension mm	*IT*																			
	0.1	*0*	*1*	*2*	*3*	*4*	*5*	*6*	*7*	*8*	*9*	*10*	*11*	*12*	*13*	*14*	*15*	*16*	*17*	*18*
from 1 upto 3	0.3	0.5	0.8	1.2	2	3	4	5	10	14	25	40	60	100	140	250	400	600	–	–
over 3 upto 6	0.4	0.6	1	1.5	2.5	4	5	8	12	18	30	48	75	120	180	300	480	750	–	–
over 6 upto 10	0.4	0.6	1	1.5	2.5	4	5	9	15	22	36	58	90	150	220	360	580	900	1500	–
over 10 upto 18	0.5	0.8	1.2	2	3	5	8	11	18	27	43	70	110	180	270	430	700	1100	1800	2700
over 18 upto 30	0.6	1	1.5	2.5	4	5	9	13	21	33	52	84	130	210	330	520	840	1300	2100	3300
over 30 upto 50	0.6	1	1.5	2.5	4	7	11	16	25	39	62	100	160	250	390	620	1000	1600	2500	3900

over 50 upto 80	0.8	1.2	2	3	5	8	13	19	30	46	74	120	190	300	460	740	1200	1900	3000	4600
over 80 upto 120	1	1.5	2.5	4	6	10	15	22	35	54	87	140	220	350	540	870	1400	2200	3500	5400
over 120 upto 180	1.2	2	3.5	5	8	12	18	25	40	63	100	160	250	400	630	1000	1600	2500	4000	6300
over 180 upto 250	2	3	4.5	7	10	14	20	29	46	72	115	185	290	460	720	1150	1850	2900	4600	7200
over 250 upto 315	2.5	4	6	8	12	16	23	32	52	81	130	210	320	520	810	1300	2100	3200	5200	8100
over 315 upto 400	3	5	7	9	13	18	25	36	57	89	140	230	360	570	890	1400	2300	3600	5700	8900
over 400 upto 500	4	5	8	10	15	20	27	40	63	97	155	250	400	630	970	1550	2500	4000	6300	9700

Referring to Table 7.1, it can be seen that the basic sizes from 1 mm to 500 mm have been subdivided into 13 steps or ranges. There are 8 steps in the basic sizes over 500 mm to 3150 mm. For each step, there are 20 grades of tolerance designated IT 01 to IT 18 as per the international codes. These are known as the "Standard Tolerances", where IT stands for "International Tolerance" or ISO Tolerance Series. ISO is the abbreviation for the International Organisation for Standardisation.

The standard tolerance is a function of the nominal size and its unit is given by the empirical relation:

$$\text{Standard tolerance unit, } i = 0.45\sqrt[3]{D} + 0.001\,D,$$

where "i" is expressed in micrometers and D in mm. D is the geometrical mean of the limiting values of the steps of size. The above relation stems from the fact that under the same manufacturing conditions, the practical relation between the manufacturing error and the nominal size follows approximately a parabolic pattern. This relation holds good for grades 5 to 16 and sizes from 3 to 500 mm. Other empirical relations are used for grades below 5 and for sizes above 500 mm. From grade 5 onwards, the IT values are multiples of the unit i. These are stepped in geometrical progression as per the Series R 5 of the Preferred Number Series given in Section 10.7. As per this series, the value of each grade is 1.6 times higher than that of the previous grade. This is apparent from the relative magnitude of each grade as given in Table 7.1, for sizes from 3 to 500 mm.

The discussion, regarding the determination of the numerical value of a tolerance is illustrated below: For the step 120-180 mm of nominal size, the geometrical mean is given by:

$$D = \sqrt{120 \times 180} \approx 147 \text{ mm}$$

$$i = 0.45\sqrt[3]{147} + 0.001 \times 147 \approx 2.522\ \mu\text{m}$$

From Table 7.2, for grade 7 and step 120-180 mm, the tolerance value = $16 \times 2.522 \approx 40\ \mu$m. This conforms to the value given in Table 7.1 for quality 7, the range being 120-180 mm.

Table 7.2 Relation between different grades of tolerances (for sizes from 3 to 500 mm)

IT 5	*IT 6*	*IT 7*	*IT 8*	*IT 9*	*IT 10*	*IT 11*	*IT 12*	*IT 13*	*IT 14*	*IT 15*	*IT 16*
7 *i*	10 *i*	16 *i*	25 *i*	40 *i*	64 *i*	100 *i*	160 *i*	250 *i*	400 *i*	640 *i*	1000 *i*

It should be noted here that the IT value of a nominal size ranges and quality grade represents the absolute value of the difference between the upper and the lower limits of a tolerance. For example, consider the following tolerances :

$$150^{H7} = 150^{+0.040}_{\ 0}, \ 150_{u7} = 150^{+0.230}_{+0.190}, \ 150_{f7} = 150^{-0.043}_{-0.083}, \ 150^{J7} = 150^{+0.026}_{-0.014}$$

In the above example, the nominal size is 150 mm and the quality is 7 in each case. Depending on the zone represented by the letter symbol, the positions of the limits of tolerances vis-a-vis the basic size represented by the zero line, are at varying distances in each case, but the absolute value of the range, *i.e.*, the difference between the limits is the same in each case, namely, 0.040 mm or 40 μm.

The standard tolerance unit for sizes over 500 upto 3150 mm is given by the empirical relation:

$$I = 0.004\,D + 2.1,$$

where I is expressed in μm and D in mm as before. The relative magnitude of each grade is given in Table 7.3.

Table 7.3 Relation between different grades of tolerances (for sizes from 500 to 3150 mm)

IT 6	*IT 7*	*IT 8*	*IT 9*	*IT 10*	*IT 11*	*IT 12*	*IT 13*	*IT 14*	*IT 15*	*IT 16*
10 *I*	16 *I*	25 *I*	40 *I*	64 *I*	100 *I*	160 *I*	250 *I*	400 *I*	640 *I*	1000 *I*

Unilateral and bilateral tolerances : The unilateral and bilateral tolerance systems have been defined in Section 7.1. The examples of those tolerances are:

Unilateral tolerance : $50^{+0.020}_{\ 0}$, Bilateral tolerance: 50.01 ± 0.010

In both the cases, the final limiting values are 50 and 50.02 mm. Though both the above tolerance systems along with the basic hole and the basic shaft systems are in vogue in different countries, in India the basic hole system is generally followed along with the hole tolerance *H7* or *H8* as standard which is a unilateral tolerance. Note that the lower limit of zone *H* with any quality is always zero. As stated earlier, a unilateral tolerance system allows deviation in one direction only, (either plus or minus, but not both at the same time), from the design size. Consequently, one limit of tolerance must always be zero, so that the tolerancing begins from the design size. In manufacturing industries, unilateral system is generally preferred for certain advantages it offers as compared to the bilateral system.

Cumulative tolerances: While assigning tolerances to the dimensions of a part, it is extremely important to consider the inter-relation and the effect of one tolerance on another. When a particular feature in a drawing, *e.g.*, the side of a work-piece or the centre distance of a hole, is controlled by more than one tolerance, it may result in an accumulation of tolerances. This is known as "cumulative tolerances" and it may lead to misinterpretation and confusion. Referring to Fig. 7.3(a), the dimensions of the sides marked 2, 3, and 4, are all toleranced. Besides, the dimension of the side 4 is also individually toleranced with respect to side 1. Clearly, the drawing is over-dimensioned. If all the maximum individual tolerances of the sides 2, 3, and 4 are added up, the resulting dimension of the side 4 with respect to the side 1 becomes 150 + 0.3 = 150.3 mm. This is obviously inconsistent with the maximum allowable dimension of side 4 which is 150 + 0.1 = 150.1 mm.

The above situation can be avoided if the dimensioning is carried out with respect to one position which is used as a datum or reference surface for all the dimensions. This is illustrated in Fig. 7.3(b). Note that the dimension spanning the sides 3 and 4 is left out. This is known as the base-line dimensioning system and it eliminates the possibility of superfluous dimensioning.

Positional tolerances : For locating the centre of a hole, the traditional method is to use rectangular co-ordinates which define the position of the centre with respect to one or more reference features. Each of these co-ordinates or dimensions are assigned tolerances. Fig. 7.4 (a) illustrates this method.

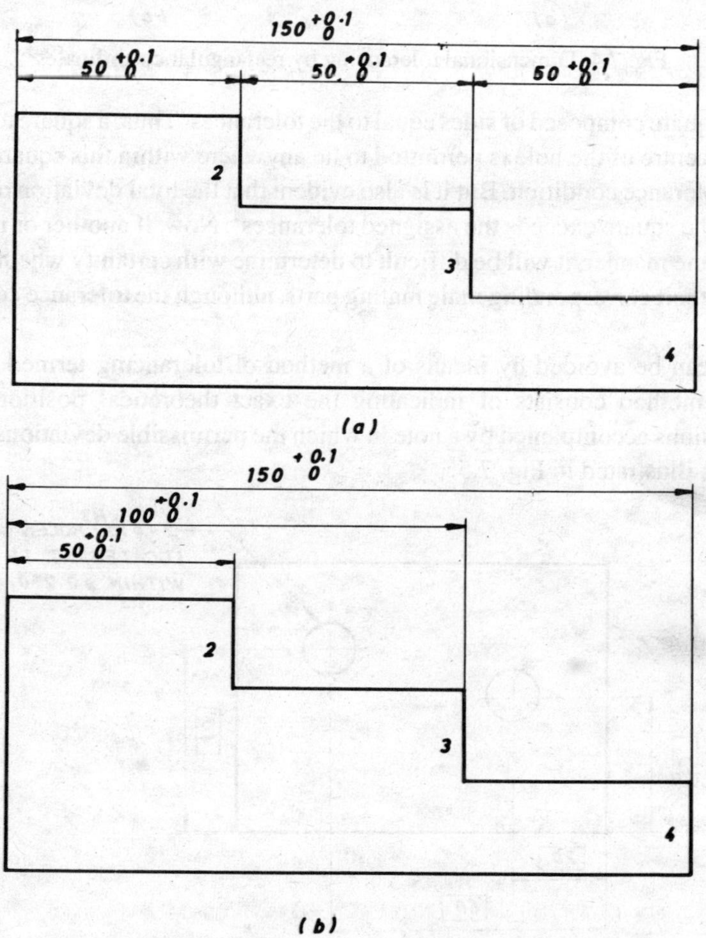

Fig. 7.3 Cumulative tolerances and base line dimensioning

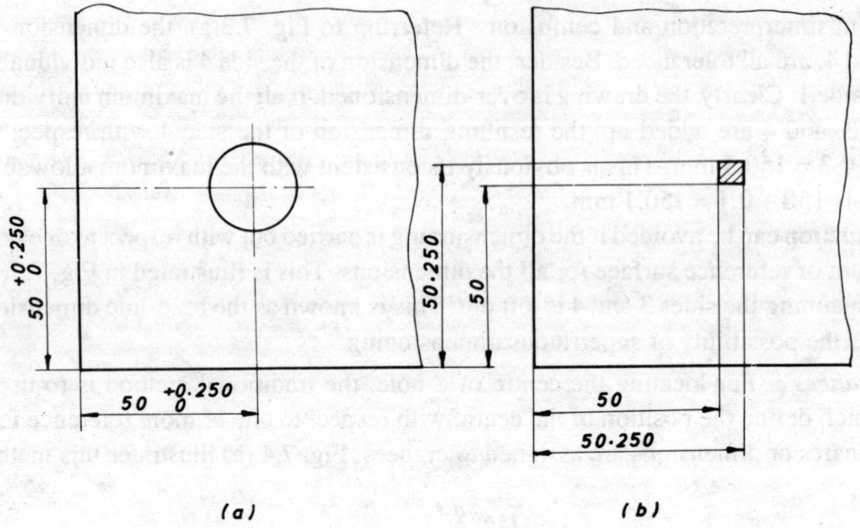

Fig. 7.4 Dimensional tolerancing by rectangular co-ordinates

Fig. 7.4(b) shows a square composed of sides equal to the tolerances. Thus, a square tolerance zone ensues. It is evident that the centre of the hole is permitted to lie anywhere within this square which is in keeping with the stipulated tolerance condition. But it is also evident that the total deviation of the centre positions along a diagonal of the square exceeds the assigned tolerances'. Now, if another or more hole-centres are demarcated in the same manner, it will be difficult to determine with certainty whether the resulting holes will actually fit with their corresponding male mating parts, although the tolerance conditions are satisfied as per the drawing.

Such a situation can be avoided by means of a method of tolerancing termed as the "true-position dimensioning. The method consists of indicating the exact theoretical positions of the centres by untoleranced dimensions accompanied by a note in which the permissible deviations from these positions are specified. This is illustrated in Fig. 7.5.

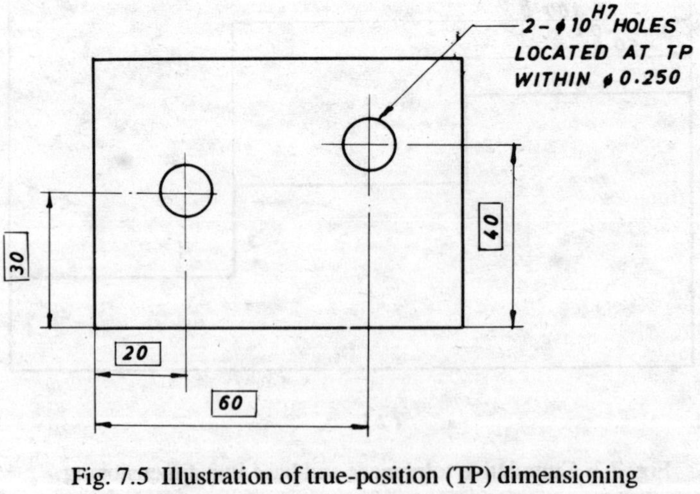

Fig. 7.5 Illustration of true-position (TP) dimensioning

Here, the tolerance zone of each hole is a circle. That is, the centre of a hole is allowed to lie anywhere within a circular tolerance zone of diameter 0.250 mm. The true position denotes the basic or theoretically exact location of a feature and the TP should always be established with reference to a datum to eliminate any scope for misunderstanding. Moreover, it is important to note here that TP represents an exact location and though untoleranced, it is not subjected to the application of general tolerances given in Section 7.6. A note to this effect can be inserted in the drawing to prevent any ambiguity, if necessary. A true-position dimension is normally indicated within a frame as shown in Fig. 7.5.

Since the part in which a hole is bored has a thickness, the "circular tolerance zone" is actually a "cylindrical tolerance zone". This is apparent from the pictorial view given in Fig. 7.6. The central cylinder

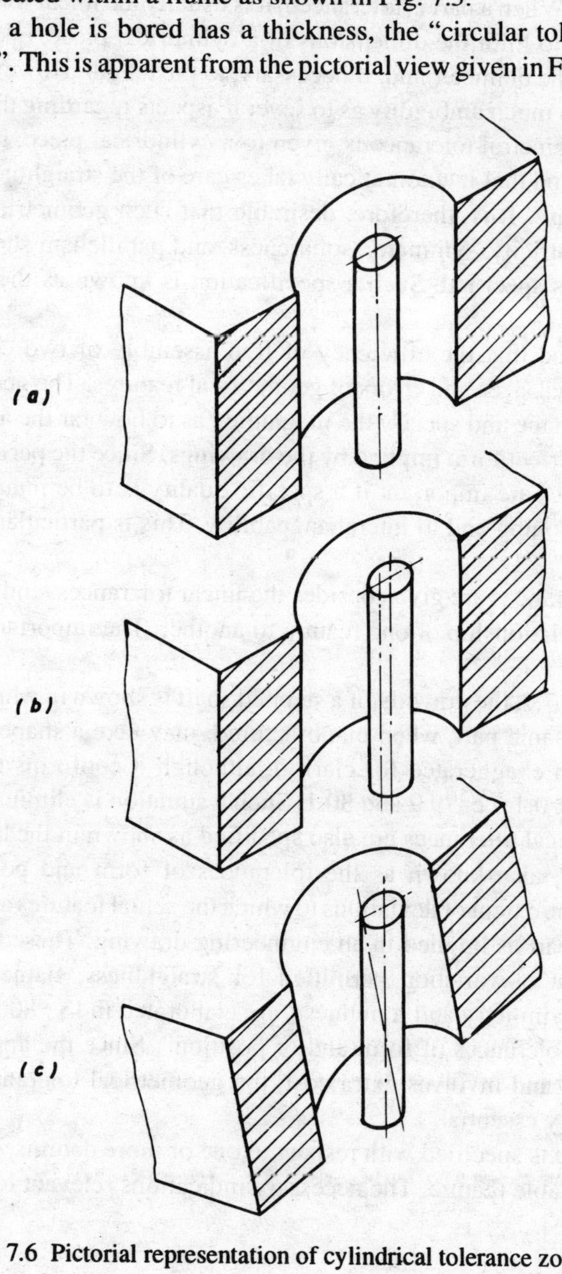

Fig. 7.6 Pictorial representation of cylindrical tolerance zone

of diameter, say 0.250 mm, must contain the axis of the hole. The hole-axis may exactly coincide with the axis of the tolerance cylinder as shown in Fig. 7.6 (a) which is an idealised case seldom attainable in the practical field, or it may be parallel to the cylinder axis while remaining within the cylinder as in Fig. 7.6 (b) or the two axes may make an angle with each other as shown in Fig. 7.6 (c). One of the advantages of the true-position tolerancing system in the elimination of cumulative tolerances which often arise in case of the conventional tolerancing systems.

Geometrical tolerances : When a part is toleranced, it is usually the linear tolerance which is considered. For example, if one wants to limit the dimensions of a cylindrical piece, one thinks that it is enough to assign tolerance values to the diameter and, if necessary, to the length. However, these linear tolerancing is not enough as it leaves as much ambiguity as to several aspects regarding the shape and form and other features of the part. If a diametral tolerance is given to a cylindrical piece, it is assumed that this holds good for the entire length and that it automatically takes care of the straightness of the piece. In practice, however, this is not the case. It is, therefore, desirable that such geometrical features as straightness, flatness, angularity, concentricity, symmetry, squareness, and parallelism should also be considered and their allowable magnitudes specified. Such a specification is known as the "Geometrical Tolerancing System".

It must be borne in mind that the efficiency of free assembly of two components depends on the combined effect of the actual sizes and errors of geometrical features. The geometrical tolerances signify the tolerances of form or shape and specify the magnitude as to how far the actual surfaces are permitted to vary from the ideal or perfect form implied by the drawings. Since the perfect form cannot be attained, these geometrical tolerances are important if a specific quality is to be maintained for all requirements which are critical to functioning and to interchangeability. This is particularly relevant where close fits are involved.

Hence, geometrical tolerances are given besides the linear tolerances, and those are mainly concerned with the accuracy of the relationship of one feature to another. The importance of this inter-relationship can be readily seen in Fig. 7.7.

In the top picture of Fig. 7.7, the drawing of a stepped shaft is shown in which only the linear tolerances have been specified. The same part, when manufactured, may take a shape as illustrated in the middle picture, (deviations shown exaggerated for clarity), although it conforms to be linear tolerance limits assigned to it, namely 80 ± 0.1, i.e. 79.9 and 80.1. Such a situation is eliminated when, in addition to the linear tolerances, geometrical tolerances are also specified as shown in the bottom picture.

Geometrical tolerances, also known as the tolerances of form and position, are, therefore, those tolerances which specify the extent of deviations to which the actual features are permitted from the perfect or ideal geometry of the part as implied in an engineering drawing. These tolerances, which define the condition and the amount of variation permitted for straightness, flatness, squareness, parallelism, angularity concentricity, symmetry and roundness, are elaborated in IS : 8000 - 1976 which conforms to ISO/R 1101 - 1969 on "Tolerances of form and of position". Since the application of these tolerances entails further constraints and involves extra cost, the geometrical tolerances are to be specified only where they are functionally essential.

A geometrical tolerance is specified with respect to one or more datums. A datum may be an axis or a surface or some other suitable feature. The necessary indications relevant to a geometrical tolerance are

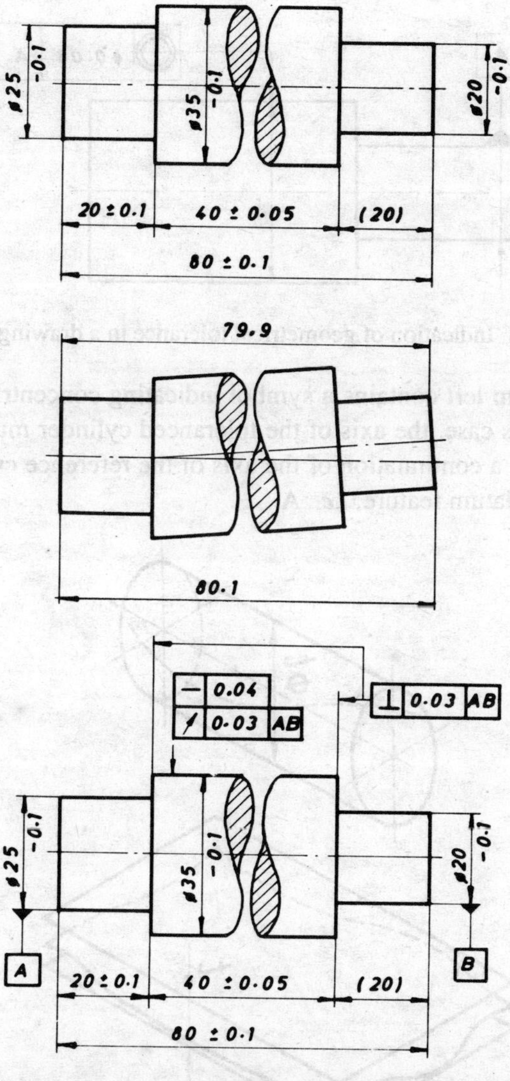

Fig. 7.7 Comparison of linear tolerance and geometrical tolerance
Based on *Fachzeichnen Metalltechnik*, Fachstufe, 1982, Fig. No. 79.1 - 79.2, P. 79, Verlag Europa - Lehrmittel, Wuppertal, Germany.

contained within a rectangular frame which is divided into several compartments. These compartments are filled in with:

(i) a symbol for the characteristic to be toleranced,

(ii) the numerical value of the tolerance, and

(iii) a letter symbol identifying the datum feature.

The above system is illustrated in Fig. 7.8.

Here, the datum feature is the axis of the reference cylinder and is indicated by the framed letter symbol "A". The leader line from this frame ends in a black full triangle as shown. The other frame has three

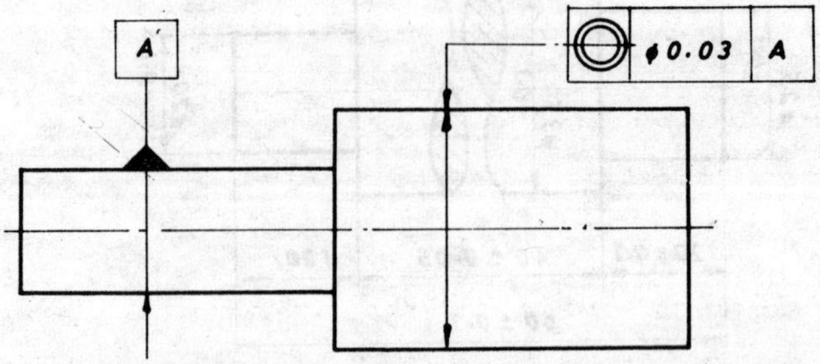

Fig. 7.8 Indication of geometrical tolerance in a drawing

compartments. The first one from left contains a symbol indicating concentricity. The second contains the value of the tolerance. In this case, the axis of the toleranced cylinder must lie within a cylinder of diameter 0.03 mm whose axis is a continuation of the axis of the reference cylinder. The symbol in the third compartment refers to the datum feature, *i.e.*, A.

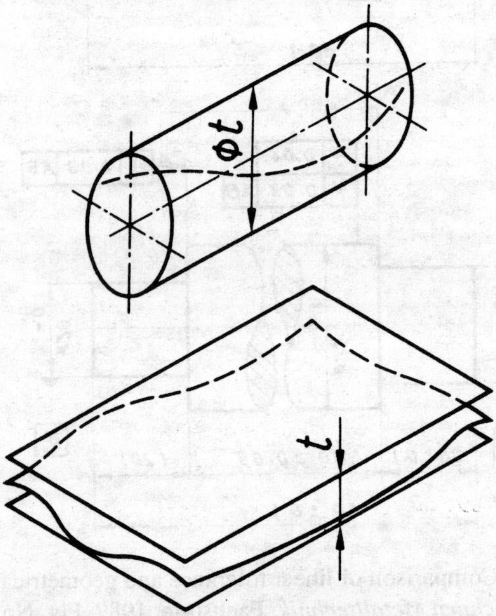

Fig. 7.9 Cylindrical and plane tolerance zones
Based on Fachzeichnen Metalltechnik, Fachstufe, 1982, Fig.No. 80.1 - 80.2, P. 80.
Verlag Europa - Lehrmittel, Wuppertal, Germany.

Fig. 7.9 shows the representation of the most common geometrical tolerances, namely, those applied for cylindrical objects and planes. The picture at the top shows the tolerance zone cylinder of diameter "*t*" within which the axis of the work-piece must lie. The bottom picture shows two plane surfaces which are "*t*" mm apart. The surface of the work-piece must lie within these planes.

In Table 7.4 the common datum features, symbols, characteristics, tolerance zone and examples of typical applications are shown.

Table 7.4 Geometrical Tolerances

Symbol and Characteristic		*Tolerance Zone*	*Examples*	*Remarks*
—	Straightness	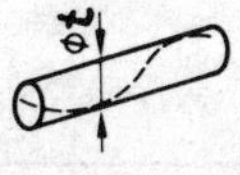	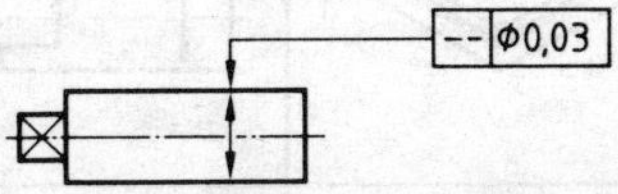	The axis of cylinder must lie within a cylinder of diameter $t = 0.03$ mm.
▱	Flatness	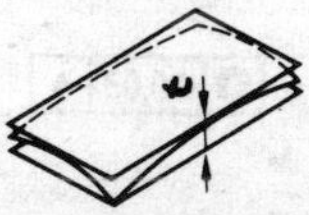	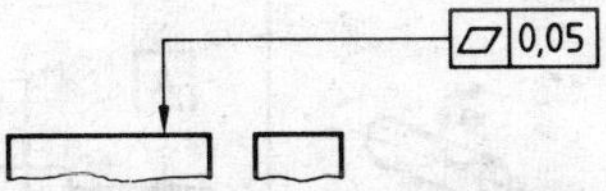	The toleranced planes must lie between two parallel planes which are $t = 0.05$ mm apart.
○	Circularity		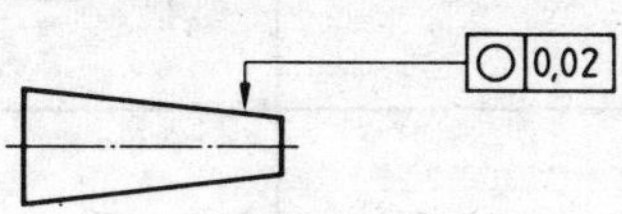	The circumference of each cross section must be contained in a circular ring of width $t = 0.02$.
//	Parallelism	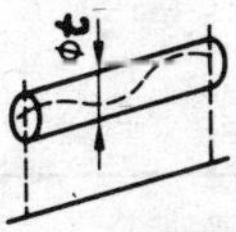	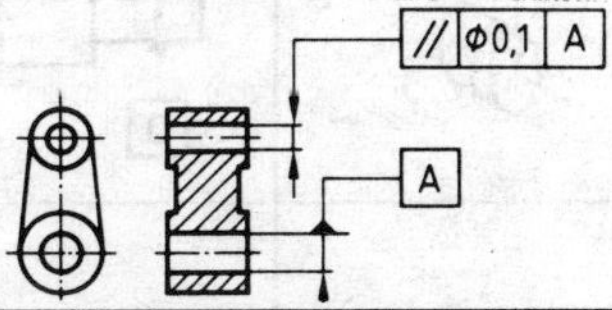	The toleranced axis must be within a cylinder of diameter $t = 0.1$ mm whose axis is parallel to the reference axis.
			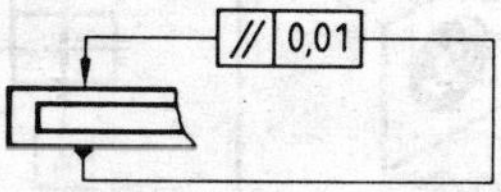	The toleranced plane must be between two parallel planes $t = 0.01$ apart, and parallel to the reference plane.
⊥	Pependicularity	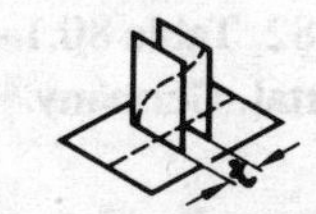	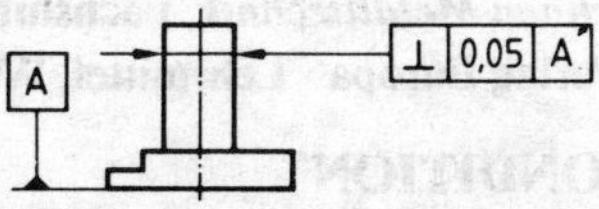	The toleranced axis must lie between two parallel planes, $t = 0.05$ mm apart and at *right.* angles to the reference plane *A*.
∠	Angularity	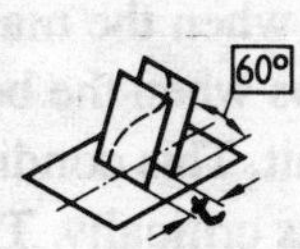	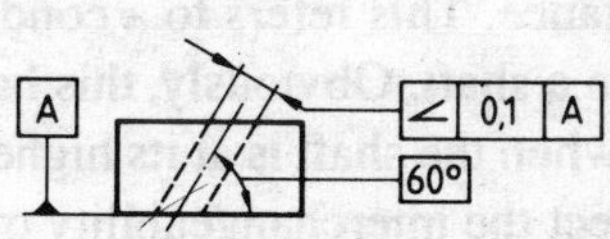	Toleranced axis must lie between two parallel planes, $t = 0.1$ mm apart, and inclined at 60° to ref. plane *A*
⌖	Position	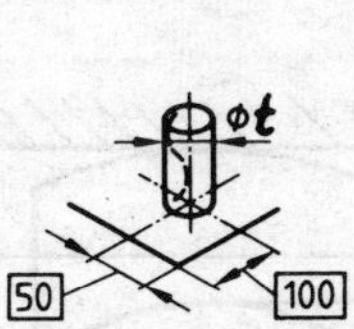	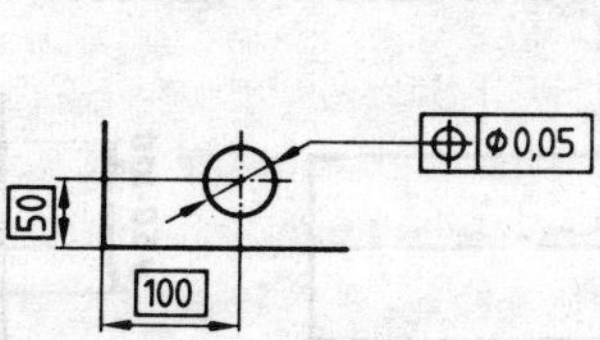	The toleranced axis must lie within a cylinder of diameter, $t = 0.05$ mm whose axis is situated at the true position (TP) indicated by framed dimensions.

⌯	Symmetry	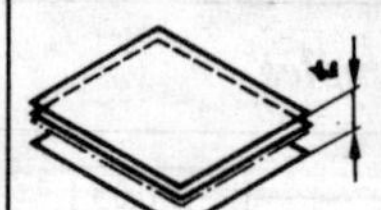	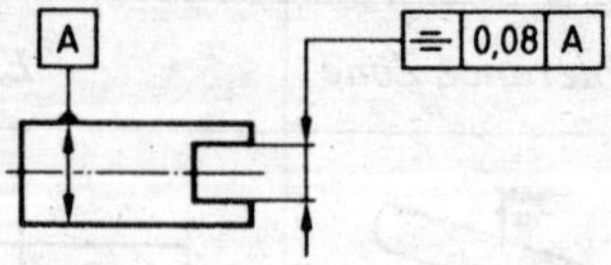	The median plane of the slot must lie between two parallel planes, t = 0.08 mm apart, which are symmetrically disposed about the median plane of the reference plane *A*.
◎	Concentricity	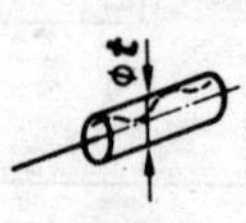	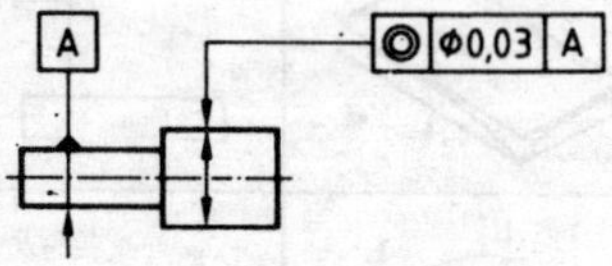	The axis of the toleranced cylinder must lie within a cylinder of diameter t = 0.03 mm whose axis is a continuation of the axis of the reference cylinder.
	Axial run-out	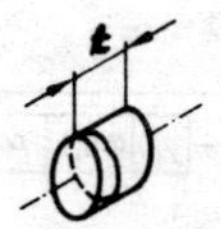	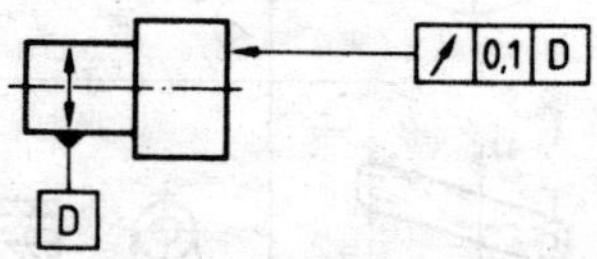	By turning around the reference axis *D*, the axial run-out must not be greater than 0.1 mm in any measuring cylinder.
↗	Radial run-out	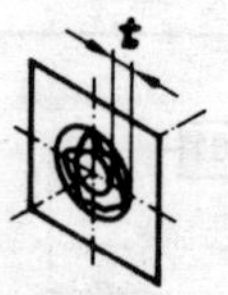	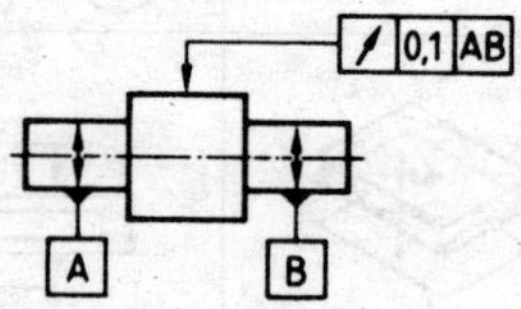	The radial run-cut must not be greater than 0.1 mm in any measuring plane, perpendicular to the axis, during one complete revolution about the common axis of *A* and *B*.

Based on *Fachzeichnen Metalltechnik*, Fachstufe, 1982, Table 80.1 - 81.1 P. 80 - 81, Verlag Europa - Lehrmittel, Wuppertal, Germany.

MAXIMUM MATERIAL CONDITION

In specifying geometrical tolerances, the condition known as the "maximum material condition" (or MMC for short) is of utmost importance. This refers to a condition when the maximum amount of material is left on a part with a bore or on a shaft. Obviously, this happens when the bore is at the lower limit of the tolerances assigned to it and when the shaft is at its higher limit. This condition is important because it is always these limits which affect the interchangeability of parts critically. The importance of geometrical tolerance vis-a-vis the MMC is illustrated in Fig. 7.10.

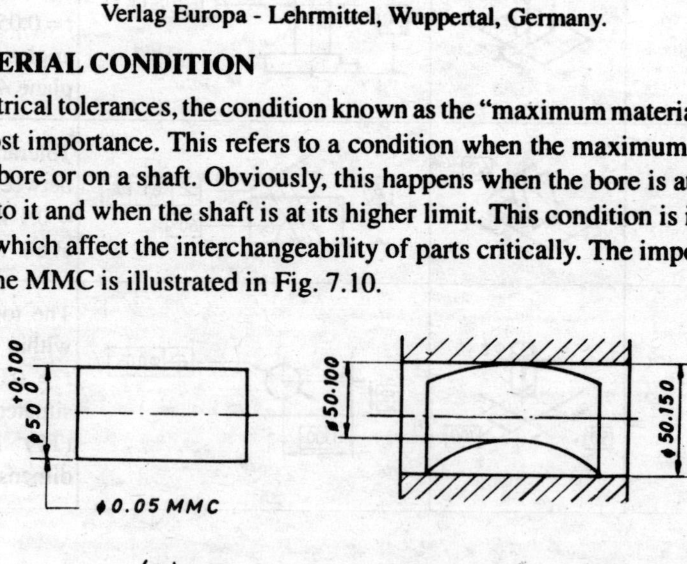

Fig. 7.10 Geometrical tolerance and maximum material condition

In Fig. 7.10 (a), a plain cylinder is shown having a tolerance as specified. Besides, a geometrical tolerance of 0.05 mm has been specified to take care of straightness. In Fig. 7.10 (b), the conditions at MMC are shown. The diameter of the cylinder is at its highest value, namely, 50.000 + 0.100 = 50.100 mm.

The cylinder has entered a hole of which the diameter is a summation of the nominal dimension, the linear tolerance at MMC, and the permissible geometrical tolerance, namely.

$$50.000 + 0.100 + 0.050 = 50.150 \text{ mm.}$$

In this condition, the above cylinder will just enter the hole.

From the point of view of assembly and interchangeability, the above arrangement is in the worst possible condition. This is because of the constraints involved. If, however, all the linear tolerance is not taken up by the cylinder, that is, the diameter of the cylinder measures less than 50.100 (say, 50 or thereabouts), then obviously a greater straightness tolerance is available without endangering free assembly.

In other words, assembly becomes easier and the scope for interchangeability increases.

7.3 TYPES OF FIT

In Section 7.1, we have defined the term "fit" as applied in engineering parlance. Mating parts can be made to be completely interchangeable if proper tolerances and allowances are assigned to the individual components comprising the assembly. In an assembly, each working part of an equipment or machine bears some sort of definite relationship, with its corresponding mating part. When the parts are assembled, this relationship may result in free rotation, free linear movement, permanently fixed position with no relative movement between the parts, *etc.*, as the case may be. As mentioned elsewhere in this Chapter, before the advent of the modern, systematized fit systems, the ultimate efficiency of fit of components in actual, assembled condition depended by and large on the skill and experience of the machinist who did his job solely on his own interpretation of such notes in the drawings as "running" fit, "force" fit, *etc.* As it is not possible to make parts conforming to absolute sizes, elaborate tolerancing systems, specifying "Limits" of sizes of mating parts, have been drawn up to dispense with the individual judgement of the workman and thereby to ensure proper and interchangeable assembly. Since the limits, prescribing the maximum and minimum permissible sizes of parts within the ambit of the required range of accuracy, are clearly indicated in modern engineering drawings, the individual factor has largely been eliminated and the possibility of rejects has been alleviated to a great extent.

We recall that the three main classification of fits are:

(i) Clearance fit, where a clearance is always assured between the mating parts when assembled;

(ii) Interference fit, where an interference always exists between the mating parts, and

(iii) Transition fit, where either a clearance fit or an interference fit ensues, depending on the actual sizes of the mating parts.

In Fig. 7.11, the above three principal kinds of fits have been illustrated. It can be seen that in case of clearance fit, the tolerance zone of the hole is entirely above that of the shaft, while in case of interference fit, this zone is entirely below that of the shaft. In case of transition fit, the two zones overlap. Besides the above main classes, the fit system has been categorized into a number of sub-divisions which have been elaborated in Section 7.5.

For practical application of the systems of fits, two principal ones are in vogue, namely, the basic hole system and the basic shaft system. Since normally it is easier to make a shaft to a pre-specified tolerance

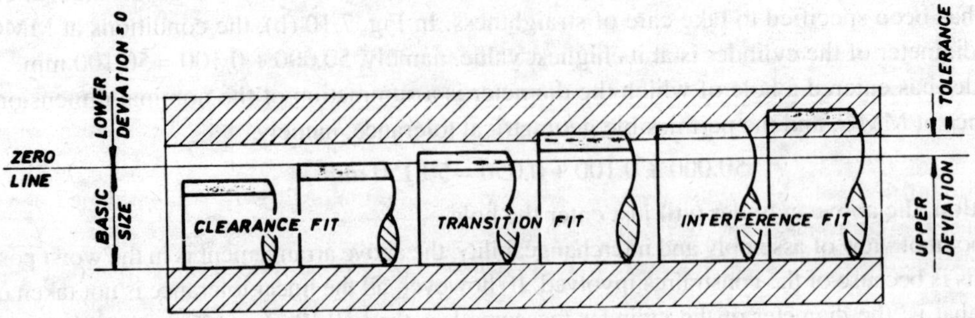

Fig. 7.11 Different types of fit

than a hole with the same tolerance, the hole basis system is most extensively used in industries from the standpoint of production economy. To produce a hole to conform to the desired toleranced size, reamers, broaches and other standard tools have to be used, while shaftings can easily be machined to any specified size. Hence, the preference for the hole-basis system. As we have seen in Section 7.1, in a basic-hole system, the minimum size of the hole is taken as the base-size from which all size variations are made. This hole can generally be made with a standard tool. Allowances and tolerances are assigned from this basic size which results in producing the different kinds of fits required. In the hole basis system, it is now an almost universal practice to take the zone as "*H*" and the quality or grade as "7" (or sometimes 8), so that the bore has a tolerance of *H* 7. It will be noted here that irrespective of grade, the lower deviation of the zone *H* is always zero, so that the minimum limit of the hole-size is the basic size of the system.

In Fig. 7.12, a pictorial view of the hole-basis system (with constant hole tolerance zone H) has been depicted. The zones of tolerance of the shafts vary from "*a*" to "*z*", producing the following different fits:

(i) *Clearance fit*: with shafts having tolerance zones from "*a*" to "*h*",

(ii) *Transition fit*: with shafts having tolerance zones from "*j*" to "*p*", and

(iii) *Interference fit*: with shafts having tolerance zones from "*p*" to "*z*".

The zone "*p*" is a border-line case which produces either transition fit or interference fit, depending upon the qualities of "*H*" and "*p*", *e.g.*, *H7/p6* is an interference fit, while *H8/p6* is a transition fit. (See Appendix A on tolerance values for comparison).

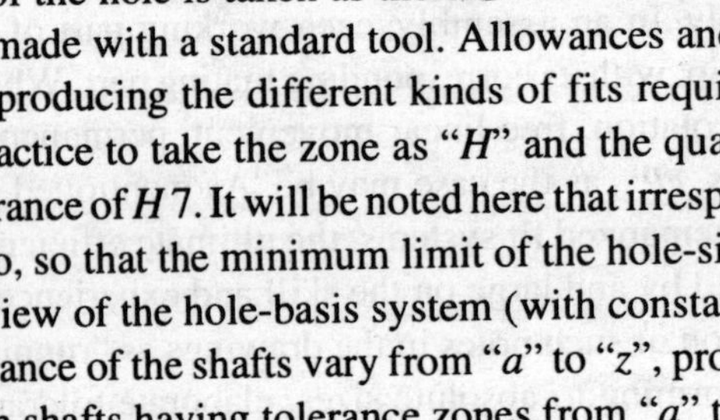

Fig. 7.12 Types of fit vis-a-vis tolerance zones.
Based on *Fachzseichnen Metalltechnik*, Fachstufe, 1982, Fig. No. 71.2 P. 71
Verlag Europa - Lehrmittel, Wuppertal, Germany.

7.4 ASPECTS OF TOLERANCE SYSTEM

Factors influencing the choice of tolerance and the cost angle: In Sections 7.1 and 7.2, the main aspects of the systems governing the tolerancing of parts have been discussed. In olden days, no systematic tolerancing system existed and the efficiency of fitting parts depended largely on the skill of the workshop personnel who made and fitted these parts. Before the limit dimensioning system came into vogue, only the nominal dimensions of the parts to be mated were given, and the desired fit was indicated by a note in the drawing, viz., running fit, drive fit, sliding fit, force fit, *etc.*

The interchangeability aspect of modern manufacturing technology has been emphasised before. An essential element of mass production is the ability to fit parts successfully which are made in widely separated places, and then brought together for assembly. Modern industry cannot function without interchangeable manufacturing which in turn cannot be effective unless there is a systematised size control in practice. Often the main manufacturer of an equipment off-loads many components comprising the equipment to sub-contractors situated at far-flung localities. It is obvious that size control is imperative if the whole equipment is to be assembled and function properly.

Since it is virtually impossible to make parts to the exact sizes required or specified and since it is not economically viable even if it is possible as the cost will be prohibitive, only varying degrees of accuracy, depending on functional and operational requirements, will suffice. This is the fundamental aspect and consideration from which the modern tolerancing systems have emerged. To summarise: tolerance in an engineering drawing is to be provided,

(i) If the functional or the economic aspect of the manufacturing process of the job warrants that certain limiting dimensions be adhered to;

(ii) If the work-pieces are to be mated in a specified manner to have a desired fit;

(iii) If the components are separately finished and are to be assembled without any further machining;

(iv) If the parts are to be interchangeable, as in the case of spare parts of an equipment, and

(v) If jigs and fixtures for finishing operations require any particular tolerance to be assigned.

In Section 7.6, general tolerance (also known as free measure tolerances) have been discussed. These tolerances will suffice in ordinary cases.

We have seen that the proper choice of tolerance should be commensurate with the relevancy of application of the work-pieces for which it is meant as regards functionability and interchangeability. The specified tolerance should just serve the purpose. In Fig. 7.13, a comparative picture of the tolerance grades or quality has been shown. It can be seen that the amount of tolerance varies widely throughout the whole gamut of the tolerance grades.

An idea of the relative cost versus tolerance grade can be obtained from Fig. 7.14 which is based on practical considerations and encompasses all types of costs, viz., material cost, labour cost, overhead cost, cost due to rejects, *etc.* From the figure, it is clear that the cost increases sharply as the quality becomes finer.

Figure 7.13 shows the classification of tolerance grades and the different purposes for which these grades are applicable. Grades 0 1 to 5 are mainly meant for gauges. Grades 0 1 to 5 are obtained through honing, lapping and finest grinding. Grades 6 to 14 are used generally for work-pieces and fitting parts and are obtained by common machining processes like turning, milling, grinding, *etc.* Grades 12 to 18 belong to the coarse category and are primarily meant for forged, rolled, stamped parts, *etc.*

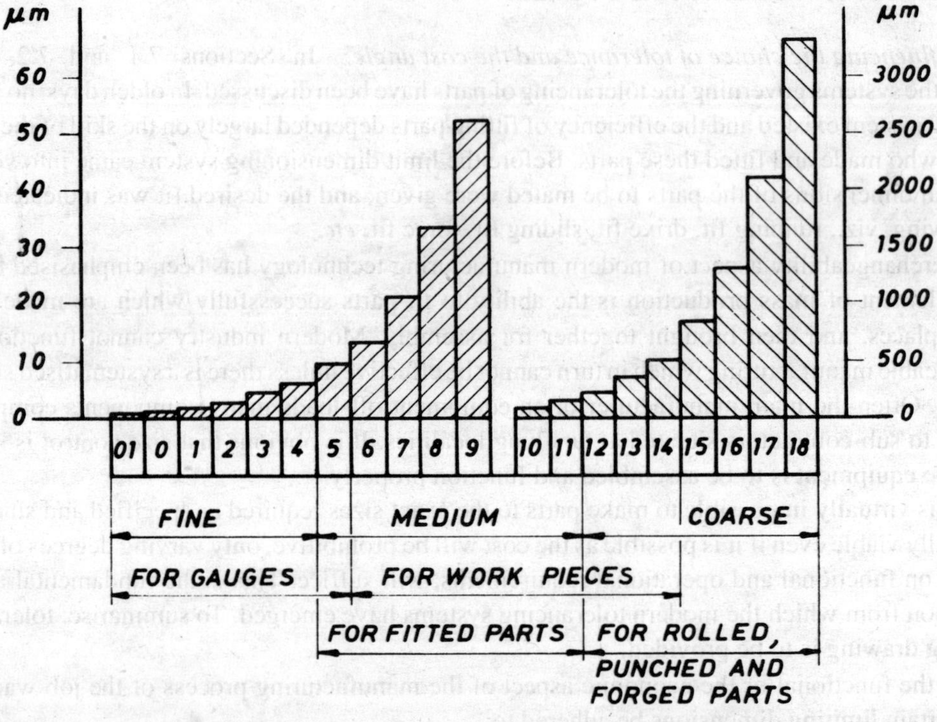

Fig. 7.13 Diagramatic representation of tolerance grades. (Tolerance values shown correspond to a nominal dimension of 30 mm)

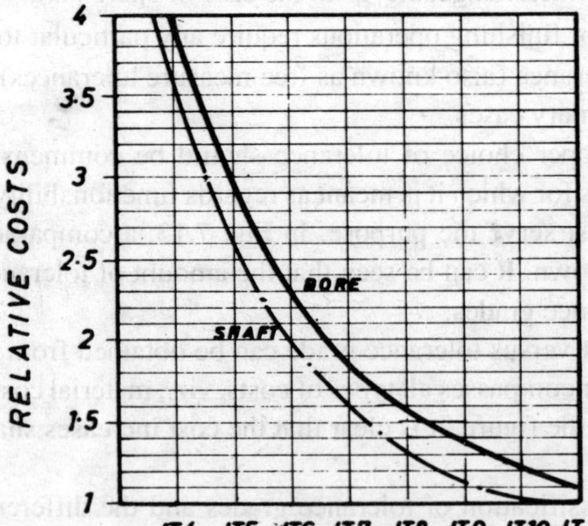

Fig. 7.14 Tolerance grade vs. relative cost

Limit gauging : To ensure that a finished part conforms to the tolerances and specifications given in the drawing, some form of inspection has to be resorted to. This can be achieved either by measurement or by using and comparing with some sort of standard gauges. The main distinction between measurement and gauging techniques lies in the fact that by measurement, one can ascertain the exact size and hence the error, if any, of the work-piece, whereas gauging can only check whether the piece is made within the limits prescribed, but does not reveal the amount of any error which may be present after finishing the job. For quantity production work, limit gauging (or simply gauging as it is commonly called) is extensively used and is also preferred because the method is much quicker than direct measurement. Besides, for ordinary shop practices, the magnitude of error is not usually required, but only the knowledge that the piece is acceptable or a reject is sufficient. However, if for statistical quality control techniques, it becomes necessary to know the extent of errors and size variations of components, then the direct measurement method has to be carried out. Also, the gauges themselves are vulnerable to wear due to continued use and are subject to manufacturing errors.

Though complex receiver gauges and gauging fixtures are used for checking sophisticated parts, ordinary hand gauges are mostly used in work-shops. To these categories belong the plug gauge, caliper gauge or snap gauge which is in the form of a hardened and ground plate having the high and the low limits of the work at the two ends, ring gauge, *etc.*

A plug gauge, which is most commonly used gauge for checking machined bores, has a "GO" and conforming to the lower limit of tolerance prescribed for the diameter of the workpieces bore and a "NOT GO" end conforming to the upper limit. "GO" end should enter the bore and "NOT GO" end should not enter for acceptability of the piece. The "GO" end is longer than the "NOT GO" end. Besides, the "GO" end must be perfectly round and straight within the specified geometrical tolerances. A gauge must also have a manufacturing tolerance which is generally equal to 10% of the tolerance of the work-piece. The initial tolcrances of the two ends of a gauge are disposed above the respective tolerances of the work-piece to compensate for the gradual wearing down of the gauge due to continued use. Figure 7.15 shows limit gauges for checking bore and shaft. The "GO" limit corresponds to the maximum material condition in each case, while the "NOT GO" limit corresponds to the minimum material condition.

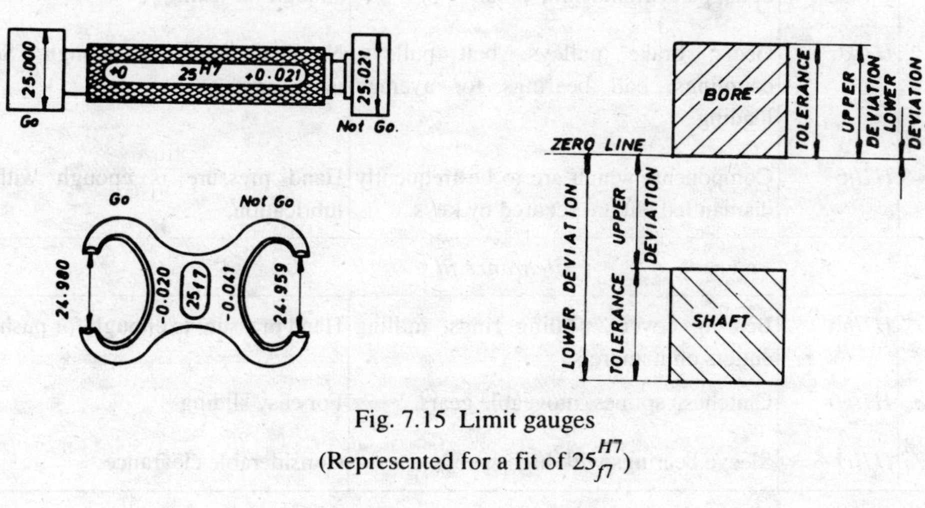

Fig. 7.15 Limit gauges

(Represented for a fit of $25\frac{H7}{f7}$)

7.5 SELECTION OF FITS AND GRADES

In Section 7.3, we have discussed the principal types of fit and their peculiarities. Fits have been classified into clearance fit, interference fit, and transition fit. For practical application purposes, the system of fit has been further categorised into several sub-divisions conforming to the degree of tightness or looseness actually required in the different types of applications in the industrial fields. The fits which are in common usage in manufacturing technology and which the draughtsman or engineers in the drawing office are normally required to specify are given in a tabular form for ready reference in Table 7.5, which also shows their typical scope of application with examples. The fits are based on the hole basis system (*H*7 and *H*8) which is now the most commonly accepted system.

Table 7.5 Common fits and their applications

Type of fit	*Symbol of fit*	*Examples of application*	*Remarks.*
		Interference fit	
Shrink fit	*H*8/*x*8 *H*8/*u*8 *H*7/*u*6	Gears fitted on shafts without keys, tyres fitted on traction wheels, bronze crowns on work-wheel hubs.	The degree of heating or cooling depends on the coefficient of thermal expansion
Press fit	*H*7/*r*6	Bearing bushes in mountings, couplings on shaft ends, gears on shafts.	Adequate pressure required for assembly
		Transition fit	
Light press fit	*H*7/*n*6	Shaft and wheel assembly with parallel key, bearing bushes in hubs.	High pressure required, securing against turning necessary.
Force fit	*H*7/*m*6	Inner ring of ball bearing on shafts, fit bolts, parts on machine tools.	Hand hammer with adequate force is enough for fitting.
Push fit	*H*7/*k*6	Gears, brake pulleys, belt pulleys, couplings, ball bearings for average loading.	Nominal force is enough. Securing necessary.
Easy push fit	*H*7/*j*6	Components which are to be frequently dismantled but are secured by keys.	Hand pressure is enough with good lubrication.
		Clearance fit	
Sliding fit	*H*7/*h*6	Bearing covers, sealing rings, milling cutters on mandrels.	Hand pressure is enough for pushing
	*H*7/*g*6	Clutches, splines, moveable gears.	For easy sliding.
	*H*7/*i*7	Sleeve bearings with high speeds.	Considerable clearance.

Clearance fit			
Sliding fit	*H8/e8*	Gears sliding on shafts, journal bearings with medium speed.	Enough clearance for free running.
Running fit	*H8/d9*	Journal bearings with low speeds, bearings made of plastic materials	High clearance between mating components.
	H8/h11	Rotary shaft seals, seals with metal housing, multispline shafts.	Special purposes application

7.6 GENERAL TOLERANCES

For the attainment of economic advantages, an engineering component is usually toleranced as per the relevant parameters, such as, interchangeability, manufacturing processes, nature of fit and other technical considerations. Only the correct and appropriate tolerancing commensurate with the purpose brings about the economic advantages. It is, however, not necessary to provide all dimensions with tolerances in a drawing. Tolerance is required only for those dimensions which play a role in fixing and maintaining certain limiting conditions based on certain specific reasons. It has been emphasised before that assignment of rigorous tolerancing, which is unnecessary to serve the purpose, results in making the part costlier and thereby rendering it uncompetitive in market and loosing its economic viability. Parts or dimensions which are not toleranced make the work of the designer easier. Besides, since only the toleranced dimensions are specially checked or controlled during inspection, this work is dispensed with in case of untoleranced components.

However, dimensions which are untoleranced cannot be left to the discretion of the workshop people as regards their final values in the finished conditions and therefore, some sort of limiting dimensions has to be provided in such cases to restrict these dimensions within reasonable values. These are covered by what is known as the "General tolerances". These tolerances were previously termed as "Free measure tolerances" and "Allowable deviations for dimensions without specified tolerances". These are permissible deviations which correspond to the standards of accuracy normally attainable in workshop practice, and are to be used for nominal sizes where no tolerance has been specified in the engineering drawings.

To avoid controversy, ambiguity or enquiry, the values of the general tolerances have been laid down in various Indian Standards along with the specific fields in which these tolerances are applicable. IS:2102 specifies the permissible machining variations for linear and angular dimensions where no tolerances have been indicated in the drawing. These are given in Tables 7.6 to 7.8. The Standard is applicable for all chip-forming machining processes, *e.g.*, turning, milling, *etc.*, and also for processes where no chips are formed, such as, drawing, stamping, embossing, *etc.*

In the drawing, either the class of deviation followed should be indicated in the title block (*e.g.*, Medium IS:2102, Coarse IS:2102, *etc.*) or the permissible values should be entered in a small box in the title block. IS:2102 is not applicable for production methods, such as, casting, forging, welding, *etc.*

For values of general tolerances on iron and steel castings, see Appendix - B.

Table 7.6 Deviations for linear dimensions

All dimensions in millimeters

Class of deviation	*Range of Nominal Dimensions*											
	Above	*0.5*	*3*	*6*	*30*	*120*	*315*	*1000*	*2000*	*4000*	*8000*	*12000*	*16000*
	Up to and including	*3*	*6*	*30*	*100*	*315*	*1000*	*2000*	*4000*	*8000*	*12000*	*16000*	*20000*
Fine		± 0.05	± 0.05	± 0.1	± 0.15	± 0.2	± 0.3	± 0.5	± 0.8	–	–	–	–
Medium		± 0.1	± 0.1	± 0.2	± 0.3	± 0.5	± 0.8	± 1.2	± 2	± 3	± 4	± 5	± 6
Coarse		–	± 0.2	± 0.5	± 0.8	± 1.2	± 2	± 3	± 4	± 5	± 6	± 7	± 8
Extra coarse		–	± 0.5	± 1	± 1.5	± 2	± 3	± 4	± 6	± 8	± 10	± 12	± 12

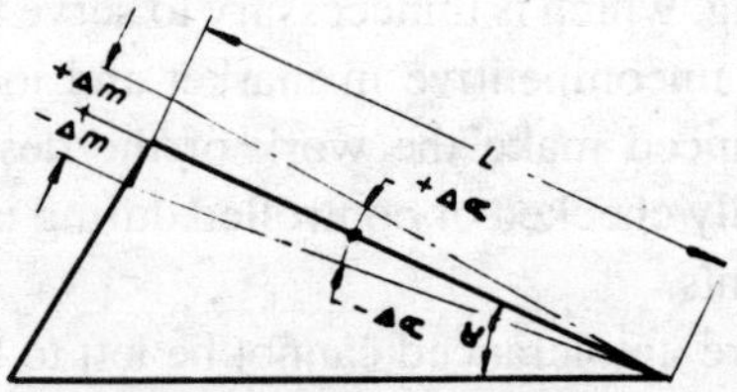

$\Delta\alpha$ = angle tolerance in angular units

Δm = angle tolerance in linear units

Table 7.7 Deviations for angular dimensions

All dimensions in millimetres

Class of Deviation	*Permissible Variation on Length of Shorter Side of Angle*							
	Up to 10		*Over 10 to 50*		*Over 50 to 120*		*Over 120*	
	Degree	*mm per 100 mm*	*Degree*	*mm per 100 mm*	*Degree*	*mm per 100 mm*	*Degree*	*mm per 100 mm*
Fine and medium	± 1°	± 1.8	± 30′	± 0.9	± 20′	± 0.6	± 10′	± 0.3
Coarse	± 1°30′	± 2.6	± 50′	± 1.5	± 25′	± 0.7	± 15′	± 0.4
Extra coarse	± 3°	± 5.2	± 2°	± 3.5	± 1°	± 1.8	± 30′	± 0.9

Table 7.8 Deviations for radii and chamfers

All dimensions in millimetres

Class of Deviation	Range of Nominal Dimensions					
	Above	*0.5*	*3*	*6*	*30*	*120*
	Upto and including	*3*	*6*	*30*	*120*	*315*
Fine and medium		±0.2	±0.5	±1	±2	±4
Coarse and extra coarse		±0.2	±1	±2	±4	±8

7.7 SURFACE ROUGHNESS AND SURFACE FINISH SYMBOLS

The surface finish of an engineering component is directly related to the functionability, wear characteristics and operational life of the component. Accurate control of surface quality has become imperative to meet the ever-increasing demands on modern design of parts as regards heavier loads, higher speeds, lesser wear due to lesser friction. In practice, it is not possible to attain a geometrically ideal surface of a work-piece. Therefore, an engineering drawing must also specify the surface texture which is permissible under the circumstances in question. The surface roughness can be represented in several ways in a drawing and these methods along with the relevant nomenclatures will be discussed in this section.

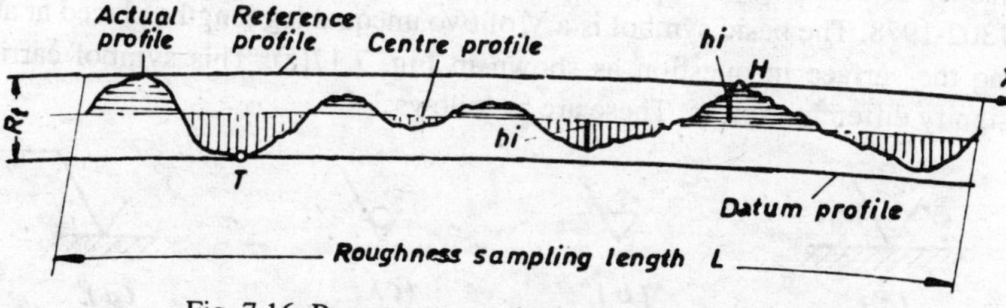

Fig. 7.16 Parameters pertaining to surface roughness

Nomenclatures : Referring to Fig. 7.16, the relevant terms are defined below:

Actual profile: It is the profile of the actual surface.

Reference profile: It is the profile to which the irregularities of the surface are referred to. It passes through the highest point H of the actual profile.

Datum profile: It is the profile which is parallel to the reference profile and passes through the farthest point *T* of the actual profile.

Centre profile or mean profile: It is the profile within the sampling length so placed that the sum of the material-filled areas enclosed above it by the actual profile is equal to the sum of the material-void areas enclosed below it by the actual profile. The term mean profile will also be used synonymously as mean surface in this section.

Roughness sampling length: It is the test length in question of which the roughness is measured.

Peak-to-valley height R_t: It is the distance from the reference profile to the datum profile. In a sampling length, it represents the maximum peak-to-valley height.

Mean roughness index R_a: It is the arithmetic mean of the absolute values of the distances h_i between the actual and the centre profiles. The significance of R_a will be explained at length later in this section

Representation of surface roughness in drawings: For general application, the degree of surface roughness is automatically controlled by the machining methods adopted, namely, turning, grinding, milling, lapping, *etc*. The type of surface roughness plays a secondary role in case of non-working surfaces, and hence these surfaces do not require close control of surface finish except in those cases where constraints are imposed by process control or for the sake of aesthetic reasons concerning outward appearance.

In previous practice, instructions on drawings regarding the degree of surface finish required were rather vague. Expressions like "smooth surface", "very fine surface", *etc*. were used on the drawings indicating the surface finish required. In English-speaking countries, the letter "*f*" was used to indicate that some sort of finish was required for the surface in question. Subscripts were used to indicate the degree of roughness in conjunction with the letter "*f*", *viz*., f_1, f_2, f_3, *etc*., indicating finer and finer surfaces. When all the surfaces were to be finished, the expression *F A O*, meaning "Finish All Over", was used as a general instruction written by the side of the drawing. As per modern practice, surface roughness can be represented in several ways in an engineering drawing. We will first discuss the latest ISO method of incorporating the surface finish symbols in drawings. Later, the finish symbols involving triangles as given in DIN 3141 will be dealt with. Both the systems are concurrently in vogue in present-day industrial arenas.

The latest practice of denoting the surface texture has been standardised in IS:10719-1983 which is based on ISO 1302-1978. The basic symbol is a V of two unequal leg lengths placed at about 60° to the line representing the surface in question as shown in Fig. 7.17(a). This symbol carries appropriate indications to signify different aspects. These are as follows:

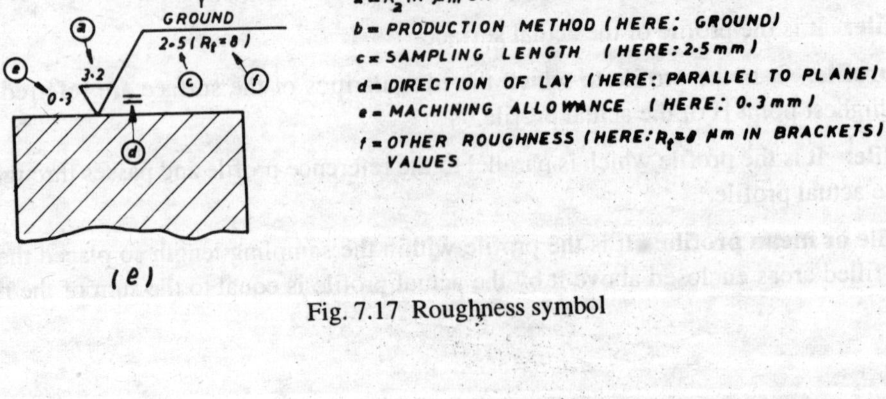

Fig. 7.17 Roughness symbol

Table 7.9 Roughness values and grade members

Roughness value R_a μm	*Roughness grade numbers*
50	N12
25	N11
12.5	N10
6.3	N9
3.2	N8
1.6	N7
0.8	N6
0.4	N5
0.2	N4
0.1	N3
0.05	N2
0.025	N1

1. If it is intended to show removal of material by machining, a bar is added to the basic symbol, Fig. 7.17 (b).
2. If it is not permitted to remove material, then a circle is added, Fig. 7.17 (c).
3. A horizontal line is added to the larger arm of the basic symbol when some special surface characteristics have to be indicated, Fig. 7.17(d).
4. The roughness values R_a is entered as shown in "*a*" in Fig. 7.17(e). In Table 7.9, the roughness value are also indicated by grade numbers to avoid misinterpretation of numerical values. Either numerical values in micrometer or the corresponding grade number can be inserted in the symbol to indicate the roughness value.
5. If only one value of R_a is specified, then it stands for the maximum permissible value of surface roughness. In case of two values, *viz*., maximum and minimum, these values should be shown one above the other.
6. If it is desired that the final surface texture should be produced by any particular production method, then this should be indicated as shown in "*b*" in Fig. 7.17(e).
7. Sampling length and direction of lay should be shown as in "*c*" and "*d*" respectively in Fig. 7.17(e).
8. In Fig. 7.17(e), "*e*" indicates the value of the machining allowance in mm, when it is necessary to be specified.
9. Other roughness values, viz., R_t, R_z, should be entered within brackets as shown in "*f*" in Fig. 7.17(e).

Methods of indicating surface finish symbols on drawings: Guidelines as to the incorporation of the surface texture symbols discussed above have been laid down in IS:10719-1983 in conformity with ISO/R 129. These are summarised below:

1. The symbols should be so oriented that they are readable from the right-hand side of the drawing or from the bottom, Fig. 7.18.

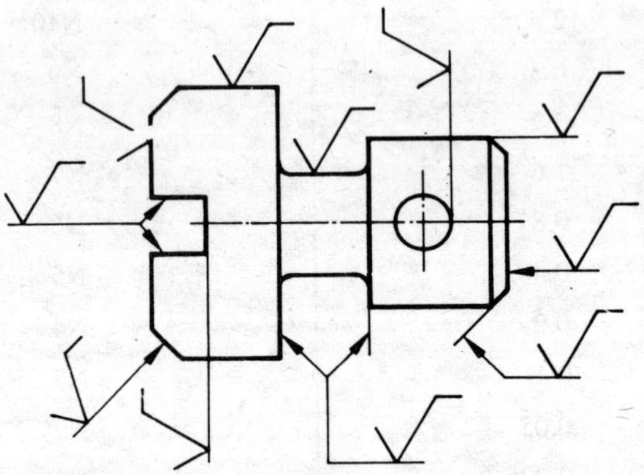

Fig. 7.18

2. The symbol is to be used once only for a given surface, and preferably on the view which carries the dimension pertaining to the surface in question.
3. If all the surfaces of a part have the same texture, the symbol carrying the roughness value is placed outside the drawing by the side of the part number, as shown in Fig. 7.19.

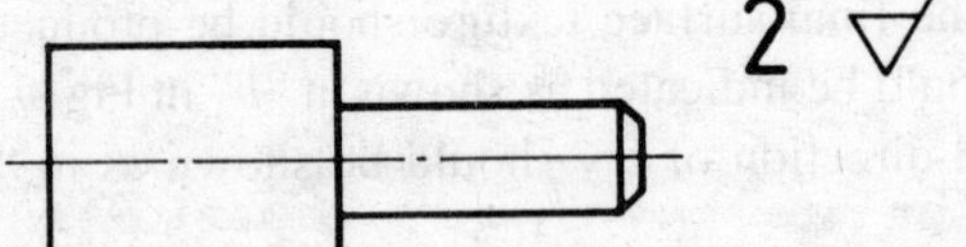

Fig. 7.19

4. In cases where a particular surface texture is required on the majority of the surfaces with a few exceptions, then those exceptions are shown on the surface for which they are meant on the body of the drawing. The symbol meant for the majority of the surfaces is shown outside the drawing and the exceptions are indicated inside the parentheses as shown in Fig. 7.20, where a_1, a_2, *etc.*,

stand for the R_a values. In case of a number of exceptions these can be indicated by the basic symbol enclosed within brackets as shown in Fig. 7.21. Here, the general finish symbol is a_1 and the basic symbol stands for all the exceptions, *i.e.*, a_2 and a_3.

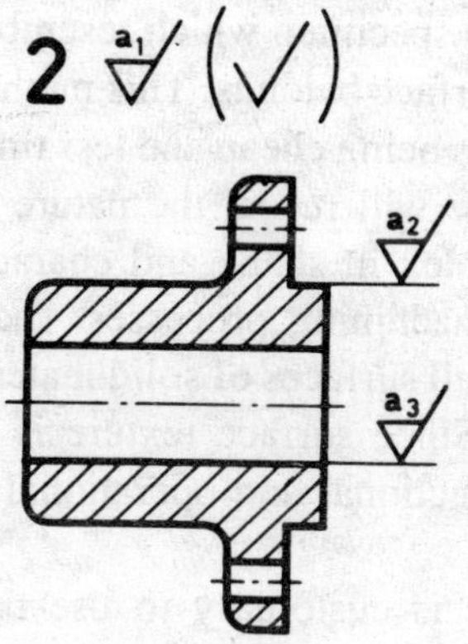

Fig. 7.20

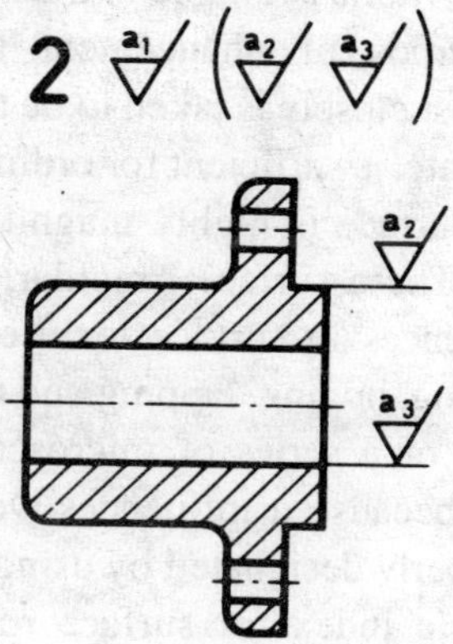

Fig. 7.21

5. Complicated specifications when required for a number of surfaces need not be repeated. Instead, they may be shown as in Fig. 7.22.

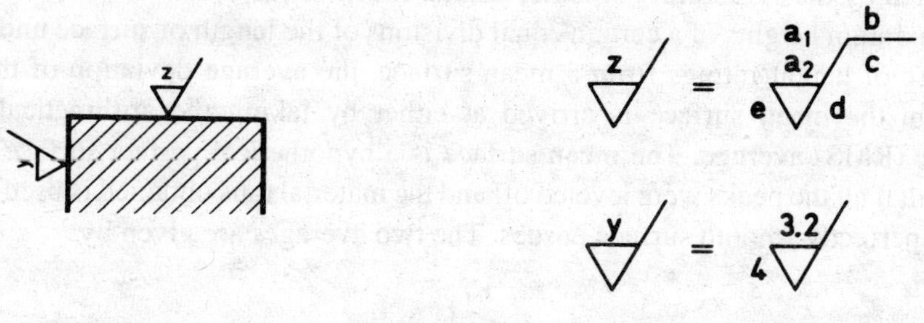

Fig. 7.22

6. When the same type of surface texture is required on a number of surfaces, one simplified, appropriate and relevant symbol only is required to be shown as such surfaces and its meaning can be inserted elsewhere near the drawing. This is illustrated in Fig. 7.23 to 7.25.

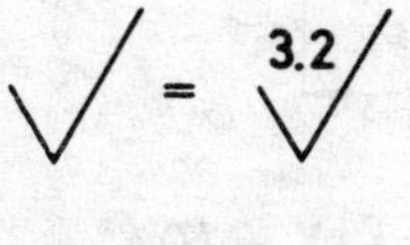

Fig. 7.23

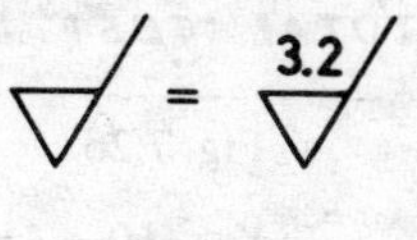

Fig. 7.24

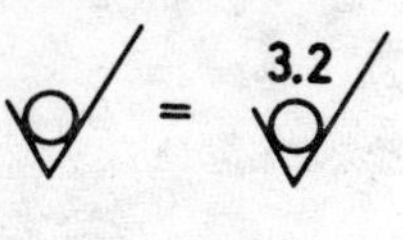

Fig. 7.25

7. No surface symbol should be shown when the ordinary manufacturing processes by themselves ensure as acceptable surface finish.
8. Roughness, method of production, machining allowance, *etc.*, should be indicated only when the surfaces really need them. Otherwise the cost is unnecessarily increased.

Methods of evaluation and measurement of surface roughness : One of the principal methods employed in shops for the evaluation of surface finish of a machined part is by subjective comparison. In this method, the surface to be inspected is visually compared with the specimens having standard surface-finish criteria available in the inspection department of the shop. Usually magnifying glasses are used for the purpose of enhancement. The quality of the standard specimen which resembles the surface in question most closely is taken to be the quality to which the surface belongs. This method, though not the most accurate, is sufficient for ordinary shop practices, besides being cheap and less time-consuming.

If a given surface is highly magnified, the contour or profile will reveal the nature of the surface irregularities. These surface irregularities are minute and complex in shape and character. There are distinct differences in surface textures produced by various machining processes, such as, turning, grinding, milling, honing, lapping and other methods. However, all surfaces of solid materials in general are composed of a series of microscopic peaks and valleys. Since surface texture is a vital design consideration because it influences wear, strength and other functional and operational parameters, it should be properly designated by using an index of some sort.

As a suitable index for surface roughness measurement, it is customary to use one of the two interpretations of the measured values of the surface texture. These are: (i) the maximum peak-to-valley height which is given by the vertical distance between the top of the highest peak and the bottom of the lowest valley, measured over a particular stretch of surface, and (ii) the average deviation from the mean surface. The maximum peak-to-valley height, measured over the whole length of the inspected stretch in question, is denoted by the symbol: R_t. Another similar index, R_z, is used which is the average value of the individual maximum heights of a certain equal divisions of the length of surface under reference (see Fig. 7.26). In case of height ratings from a mean surface, the average deviation of the surface of the irregularities from the mean surface is arrived at either by taking the arithmetical average or the root-mean-square (RMS) average. The mean surface is a hypothetical perfect surface and is a surface which would result if all the peaks were leveled off and the material thus obtained is used to fill the valleys exactly, so that a perfectly smooth surface ensues. The two averages are given by:

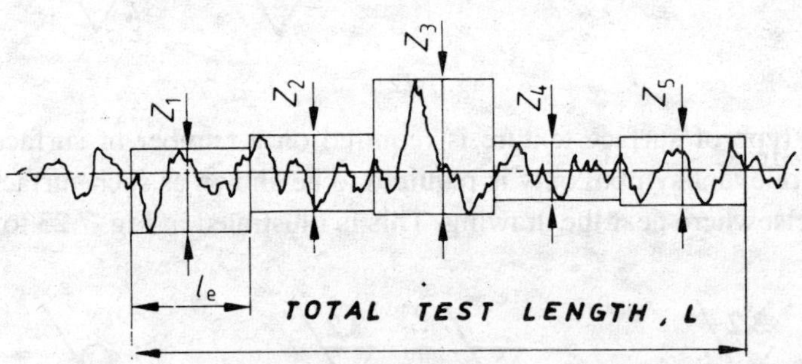

Fig. 7.26

$$\text{Arithmetical average} = \frac{(Y_1 - H) + (Y_2 - H) + ... + (Y_n - H)}{n}$$

$$\text{RMS average} = \sqrt{\frac{(Y_1 - H)^2 + (Y_2 - H)^2 + \ldots + (Y_n - H)^2)}{n}}$$

For meanings of Y_1, Y_2, H, *etc.*, see Fig. 7.27

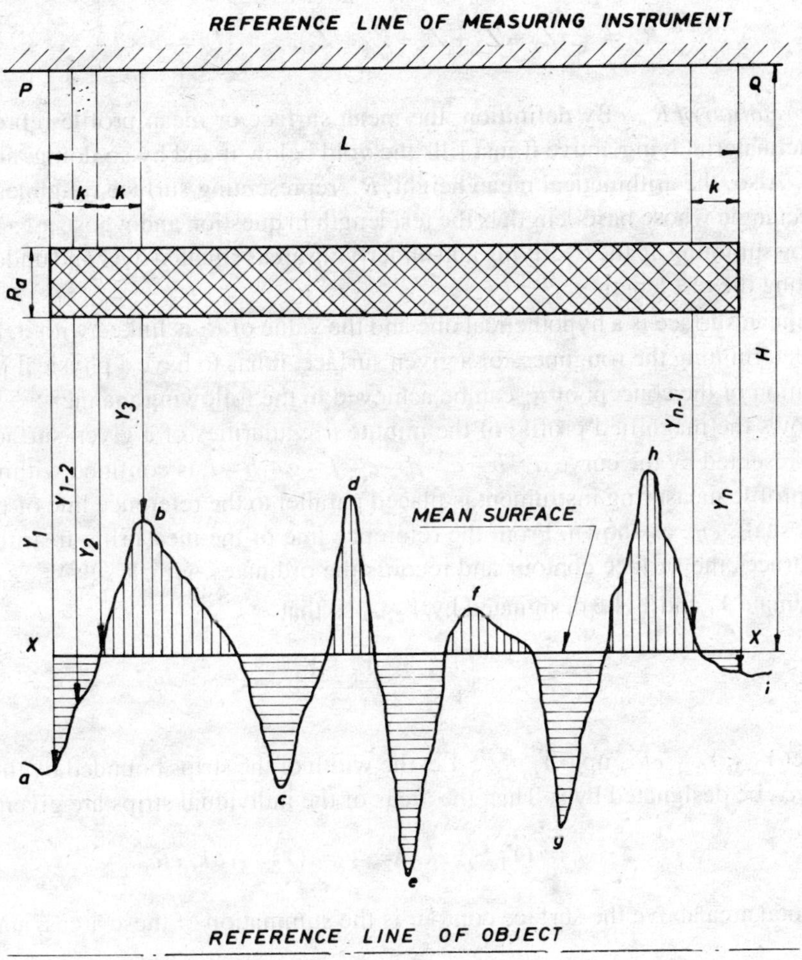

Fig. 7.27 Physical interpretation of R_a

The arithmetical average is somewhat less than the RMS average. The arithmetical average system is used in preference to the other because the commercially available roughness measuring instruments are calibrated for the arithmetical average. Besides, it is also simple and leads to quicker calculations.

For actual measurement of surface roughness, various precision instruments are commercially available. Such instruments are profilemeter, Brush Surface Analyser, Surfagage, *etc.* In such instruments, a finely pointed diamond moves up and down over the irregularities of the surface whose roughness is being measured. As the diamond point moves over the peaks and valleys of the surface, a suitable tracer head, to which the diamond is flexibly suspended, actuates an electro-magnetic device which in turn indicates the surface roughness. The form, shape and the amplitude of the surface irregularities are traced by a stylus having a point radius of about 10 μm and recorded in ink on a moving graduated paper chart. Once a magnified picture of the contour, traced by the stylus, is obtained, the average roughness values can be calculated from the record.

$$R_z = \frac{1}{5}(Z_1 + Z_2 + Z_3 + Z_4 + Z_5). \qquad (L = 5 \times l_e) \qquad \text{(see Fig. 7.26)}$$

Physical interpretation of R_a: By definition, the mean surface or mean profile represents a surface so situated that the material lying above it just fills the void below it and by so doing, an idealised, perfect surface ensues. Also, the arithmetical mean height, R_a, representing surface roughness is defined as the height of the rectangle whose base-length is the test length in question and whose area is equal to the total area arrived at by summing up the overlying and underlying areas situated over and under the mean surface respectively along the test length.

Clearly, the mean surface is a hypothetical one and the value of R_a is linked with it. Since R_a is a given parameter for determining the roughness of a given surface, it has to have a physical interpretation. This actual visualisation of the concept of R_a can be achieved in the following manner :

Fig. 7.27 shows the magnified profile of the minute irregularities of a given surface. The contour of this profile, represented by the curve $a - b - c - d - e - f - g - h - i$, is confined within the test length in question. L. A profile measuring instrument is placed parallel to the reference line of the object, *e.g.*, the centre line of a shaft, *etc.* as shown. From the reference line of the measuring instrument, the stylus of the instrument traces the surface contour and records the ordinates, *viz.*, Y_1, Y_2 Y_3, ... Y_{n-1}, Y_n.. Let the mean of the ordinate Y_1 and Y_2 be designated by Y_{1-2}, so that

$$Y_{1-2} = \frac{Y_1 + Y_2}{2}$$

Similarly, we get Y_{2-3}, Y_{3-4}, *etc.*, upto $Y_{(n-1)-n}$. Let the width of the strips bounded by the ordinates Y_1 and Y_2, Y_2 and Y_3, *etc.*, be designated by k. Then the areas of the individual strips are given by:

$$A_{1-2} = (Y_{1-2}) \cdot k,\ A_{2-3} = (Y_{2-3}) \cdot k,\ etc.$$

Therefore, the total area above the surface contour is the summation of these areas, and is given by

$$\text{Total area, } A_T = [Y_{1-2}] \cdot k + [Y_{2-3}] \cdot k + \ldots + [Y_{(n-1)} - n] \cdot k.$$

Let H be the height of the mean surface which is yet to be determined. By definition, we have,

$$A_T = L.H. = k.n.H,$$

where n = Number of strips, so that $L = k.n$.
Since the values of Y_1, Y_2 ... Y_n are known, A_T can be calculated, and since the test length L is known, H can be calculated. Thus, the height H of the mean surface $X - X$ can be established.

Now, to determine the sum of the overlying areas (vertically hatched) and the underlying areas (horizontally hatched) with reference to the mean surface $X - X$, we proceed as follows: the area of the portion of the strip confined between the mean surface $X - X$ and the surface contour a...*i* is obviously given by the algebraic sum of the mean ordinate and H multiplied by k. That is, the area of the portion A_p, is given by,

$$A_p = (Y_m - H).k \text{ or } (H - Y_m).k$$

as the case may be, depending on whether Y_m is greater than H or H is greater than Y_m. Here, Y_m is a generalised symbol for mean ordinate representing Y_{1-2}, Y_{2-3}, Y_{3-4}, *etc.* Therefore, the total area over and below $X - X$ is given by,

$$A = A_P \,.\, n = |Y_m - H| \,.\, k.n,$$

where $|Y_m - H|$ represents the absolute numerical value of the height of the portion of the individual strips bounded by the contour and $X - X$. This area A can be represented by a rectangle of equal area (shown cross-hatched) which is given by,

$$A = L \,.\, R_a, \text{ or } |Y_m - H|\, k.n = k.n.R_a \quad \therefore\ R_a = |\, Y_m - H\,|$$

or,
$$R_a = \frac{[|Y_{1-2} - H| + |Y_{2-3} - H| + \ldots + |Y_{(n-1)-n} - H|]}{n}$$

Since the parameters Y_{1-2}, Y_{2-3}, *etc*., as well as the values of H and n are known, R_a can be calculated. This calculated value of R_a of the given surface is now compared with the required surface index R_a given in the drawing (say $R_a = 6.3$ μ m or N 9) to decide whether the actual calculated value conforms to the limit set by the desired value given in the drawing. The conclusion as to the acceptance or rejection of the surface can then be arrived at.

Table 7.10 shows a comparison between the different methods used for indicating surface roughness or drawings. An exact conversion of R_a values to R_t values or vice versa is not possible. As regards the relation between roughness and base tolerance, the following rule of thumb may be used for general purposes:

$$\text{Average depth of roughness, } R_z = \leq \frac{1}{2}\, IT,$$

where IT is the base tolerance as explained in Section 7.2.

Table 7.10 Comparison of surface roughness values

Roughness symbol	▽				▽▽				▽▽▽				▽▽▽▽			
Series	*1*	*2*	*3*	*4*	*1*	*2*	*3*	*4*	*1*	*2*	*3*	*4*	*1*	*2*	*3*	*4*
	MAXIMUM ALLOWABLE VALUES IN μ m															
R_t	160	100	63	25	40	25	16	10	16	6.3	4	2.5	–	1	1	0.4
R_a	25	12.5	6.3	3.2	6.3	3.2	1.6	1.6	1.6	0.8	0.4	0.2	–	0.1	0.1	0.025
Roughness grade	N11	N10	N9	N8	N9	N8	N7	N7	N7	N6	N5	N4	–	N3	N3	N1

Based on *Fachzeichnon Metalltechnik, Grundstufe*, 1984, Table 32/1 – 32/2, P. 32, Verlag Europa – Lehrmittel, Wuppertal, Germany.

Lay : When a part is made using any particular production method, it leaves a certain surface pattern. The term "lay" is used in connection with the direction of the predominant surface which is normally determined by the production method employed. Lay symbols are given in Table 7.11.

Table 7.11 Symbols for the direction of Lay

Symbol	*Interpretation*	
=	Parallel to the plane of projection of the view in which the symbol is used	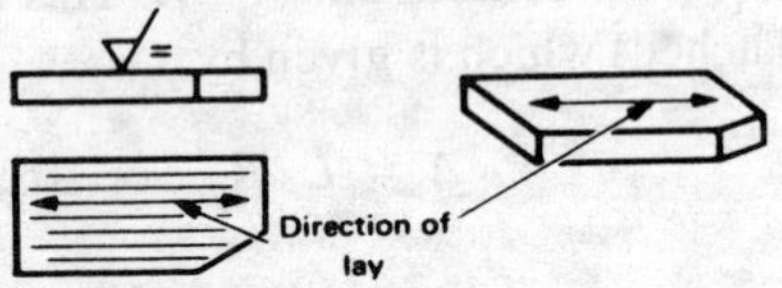
⊥	Perpendicular to the plane of projection of the view in which the symbol is used	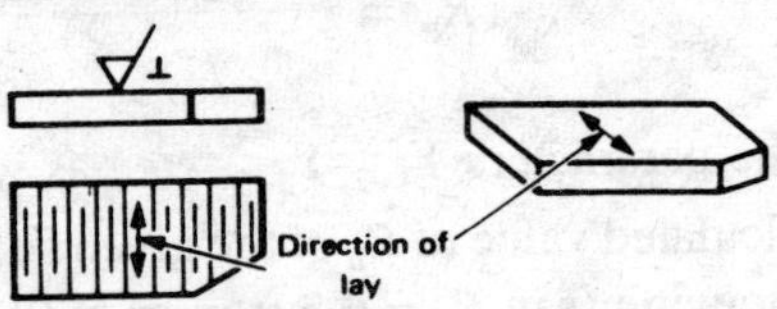
X	Crossed in two slant directions relative to the plane of projection of the view in which the symbol is used	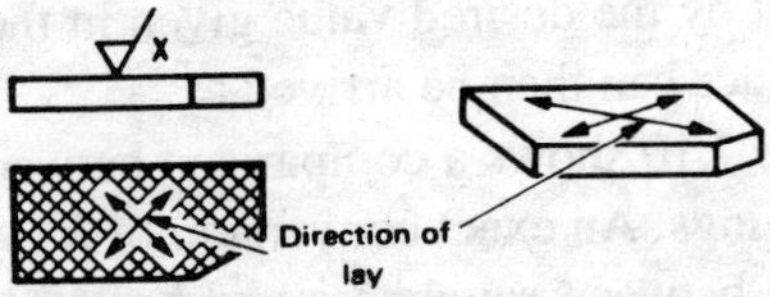
M	Multi–direction	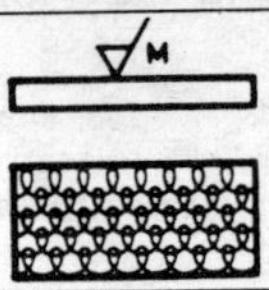
C	Approximately circular relative to the centre of the surface to which the symbol is applied	
R	Approximately radial relative to the centre of the surface to which the symbol is applied	

Note : Should it be necessary to specify a direction of lay not clearly defined by these symbols then this must be achieved by a suitable note on the drawing.

If necessary, the relevant lay symbol is to be inserted along with the surface finish symbol as indicated in "*d*" in Fig. 7.17(e).

Table 7.12 Relation between surface roughness, tolerance grade and manufacturing process

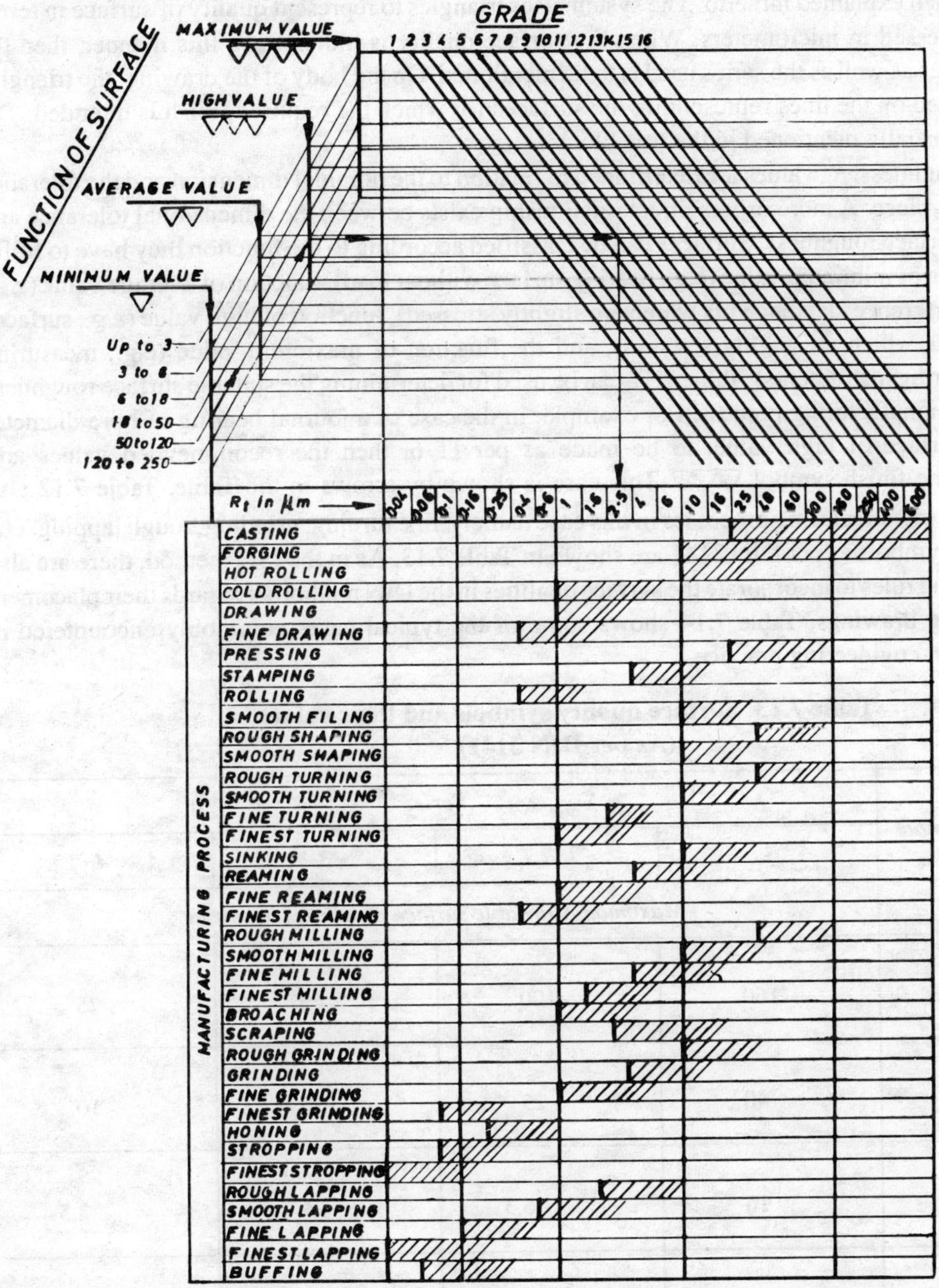

Based on *DUBBEL Taschenbuch fuer den Maschinenbau,* vol. 1. p. 625, Springer Verlag, Berlin, Germany, 1970

Indication of surface finish on drawing as per DIN 3141: Surface finish symbols are also incorporated in the engineering drawings as per DIN 3141 and this system is widely followed concurrently with the ISO system explained hitherto. The system uses triangles to represent quality of surface in terms of R_t values expressed in micrometers. When the surface quality is indicated in this manner, then the number of triangles as well as the series used is to be mentioned. On the body of the drawing, the triangles are directly inserted on the lines representing the surfaces for which the required finish is intended. The series number is usually mentioned in the title block.

The surface roughness, R_t, values are determined in relation to the nominal dimension and the tolerance grade of the work-piece. A more or less direct relationship exists between the dimensional tolerance and the permissible surface roughness. Surfaces can be classified according to the function they have to fulfil. These are: function of minimum value (e.g., bearing surface without load), function of average value (*e.g.*, surfaces with interference clearance fits which are slightly stressed), function of high value (*e.g.*, surfaces with clearance fits which are highly stressed), and the function of maximum value (e.g., measuring surfaces of measuring instruments). Table 7.12 can be used for determining the suitable surface roughness and the relevant manufacturing process. For example: in the case of a journal bearing of bore diameter 60 mm and a surface of high value, to be made as per IT 6, then the recommended values are: $R_t = 2.5$ μm and the finish symbol ▽▽▽ This case is shown by arrows in the Table. Table 7.12 also shows the suitable manufacturing processes in this case namely, fine turning, grinding, rough lapping, etc.

Surface finish symbols as per DIN 3141 are shown in Table 7.13. As in the ISO method, there are also prescribed ways and rules to incorporate the surface qualities in the DIN method as regards their placement on the engineering drawings. Table 7.14 shows some of the typical cases commonly encountered in general mechanical engineering practice.

Table 7.13 Surface quality symbols and their values (As per DIN 3141)

Surface quality symbol	*Series*			
	1	*2*	*3*	*4*
	Maximum Allowable surface roughness R_t in μ m			
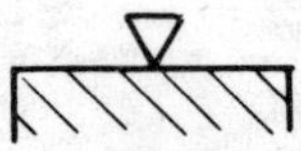	160	100	63	25
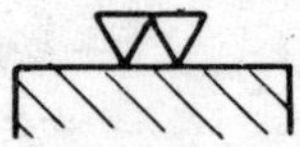	40	25	16	10
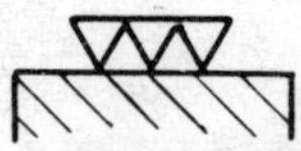	16	6.3	4	2.5
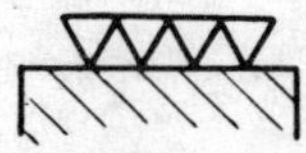	—	1	1	0.4

Table 7.14 Representation of surface finish symbols on drawings (As per DIN 140)

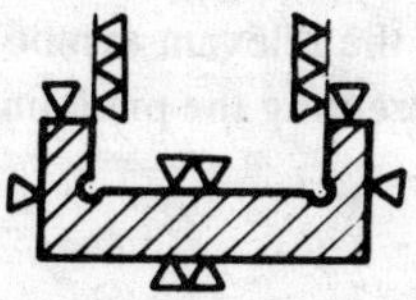	Surface quality symbols should be normally indicated on the lines representing the surfaces to be machined. When there is a lack of space, these symbols may be shown on lines extended from the body lines.
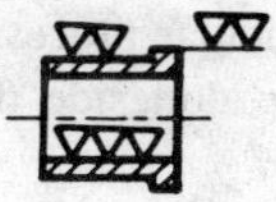	For work-pieces which are machined by turning process, the symbols should be indicated on one boundary line only.
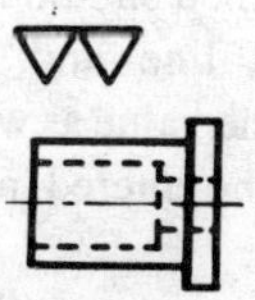	When the same surface quality is required all over, the relevant symbol may be placed alongside the illustration and should be of bigger than normal size.
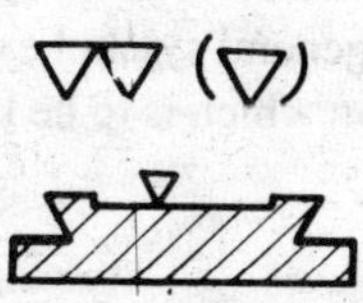	Here, the symbol ∇∇ is meant for all surfaces except the surface having ∇ symbol. The exception is placed within brackets. The same rule is applied if two or more surface quality symbols are intended to be exceptions to the general finish, namely: (∇, ∇∇).
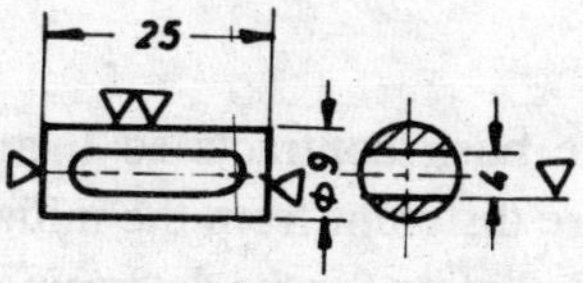	It is desirable to show the symbol for a surface in view or section in which the surface concerned is dimensioned.
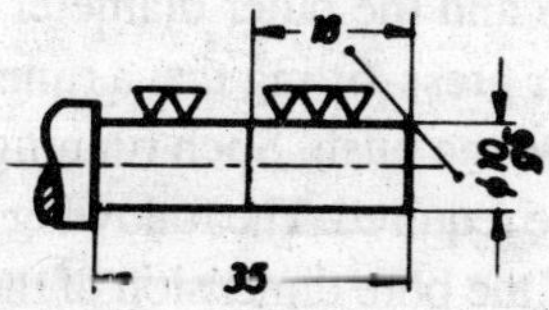	In case only a portion of a work-piece is meant to have a particular finish or carry a specified tolerance, this should be indicated as shown. Here, ∇∇∇ is valid for a length of 18, the diameter having a tolerance of *g*6.

Table 7.14 (*Contd.*)

	To indicate the finish of gear teeth surfaces on a gear drawing, the relevant symbol is placed on the line representing the pitch line.
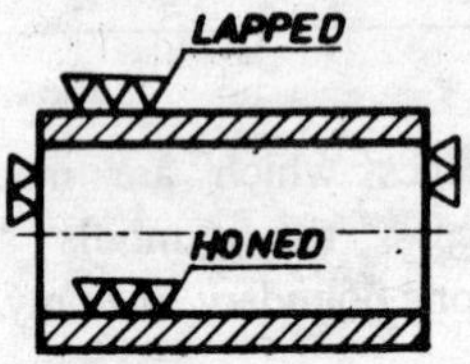	If the required surface finish is desired to be achieved by a particular process, this can be written specifically alongside the symbol as shown.
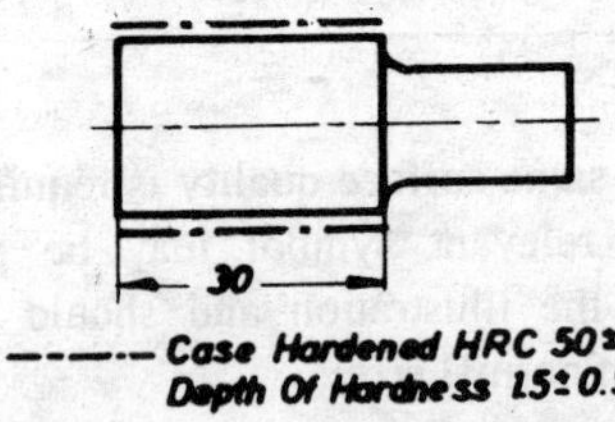	Surface to be hardened should be indicated by thick chain-dotted line and the extent of hardened surface, the value as well as the depth of hardness should be entered in the drawing as shown.
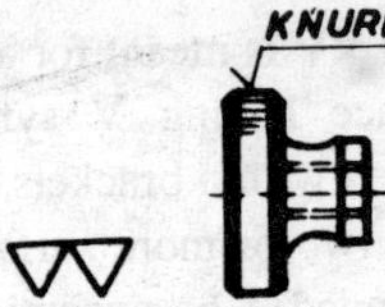	Here, all the surfaces are smooth-finished as indicated by the general symbol except for the finger-grip portion which is to be knurled with a pitch of 0.6.

7.8 PRESS AND SHRINK FIT CALCULATIONS

Press fit: When a bush is press-fitted inside a housing, the bush contracts so that the final actual dimensions of the outer and the inner diameters of the bush are different from the initial values prior to pressing. The final values are obviously somewhat smaller. Calculation for the determination of tolerance before press-fitting is discussed here.

For a press-fit between the bush and the housing bore into which the bush is introduced, the tolerance combination of $H\,7/r\,6$ is normally used between housing bore and the outer diameter of the bush. The inner diameter of the bush should have such an actual value after press-fitting that a running fit is ensured between the inner bore of bush and the shaft which is held inside the bush. Such running fits are $H\,7/g\,6$, $H\,7/e\,8$, $H\,7/f\,7$, $H\,8/d\,9$, etc., depending on the degree of clearance required. The following three possibilities may arise as regards the steps to be taken for the determination of the bore dimension of the bush:

(i) The housing along with the press-fitted bush can be mounted on a suitable machine tool where the bush, which is usually provided with a pilot bore, is bored to the required toleranced dimension, *e.g.*, $H\,7$. This arrangement, however, is possible only in case of small housings.

(ii) Due to practical difficulties, big housings cannot be easily mounted as in case (i) above. In such cases, the bore of the bush is so toleranced before assembly that ample clearance is ensured after press-fitting. Such tolerances are usually $C\,11, E\,6, E\,9, F\,7$, *etc*., depending on the type of running fit needed for the purpose.

(iii) In case (ii) above, the requisite pre–assembly tolerance for the bore of the bush can also be calculated as illustrated in example 7.8a. The assumptions are as per the experimental results.

Example 7.8a *Given, housing bore* $D_H = 135^{H7}$, *outer diameter of bush* $d_B = 135_{r6}$, *shaft diameter* $d_S = 120_{g6}$. *To find the pre-assembly tolerance of the inner diameter of the bush* D_B, *having nominal dimension of 120, as that the final dimension is around* 120^{H7}.

Solution :

$$D_H = 135^{H7} = 135^{+0.040}_{\;\;0}, \quad d_B = 135_{r6} = 135^{+0.088}_{+0.063}$$

Max. interference, $I_{max} = 0.088 - 0 = 0.088$, Min. interference, $I_{min} = 0.063 - 0.04 = 0.023$

From experimental results, we assume a contraction of the inside diameter of the bush to be around 75% of the interference after press-fitting.

$\therefore$ Max. contraction $C_{max} = 0.75 \times 0.088 = 0.066$, Min. contraction $C_{min} = 0.75 \times 0.023 = 0.018$

Final dimension required for the bore of the bush = 120^{H7}.

$$= 120^{+0.035}_{\;\;0}$$

Upper limit of tolerance of D_B = Lower tolerance limit of $120^{H7} + C_{max} = = 0 + 0.066 = 0.066$.
Lower limit of tolerance of D_B = Upper tolerance limit of $120^{H7} + C_{min} = 0.035 + 0.018 = 0.053$

So, the dimensions of the bore of the bush before press-fitting to effect a final tolerance of around $H7$ after assembly are:

$$D_B = 120^{+0.066}_{+0.053}$$

This tallies with the experimental results.

Shrink fit: Shrink fits are used in applications where heavy force fits are not possible from the practical point of view or where there are other kinds of limitations and constraints. These fits cause excessive stress in cast iron bores. Shrink fits are mainly used for steel holes where the material can be stressed to a high degree within the elastic limit. Tyres on wheels of locomotives and rolling stocks are shrink-fitted. In case of a shrink-fitted assembly, the female or the external part may be expanded by the application of heat through oil-bath, *etc.* or the male part may be cooled by liquid nitrogen, etc., before assembly. In some cases, the external part is heated and the internal part is cooled as well. In this way, the normal interference resulting from the tolerances of the mating parts is eliminated and dimensional change is obtained before assembly to effect the shrink-fit.

A shrink-fit assures high joint strength and is useful in heavy construction where high holding power is imperative. The usual tolerance combinations for shrink fits are:

H 8/x 8, H 8/u 8, H 7/u 6, etc. In many cases, the torque is transmitted through shafts on which wheels, gears, etc., are shrink-fitted. Often, the drawing office personnel are asked to examine whether a certain shrink-fitted combination has enough resistance against sliding or turning between mating parts or whether it is suitable for transmission of a specified amount of torque. One such case is illustrated in example 7.8b given below.

Example 7.8b *A pinion is shrink-fitted on a shaft as shown in Fig. 7.28. The materials of pinion and shaft are 40 Cr 4 and 45 C 8 respectively. Using the data given in Figs. 7.28 and 7.29, and taking a service factor of 1.25, determine whether the system is capable of transmitting a nominal torque of 3000 N m.*

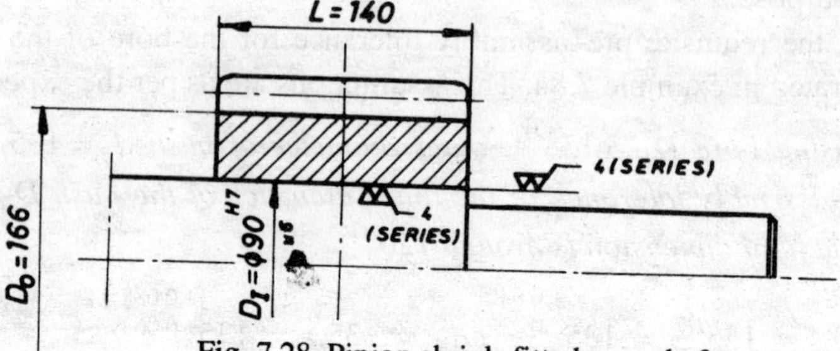

Fig. 7.28 Pinion shrink-fitted on a shaft

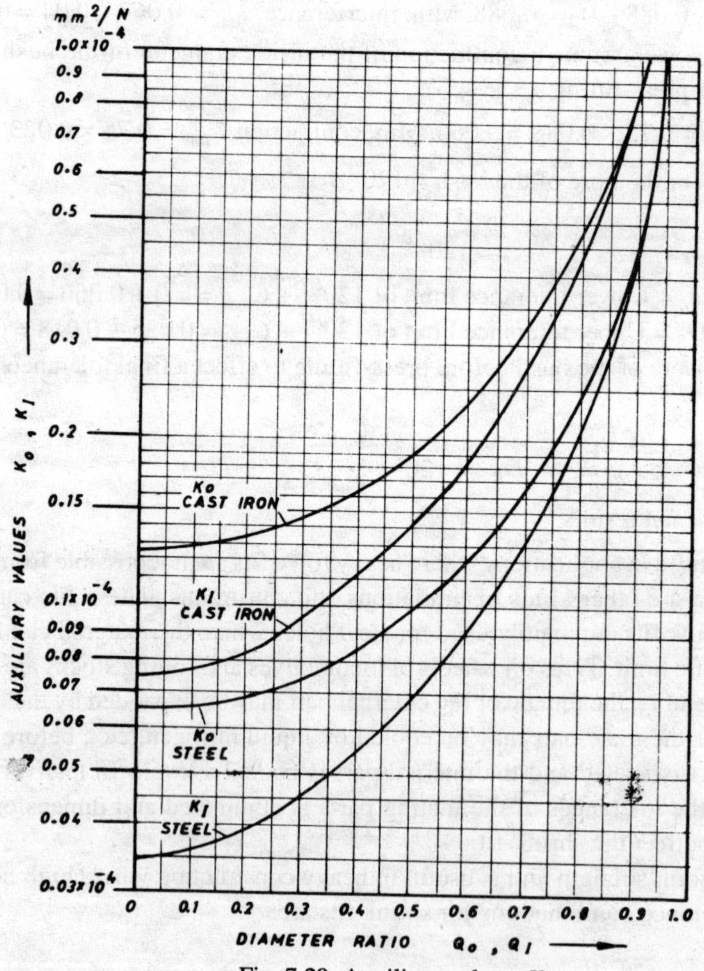

Fig. 7.29 Auxiliary values, *K*

Based on *Neue Festigkeitsberechnung fuer den Maschinenbau,* Haenchen u. Decker, 3rd Edition, 1967, Fig. No. 210, P. 167, Carl Hanser, Verlag, Munich, Germany.

Solution From Fig. 7.28 the shrink-fit chosen is 90 *H7*/*u6*. The relevant values of the tolerances are:

$$90^{H7} = 90^{+0.035}_{0}, \quad 90_{u6} = 90^{+0.146}_{+0.124}$$

The maximum and the minimum interferences are:

$$I_{max} = (146 - 0) = 146\ \mu m, \text{ and } I_{min} = (124 - 35) - 89\ \mu m$$

From Section 7.7, the surface roughness value is 10μm for series 4 and 2 triangle finish. The contact pressure at the mating surfaces is given by:

$$p = \frac{Z}{(K_O + K_I)\, D_I\, 10^3},\ \text{N/ mm}^2 \qquad \text{(or Megapascal, MPa)}$$

The factors K_O and K_I are auxiliary values to be taken from Fig. 7.29. The subscripts *O* and *I* stand for the outer and inner part or diameter respectively. The diameter ratio is given by:

$$Q_O = \frac{D_I}{D_O}, \text{ and } Q_I = \frac{D_H}{D_I},$$

where, D_O = Outside diameter of the outer part, and D_I = Diameter at the joint = Inside diameter of the outer part = Outside diameter of the inner part.

D_H = Inside diameter of the inner part in case of a hollow cylinder.

In this case, D_I = 90 mm and since it is a solid shaft, $D_H = 0$, therefore $Q_I = 0$. Dimension of adhesion is given by,

$$Z(\mu m) = I - 2\,(G_O + G_I),$$

where *I* stands for interference and G_O and G_I are magnitude of smoothness in case of the outer and inner parts respectively. These are given by:

$$G_O = 0.6\, R_{tO} \quad \text{and} \quad G_I = 0.6\, R_{tI}$$

Here, R_{tO} and R_{tI} are roughness values of outer and inner surfaces respectively. Since in this example, R_{tO} and R_{tI} are equal to 10 μm, $G_O = G_I = 0.6 \times 10 = 6$ μm.

Corresponding to the maximum and minimum values of interference, there will be two values of *Z*,

$$Z_{max} = 146 - 2(6 + 6) = 122\ \mu\ m, \quad \text{and} \quad Z_{min} = 89 - 2(6 + 6) = 65\ \mu\ m.$$

Corresponding to $Q_O = 90/166 = 0.54$ and $Q_I = 0$, the *K* values are found from Fig. 7.29. These are:

$$K_O = 0.1 \times 10^{-4}\ \text{mm}^2/\text{N}, \quad \text{and} \quad K_I = 0.03 \times 10^{-4}\ \text{mm}^2/\text{N}.$$

The maximum and the minimum contact pressures are found to be:

$$p_{max} = \frac{122}{(0.1 + 0.03) \times 10^{-4} \times 90 \times 10^3} = 104\ \text{N/mm}^2$$

$$p_{\min} = \frac{65}{(0.1 + 0.03) \times 10^{-4} \times 90 \times 10^3} = 56 \text{ N/mm}^2$$

The maximum torque to be transmitted, T_{max} = Nominal torque × Service factor = 3000 × 1.25 = 3750 Nm.

The area of contact, $A = D_I \times \pi \times L = 90 \times 3.14 \times 140 = 39564 \text{ mm}^2$.

The sliding torque, i.e., the torque required to overcome the frictional resistance at the surface of contact is given by:

$$T_s = \frac{1}{2} \times D_I \times A \times p \times F$$

where F = The coefficient of adhesion given in Table 7.15.

Taking $F = 0.12$ and the minimum value of p, i.e., 56 N/mm², we have

$$T_s = \frac{1}{2} \times 90 \times 39564 \times 56 \times 0.12 \times 10^{-3} \text{ N m.} = 11964 \text{ N m.}$$

Since this value is much larger than the value of T_{max}, the shrink-fitted system is capable of transmitting the required torque.

Table 7.15 Coefficient of adhesion, *F*

Material	*Coefficient, F*	*Shrink-fitted*
Steel/Steel	0.10 to 0.15	in oil
Steel/Steel (hardened)	0.12	dry
Steel/Cast iron	0.07 to 0.09	dry
Steel (hardened)/Steel (cast)	0.10	in oil
Steel/Brass or bronze	0.07 to 0.14	dry

CHAPTER 8

COMMON STANDARDS AND MACHINE ELEMENTS

8.1 SCREW THREADS AND THREADED FASTENERS

Threaded fasteners are extensively used in all types of engineering applications. There are many standardised forms of screw threads used in bolts, nuts, shaft ends and other parts, *viz.*, the metric thread, Whitworth thread, and the different forms of American thread profiles. Mainly two varieties of pitch are used in common threads: coarse and fine. In metric system, (which has now largely replaced all other systems), the nominal size of the thread is preceded by the prefix "*M*", indicating metric. In case of coarse thread profile, the pitch is not mentioned, while in case of fine thread, the relevant pitch is indicated in the designation, *e.g.*,

Coarse thread: M 10

Fine thread: M 10 × 1.25

Terminology: The definitions relevant to screw threads are as follows:

Screw thread: A ridge of uniform section in the form of a helix wrapped on the external or internal surface of a cylinder. The section most commonly used is in the form of a " V ". Other sections are: square, acme, buttress, knuckle, *etc.*

External thread: A thread on the outside surface of a member.

Internal thread: A thread on the internal surface of a member.

Major diameter: The largest diameter of a screw thread. It is the outside diameter of the thread profile in case of an external member, and the root diameter of an internal member.

Minor diameter: It is the smallest diameter of a screw thread, namely, the root diameter of an external thread and the inside diameter of an internal thread.

Pitch diameter: The diameter of an imaginary cylinder passing through the threads in which the widths of threads equal to widths of gaps where this cylinder cuts the thread profile.

Pitch: It is the distance from a point on a screw thread to a corresponding point on the next threa. measured parallel to the axis of the threaded member.

Lead: It is the distance a screw thread advances axially in one complete turn (360°). In case of a single thread or "start", the pitch equals the lead. In multi-start screw, lead = pitch × number of starts.

Thread form: It is the form or shape of the thread. Both metric and Whitworth threads have " V " forms with thread angles 60° and 55° respectively. For general applications in inch system, the British Standard Whitworth (BSW) form is used. The British Standard Pipe (BSP) or simply "pipe thread" is used in specific cases. Details of forms are given later.

In this section and in Section 8.2, the thread profiles of common forms, some fasteners in general use, standards pertaining to threads and fasteners, accessories to fasteners and other relevant topics pertinent to drawing making will be discussed.

Presentation of threaded parts on technical drawings: Code for representation of threaded parts on technical drawings has been laid down in IS:10715 - 1983 which is identical with ISO 6410 - 1981. The Standard specifies the conventional method of simplified representation of threaded parts and is valid for all types of screw threads. Type, dimensions and other data pertaining to the screw thread are to be indicated as per the relevant Standards. While drawing threaded parts, the following rules should be observed:

1. In case of external or male threads, the lines representing the major diameter (outside diameter) should be defined by continuous thick lines and the minor (rest) diameter by continuous thin lines. Previously, the minor diameter was represented by dotted lines.
2. In case of internal or female threads, the representation should be the reverse of that of the external threads.
3. For threaded parts in sections, hatching should be extended to the lines representing the crests of the threads, that is, the outside diameter of the external threads and inside (minor) diameter of the internal threads.
4. On end views, the thread roots should be represented by a portion of a circle (about three-quarters of the circumference), drawn with a thin line.
5. Thread run-outs, which were shown in previous drawing practice, are no longer to be shown, unless they serve some specific or useful purpose and are a functional necessity.
6. In case of threaded parts in assembled condition, the threaded portion should be shown as in the case of external threads.

All the above conventions have been illustrated in Fig. 8.1.

Detailed drawing of a hexagonal bolt and nut

In Fig. 8.2, the proportional drawing parameters of a hexagonal belt and nut have been illustrated. These are in terms of the corner dimension "*e*" and the across flat dimension (A/F) "*s*" of the bolt head and nut. A/F also determines the relevant spanner size. Later in this section, the actual dimensions, such as, the nominal diameter "*d*", the heights of bolt heads and nuts, and values of "*e*" and "*s*" are given.

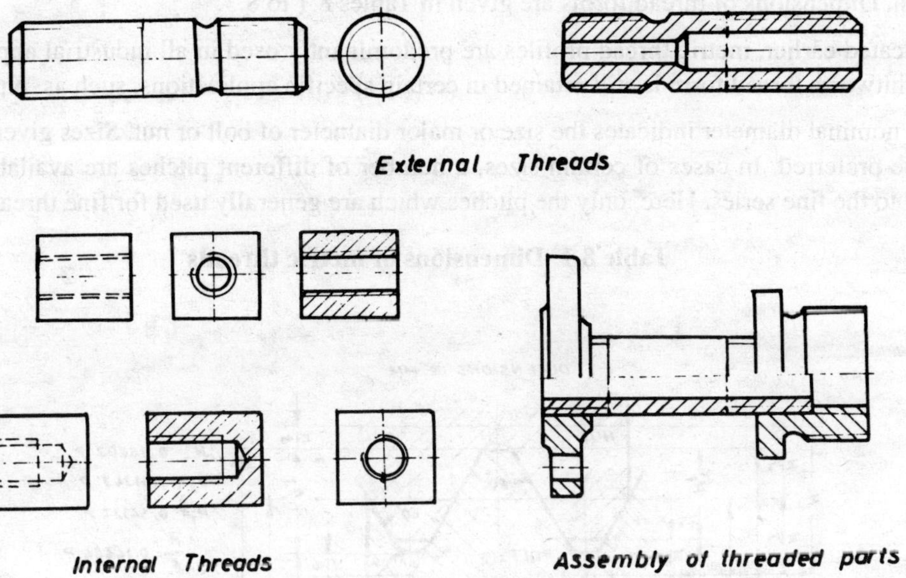

Fig. 8.1 Drawing representations of threaded parts

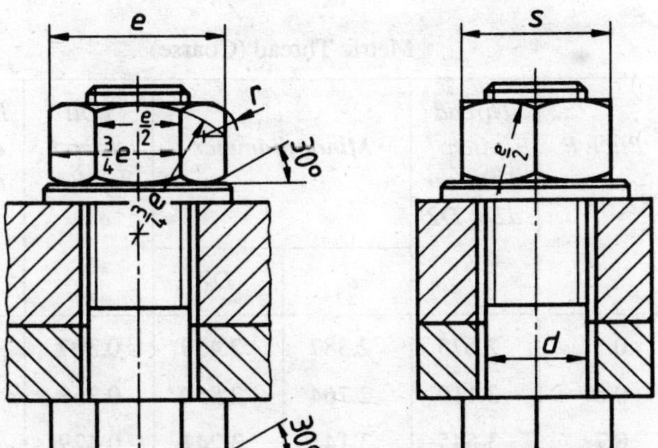

Fig. 8.2 Detailed drawing of a hexagonal bolt and nut
Based on *Fachzeichnen Metalltechnik, Grundstufe*, 1984, Fig. No. 72/2, P. 72,
Verlag Europa - Lehrmittel, Wuppertal, Germany.

Screw thread profiles : Thread profiles which are in vogue are mainly of two types: Metric and Whitworth. Dimensions of thread forms are given in Tables 8.1 to 8.3.

As indicated earlier, metric thread profiles are predominantly used in all industrial applications now, but the Whitworth thread form is still retained in certain specific applications, such as, pipe threads.

Thread nominal diameter indicates the size or major diameter of bolt or nut. Sizes given under Group 1 are to be preferred. In cases of certain sizes, a number of different pitches are available for threads belonging to the fine series. Here, only the pitches which are generally used for fine threads are given.

Table 8.1 Dimensions of metric threads

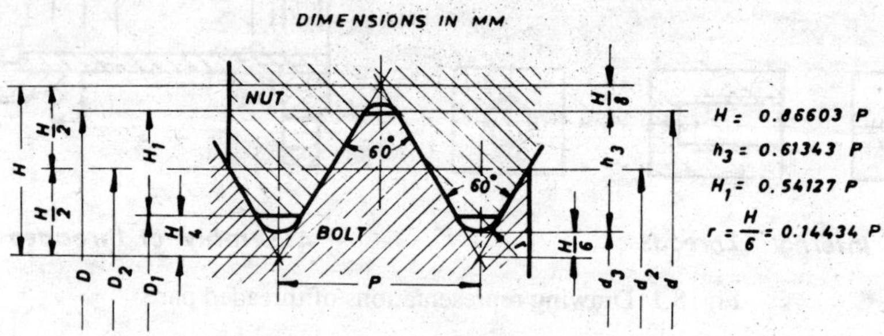

Metric Thread (Coarse)

Thread nominal diameter d = D		*Pitch P*	*Thread pitch diameter d2 = D2*	*Minor diameter*		*Bolt thread depth h_3*	*Depth of engagement H_1*	*Radius r*	*Stress Area A_S mm^2*
Group-1	*Group-2*			d_1	D_3				^
3		0.5	2.675	2.387	2.459	0.307	0.271	0.072	5.03
	3.5	0.6	3.110	2.764	2.850	0.368	0.325	0.087	6.78
4		0.7	3.545	3.141	3.242	0.429	0.379	0.101	8.78
	4.5	0.75	4.013	3.580	3.688	0.460	0.406	0.108	11.3
5		0.8	4.480	4.019	4.134	0.491	0.433	0.115	14.2
6		1	5.350	4.773	4.917	0.613	0.541	0.144	20.1
	7	1	6.350	5.773	5.917	0.613	0.541	0.144	28.9
8		1.25	7.188	6.466	6.647	0.767	0.677	0.180	36.6
10		1.5	9.026	8.160	8.376	0.920	0.812	0.217	58.0
12		1.75	10.863	9.853	10.106	1.074	0.947	0.253	84.3
	14	2	12.701	11.546	11.835	1.227	1.083	0.289	115

16		2	14.701	13.546	13.835	1.227	1.083	0.289	157
	18	2.5	16.376	14.933	15.294	1.534	1.353	0.361	192
20		2.5	18.376	16.933	17,294	1.534	1.353	0.361	245
	22	2.5	20.376	18.933	19.294	1.534	1.353	0.361	303
24		3	22.051	20.319	20.752	1.840	1.624	0.433	353
	27	3	25.051	23.319	23.752	1.840	1.624	0.433	459
30		3.5	27.727	25.706	26.211	2.147	1.894	0.505	561
	33	3.5	30.727	28.706	29.211	2.147	1.894	0.505	694
36		4	33.402	31.093	31.670	2.454	2.165	0.577	817
	39	4	36.402	34.093	34.670	2.454	2.165	0.577	976

Metric Thread (Fine)

8		1	7.350	6.773	6.917	0.613	0.547	0.144	39.2
10		1.25	9.188	8.465	8.647	0.767	0.677	0.180	61.2
12		1.25	11.188	10.466	10.647	0.767	0.677	0.180	92.1
	12	1.5	11.026	10.160	10.376	0.920	0.812	0.217	88.1
	14	1.5	13.026	12.160	12.376	0.920	0.812	0.217	125
16		1.5	15.026	14.160	14.376	0.920	0.812	0.217	167
	18	1.5	17.026	16.160	16.376	0.920	0.812	0.217	216
20		1.5	19.026	18.160	18.376	0.920	0.812	0.217	272
	22	1.5	21.026	20.160	20.376	0.920	0.812	0.217	333
24		2	22.701	21.546	21.835	1.227	1.083	0.289	384
	27	2	25.701	24.546	24.835	1.227	1.083	0.289	496
30		2	28.701	27.546	27.835	1.227	1.083	0.289	621
	33	2	31.701	30.546	30.835	1.227	1.083	0.289	761
36		3	34.051	32.319	32.752	1.840	1.624	0.433	865
	39	3	37.051	35.319	35.752	1.840	1.624	0.433	1030

The nominal stress is calculated from the formula:

$$A_s = \frac{\pi}{4}\left(\frac{d_2 + d_3}{2}\right)^2$$

The tensile strength of a bolt would be the same as that of an unthreaded rod of cross sectional area equal to A_s.

Table 8.2 Dimensions of Whitworth pipe thread

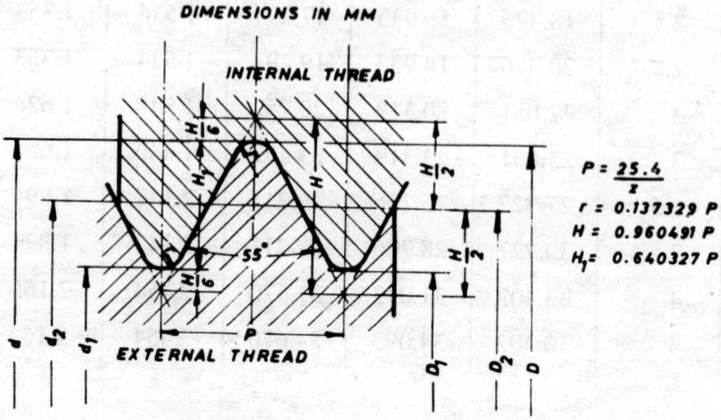

Thread size inch	Thread dimensions						
	Major diameter $d = D$	Effective diameter $d_2 = D_2$	Minor diameter $d_1 = D_1$	Pitch p	Threads per in. z	Depth of thread H_1	Radius $r \approx$
R 1/8	9.728	9.147	8.566	0.907	28	0.581	0.125
R 1/4	13.157	12.301	11.445	1.337	19	0.856	0.184
R 3/8	16.662	15.806	14.950	1.337	19	0.856	0.184
R 1/2	20.955	19.793	18.631	1.814	14	1.162	0.249
(R 5/8)	22.911	21.749	20.587	1.814	14	1.162	0.249
R 3/4	26.441	25.279	24.117	1.814	14	1.162	0.249
(R 7/8)	30.201	29.039	27.877	1.814	14	1.162	0.249
R1	33.249	31.770	30.291	2.309	11	1.479	0.317
(R1 1/8)	37.897	36.418	34.939	2.309	11	1.479	0.317
R1 1/4	41.910	40.431	38.952	2.309	11	1.479	0.317
(R1 3/8)	44.323	42.844	41.365	2.309	11	1.479	0.317
R1 1/2	47.803	46.324	44.845	2.309	11	1.479	0.317
(R1 3/4)	53.746	52.267	50.788	2.309	11	1.479	0.317
R2	59.614	58.135	56.656	2.309	11	1.479	0.317
(R2 1/4)	65.710	64.231	62.752	2.309	11	1.479	0.317
R2 1/2	75.184	73.705	72.226	2.309	11	1.479	0.317
(R2 3/4)	81.534	80.055	78.576	2.309	11	1.479	0.317

R3	87.884	86.405	84.926	2.309	11	1.479	0.317
(R3 1/4)	93.980	92.501	91.022	2.309	11	1.479	0.317
R3 1/2	100.330	98.851	97.372	2.309	11	1.479	0.317
(R3 3/4)	106.680	105.201	103.722	2.309	11	1.479	0.317
R4	113.030	111.551	110.072	2.309	11	1.479	0.317
(R4 1/2)	125.730	124.251	122.772	2.309	11	1.479	0.317
R5	138.430	136.951	135.472	2.309	11	1.479	0.317
(R5 1/2)	151.130	149.651	148.172	2.309	11	1.479	0.317
R6	163.830	162.351	160.872	2.309	11	1.479	0.317

Pipe threads are used on ends of screwed pipes. Here, only parallel internal and external threads are shown, but screwed pipe ends can be tapered also. The size is designated by the prefix "*R*" to indicate pipe thread. The prefix "*P*" is also used to indicate pipe thread. The size dimension is given in inch, *e.g.*, 1/4, 2, 3 1/2, etc., and these dimensions correspond to the nominal bore (NB) of the pipes.

Table 8.3. Metric trapezoidal screw threads

All dimensions in mm

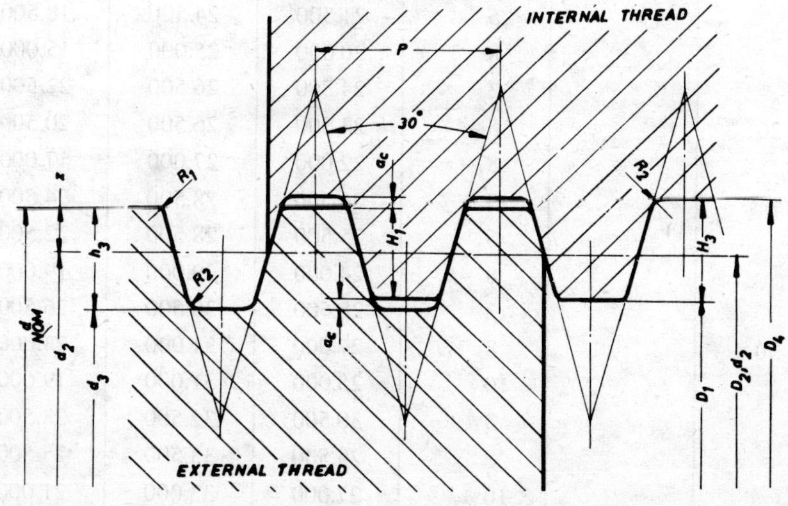

Nominal Diameter d			*Pitch P*	*Pitch diameter* $d_2 = D_2$	*Major Diameter* D_1	*Minor Diameter*	
Column 1	*Column 2*	*Column 3*				d_3	D_1
8			1.5	7.250	8.300	6.200	6.500
	9		1.5	8.250	9.300	7.200	7.500
			2	8.000	9.500	6.500	7.000
			1.5	9.250	10.300	8.200	8.500

10			2	9.000	10.500	7.500	8.000
	11		2	10.000	11.500	8.500	9.000
			3	9.500	11.500	7.500	8.000
12			2	11.000	12.500	9.500	10.000
			3	10.500	12.500	8.500	9.000
	14		2	13.000	14.500	11.500	12.000
			3	12.500	14.500	10.500	11.000
16			2	15.000	16.500	13.500	14.000
			4	14.000	16.500	11.500	12.000
	18		2	17.000	18.500	15.500	16.000
			4	16.000	18.500	13.500	14.000
20			2	19.000	20.500	17.500	18.000
			4	18.000	20.500	15.500	16.000
	22		3	20.500	22.500	18.500	19.000
			5	19.500	22.500	16.500	17.000
			8	18.000	23.000	13.000	14.000
24			3	22.500	24.500	20.500	21.000
			5	21.500	24.500	18.500	19.000
			8	20.000	25.000	15.000	16.000
	26		3	24.500	26.500	22.500	23.000
			5	23.500	26.500	20.500	21.000
			8	22.000	27.000	17.000	18.000
28			3	26.500	28.500	24.500	25.000
			5	25.500	28.500	22.500	23.000
			8	24.000	29.000	19.000	20.000
	30		3	28.500	30.500	26.500	27.000
			8	27.000	31.000	23.000	24.000
			10	25.000	31.000	19.000	20.000
32			3	30.500	32.500	28.500	29.000
			8	29.500	33.500	25.500	26.000
			10	27.000	33.000	21.000	22.000
	34		3	32.500	34.500	30.500	31.000
			6	31.000	35.000	27.000	28.000
			10	29.000	35.000	23.000	24.000
36			3	34.500	36.500	32.500	33.000
			6	33.000	37.000	29.000	30.000
			10	31.000	37.000	25.000	26.000
			3	36.500	38.500	34.500	35.000

	38		7	34.500	39.000	30.000	31.000
			10	33.000	39.000	27.000	28.000
			3	38.500	40.500	36.500	37.000
40			7	36.500	41.000	32.000	33.000
			10	36.000	41.000	29.000	30.000
			3	40.500	42.500	38.500	39.000
	42		7	38.500	43.000	34.000	35.000
			10	37.000	43.000	31.000	32.000
			3	42.500	44.500	40.500	41.000
44			7	40.500	45.000	36.000	37.000
			12	38.000	45.000	31.000	32.000
			3	44.500	46.500	42.500	43.000
	46		8	42.000	47.000	37.000	38.000
			12	40.000	47.000	33.000	34.000
			3	46.500	48.500	44.500	45.000
48			8	44.000	49.000	39.000	40.000
			12	42.000	49.000	35.000	36.000
			3	48.500	50.500	46.500	47.000
	50		8	46.000	51.000	41.000	42.000
			12	44.000	51.000	37.000	38.000
			3	50.500	52.500	48.500	49.000
52			8	48.000	53.000	43.000	44.000
			12	46.000	53.000	39.000	40.000
			3	53.500	55.500	51.500	52.000
	55		9	50.500	56.000	45.000	46.000
			14	48.000	57.000	39.000	41.000
			3	58.500	60.500	56.500	57.000
60			9	55.500	61.000	50.000	51.000
			14	53.000	62.000	44.000	46.000
			4	63.000	65.500	60.500	61.000
	65		10	60.000	66.000	54.000	55.000
			16	57.000	67.000	47.000	49.000
			4	68.000	70.500	65.500	66.000
70			10	65.000	71.000	59.000	60.000
			16	62.000	72.000	52.000	54.000

			4	73.000	75.500	70.500	71.000
	75		10	70.000	76.000	64.000	66.000
			16	67.000	77.000	57.000	59.000
			4	78.000	80.500	75.500	76.000
80			10	75.000	81.000	69.000	70.000
			16	72.000	82.000	62.000	54.000
			4	83.000	85.500	80.500	81.000
	85		12	79.000	86.000	72.000	73.000
			18	76.000	87.000	65.000	67.000
			4	88.000	90.500	85.500	86.000
90			12	84.000	91.000	77.000	78.000
			18	81.000	92.000	70.000	72.000
			4	93.000	95.500	90.500	91.000
	95		12	89.000	96.000	82.000	83.000
			18	86.000	97.000	75.000	77.000
			4	98.000	100.500	85.000	90.000
100			12	94.000	101.000	87.000	88.000
			20	90.000	102.000	78.000	80.000
			4	103.000	105.500	100.500	101.000
		105	12	99.000	106.000	92.000	93.000
			20	95.000	107.000	83.000	85.000
			4	108.000	110.500	105.500	106.000
	110		12	104.000	111.000	97.000	98.000
			20	100.000	112.000	88.00	90.000
			6	112.000	116.000	108.000	109.000
		115	14	108.000	117.000	99.000	101.000
			22	104.000	117.000	91.000	93.000
			6	117.000	121.000	113.000	114.000
120			14	113.000	122.000	104.000	106.000
			22	109.000	122.000	96.000	98.000
			6	122.000	126.000	118.000	119.000
		125	14	118.000	127.000	109.000	111.000
			22	114.000	127.000	101.000	103.000

Table 8.4 Hexagon head bolts

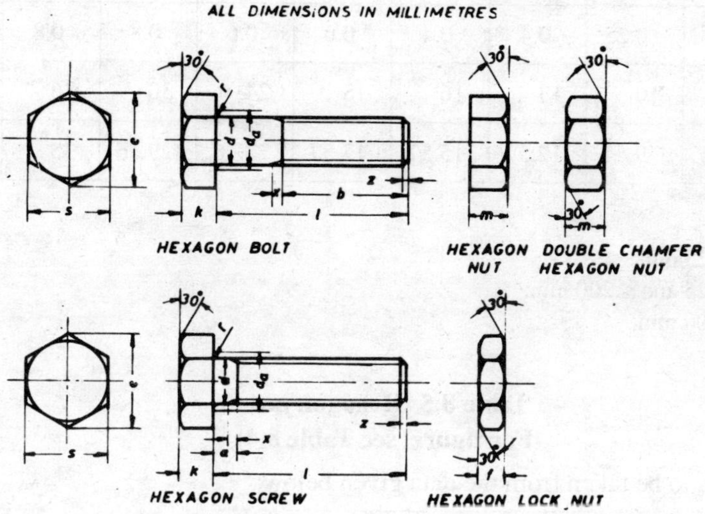

Thread Size d		M5	M6	M8	M10	M12	M16	M20	M24	M30	M36
P*		0.8	1	1.25	1.5	1.75	2	2.5	3	3.5	4
b	†	16	18	22	26	30	38	46	54	66	78
	‡	—	—	28	32	36	44	52	60	72	84
	§	—	—	—	—	—	57	65	73	85	97
d_a	Max	6	7.2	10.2	12.2	14.7	18.7	24.4	28.4	35.4	42.4
e	Min	8.63	10.89	14.20	17.59	19.85	26.17	32.95	39.55	50.85	60.79
k	Nom	3.5	4	5.3	6.4	7.5	10	12.5	15	18.7	22.5
	Min	3.12	3.62	4.92	5.95	7.05	9.25	11.6	14.1	17.65	21.45

	Max	3.88	4.38	5.68	6.85	7.95	10.75	13.4	15.9	19.75	23.55
r	*Min*	0.2	0.25	0.4	0.4	0.6	0.6	0.8	0.8	1	1
s	*Max*	8	10	13	16	18	24	30	36	46	55
	Min	7.64	9.64	12.57	15.57	17.57	23.16	29.16	35	45	53.8

*p = pitch of the thread.

† For nominal lengths ≤ 125 mm.

‡ For nominal lengths > 125 and ≤ 200 mm.

§ For nominal lengths > 200 mm.

Table 8.5 Hexagon nut
(For figure, see Table 8.4)

The height of lock nut is to be taken from the data given below:

Size	*M 6*	*M 8*	*M 10*	*M 12*	*M 16*	*M 20*	*M 24*	*M 30*	*M 36*
f_{js16}	3	4	5	7	8	9	10	12	14

The above dimensions of "*f*" are nominal dimensions in mm.

Thread Size d		*M5*	*M6*	*M8*	*M10*	*M12*	*M16*	*M20*	*M24*	*M30*	*M36*
*p**		0.8	1	1.25	1.5	1.75	2	2.5	3	3.5	4
m	*Max*	5.6	6.1	7.9	9.5	12.2	15.9	18.7	22.3	26.4	31.5
	Min	4.4	4.9	6.4	8	10.4	14.1	16.6	20.2	24.3	29.4
s	*Max*	8	10	13	16	18	24	30	36	46	55
	Min	7.64	9.64	12.57	15.57	17.57	23.16	29.16	35	45	53.8
e	*Min*	8.63	10.89	14.20	17.59	19.85	26.17	32.95	39.55	50.85	60.79

**p* = Pitch of the thread.

Table 8.6 Hexagon head bolts and nuts

ALL DIMENSIONS IN MILLIMETRES

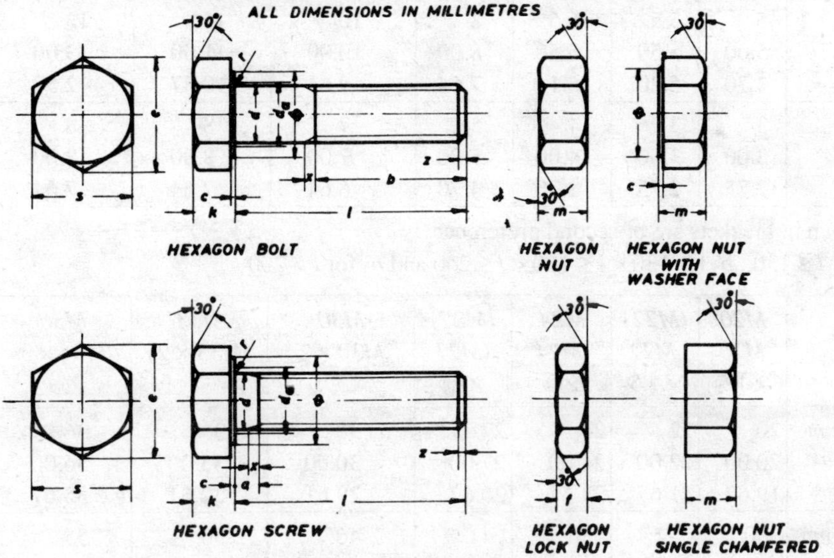

x ACCORDING TO I S:1369 DIMENSIONS OF SCREW THREAD RUN OUTS AND UNDERCUTS
z ACCORDING TO I S:1368 DIMENSIONS OF ENDS OF BOLTS AND SCREWS

Size		M6	(M7)	M8 M8 × 1	M10 M10 × 1.25	M12 M12 × 1.25	M14 (M14 × 1.5)	M16 M16 × 1.5	M18 (M18 × 1.5)
*d h*13	Nom Max. Min	6 6.00 5.82	7 7.00 6.78	8 8.00 7.78	10 10.00 9.78	12 12.00 11.73	14 14.00 13.73	16 16.00 15.73	18 18.00 17.73
s ≤ 32–*h* 13 > 32–*h* 14	Nom Max. Min	10 10.00 9.78	11 11.00 10.73	13 13.00 12.73	17 17.00 16.73	19 19.00 18.67	22 22.00 21.67	24 24.00 23.67	27 27.00 26.67
e	Min	11.05	12.12	14.38	18.90	21.10	24.49	26.30	30.14
*k js*14	Nom Max Min	4 4.15 3.85	5 5.15 4.85	5.5 5.65 5.35	7 7.18 6.82	8.0 8.18 7.82	9.0 9.18 8.82	10 10.18 9.82	12 12.22 11.78
r	Min	0.25	0.25	0.4	0.4	0.6	0.6	0.6	0.6
d_a	Max	6.8	7.8	9.2	11.2	14.2	16.2	18.2	20.2
ϕ	Min	9.0	9.9	11.7	15.3	17.1	19.8	21.6	24.3
c	Approx	0.3	0.3	0.4	0.4	0.4	0.4	0.4	0.4
a	Max	2.5	2.5	3.0	4.0	5.0	5.0	5.0	6.5
b { b_1 b_2 b_3	Nom Min Max	18 24 –	20 26 –	22 28 –	26 32 –	30 36 49	34 40 53	38 44 57	42 48 61

*m h*14	*Nom* *Max.* *Min*	5 5.00 4.70	5.5 5.50 5.20	6.5 6.50 6.14	8 8.00 7.64	10 10.00 9.64	11 11.00 10.57	13 13.00 12.57	15 15.00 14.57
*f h*14	*Nom* *Max.* *Min*	3 3.00 2.75	3 3.00 2.75	4 4.00 3.70	5 5.00 4.70	7 7.00 6.64	8 8.00 7.64	8 8.00 7.64	9 9.00 8.64

Note 1 : Sizes shown in brackets are of second preference.
Note 2 : Use b_1 for $l \le 130$, b_2 for $130 < l \le 130 < l \le 200$ and b_3 for $l > 200$

Size		*M20* *M20* × 1.5	*(M22)* *M22* × 1.5	*M24* *M24* × 2	*(M27)* *(M27* × 2)	*M30* *M30* × 2	*M33* (*M33* × 2)	*M36* *M36* × 3	*M39* (*M39* × 3)
d h 13	*Nom* *Max.* *Min*	20 20.00 19.67	22 22.00 21.67	24 24.00 23.67	27 27.00 26.67	30 30.00 29.61	33 33.00 32.61	36 36.00 35.67	39 39.00 38.61
s ≤ 32–*h* 13 > 32–*h* 14	*Nom* *Max.* *Min*	30 30.00 29.67	32 32.00 31.61	36 36.00 35.38	41 41.00 40.38	46 46.00 45.38	50 50.00 49.38	55 55.00 54.26	60 60.00 59.26
e	*Min*	33.53	35.72	39.98	45.63	51.28	55.80	61.31	66.96
*k js*14	*Nom* *Max.* *Min*	13 13.22 12.78	14 14.22 13.78	15 15.22 14.78	17 17.22 16.78	19 19.26 18.74	21 21.26 20.74	23 23.26 22.74	25 25.26 24.74
r	*Min*	0.8	0.8	0.8	1.0	1.0	1.0	1.0	1.0
d_a	*Max*	22.4	24.4	26.4	30.4	33.4	36.4	39.4	42.5
φ	*Min*	27.0	28.8	32.4	36.9	41.4	45.0	49.5	54.00
c	*Approx*	0.4	0.4	0.5	0.5	0.5	0.5	0.5	0.6
a	*Max*	6.5	6.5	7.5	11.0	13.0	13.0	14.0	14.0
b $\begin{cases} b_1 \\ b_2 \\ b_3 \end{cases}$	*Nom* *Max.* *Min*	46 52 65	50 56 69	54 60 73	60 66 79	66 72 85	72 78 91	78 84 97	84 90 103
m h 14	*Nom* *Max.* *Min*	16 16.00 15.57	18 18.00 17.57	19 19.00 18.48	22 22.00 21.48	24 24.00 23.48	26 26.00 25.48	29 29.00 28.48	31 31.00 30.38
f h 14	*Nom* *Max.* *Min*	9 9.00 8.64	10 10.00 9.64	10 10.00 9.64	12 12.00 11.57	12 12.00 11.57	14 14.00 13.57	14 14.00 13.57	16 16.00 15.57

Note 1 : Sizes shown in brackets are of second preference.
Note 2 : Use b_1 for $l \le 130$, b_2 for $130 < l \le 200$ and b_3 for $l > 200$

Table 8.7 Hexagon head bolts and nuts. Semi precision grade

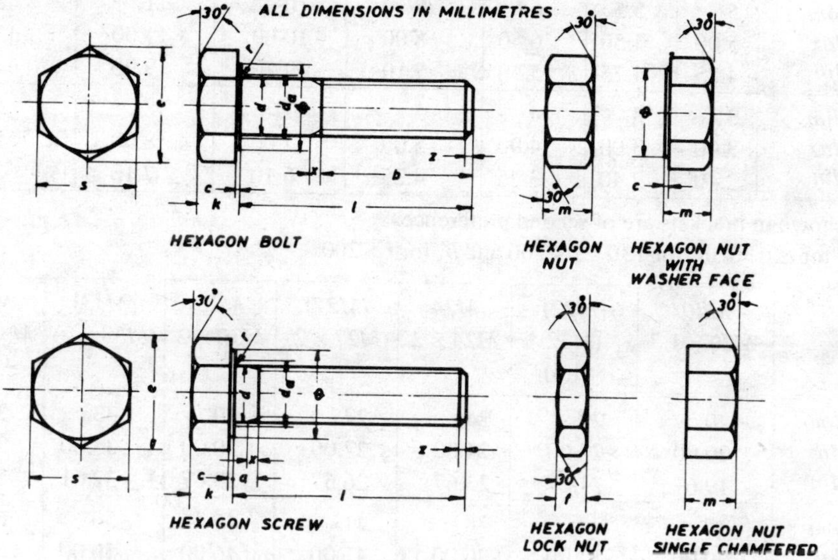

x ACCORDING TO I S:1369 DIMENSIONS OF SCREW THREAD RUN OUTS AND UNDERCUTS
z ACCORDING TO I S:1368 DIMENSIONS OF ENDS OF BOLTS AND SCREWS

Size		M6	(M7)	M8 M8 × 1	M10 M10 × 1.25	M12 M12 × 1.25	(M14) (M14 × 1.5)	M16 M16 × 1.5	(M18) (M18 × 1.5)
*d h*13	Nom Max. Min	6 6.00 5.82	7 7.00 6.78	8 8.00 7.78	10 10.00 9.78	12 12.00 11.73	14 14.00 13.73	16 16.00 15.73	18 18.00 17.73
s ≤ 19–*h* 14 > 19–*h* 15	Nom Max. Min	10 10.00 9.64	11 11.00 10.57	13 13.00 12.57	17 17.00 16.57	19 19.00 18.48	22 22.00 21.16	24 24.00 23.16	27 27.00 26.16
e	Min	10.89	11.94	14.20	18.72	20.88	23.91	26.17	29.56
*k js*15	Nom Max. Min	4 4.24 3.76	5 5.24 4.76	5.5 5.74 5.26	7 7.29 6.71	8 8.29 7.71	9 9.29 8.71	10 10.29 9.71	12 12.35 11.65
r	Min	0.25	0.25	0.4	0.4	0.6	0.6	0.6	0.6
d_a	Max	6.8	7.8	9.2	11.2	14.2	16.2	18.2	20.2
φ	Min	9.0	9.9	11.7	15.3	17.1	19.8	21.6	24.3
c	Approx	0.3	0.3	0.4	0.4	0.4	0.4	0.4.	0.4
a	Max	2.5	2.5	3.0	4.0	5.0	5.0	5.0	6.5
b $\begin{cases} b_1 \\ b_2 \\ b_3 \end{cases}$	Nom Max Min	18 24 –	20 26 –	22 28 –	26 32 –	30 36 49	34 40 53	38 44 57	42 48 61

m h 16	*Nom* *Max* *Min*	5 5.00 4.25	5.5 5.50 4.75	6.5 6.50 5.60	8 8.00 7.10	10 10.00 9.10	11 11.00 9.90	13 13.00 11.90	15 15.00 13.90
f h 16	*Mim* *Max* *Min*	3 3.00 2.40	3 3.00 2.40	4 4.00 3.25	5 5.00 4.25	7 7.00 6.10	8 8.00 7.10	8 8.00 7.10	9 9.00 8.10

Note 1 : Sizes shown in brackets are of second preference.
Note 2 : Use b_1 for $l \leq 130$, b_2 for $130 < l \leq 200$ and b_3 for $l > 200$

Size		*M20* *M20 × 1.5*	*(M22)* *(M22 × 1.5)*	*M24* *M24 × 2*	*(M27)* *(M27 × 2)*	*M30* *M30 × 2*	*(M33)* *(M33 × 2)*	*M36* *M36 × 3*	*(M39)* *(M39 × 3)*
d h 13	*Nom* *Max* *Min*	20 20.00 19.67	22 22.00 21.67	24 24.00 23.67	27 27.00 26.67	30 30.00 29.67	33 33.00 32.61	36 36.00 35.61	39 39.00 38.61
s ≤ 19–*h* 14 > 19–*h* 15	*Nom* *Max* *Min*	30 30.00 29.16	32 32.00 31.00	36 36.00 35.00	41 41.00 40.00	46 46.00 45.00	50 50.00 49.00	60 55.00 53.00	66 60.00 58.80
e	*Min*	32.95	35.03	39.55	45.20	50.85	55.37	60.79	66.44
k js 15	*Nom* *Max* *Min*	13 13.35 12.65	14 14.35 13.65	15 15.35 14.65	17 17.35 16.65	19 19.42 18.58	21 21.42 20.58	23 23.42 22.58	25 25.42 24.58
r	*Min*	0.8	0.8	0.8	1.0	1.0	1.0	1.0	1.0
d_a	*Max*	22.4	24.4	26.4	30.4	33.4	36.4	39.4	42.5
ϕ	*Min*	27.0	28.8	32.4	36.9	41.4	45.0	49.5	54.0
c	*Approx*	0.4	0.4	0.5	0.5	0.5	0.5	0.5	0.6
a	*Max*	6.5	6.5	7.5	11.0	13.0	13.0	14.0	14.0
b $\begin{cases} b_1 \\ b_2 \\ b_3 \end{cases}$	*Nom* *Max* *Min*	46 52 65	50 56 69	54 60 73	60 66 79	66 72 85	72 78 91	78 84 97	84 90 103
m h 16	*Min* *Max* *Min*	16 16.00 14.90	18 18.00 16.90	19 19.00 17.70	22 22.00 20.70	24 24.00 22.70	26 26.00 24.70	29 29.00 27.70	31 31.00 29.40
f h 16	*Nom* *Max* *Min*	9 9.00 8.10	10 10.00 9.10	10 10.00 9.10	12 12.00 10.90	12 12.00 10.90	14 14.00 12.90	14 14.00 12.90	16 16.00 14.90

Note 1 : Sizes shown in brackets are of second preference.
Note 2 : Use b_1 for $l \leq 130$, b_2 for $130 < l \leq$ and b_3 for $l > 200$

Table 8.8 Hexagon fit bolts

ALL DIMENSIONS IN MILLIMETRES

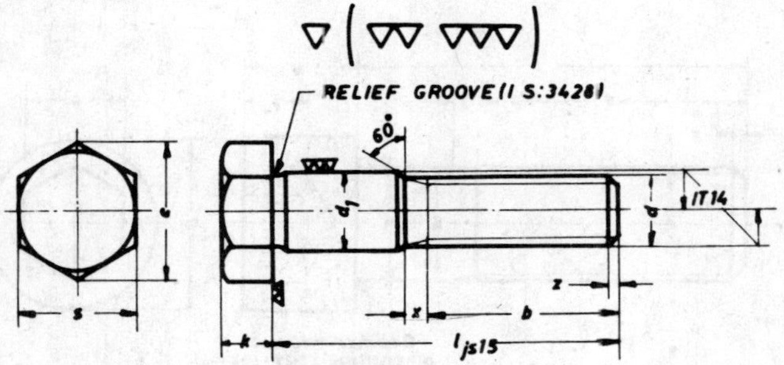

x ACCORDING TO I S:1369 DIMENSIONS OF SCREW THREAD RUNOUTS AND UNDERCUTS

z ACCORDING TO I S:1368 DIMENSIONS OF ENDS OF BOLTS AND SCREWS

Nominal Size d			M8	M10	M12	(M14)	M16	(M18)	M20	(M22)	M24	(M27)	M30	(M33)	M36	(M39)	M42	(M45)	M48	(M52)
$b + 2\underset{0}{p}$		*	14.5	17.5	20.5	22	25	27.5	28.5	32.5	—	—	—	—	—	—	—	—	—	—
		†	16.5	19.5	22.5	24	27	29.5	30.5	34.5	36.5	39.5	43	45	49	51	56	59	63	65
		‡	21.5	24.5	27.5	29	32	34.5	35.5	39.5	41.5	44.5	48	50	54	56	61	64	68	70
d_1	k6	§	9	11	13	15	17	19	21	23	25	28	32	34	38	40	44	46	50	55
e	Min		—	17.60	19.86	22.78	26.17	29.56	32.95	37.29	39.55	45.20	50.85	55.37	60.79	66.44	71.30	76.95	82.60	88.25
k js 15			5.5	7	8	9	10	11	12	13	14	15	17	19	21	23	25	26	28	30

$s > 19$	$\leq 19\ h14$ $\leq 60\ h15$ $> 60\ h16$	13	16	18	21	24	27	30	34	36	41	46	50	55	60	65	70	75	80

Note : Sizes shown in parenthesis are of second preference.

* For lengths, up to 50 mm.

† For lengths, above 50 up to 150 mm.

‡ For lengths, above 150 mm.

§ Other tolerance classes shall be specified when ordering.

Table 8.9 Hexagon socket head cap screws

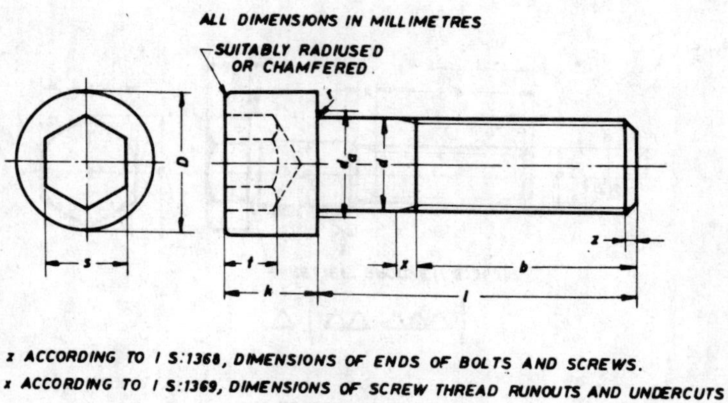

Nominal Size d		M1.6	M2	M2.5	M3	M4	M5	M6	M8	M10	*M12*	(M14)	M16	M20	M24	M30	M36
P	(Pitch)	0.35	0.4	0.45	0.5	0.7	0.8	1	1.25	1.5	1.75	2	2	2.5	3	3.5	4
b	*Ref*	15	16	17	18	20	22	24	28	32	36	40	44	52	60	72	84

D	*Max*	3	3.8	4.5	5.5	7	8.5	10	13	16	18	21	24	30	36	45	54
	Min	2.86	3.62	4.32	5.32	6.78	8.28	9.78	12.73	15.73	17.73	20.67	23.67	29.67	35.61	44.61	53.54
d_a	Max	2	2.6	3.1	3.6	4.7	5.7	6.8	9.2	11.2	14.2	16.2	18.2	22.4	26.4	33.4	39.4
k $h13/h14$	*Max*	1.6	2	2.5	3	4	5	6	8	10	12	14	16	20	24	30	36
	Min	1.46	1.86	2.36	2.86	3.82	4.82	5.70	7.64	9.64	11.57	13.57	15.57	19.48	23.48	29.48	35.38
r	*Min*	0.1	0.1	0.1	0.1	0.2	0.2	0.25	0.4	0.4	0.6	0.6	0.6	0.8	0.8	1	1
s h 13	*Nom*	1.5	1.5	2	2.5	3	4	5	6	8	10	12	14	17	19	22	27
	Min	1.52	1.52	2.02	2.52	3.02	4.02	5.02	6.02	8.025	10.025	12.032	14.032	17.05	19.065	22.065	27.065
	Max	1.56	1.56	2.06	2.58	3.071	4.084	5.084	6.14	8.175	10.127	12.146	14.159	17.216	19.275	22.275	27.275
t	*Min*	0.7	1	1.1	1.3	2	2.5	3	4	5	6	7	8	10	12	15.5	19

Table 8.10 Countersunk head screws

All dimensions in mm

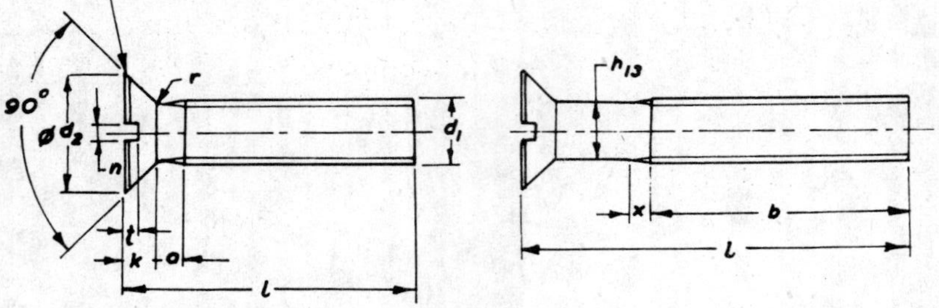

SHANK DIAMETER ≈ PITCH DIAMETER

a and x according to IS: 1369 - 1975

Nominal Size d_1		M1	M 1.2	(M 1.4)	M 1.6	(M 1.8)	M2	(M 2.2)	M 2.5	M3	(M 3.5)	M.4	(M 4.5)	M5	M6	M8	M10	M12	(M14)	M16	(M18)	M20
d_2 h13 $d < 3$, h13 $d \geq 3$	Nom	1.9	2.3	2.6	3	3.4	3.8	4.2	4.7	5.6	6.5	7.5	8.3	9.2	11	14.5	18	22	25	29	33	36
	Max	1.90	2.30	2.60	3.00	3.40	3.80	4.20	4.70	5.60	6.50	7.50	8.80	9.20	11.00	14.50	18.00	22.00	25.00	29.00	33.00	36.00
	Min	1.76	2.16	2.46	2.75	3.10	3.5	3.90	4.40	5.30	6.11	7.19	7.94	8.34	10.57	14.07	17.57	21.48	24.48	28.48	32.38	35.38
k	Max	0.6	0.72	0.84	0.95	1.08	1.2	1.32	1.5	1.05	1.93	2.2	2.25	2.5	3	4	5	6	7	8	9	10
n c13, $n \leq 1$, c14, $n > 1$	Nom	0.25	0.3	0.3	0.4	0.4	0.5	0.6	0.6	0.8	0.8	1	1	1.2	1.6	2	2.5	3	3	4	4	5
	Max	0.45	0.5	0.5	0.6	0.6	0.7	0.8	0.8	1	1	1.2	1.2	1.51	1.91	2.31	2.31	3.31	3.31	4.37	4.37	5.37
	Min	0.31	0.36	0.36	0.46	0.46	0.56	0.65	0.66	0.85	0.86	1.06	1.06	1.26	1.66	2.06	2.56	3.06	3.06	4.07	4.07	5.07
t	Max	0.3	0.35	0.4	0.5	0.55	0.6	0.65	0.73	0.85	1	1.1	1.25	1.35	1.6	2.1	2.6	3	3.5	4	4.5	5
	Min	0.2	0.24	0.28	0.32	0.36	0.4	0.44	0.5	0.6	0.7	0.8	0.9	1	1.2	1.6	2	2.4	2.8	3.2	3.6	4
r ≈		0.1	0.12	0.14	0.16	0.18	0.2	0.22	0.25	0.3	0.35	0.4	0.45	0.5	0.6	0.8	1	1.2	1.4	1.6	1.8	2
b	+2p	*	*	*	15	15	16	17	18	19	20	22	24	25	28	34	40	46	52	58	64	70

Table 8.11 Slotted cheese head screws

All dimensions in mm

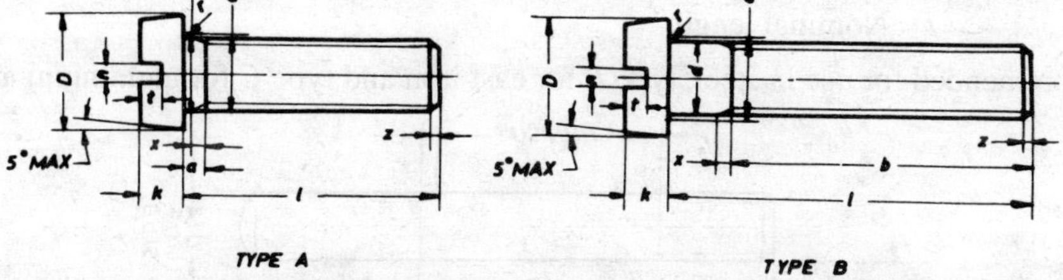

z ACCORDING TO I S:1368, DIMENSIONS OF ENDS OF BOLTS AND SCREWS.

x ACCORDING TO I S:1369, DIMENSIONS OF SCREW THREAD RUNOUTS AND UNDERCUTS.

Size		M1.6	M2	(M2.2)	M2.5	M3	(M3.5)	M4	(M4.5)	M5	M6
d	Nom	1.6	2	2.2	2.5	3	3.5	4	4.5	5	6
	Max	1.60	2.00	2.20	2.50	3.00	3.50	4.00	4.50	5.00	6.00
	Min	1.46	1.86	2.06	2.36	2.86	3.32	3.82	4.32	4.82	5.82
D	Nom	3	3.8	4	4.5	5.5	6	7	8	8.5	10
	Max	3.00	3.80	4.00	4.50	5.50	6.00	7.00	8.00	8.50	10.00
	Min	2.60	3.32	3.52	4.02	5.02	5.52	6.42	7.42	7.92	9.42
k	Nom	1	1.3	1.5	1.6	2	2.4	2.6	3.1	3.3	3.9
	Max	1.00	1.30	1.50	1.60	2.00	2.40	2.60	3.10	3.30	3.90
	Min	0.86	1.05	1.25	1.35	1.75	2.15	2.35	2.80	3.00	3.60
n	Nom	0.4	0.5	0.6	0.6	0.8	0.8	1	1	1.2	1.6
	Max	0.60	0.70	0.80	0.80	1.00	1.00	1.20	1.20	1.51	1.91
	Min	0.46	0.56	0.66	0.66	0.86	0.86	1.06	1.06	1.26	1.66
t	Max	0.65	0.85	1.0	1.0	1.3	1.4	1.6	1.8	2	2.3
	Min	0.45	0.6	0.7	0.7	0.9	1.0	1.2	1.4	1.5	1.8
r	Min	0.1	0.1	0.1	0.1	0.1	0.1	0.2	0.2	0.2	0.25
d_a	Max	2.0	2.6	2.8	3.1	3.6	4.1	4.7	5.2	5.7	6.8
a	Max	0.7	0.8	0.8	0.9	1	1.2	1.4	1.5	1.6	2
b	Min	15	16	17	18	19	20	22	24	25	28

Table 8.12 Studs

In the following table :

d = Diameter of stud
b = Nut end. This end is not screwed into the component.
b_1 = Metal end. This end is screwed into the component.
l = Nominal length.

Type *A* is recommended for use in steel, type *B* for cast iron and type *C* for aluminium alloys.

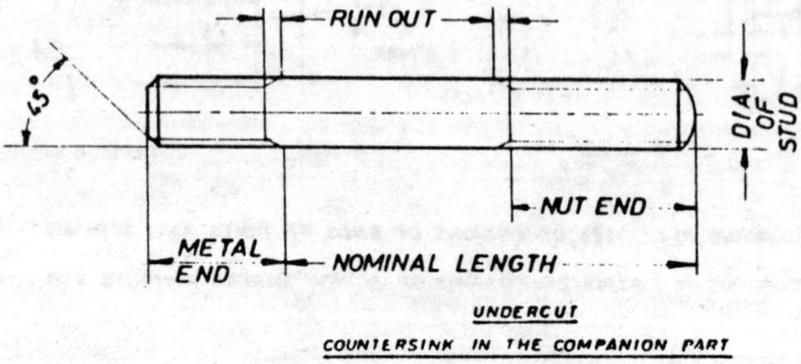

All dimensions in mm

Nominal size d h 13		M3	(M3.5)	M4	(M4.5)	M5	M6	M8 M8 × 1	M10 M10 × 1.25	M12 (M12 × 1.25)	(M14) (M14 × 1.5)
$b + 2P$ – 0	For $l \le 125$	12	13	14	15	16	18	22	26	30	34
	$125 < l < 200$	–	–	–	–	–	–	–	32	36	40
	Type A Nom.	3	3.5	4	4.5	5	6	8	10	12	14
	Max.	3.3	3.88	4.38	4.88	5.38	6.38	8.45	10.45	12.55	14.55
	Min.	2.7	3.12	3.62	4.12	4.62	5.62	7.55	9.55	11.45	13.45
$b_1(js16)$	Type B Nom.	4.5	5	6	7	7.5	9	12	15	18	21
	Max.	4.88	5.38	6.38	7.45	7.95	9.45	12.55	15.55	18.55	21.65
	Min.	4.12	4.62	5.62	6.55	7.05	8.55	11.45	14.45	17.45	20.35
	Type C Nom.	–	–	8	–	–	–	16	–	24	–
	Max.	–	–	8.45	–	–	–	16.55	–	24.65	–
	Min.	–	–	7.55	–	–	–	15.45	–	23.35	–

Nominal size d		M16 M16 × 1.5	(M18) (M18 × 1.5)	M20 M20 × 1.5	(M22) (M22 × 1.5)	M24 M24 × 2	(M27) (M 27 × 2)	M30 M30 × 2	(M33) (M33 × 2)	M36 M36 × 3	(M39) (M39 × 3)
$b + 2P$ – 0	For $l \le 125$	38	42	46	50	54	60	66	72	78	84
	$125 \le l \le 200$	44	48	52	56	60	66	72	78	84	90
	For $l > 200$	57	61	65	69	73	79	85	91	97	103
	Type A Nom.	16	18	20	22	24	27	30	33	36	39
	Max.	16.55	18.55	20.65	22.65	24.65	27.65	30.65	33.8	36.8	39.8
	Min.	15.45	17.45	19.35	21.35	23.35	26.35	29.35	32.2	35.2	38.2

$b_1(js16)$	Type B	*Nom.* *Max.* *Min.*	24 24.65 23.35	27 27.65 26.35	30 30.65 29.35	33 33.8 32.2	36 36.8 35.2	40.5 41.3 39.7	45 45.8 44.2	49.5 50.3 48.7	54 54.95 53.05	57 57.95 56.05
	Type C	*Nom.* *Max.* *Min.*	32 32.8 31.2	– – –	40 40.8 39.2	– – –	– – –	– – –	– – –	– – –	– – –	– – –

Note : Sizes shown within brackets are of second preference.

Table 8.13 Castle nuts

ALL DIMENSIONS IN MILLIMETRES

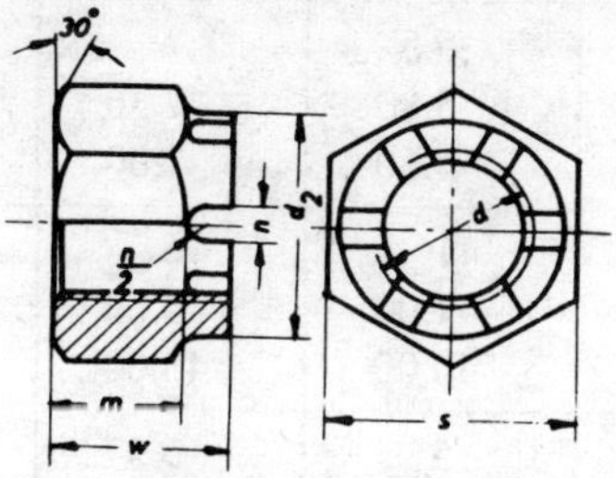

Size, d		*M12*	*(M14)*	*M16*	*(M18)*	*M20*	*(M22)*
d_2 *h*16	Nom Max Min	17 17.00 15.90	19 19.00 17.70	22 22.00 20.70	25 25.00 23.70	28 28.00 26.70	30 30.00 28.70
*w js*16	Nom Max Min	15 15.55 14.45	16 16.55 15.45	19 19.65 18.35	21 21.65 20.35	22 22.65 21.35	26 26.65 25.35
*m h*16	Nom Max Min	10 10.00 9.10	11 11.00 9.90	13 13.00 11.90	15 15.00 13.90	16 16.00 14.90	18 18.00 16.90
*n h*14	Nom Max Min	3.5 3.80 3.50	3.5 3.80 3.50	4.5 4.80 4.50	4.5 4.80 4.50	4.5 4.80 4.50	5.5 5.80 5.50
$s \leq 19 - h14$ $> 19 - h15$	Nom Max Min	19 19.00 18.48	22 22.00 21.16	24 24.00 23.16	27 27.00 26.16	30 30.00 29.16	32 32.00 31.00
Size of split cotter pin		3.2	3.2	4	4	4	5

Size d		M24	(M27)	M30	(M33)	M36	(M39)
d_2 h16	Nom	34	38	42	46	50	55
	Max	34.00	38.0C	42.00	46.00	50.00	55.00
	Min	32.40	36.40	40.40	44.40	48.40	53.10
w js16	Nom	27	30	33	35	38	40
	Max	27.65	30.65	33.80	35.80	38.80	40.80
	Min	26.35	29.35	32.20	34.20	37.20	39.20
m h16	Nom	19	22	24	26	29	31
	Max	19.00	22.00	24.00	26.00	29.00	31.00
	Min	17.70	20.70	22.70	24.70	27.70	29.40
n h14	Nom	5.5.	5.5	7	7	7	7
	Max	5.80	5.80	7.36	7.36	7.36	7.36
	Min	5.50	5.50	7.00	7.00	7.00	7.00
s h15	*Nom*	36	41	46	50	55	60
	Max	36.00	41.00	46.00	50.00	55.00	60.00
	Min	35.00	40.00	45.00	49.00	53.80	58.80
Size of spit cotter pin		5	5	6.3	6.3	6.3	6.3

Size, d		M42	(M45)	M48	(M52)	M56	(M60)	M64	(M68)
d_2 h16	*Nom*	58	62	65	70	75	80	85	90
	Max	58.00	62.00	65.00	70.00	75.00	80.00	85.00	90.00
	Min	56.10	60.10	63.10	68.10	73.10	78.10	82.80	87.80
w js16	*Nom*	46	48	50	54	57	63	66	69
	Max	46.80	48.80	50.80	54.95	57.95	63.95	69.95	69.95
	Min	45.20	47.20	49.20	53.05	56.05	62.05	65.05	68.05
m h16	Nom	34	36	38	42	45	48	51	54
	Max	34.00	36.00	38.00	42.00	45.00	48.00	51.00	54.00
	Min	32.40	34.40	36.40	40.40	43.40	46.40	49.10	52.10
n h14	*Nom*	9	9	9	9	9	11	11	11
	Max	9.36	9.36	9.36	9.36	9.36	11.43	11.43	11.43
	Min	9.00	9.00	9.00	9.00	9.00	11.00	11.00	11.00
s h15	*Nom*	65	70	75	80	85	90	95	100
	Max	65.00	70.00	75.00	80.00	85.00	90.00	95.00	100.00
	Min	63.80	68.80	73.80	78.80	83.60	88.60	93.60	98.60
Size of split cotter pin		8	8	8	8	8	10	10	10

8.2 STANDARDS, DATA AND ACCESSORIES FOR THREADED FASTENERS

Materials for fasteners: In previous Standards for fasteners, the relevant materials for bolts, nuts, studs and other fasteners were indicated by the designation of the fastener materials, and these were entered into the bill of materials in the drawings accordingly, *e.g.*, St 42, C 45, *etc*. In view of the fact that in case of fasteners, the strength properties are more important and revealing than the material designations, this practice has now been discarded and replaced by the indication of strength properties by means of appropriate numerical symbols representing the relevant strength properties.

In IS:1367, Technical supply conditions for threaded steel fasteners - the property classes product grades, mechanical properties, testing, acceptance conditions and other informations have been stipulated. In Table 8.14, the mechanical properties of the fastener materials represented by numerical symbols have been shown. Heat treatment, chemical limiting conditions and other data relating to the different property classes are given in Table 8.15. Table 8.16 shows the previous and revised specifications of property classes of the numerical systems. (The previous numerical system is still used in some cases). In

Table 8.14 Data of mechanical characteristics of property class

Mechanical Property			*4.6*		*4.8*		*5.6*		*5.9*	*6.8*	*8.8* < *M16*	*8.8* > *M16*	*9.8†*	*10.9*	*12.9*
Tensile strength, σ_{Ut}	N/mm²	Nom		400				500		600	800	800	900	1000	1200
		Min.	400		420		500		520	600	800	800	900	1040	1220
Vickers hardness	HV	Min.	120		130		155		160	190	230	255	280	310	372
		Max.				220				250	300	336	360	382	434
Brinell hardness	HB	Min.	114		124		147		152	181	219	242	266	295	353
		Max.				209				238	285	319	342	362	412
Rockwell hardness	Min. HR	HRB	67		71		79		82	89	–	–	–	–	–
		HRC	–		–		–		–	–	20	23	27	31	38
	Max.	HRB				95				99	–	–	–	–	–
		HRC		–		–				–	30	34	36	39	44
Surface hardness $HV_{0.3}$	Max.										320	356	380	402	454
Yield stress, σ_Y N/mm²		Nom	240		320		300		400	480	–	–	–	–	–
		Min.	240		340		300		420	480	–	–	–	–	–
Stress at permanent set limit, $\sigma_{0.2}$	N/mm²	Nom.				–					640	640	720	900	1080
		Min.				–					640	660	720	940	1100
Stress under proof load S_p		S_p/σ_y	0.94		0.91		0.94		0.91	0.91	0.91	0.91	0.91	0.88	0.88
		N/mm²	225		310		280		380	440	580	600	650	830	970
Elongation after fracture A_s%		Min.	22		14		20		10	8	12	12	10	9	8

Table 8.17, the property classes for nuts and those of mating bolts are given. The property class of nuts is indicated by a numeral which is 1/10th of the proof load stress in kgf/mm^2. This proof load stress is equal to the minimum tensile strength of the bolt with which the nut is paired. Thus a nut of property class 4 has a proof load of 40 kgf/mm^2 which is also equal to the minimum tensile strength of the bolts 4.6 or 4.8 with which the nut is mated. Nuts of higher property class may substitute the nuts of lower property class but not vise versa. The surface finish for the fasteners are given in the Table 8.18.

Explanation of Table 8.14

The property class consists of two numerals separated by a point, as for example, 4.6. The 1st numeral indicates 1/10th of the value of the minimum tensile strength of the material in kgf/mm^2. The 2nd numeral indicates 10 times the value of the ratio between the minimum yield strength and the minimum tensile strength. Thus a bolt belonging to the property class 4.6 will have the following strength values:

$$\text{Min. tensile strength} = 4 \times 10 = 40 \text{kgf/mm}^2$$

$$\frac{\text{Min. yield strength } \sigma_y}{\text{Min. tensile strength } 6_{ut}} \times 10 = 6$$

or, $$\text{min. yield strength } \sigma_y = \frac{6}{10} \cdot \sigma_{ut} = \frac{6}{10} \times 40 = 0.6 \times 40 = 24 \text{ kgf/mm}^2$$

It is to be noted that the multiplication of the two numerals of the property class gives the yield strength, *e.g.*, $4 \times 6 = 24$ kgf/mm^2.

When no well-defined yield point is available as, for example, in the case of a high tensile steel, *etc.*, 0.2 % proof stress, symbolised as $\sigma_{0.2}$, is taken to represent yield strength and is considered as such for the purpose of mechanical design calculations. This stress, obtained from the stress-strain curve, is the stress corresponding to a permanent set of 0.2%. In other words, the strain in this case is given by:

$$\text{Strain}_{0.2} = 0.2\% = \frac{0.2}{100} = \frac{2}{1000}$$

It is presumed that when a non-recoverable strain or permanent set of 0.2% has taken place, the material has entered the plastic range analogous to yield point.

Elongation after fracture or percentage elongation is a measure of the ductility of the material. There is no hard and fast line of demarcation between brittleness and ductility, but normally a material having less than 5% elongation at fracture is said to be brittle whereas a material having more elongation than this value is said to be ductile. The percentage elongation of different materials are given in books dealing with materials. In order to obtain comparable elongation, the ratio between the gauge length and diameter of the test specimen shall have a certain definite value. As a rule only proportional test bars are used, *e.g.*, specimen having a gauge length $L = 5\ d \approx (5.65 \sqrt{A})$ with short proportional bar or $L = 10\ d \approx (11.3 \sqrt{A})$ with long proportional bar, L being the G.L., d the diameter and A the cross-sectional area of the test specimen.

Table 8.15 Property class, material and treatment for bolts, screws and studs

Property Class	*Material and Treatment*	*Chemical Composition Limits (Check Analysis) Percent*				*Tempering*
		C	+	*P*	*S*	*°C Min.*
		Min.	*Max.*	*Max.*	*Max*	
4.6 4.8	Low or medium carbon steel	–	0.55	0.05	0.06	–
5.6 5.8 6.8	Low or medium carbon steel	–	0.55	0.05	0.06	–
8.8	Low carbon steel with additives (for example. boron or Mn or Cr), quenched and tempered or Medium carbon steel, quenched and tempered	0.15 0.25	0.35 0.55	0.04	0.05	425 450
9.8	Low carbon steel with additives (for example, boron or Mn or cr) quenched and tempered or Medium carbon steel quenched and tempered	0.15 0.25	0.35 0.55	0.04	0.05	410
10.9	Low carbon steel with additives (for example boron or Mn or Cr), quenched and tempered or Medium carbon steel, quenched and tempered or Medium carbon steel with additives (for example, boron or Mn or Cr), quenched and tempered or Alloy steel	0.15 0.25 0.20 0.20	0.35 0.55 0.55 0.55	 0.04 0.035	 055 0.035	340 425
12.9	Alloy steel	0.20	0.50	0.035	0.035	380

Material designations of some common steels which correspond to the different property classes of bolts, screws and studs are given below. The suggested steels are for guidance only.

Property class	*Material designation of steel (IS symbol)*
4.6	10 C 4, 15 C 4, C 15 Mn 75, 14 Mn 1 Si 4
4.8	15 C 4, 14 Mn 1 Si 4, C 25 Mn 75, C 15 Mn 75
5.6 and 5.8	15 C 4, 25 Mn 1 Si 4, C 35 Mn 75, 40 C 8
8.8	40 C 8, 40 Cr 4, 37 Mn 2, 35 Ni 1 Cr 60
10.9	40 Cr 4, 35 Mn 2 Mo 28, 40 Ni 2Cr 1 Mo 28
12.9	40 Cr 1 Mo 28, 40 Ni 3 Cr 65 Mo 55, 40 Cr 4

In case of above-mentioned steels, the different strength properties are obtained by suitable heat treatment.

Table 8.16 Previous and revised property class specifications

Previous specifications	*4A*	*4D*	*4S*	*5D*	*5S*	*6 E* *6 S*	*8G*	*10K*	*12K*
Revised specifications	4.6		4.8	6.6			8.8	10.9	12.9

Table 8.17 Property class for nuts (nominal height ≥ 0.8*d*)

Property Class of Nut	*4*	*5*	*6*	*8*	*9*	*10*	*12*
Property Class of mating bolt	4.6 4.8	5.6 5.8	6.8	8.8	8.8 9.8	10.9	12.9

Grades : The threaded fasteners are classified into three grades, namely,

Precision (P)
Semi-precision (S)
Black (B)

The surface finish for fasteners is given in Table 8.18 in relation to the different grades. The DIN grades are also given for comparison purposes.

Table 8.18 Surface finish for fasteners

Fastener grade		*Surface*		*Surface finish symbol*
IS	*DIN*	*Bolt*	*Nut*	
P	*m*	All surfaces and threads except points	All surfaces and threads	▽▽
		Point of bolt	Minor dia of thread	∼
S	*mg*	Thread shank and bearing faces	Thread and bearing face	▽▽
		All other faces	Flats and minor dia. of thread	∼
B	*g*	Flank and minor dia. of thread	Flank and major dia. of thread	▽▽
		Major dia. of thread	Minor dia. of thread	∼
		All other faces	All other faces	∼

Strength calculations of bolts, studs and other similar threaded fasteners are based on the nominal stress area of the fasteners. The core areas are also sometimes required for these calculations. Data regarding these areas of common bolt sizes are given in Table 8.19.

Table 8.19 Nominal stress area and core area of the bolt

Coarse Thread				*Fine Thread*			
Nom. Size	*Pitch P*	*Nom. Stress Area A_s*	*Core Area A_c*	*Nom. Size*	*Pitch P*	*Nom. Stress Area, A_s*	*Core Area A_c*
mm	*mm*	*mm²*	*mm²*	*mm*	*mm*	*mm²*	*mm²*
M4	0.7	8.78	7.75	M8	1	39.2	36.0
M5	0.8	14.2	12.7	M10	1.25	61.2	56.3
M6	1.0	20.1	17.9	M12	1.25	92.1	86
M7	1.0	28.9	26.2	M12	1.5	88.1	81.1
M8	1.25	36.6	32.8	M14	1.5	125	116
M9	1.25	48.1	43.8	M16	1.5	167	157
M10	1.5	58.0	52.3	M18	1.5	216	205
M12	1.75	84.3	76.2	M20	1.5	272	259
M14	2.0	115	105	M22	1.5	333	319
M16	2.0	157	144	M24	2	384	365
M18	2.5	192	175	M27	2	496	473
M20	2.5	245	225	M30	2	621	596
M22	2.5	303	282				
M24	3.0	353	324				
M27	3.0	459	427				
M30	3.5	561	519				

A_s = Nom. stress area $= \frac{\pi}{4}\left[\frac{d_2 + d_3}{2}\right]^2$. A_c = Core area $= \frac{\pi}{4} d_3^2$

d_2 = Pitch dia. of the bolt in mm.
d_3 = Core dia. of the bolt in mm.

Designation of fasteners: In the bill of materials of a drawing, the proper designations of fasteners (and other items) are to be entered by the draughtsman in the relevant columns provided for the purpose (see Fig. 4.39). In Table 8.20, an example is shown illustrating the designation system - in this case, that of precision and semi-precision bolts and nuts according to IS:1364. Similarly, designations of other fasteners and items are to be written as per the relevant IS Specifications.

Table 8.20 Designation system for bolts, screws, nuts and lock nuts
(All dimensions in millimetres)

Description						
Fastener	*Size*	*Pitch for Fine Thread Only*	*Length*	*Grade*	*Property Class*	*Dimension*
Bolt with a nut and a lock nut	M20	1.5	75	S	4.6	Hex Bolt M20 × 1.5 × 75 NL-IS : 1364-S-4.6
Bolt with a nut	M20	1.5	75	S	4.6	Hex Bolt M20 × 1.5 × 75N-IS : 1364 S-4.6
Bolt with two nuts	M20	1.5	75	S	4.6	Hex Bolt M20 × 1.5 × 75 NN-IS : 1364 S-4.6
Bolt	M20	1.5	75	S	4.6	Hex Bolt M20 × 1.5 × 75 IS : 1364 S-4.6
Screw	M8	—	30	P	8.8	Hex Screw M8 × 30–IS: 1364-P-8.8
Nut	M10	—	—	S	4	Hex Nut M10: IS 1364-S-4
Lock Nut	M10	1.25	—	S	4	Hex Lock Nut M10 × 1.25-IS : 1364-S-4
Nut single chamfered	M10	—	—	S	4	Hex Nut Single Chamfered M10-IS : 1364-S-4
Nut with washer face	M10	—	—	P	4	Hex Nut Washer Face M10-IS : 1364-P-4

Holes on mountings for fasteners: The engineering fasteners, the draughtsman is required to provide dimensioned holes on mountings into which the fasteners are inserted. These holes are standardised to accept standard fasteners.

Table 8.21 shows the clearance holes in the mountings meant for receiving hexagonal bolts. The grades of holes are to be decided upon as per the requirements and design considerations.

Table 8.22 shows the hole dimensions corresponding to the various types of bolts and screws as indicated. The first figure from left is meant for countersunk head screw as per IS:1365, the second for hexagonal socket head cap screw (IS: 2269) and slotted cheese head screw (IS:1366). In the second case, while the diameters d_1 and d_3 remain the same, the diameter d_2 will depend on whether or not spring washers are used, and the depth "t" will depend on the type of screw as well as whether or not spring washers are used. The third figure is for hexagonal bolt and nut while the fourth one is also for bolt and nut with provision for the use of spanners (depth t_1 for bolt and t_2 for nut). Spot-facing, the depth of which is optional, may be provided to accommodate washers.

Table 8.21 Clearance holes for metric bolts

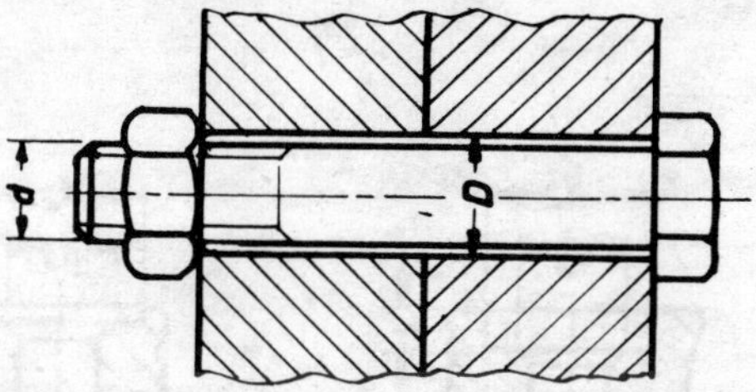

Bolt dia	*Clearance hole diameter D*		
d	*fine*	*medium*	*coarse*
1.6	1.7	1.8	2.0
2.0	2.2	2.4	2.6
2.5	2.7	2.9	3.1
3.0	3.2	3.4	3.6
4.0	4.3	4.5	4.8
5.0	5.3	5.5	5.8
6.0	6.4	6.6	7.0
8.0	8.4	9.0	10.0
10.0	10.5	11.0	12.0
12.0	13.0	14.0	15.0
16.0	17.0	18.0	19.0
20.0	21.0	22.0	24.0
24.0	25.0	26.0	28.0
30.0	31.0	33.0	35.0

Bolt dia	*Clearance hole dia D*		
d	*fine*	*medium*	*coarse*
36	37	39	42
42	43	45	48
48	50	52	56
56	58	62	66
64	68	70	74
72	74	78	82
80	82	86	91
90	93	96	101
100	104	107	112
110	114	117	122
125	129	132	137
140	144	147	155

Note: If special reasons make it necessary to indicate tolerances when specifying clearance holes, the following tolerance zones corresponding to ISO recommendation R 273 - 1962 are recommended :

H 12 for fine clearance holes

H 13 for medium clearance holes

H 14 for coarse clearance holes

Table 8.22 Countersunk and counterbore holes for fasteners

(Dimension in mm)

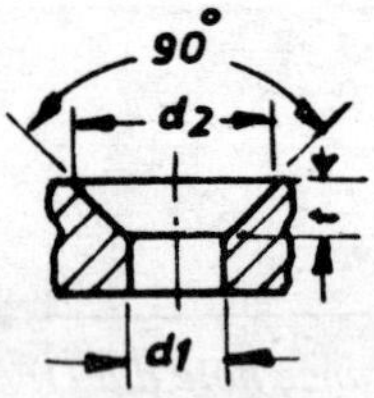

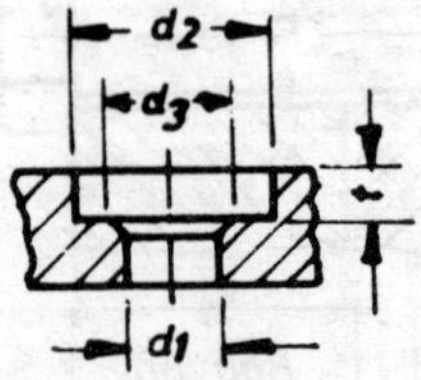

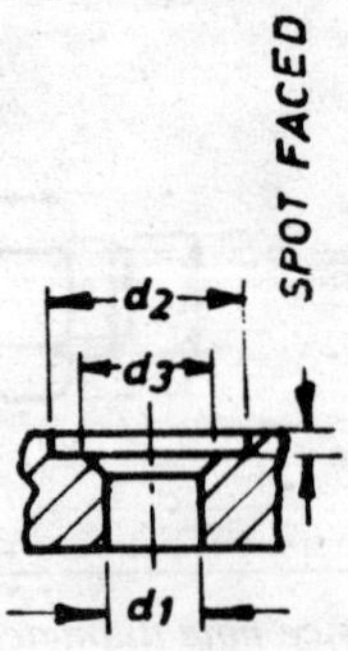

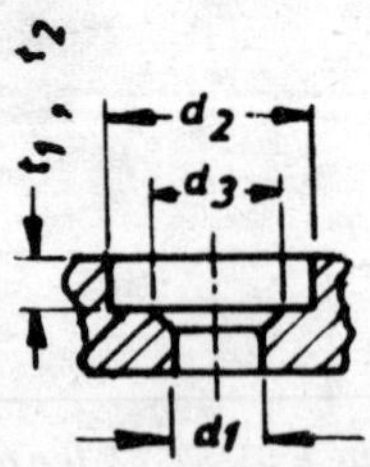

Metric thread	*Countersunk head screw. IS: 1365*			*Hexagonal socket and slotted cheese head.*		*With spring washer.*		*Slotted cheese head screw. IS: 1366*	*With spring washer.*	*Hexagonal socket screw. IS: 2269*	*With spring washers, IS: 3063*		*Hexagonal bolt and nut*		*Box spanner*	*Ring spanner*	*Hexagonal bolt*	*Hexagonal nut*
M	d_1	d_2	*t*	d_1	d_2		d_3	*t*		*t*		d_1	d_2	d_3	d_2	d_2	t_1	t_2
3	3.4	6.5	1.6	3.4	6	7		2.4	3.3	3.4	4.3	3.4	9		11	11	2.4	2.8
4	4.5	8.6	2.1	4.5	8	9		3.2	4.5	4.6	5.5	4.5	10		13	15	3.4	3.8
5	5.5	10.4	2.5	5.5	10	11		4.1	5.5	5.7	7	5.5	11		15	18	4.2	4.7
6	6.6	12.4	2.9	6.6	11	13		4.8	6.5	6.8	8.5	6.6	13		18	20	4.8	5.8
8	9	16.4	3.7	9	15	18		6	8	9	11	9	18		24	26	6.5	7.5
10	11	20.4	4.7	11	18	20		7.3	9.5	11	13.5	11	22		28	33	8	9
12	14	24.4	5.2	14	20	24		8.3	11	13	16	14	26		33	36	9	11
14	16	27.4	5.7	16	24	26		9.3	12.5	15	18.5	16	30		36	43	10	12
16	18	32.4	7.2	18	26	30		10.5	14	17.5	21	18	33		40	46	11.5	14.5
18	20	36.4	8.2	20	30	33	*Deburred*	11.5	15	19.5	23	20	36		43	50	13.5	16.5
20	22	40.4	9.2	22	33	36		12.5	16.5	21.5	25.5	22	40		46	53	14.5	17.5
22				24	36	40		13.5	17.5	23.5	27.5	24	43		50	57	15.5	19.5
24				26	40	43	30	14.5	19.5	25.5	30.5	26	48	26	57	71	16.5	20.5
27				30	43	46	33			28.5	33.5	30	53	33	61	76	18.5	23.5
30				33	48	53	36			32	38	33	61	36	71	82	21	26
33				36	53	57	39			35	41	36	66	39	76	89	23	28
36				39	57	61	42			38	44	39	71	42	82	92	25	31
39												42	76	45	89	98	27	33

·42				45	66	71	48			44	52	45	82	48	98	107	28	36
45												48	89	51	107	112	30	38
48				52	76	78	56			50	59	52	98	56	112	125	32	40
52												56	107	60	118	132	35	44
56												62	112	66				
60												66	118	70				
64												70	125	74				
68												74	132	78				
72												78	132	82				
76												82	140	86				
80												86	150	90				
90												96	170	101				
100												107	190	111				
110												117	200	122				
120												127	224	132				

Example: *For hexagonal socket head cap screw of size M 10, the hole dimensions are:*

Direct mounting: $d_2 = 18$, $t = 11$.

With spring washer: $d_2 = 20$, $t = 13.5$.

Tolerances: H13 for d_1 in each case, H 13 for d_2 in case of CSK, hex. socket head and cheese head screws, H 15 for d_2 in case of spot-faced recess and holes for hex, bolts and nuts meant for spanner insertion.

Table 8.23 Dimensions of ends of bolts and screws

(All dimensions in mm)

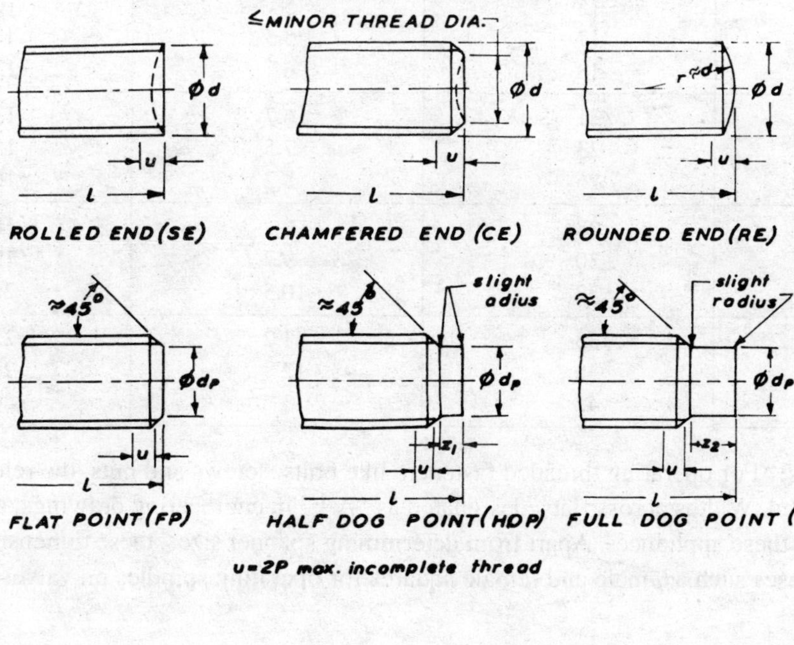

Nominal Diameter	d_p h 14	z_1 + IT14	z_2 + IT14
M1	0.5	–	–
M1.2	0.6	–	–
M1.4	0.7	0.36	0.7
M1.6	0.8.	0.4	0.8
M1.8	0.9.	0.45	0.9
M2	1	0.5	1
M2.2	1.2	0.55	1.1
M2.5	1.5	0.63	1.25
M3	2	0.75	1.5
M3.5	2.2	0.88	1.75
M4	2.5	1	2
M4.5	3	1.12	2.25
M5	3.5	1.25	2.5
M6	4	1.5	3
M7	5	1.75	3.5
M8	5.5	2	4
M10	7	2.5	5
M12	8.5	3	6
M14	10	3.5	7
M16	12	4	8
M18	13	4.5	9
M20	15	5	10
M22	17	5.5	11
M24	18	6	12
M27	21	6.7	13.5
M30	23	7.5	15
M33	26	8.2	16.5
M36	28	9	18
M39	30	9.7	19.5
M42	32	10.5	21
M45	35	11.2	22.5
M48	38	12	24
M52	42	13	26

Width across flats: For operating threaded fasteners like bolts, screws and nuts, the relevant spanners or wrenches are used. Widths across flats, designated as "A/F" in engineering drawings, are dimensions which characterise these appliances. Apart from determining spanner sizes, these dimensions are also of use for other purposes such as, male and female squares for operating spindles on valves, hand-wheels, crank handles, tools and holders. Parts in which fasteners are filled should have adequate provision of space to accommodate spanners for operating on bolt heads, nuts, *etc.* Table 8.24 shows the A/F sizes and across corner sizes for 2 - flats, squares, hexagons and octagons.

Table 8.24 Widths across flats and corners
(Dimensions in mm)

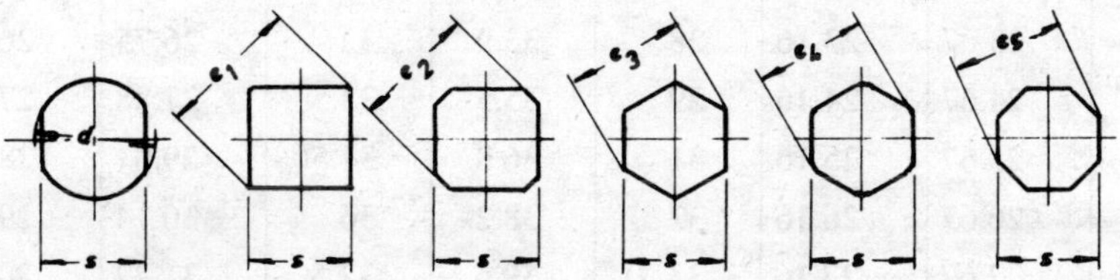

Nominal size A/F	Width across flats s			Width across corners						
	Max.	*Min.*		*d*	e_1	e_2 *min*	e_3 *min.*		e_4	e_5 *min.*
		Series 1	*Series 2*				*series 1*	*series 2*		
2	2	1.91		2.5	–	–	–			
2.5	2.5	2.41		3	–	–	–			
3	3	2.91		3.5	4.2	4.1	3.29			
3.2	3.2	3.08		3.7	4.5	4.3	3.48			
3.5	3.5	3.38		4	4.9	4.6	3.82			
4	4	3.88		4.5	5.7	5.3	4.38			
4.5	4.5	4.38		5	6.4	5.9	4.95			
5	5	4.88		6	7.1	6.5	5.51			
5.5	5.5	5.38		7	7.8	7.1	6.08			
6	6	5.88		7	8.5	8	6.64			
7	7	6.85		8	9.9	9	7.74			
8	8	7.85	7.64	9	11.3	10	8.87	8.63		
9	9	8.78	8.64	10	12.7	12	9.92	9.76		
10	10	9.78	9.64	12	14.1	13	11.05	10.89		
11	11	10.73	10.57	13	15.6	14	12.12	11.94		
12	12	11.73	11.57	14	17.0	16	13.25	13.07		
13	13	12.73	12.57	15	18.4	17	14.38	14.20		
14	14	13.73	13.57	16	19.8	18	15.51	15.33		
15	15	14.73	14.57	17	21.2	20	16.64	16.46		
16	16	15.73	15.57	18	22.6	21	17.77	17.59		
17	17	16.73	16.57	19	24	22	18.90	18.72		

18	18	17.73	17.57	21	25.4	23.5	20.03	19.85		
19	19	18.67	18.48	22	26.9	25	21.10	20.88		
20	20	19.67	19.16	23	28.3	26	22.23	21.65		
21	21	20.67	20.16	24	29.7	27	23.36	22.78		22.7
22	22	21.67	21.16	25	31.1	28	24.49	23.91		23.8
23	23	22.67	22.16	26	32.5	30.5	25.62	25.04		24.9
24	24	23.67	23.16	28	33.9	32	26.75	26.17		26
25	25	24.67	24.16	29	35.5	33.5	27.88	27.30		27
26	26	25.67	25.16	31	36.8	34.5	29.01	28.43		28.1
27	27	26.67	26.16	32	38.2	36	30.14	29.56		29.1
28	28	27.67	27.16	33	39.6	37.5	31.27	30.69		30.2
30	30	29.67	29.16	35	42.4	40	33.53	32.95		32.5
32	32	31.61	31.00	38	45.3	42	35.72	35.03		34.6
36	36	35.38	35.00	42	50.9	48	39.98	39.55		39
41	41	40.38	40.00	48	58	54	45.63	45.20		44.4
46	46	45.38	45.00	52	65.1	60	51.28	50.85		49.8
50	50	49.38	49.00	58	70.7	65	55.80	55.37		54.1
55	55	54.26	53.80	65	77.8	72	61.31	60.79		59.5
60	60	59.26	58.80	70	84.8	80	66.96	66.44		64.9
65	65	64.26	63.80	75	91.9	85	72.61	72.09		70.3
70	70	69.26	68.80	82	99	92	78.26	77.74		75.7
75	75	74.26	73.80	88	106	98	83.91	83.39		81.2
80	80	79.26	78.80	92	113	105	89.56	89.04		86.6
85	85	84.13	83.60	98	120	112	95.07	94.47		92.0
90	90	89.13	88.60	105	127	118	100.72	100.12		97.4
95	95	94.13	93.60	110	134	125	106.37	105.77		103
100	100	99.13	98.60	115	141	132	112.02	111.42		108
105	105	104.13	103.60	122	148	138	117.67	117.07		114
110	110	109.13	108.60	128	156	145	123.32	122.72		119
115	115	114.13	113.60	132	163	152	128.97	128.37		124
120	120	119.13	118.60	140	170	160	134.62	134.02		130
130	130	129.00	128.40	150	184	170	145.77	145.09		141
135	135	134.00	133.40	158	191	178	151.42	150.74		146
145	145	144.00	143.40	168	205	190	162.72	162.04		157
150	150	149.00	148.40				168.37	167.69	165	162
155	155	154.00	153.40				174.02	173.34	170	168

165	165	164.00	163.40				185.32	184.64	180	179
170	170	169.00	168.40				190.97	190.29	186	184
175	175	174.00	173.40				196.62	195.94	192	189
180	180	179.00	178.40				202.27	201.59	198	195
185	185	183.85	183.15				207.75	206.96	205	200
190	190	188.85	188.15				213.40	212.61	210	206
200	200	198.85	198.15				224.70	223.91	220	216
210	210	208.85	208.15				236.00	235.21	232	227
220	220	218.85					247.30		242	238
230	230	228.85					258.60		255	249
235	235	233.85					264.25		260	254
245	245	243.85					275.55		270	265
255	255	253.70					286.68		280	276
265	265	263.70					297.98		290	287
270	270	268.70					303.63		298	292
280	280	278.70					314.93		308	303
290	290	288.70					326.23		320	314
300	300	298.70					337.53		330	325
310	310	308.70					348.83		340	335
320	320	318.60					360.02		352	346
330	330	328.60					371.32		362	357
340	340	338.60					382.62		375	368
350	350	348.60					393.92		385	379
365	365	363.60					410.87		400	395
380	380	378.60					427.82		420	411
395	395	393.60					444.77		435	427
410	410	408.45					461.55		452	444
425	425	423.45					478.50		470	460
440	440	438.45					495.45		485	476
455	455	453.45					512.40		500	492
470	470	468.45					529.35		518	509
480	480	478.45					540.65		528	519
495	495	493.45					557.60		545	536
510	510								560	552
525	525								580	568

Table 8.25 Machined washers
(Dimensions in mm)

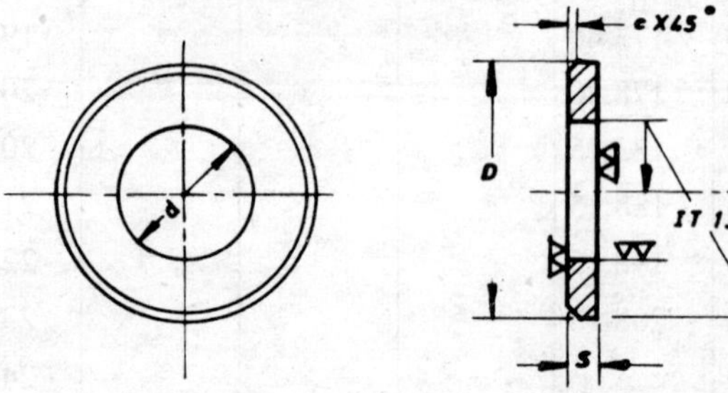

Size d H12	D		s		e Nom	For bolt or screw size
	Basic	Tol	Basic	Tol		
5.3	10	+0 –0.3	1.0	± 0.1	0.4	M5
6.4	12.5	+0 –0.4	1.6	± 0.2	0.6	M6
(7.4)	14	+0 –0.4	1.6	± 0.2	0.6	(M7)
8.4	17	+0 –0.4	1.6	± 0.2	0.6	M8
10.5	21	+0 –0.5	2	± 0.2	0.6	M10
13	24	+0 –0.5	2.5	±0.3	0.6	M12
(15)	28	+0 –0.5	2.5	±0.3	0.6	(M14)
17	30	+0 –0.5	3	±0.3	0.6	M16
(19)	34	+0 –0.8	3	±0.3	1.0	(M18)
21	37	+0 –0.8	3	±0.3	1.0	M20
(23)	39	+0 –0.8	3	+0.3	1.0	M22
25	44	+0 –0.8	4	±0.3	1.0	M24

(28)	50	+0 –0.8	4	±0.3	1.0	(M27)
31	56	+0 –1.0	4	±0.3	1.0	M30
(34)	60	+0 –1.0	5	±0.6	1.0	(M33)
37	66	+0 –1.0	5	±0.6	1.6	M36
(40)	72	+0 –1.0	6	±0.6	1.6	(M39)
43	78	+0 –1.0	7	±1.0	1.6	M42
(46)	85	+0 –1.5	7	±1.0	1.6	(M45)
50	92	+0 –1.5	8	±1.0	1.6	M48
(54)	98	+0 –1.5	8	±1.0	1.6	(M52)
58	105	+0 –1.5	9	±1.0	1.6	M56
(62)	110	+0 –1.5	9	±1.0	2.0	(M60)
66	115	+0 –1.5	9	±1.0	2.0	M64
(70)	120	+0 –1.5	10	±1.0	2.0	(M68)
74	125	+0 –1.8	10	±1.0	2.0	M72
(78)	135	+0 –1.8	10	±1.0	2.5	(M76)
82	140	+0 –1.8	12	±1.2	2.5	M80
(87)	145	+0 –1.8	12	±1.2	2.5	(M85)
93	160	+0 –1.8	12	±1.2	3.0	M90

(98)	165	+0 −1.8	12	±1.2	3.0	(M95)
104	175	+0 −1.8	14	±1.2	3.0	M100
(109)	180	+0 −1.8	14	±1.2	3.0	(M105)
114	185	+0 −2	14	±1.2	3.0	M110
(119)	200	+0 −2	14	±1.2	3.0	(M115)
(124)	210	+0 −2	16	±1.2	3.0	(M120)
129	220	+0 −2	16	±1.2	3.0	M125
(134)	230	+0 −2	16	±1.2	3.0	(M130)
144	240	+0 −2	18	±1.2	4.0	M140
(155)	250	+0 −2	18	±1.2	4.0	(M150)

Note : Sizes in brackets are of second preference.

Table 8.26 Punched washers, type A
(For hexagonal bolts and screws)

(Dimensions in mm)

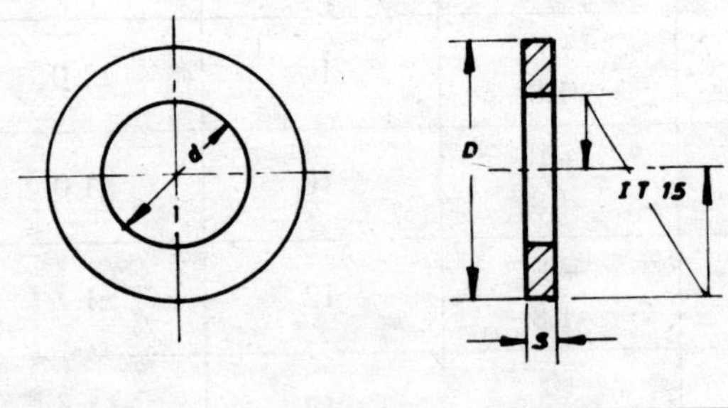

Size d	*D Nom*	*s Nom*	*For bolt or screw size*
1.8	4	0.4	M1.6
(2.1)	5	0.4	(M1.8)
2.4	5	0.4	M2

(2.6)	6	0.5	(M2.2)
2.9	6.5	0.5	M2.5
3.4	7	0.5	M3
(4.0)	8	0.8	(M3.5)
4.5	9	0.8	M4
(5.0)	10	1	(M4.5)
5.5	10	1	M5
6.6	12.5	1.6	M6
(8.6)	14	1.6	(M7)
9	17	1.6	M8
11	21	2	M10
14	24	2.5	M12
(16)	28	2.5	(M14)
18	30	3.15	M16
(20)	34	3.15	(M18)
22	37	3.15	M20
(24)	39	3.15	(M22)
26	44	4	M24
(30)	50	4	(M27)
33	56	4	M30
(36)	60	5	(M33)
39	66	5	M36
(42)	72	6	(M39)
45	78	6	M42
(48)	85	8	(M45)
52	92	8	M48

Note : Sizes in brackets are of second prefernece.

Table 8.27 Punched washers, type B
(For round and cheese head screws)
(Dimensions in mm)

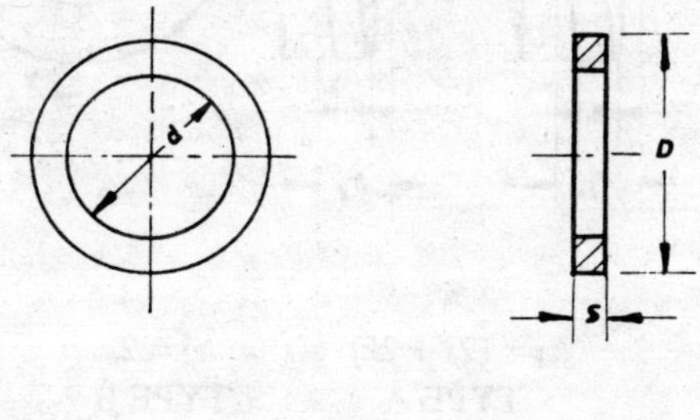

Size d	*D Nom*	*s Nom*	*For screw size*
1.8 (2.1) 2.4	3.5 3.5 4.5	0.4 0.4 0.4	M1.6 (M1.8) M2
(2.6) 2.9 3.4	4.5 5.0 6.0	0.5 0.5 0.5	(M2.2) M2.5 M3
(4.0) 4.5 (5.0)	7.0 8.0 9.0	0.5 0.8 0.8	(M3.5) M4 (M4.5)
5.5 6.6 (7.6)	9.5 11 13	1.0 1.6 1.6	M5 M6 (M7)
9 11 14	14 18 20	1.6 2 2.5	M8 M10 M12
(16) 18 (20)	24 27 30	2.5 3.15 3.15	(M14) M16 (M18)
22	33	3.15	M20

Note: Sizes in brackets are of second preference.

Table 8.28 Dimensions for spring washers

All dimensions in millimeters

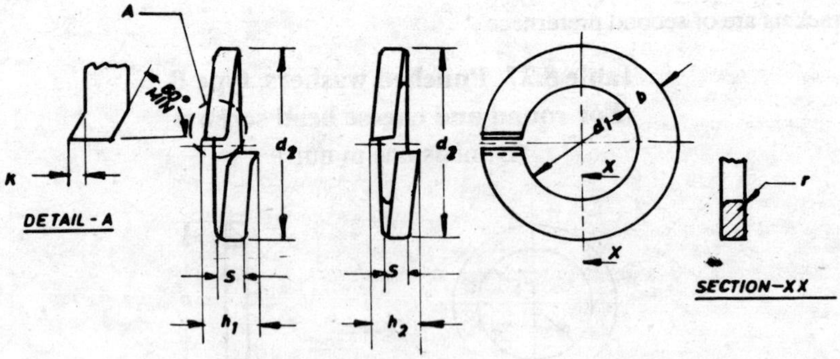

$h_1 = (2s + 2k)$ TYPE A Washer With Best Ends

$h_2 = 2s$ TYPE B Washer With Flat Ends

Nominal Size	d_1		d_2 *Max*	*b*		*s*		*r Nom*	k*	*Mass Approx (7.85 kg/dm³) kg/1000 pcs*	*For Bolt, Nut or screw size*
	Basic	*Tol +*		*Basic*	*Tol ±*	*Basic*	*Tol ±*				
2	2.1	0.3	4.4	0.9	0.1	0.5	0.1	0.1	—	0.033	M2
2.2	2.3	0.3	4.9	1	0.1	0.6	0.1	0.1	—	0.050	M2.2
2.5	2.6	0.3	5.1	1	0.1	0.6	0.1	0.1	—	0.053	M2.5
3	3.1	0.3	6.2	1.3	0.1	0.8	0.1	0.2	—	0.112	M3
(3.5)	3.6	0.3	6.7	1.3	0.1	0.8	0.1	0.2	0.15	0.12	M3.5
4	4.1	0.3	7.6	1.5	0.1	0.9	0.1	0.2	0.15	0.18	M4
5	5.1	0.3	9.2	1.8	0.1	1.2	0.1	0.2	0.15	0.36	M5
6	6.1	0.4	11.8	2.5	0.15	1.6	0.1	0.3	0.2	0.83	M6
(7)	7.1	0.4	12.8	2.5	0.15	1.6	0.1	0.3	0.2	0.93	M7
8	8.2	0.4	14.8	3	0.15	2	0.1	0.5	0.3	1.60	M8
10	10.2	0.6	18.1	3.5	0.2	2.2	0.15	0.5	0.3	2.53	M10
12	12.2	0.8	21.1	4	0.2	2.5	0.15	1.0	0.4	3.82	M12
(14)	14.2	0.8	24.1	4.5	0.2	3	0.15	1.0	0.4	6.01	M14
16	16.2	0.8	27.4	5	0.2	3.5	0.2	1.0	0.4	8.91	M16
(18)	18.2	0.8	29.4	5	0.2	3.5	0.2	1.0	0.4	9.73	M18
20	20.2	1	33.6	6	0.2	4	0.2	1.0	0.4	15.2	M20
(22)	22.5	1	35.9	6	0.2	4	0.2	1.0	0.4	16.5	M22
24	24.5	1	40	7	0.25	5	0.2	1.6	0.5	26.2	M24
(27)	27.5	1	43	7	0.25	5	0.2	1.6	0.5	28.7	M27
30	30.5	1.2	48.2	8	0.25	6	0.2	1.6	0.8	44.3	M30
(33)	33.5	1.2	55.2	10	0.25	6	0.2	1.6	0.8	63.0	M33
36	36.5	1.2	58.2	10	0.25	6	0.2	1.6	0.8	67.3	M36
(39)	39.5	1.2	61.2	10	0.25	6	0.2	1.6	0.8	71.7	M39
42	42.5	1.2	68.2	12	0.25	7	0.25	2	0.8	111	M42
(45)	45.5	1.2	71.2	12	0.25	7	0.25	2	0.8	117	M45
48	49	1.5	75	12	0.25	7	0.25	2	0.8	123	M48
52	53	1.5	83	14	0.25	8	0.25	2	1	182	M52
56	57	1.5	87	14	0.25	8	0.25	2	1	193	M56
60	61	1.5	91	14	0.25	8	0.25	2	1	203	M60
64	65	1.5	95	14	0.25	8	0.25	2	1	218	M64

72	73	1.5	103	14	0.25	8	0.25	2	1	240	M72
76	71	1.5	109	14	0.25	8	0.25	2	1	253	M76
80	81	1.5	111	14	0.25	8	0.25	2	1	262	M80
90	91	1.5	121	14	0.25	8	0.25	2	1	290	M90
100	101	1.5	131	14	0.25	8	0.25	2	1	318	M100

Note : Sizes in brackets are of second preference.

*The bend *k* shall be made on the last tenth of the washer circumference without any sharp angle.

8.3 KEYS AND SPLINES

Keys and splines are machine elements to transmit motion and torque between shafts and hubs or *vice versa*. Unless special requirements are called for, it is not generally necessary to calculate the dimensions of keys or splines as these are standardised items, and the draughtsman has only to select the proper key or spline from these Standards commensurate with the shaft diameter and other determining parameters. The Standards also specify the appropriate tolerances for the keyways in the shafts and hubs which the draughtsman should enter in the drawings of these items.

Parallel key or feather key is the most commonly used key. Other types of keys are: gib-head key, tangent key, taper key and Woodruff key.

Splines are multi-keys which are machined integral with the shaft. Common splines have straight sides. For heavy duty, involute splines are used. In general, splines are superior to key joint because of their greater strength, bigger contact area, ability to withstand shock loads and to take reversible loads. Centering of splines is very important for smooth operation. Three kinds of centering are usually employed, namely, centering by the major diameter, by the minor diameter or by the spline sides or flanks. Data concerning common keys and splines are given in this section.

For parallel keys, the tolerances are as follows:

For keys of square section: h_9 on both sides. For keys of rectangular section; h_9 on width and h_{11} on height.

Tolerances on width of parallel keyways are given in Table 8.29. Tolerances on length of parallel keyways are as follows:

Length (mm)	Tolerance (mm)
Upto 28	0 to + 0.2
32 to 80	0 to + 0.3
90 and above	0 to + 0.5

The tolerances on width and length of keyways are to be entered in the relevant drawings. Material of key is steel with a tensile strength of at least 600 N/mm^2.

Gib-head keys are easily dismentable and they transmit more torque than parallel keys. Tangent keys can take a large amount of impact type load and reversible torque.

Table 8.29 Parallel keys and keyways

Range of Shaft dia d		Key	key way											
				Tol on b									r	
				Running fit		Lightdrive fit		Force fit						
Above	Upto	b × h	b	Shaft H9	Hub D 10	Shaft N 9	Hub Js9	Shaft and Hub P9	t1	Tol on t1	t2	Tol on t2	Min.	Max.
6	8	2 × 2	2	+ 0.025 0	+ 0.060 + 0.020	– 0.004 – 0.029	+ 0.012 5 0.012 5	– 0.006 – 0.031	1.2		1.0		0.08	0.16
8	10	3 × 3	3						1.8		1.4		0.08	0.16
10	12	4 × 4	4	+ 0.030 0	+ 0.078 – 0.030	0 – 0.030	+ 0.015 0 - 0.015 0	– 0.012 – 0.042	2.5	+ 0.1 0	1.8	+ 0.1 0	0.08	0.16
12	17	5 × 5	5						3.0		2.3		0.16	0.25
17	22	6 × 6	6						3.5		2.8		0.16	0.25

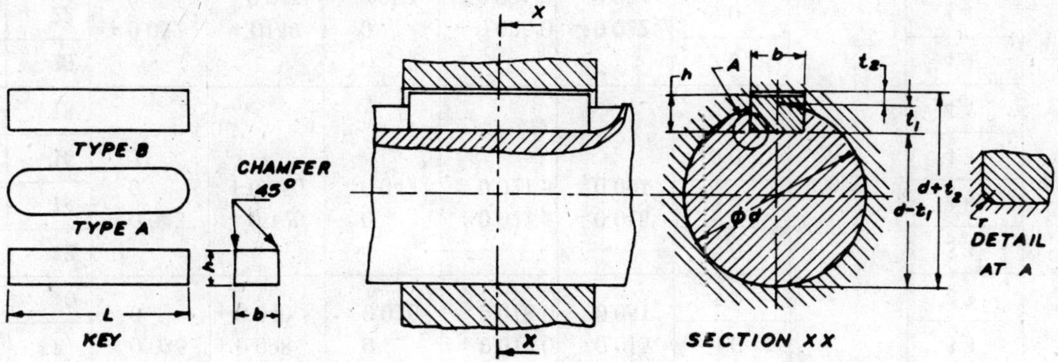

22	30	8 × 7	8	+ 0.036 0	+ 0.098 + 0.040	0 - 0.036	+ 0.018 0 - 0.018 0	- 0.015 - 0.051	4.0	+ 0.2 0	3.3	+ 0.2 0	0.16	0.25
30	38	10 × 8	10						5.0		3.3		0.25	0.40
38	44	12 × 8	12	+ 0.043 0	+ 0.120 + 0.050	0 – 0.043	+ 0.021 5 – 0.021 5	– 0.018 – 0.061	5.0		3.3		0.25	0.40
44	50	14 × 9	14						5.5		3.8		0.25	0.40
50	58	16 × 10	16						6.0		4.3		0.25	0.40
58	65	18 × 11	18						7.0		4.4		0.25	0.40
65	75	20 × 12	20	+ 0.052 0	+ 0149 + 0.065	0 – 0.052	+ 0.026 0 – 0.026 0	– 0.022 – 0.074	7.5		4.9		0.40	0.60
75	85	22 × 14	22						9.0		5.4		0.40	0.60
85	95	25 × 14	25						9.0		5.4		0.40	0.60
95	110	28 × 16	28						10.0		6.4		0.40	0.60
110	130	32 × 18	32	+ 0.062 0	+ 0.180 + 0.080	0 – 0.062	+ 0.031 0 – 0.031 0	– 0.026 – 0.088	11.0		7.4		0.40	0.60
130	150	36 × 20	35						12.0	+ 0.3 0	8.4	+ 0.3 0	0.70	1.00
150	170	40 × 22	40						13.0		9.4		0.70	1.00
170	200	45 × 25	45						15.0		10.4		0.70	1.00
200	230	50 × 28	50						17.0		11.4		0.70	1.00
230	260	56 × 32	56	+ 0.074 0	+ 0.220 + 0.100	0 – 0.074	+ 0.037 0 – 0.037 0	– 0.032 – 0.106	20.0		12.4		1.20	1.60
260	290	63 × 32	63						20.0		12.4		1.20	1.60
290	330	70 × 36	70						22.0		14.4		1.20	1.60
330	380	80 × 40	80						25.0		15.4		2.00	2.50
380	440	90 × 45	90	+ 0.087 0	+ 0.200 + 0.120	0 – 0.087	+ 0.043 5 – 0.043 5	– 0.037 – 0.124	28.0		17.4		2.00	2.50
440	500	100 × 50	100						31.0		19.5		2.00	2.50

Table 8.30 Gib-head keys and keyways

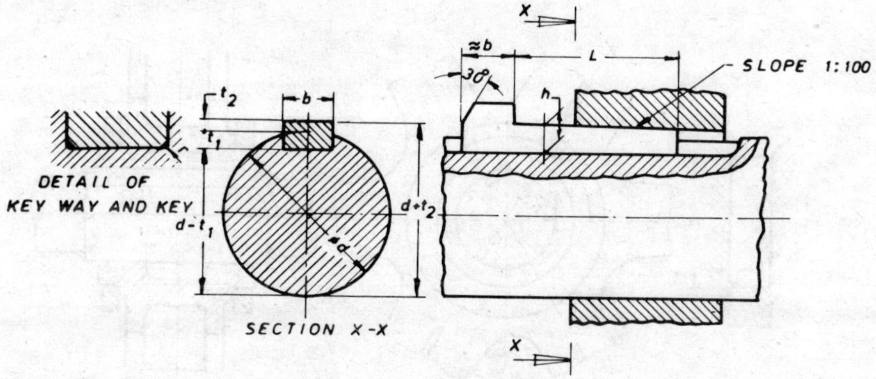

All dimensions in millimetres

Range of Shaft Dia. d		Key	key way							
Above	Upto	b × h	b	Tol on b D 10	t_1	Tol on t_1	t_2	Tol on t_2	r Min	r Max
10	12	4 × 4	4	+ 0.078 + 0.030	2.5	+ 0.1 0	1.2	+ 0.1 0	0.08	0.16
12	17	5 × 5	5		3.0		1.7		0.16	0.25
17	22	6 × 6	6		3.5		2.2		0.16	0.25
22	30	8 × 7	8	+ 0.098 + 0.040	4.0	+ 0.2 0	2.4	+ 0.2 0	0.16	0.25
30	38	10 × 8	10		5.0		2.4		0.25	0.40
38	44	12 × 8	12	+ 0.120 + 0.050	5.0		2.4		0.25	0.40
44	50	14 × 9	14		5.5		2.9		0.25	0.40
50	58	16 × 10	16		6.0		3.4		0.25	0.40
58	65	18 × 11	18		7.0		3.4		0.25	0.40
65	75	20 × 12	20	+ 0.149 + 0.065	7.5		3.9		0.40	0.60
75	85	22 × 14	22		9.0		4.4		0.40	0.60
85	95	25 × 14	25		9.0		4.4		0.40	0.60
95	110	28 × 16	28		10.0		5.4		0.40	0.60
110	130	32 × 18	32	+ 0.180 + 0.080	11.0		6.4		0.40	0.60
130	150	36 × 20	36		12.0	+ 0.3 0	7.1	+ 0.3 0	0.70	1.00
150	170	40 × 22	40		13.0		8.1		0.70	1.00
170	200	45 × 25	45		15.0		9.1		0.70	1.00
200	230	50 × 28	50		17.0		10.1		0.70	1.00
230	260	56 × 32	56	+ 0.220 + 0.100	20.0		11.1		1.20	1.60
260	290	63 × 32	63		20.0		11.1		1.20	1.60
290	330	70 × 36	70		22.0		13.1		1.20	1.60
330	380	80 × 40	80		25.0		14.1		2.00	2.50
380	440	90 × 45	90	+ 0.260 + 0.120	28.0		16.1		2.00	2.50
440	500	100 × 50	100		31.0		18.1		2.00	2.50

Table 8.31 Tangent keys and keyways

All dimensions in mm

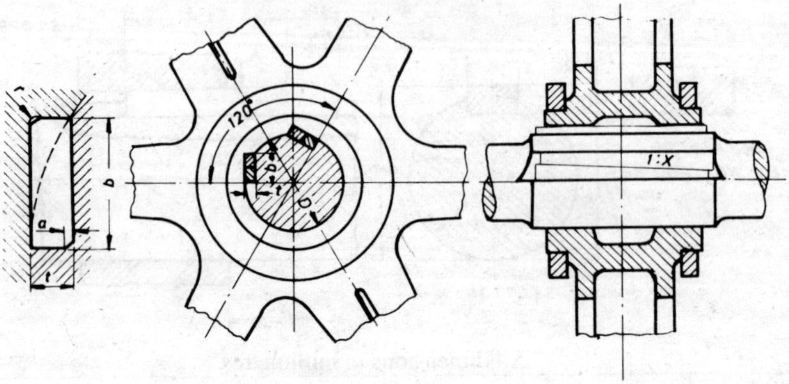

Shaft diameter D	Key Depth t	Key Width b	Key Radius r	Key Chamfer a	Shaft diameter D	Keyway Depth t	Keyway Width b	Keyway Radius r	Key Chamfer a
100	10	30	2	3	460	46	138	4	5
110	11	33	2	3	480	48	144	5	6
120	12	36	2	3	500	50	150	5	6
130	13	39	2	3	520	52	156	5	6
140	14	42	2	3	540	54	162	5	6
150	15	45	2	3	560	56	168	5	6
160	16	48	2	3	580	58	174	5	6
170	17	51	2	3	600	60	180	6	7
180	18	54	2	3	620	62	186	6	7
190	19	57	2	3	640	64	192	6	7
200	20	60	2	3	660	66	198	6	7
210	21	63	2	3	680	68	204	6	7
220	22	66	2	3	700	70	210	6	7
230	23	69	3	4	720	72	216	6	7
240	24	72	3	4	740	74	222	6	7
250	25	75	3	4	760	76	228	6	7
260	26	78	3	4	780	78	234	6	7
270	27	81	3	4	800	80	240	6	7
280	28	84	3	4	820	82	246	6	7
290	29	87	3	4	840	84	252	6	7
300	30	90	3	4	860	86	258	6	7
320	32	96	3	4	880	88	264	8	9
340	34	102	3	4	900	90	270	8	9
360	36	108	3	4	920	92	276	8	9
380	38	114	4	5	940	94	282	8	9
400	40	120	4	5	960	96	288	8	9
420	42	126	4	5	980	98	294	8	9
440	44	132	4	5	1000	100	300	8	9

Table 8.32 Straight-sided splines

All dimensions in mm

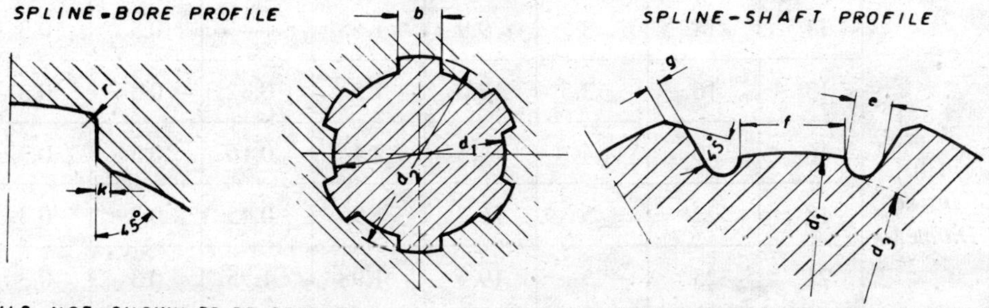

DETAILS NOT SHOWN TO BE SELECTED ACCORDING TO SUITABILITY.

THE FLANK OF EACH SPLINE KEY MUST BE PARALLEL UP TO THE MEETING POINT WITH THE INSIDE DIA.

(NO. OF SPLINE - KEYS SHOWN HERE = 6)

Light duty series

No. of spline keys	*Centering*	d_1	d_2	*b*	d_3 *min.*	*e max.*	*f*	*g max.*	*k max.*	*r* max.
6	*On Inside diameter*	23	26	6	22.1	1.25	3.54	0.3	0.3	0.2
		26	30	6	24.6	1.84	3.85	0.3	0.3	0.2
		28	32	7	26.7	1.77	4.03	0.3	0.3	0.2
8	Inside diameter or Flansk	32	36	6	30.42	1.89	2.71	0.4	0.4	0.3
		36	40	7	34.5	1.78	3.46	0.4	0.4	0.3
		42	46	8	40.4	1.68	5.03	0.4	0.4	0.3
		46	50	9	44.62	1.61	5.75	0.4	0.4	0.3
		52	58	10	49.7	2.72	4.89	0.5	0.5	0.5
		56	62	10	53.6	2.76	6.38	0.5	0.5	0.5
		62	68	12	59.82	2.48	7.31	0.5	0.5	0.5
10		72	78	12	69.6	2.54	5.45	0.5	0.5	0.5
		82	88	12	79.32	2.67	8.62	0.5	0.5	0.5
		92	98	14	89.44	2.36	10.08	0.5	0.5	0.5
		102	108	16	99.9	2.23	11.49	0.5	0.5	0.5
		112	120	18	108.8	3.23	10.72	0.5	0.5	0.5

Medium duty series

No. of spline keys	Centering	d_1	d_2	b	d_3 min.	e max.	f	g max.	k max.	r max.
6	On Inside Diameter	11	14	3	9.9	1.55	–	0.3	0.3	0.2
		13	16	3.5	12.0	1.5	0.32	0.3	0.3	0.2
		16	20	4	14.54	2.1	0.16	0.3	0.3	0.2
		18	22	5	16.7	1.95	0.45	0.3	0.3	0.2
		21	25	5	19.5	1.98	1.95	0.3	0.3	0.2
		23	28	6	21.3	2.3	1.34	0.3	0.3	0.2
		26	32	6	23.4	2.94	1.65	0.4	0.4	0.3
		28	34	7	25.9	2.94	1.70	0.4	0.4	0.3
8	On Inside Diameter or Flanks	32	38	6	29.4	3.3	0.15	0.4	0.4	0.3
		36	42	7	33.5	3.01	1.02	0.4	0.4	0.3
		42	48	8	39.5	2.91	2.57	0.4	0.4	0.3
		46	54	9	42.7	4.1	0.86	0.5	0.5	0.5
		52	60	10	48.7	4.0	2.44	0.5	0.5	0.5
		56	65	10	52.2	4.74	2.5	0.5	0.5	0.5
		62	72	12	57.8	5.0	2.4	0.5	0.5	0.5
10		72	82	12	67.4	5.43	–	0.5	0.5	0.5
		82	92	12	77.1	5.4	3.0	0.5	0.5	0.5
		92	102	14	87.3	5.2	4.5	0.5	0.5	0.5
		102	112	16	97.7	4.9	6.3	0.5	0.5	0.5
		112	125	18	106.3	5.4	4.4	0.5	0.5	0.5

Heavy duty series

No. of spline keys	*Centering*	d_1	d_2	b	d_3 *min.*	g *max.*	k *Max.*	r *max.*
10	On Inside diameter on flanks	16	20	2.5	14	0.3	0.3	0.15
		18	23	3	15.6	0.3	0.3	0.15
		21	26	3	18.44	0.3	0.3	0.15
		23	29	4	20.3	0.3	0.3	0.15
		26	32	4	23	0.4	0.4	0.15
		28	35	4	24.4	0.4	0.4	0.25
		32	40	5	28	0.4	0.4	0.25
		36	45	5	31.3	0.4	0.4	0.25
16	On flanks	42	52	6	36.9	0.5	0.4	0.4
		46	56	7	40.9	0.5	0.5	0.4
		52	60	5	47	0.5	0.5	0.4
		56	65	5	50.5	0.5	0.5	0.4
		62	72	6	56.1	0.5	0.5	0.4
		72	82	7	65.9	0.5	0.5	0.4
20		82	92	6	75.6	0.5	0.5	0.4
		92	102	7	85.5	0.5	0.5	0.4
		102	115	8	93.7	0.5	0.5	0.4
		112	125	9	103.7	0.5	0.5	0.4

Tolerances for spline hub and shaft

<table>
<tr><td rowspan="3">Spline Hub</td><td rowspan="3" colspan="2">For centering of inside diameter or on flanks</td><td>b</td><td>+</td><td>d_1</td><td>d_2</td></tr>
<tr><td>Hub unhardened</td><td>Hub hardened</td><td>Hub hardened or unhardened</td><td>Hub hardened or unhardened</td></tr>
<tr><td>D9</td><td>F10</td><td>H7</td><td>H11</td></tr>
<tr><td rowspan="4">Spline Shaft</td><td rowspan="2">For centering on Inside diameter</td><td>Shaft sliding Inside hub</td><td>h8</td><td>e8</td><td>f17</td><td rowspan="2">a11</td></tr>
<tr><td>Shaft fixed inside hub</td><td>p6</td><td>h6</td><td>j6</td></tr>
<tr><td rowspan="2">For centering on Flanks</td><td>Shaft sliding Inside hub</td><td>h8</td><td>e8</td><td>-</td><td rowspan="2">a11</td></tr>
<tr><td>Shaft fixed inside Hub</td><td>u6</td><td>k6</td><td>–</td></tr>
</table>

8.4 RIVETS AND PINS

Rivets: Rivets are permanent fasteners as distinct from removeable ones like screws, bolts and similar fasteners. Rivets are generally used to hold sheet metals or rolled steel sections together. Depending on the requirements, parts can be joined together by rivetting in the shop or they may be rivetted at the site of construction.

For general engineering purposes, hot forged steel rivets for hot closing as per IS: 1929 and cold forged steel rivets for hot closing as per IS: 2155 are used. Dimensions and other data concerning these rivets are given in Tables 8.33 to 8.40.

Material of rivets: For hot forged steel rivets for hot closing, the usual material is steel conforming to IS: 1148 and IS: 1149. For cold forged steel rivets for hot closing, the steel should conform to IS: 7557. Both the types can be made from steel conforming to IS: 226, if the steel meets certain test requirements. Rivets used for boilers and pressure vessels are of special kind and material. These are covered in other standard codes.

Pins: Pins are used for various purposes in engineering components. Cylindrical pins are used in shopwork and assemblies as dowel pins for location and alignment of mating parts. Dowel pins are usually hardened and ground to meet specific requirements.

Taper pins are effective for light work to fasten hubs or collars to shafts and for force fit applications, *e.g.*, knuckle joints.

Grooved taper pins are used as fastening elements which fit firmly in a hole made by a twist drill without any reaming being necessary.

Tables 8.41 and 8.42 show the relevant dimensions of cylindrical pins and grooved taper pins, respectively.

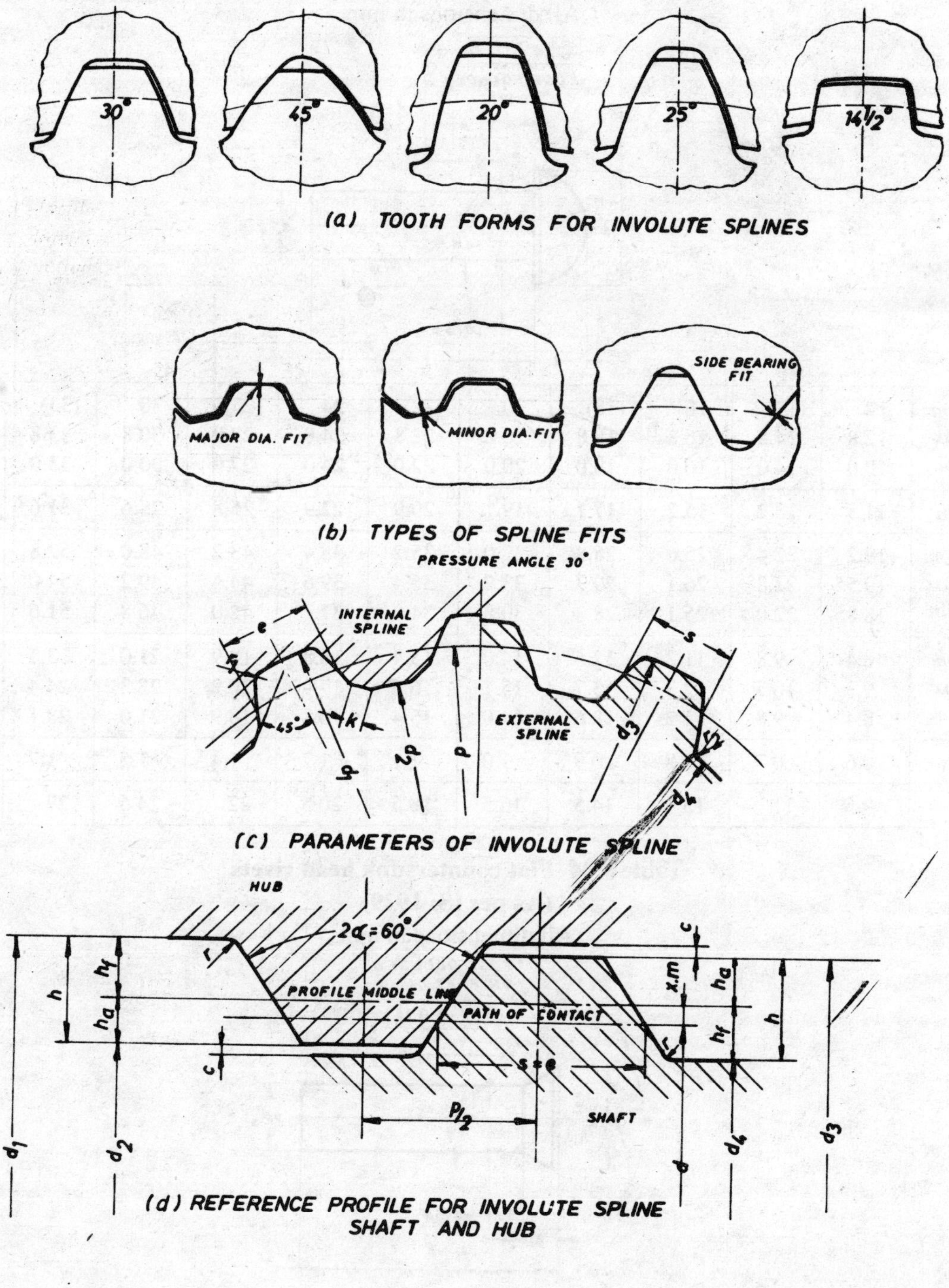

Fig. 8.3 Involute spline

Table 8.33. Snap head rivets
(As per IS: 1929)

All dimensions in mm

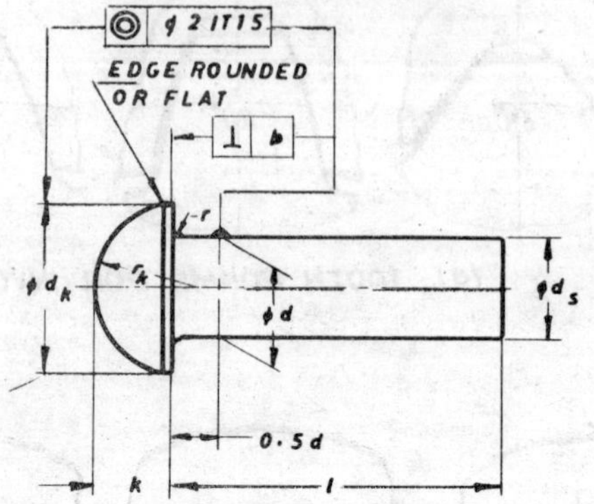

Nom	12	(14)	16	(18)	20	(22)	24	(27)	30	(33)	36
d Max	12.8	14.8	16.8	18.8	20.8	22.8	24.8	27.8	30.8	33.8	36.8
Min	12.0	14.0	16.0	18.0	20.0	22.0	24.0	27.0	30.0	33.0	36.0
d_2 *Min*	11.3	13.2	15.2	17.1	19.1	20.9	22.9	25.8	28.6	31.6	34.6
Nom	19.2	22.4	25.6	28.8	32.0	35.2	38.4	43.2	48.0	52.8	57.6
d_k *Max*	19.55	22.8	26.1	29.3	33.2	36.4	39.6	44.4	49.2	54.0	59.0
Min	18.85	22.0	25.1	28.3	30.8	34.0	37.2	42.0	46.8	51.6	56.2
Nom	8.4	9.8	11.2	12.6	14.0	15.4	16.8	18.9	21.0	23.1	25.2
k Max	9.3	10.7	12.3	13.7	15.1	16.5	17.9	20.2	22.3	24.4	26.5
Min	8.4	9.8	11.2	12.6	14.0	15.4	16.8	18.9	21.0	23.1	25.2
r Max	0.6	0.7	0.8	0.9	1.0	1.1	1.2	1.4	1.5	1.7	1.8
$r_k \approx$	9.5	11	13	14.5	16.5	18.5	20.5	22	24.5	27	30

Table 8.34 Flat countersunk head rivets
(As per IS: 1929)

All dimensions in mm

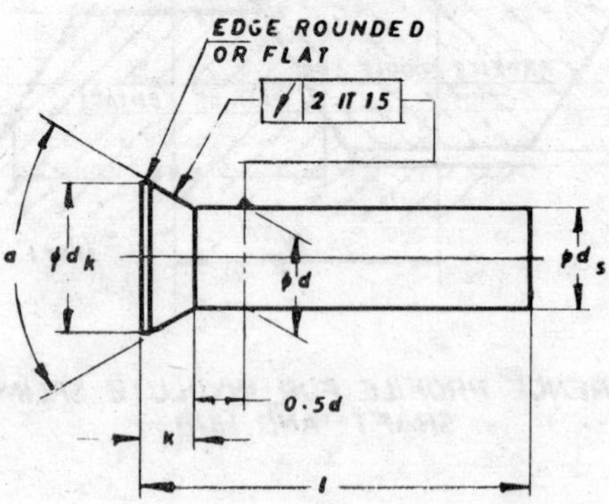

d Nom	12	(14)	16	(18)	20	(22)	24	(27)	30	(33)	36
d Max	12.8	14.8	16.8	18.8	20.8	22.8	24.8	27.8	30.8	33.8	36.8
d Min	12.0	14.0	16.0	18.0	20.0	22.0	24.0	27.0	30.0	33.0	36.0
d_s Min	11.3	13.2	15.2	17.1	19.1	20.9	22.9	25.8	28.6	31.6	34.6
d_k Nom	18	21	24	27	30	33	36	40.5	45	49.4	54
d_k Max	18.0	21.0	24.0	27.0	30.0	33.0	36.0	40.5	45.0	49.5	54.0
d_k Min	16.9	19.7	22.7	25.7	28.7	31.4	34.4	38.9	43.4	47.6	52.1
k Ref ≈	5.2	6.1	6.9	7.8	8.7	9.5	10.4	11.7	13.0	14.3	15.6
$\alpha {}^{+5^\circ}_{0}$	60°										

Table 8.35 Flat head rivets
(AS per IS:1929)
All dimensions in mm

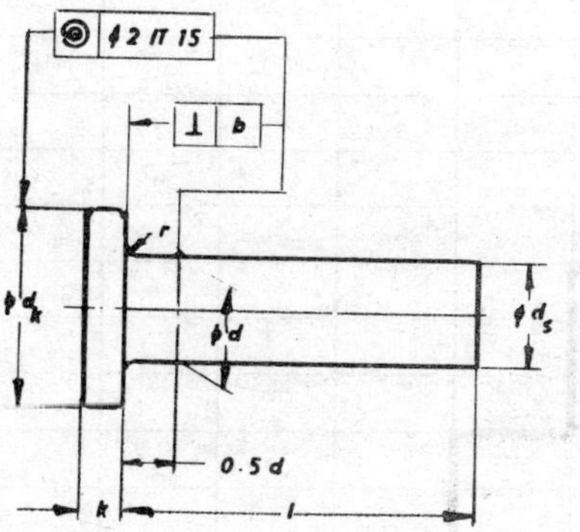

d Nom	12	(14)	16	(18)	20	(22)	24	(27)	30	(33)	36
d Max	12.8	14.8	16.8	18.8	20.8	22.8	24.8	27.8	30.8	33.8	36.8
d Min	12.0	14.0	16.0	18.0	20.0	22.0	24.0	27.0	30.0	33.0	36.0
d_s Min	11.3	13.2	15.2	17.1	19.1	20.9	22.9	25.8	28.6	31.6	34.6
d_k Nom	24	28	32	36	40	44	48	54	60	66	72
d_k Max	24.0	28.0	32.0	36.0	40.0	44.0	48.0	54.0	60.0	66.0	72.0
d_k Min	22.8	26.7	30.4	34.4	38.4	42.4	46.4	52.1	58.1	64.1	70.1
k Nom	3.0	3.5	4.0	4.5	5.0	5.5	6.0	6.75	7.50	8.25	9.0
k Max	3.6	4.25	4.75	5.25	5.75	6.25	6.75	7.65	8.40	9.15	9.9
k Min	3.0	3.50	4.0	4.5	5.0	5.5	6.0	6.75	7.50	8.25	9.0
r max	0.6	0.7	0.8	0.9	1.0	1.1	1.2	1.4	1.5	1.7	1.8

Table 8.36 Diameter-length combinations for hot forged rivets

All dimensions in millimeters

l Tol. $^{+1.5}_{0}$ For $d < 16$; $^{+3.0}_{0}$ For $d > 16$	Nominal Diameter d										
	12	*(14)*	*16*	*(18)*	*20*	*(22)*	*24*	*(27)*	*30*	*(33)*	*36*
28											
30											
32											
35											
38											
40											
45											
50											
55											
60											
65											
70											
75											
80											
85											
90											
95											
100											
105											
110											
115											
120											
125											
130											
135											
140											
145											
150											
155											
160											
165											
170											
175											

180												
185												
190												
195												
200												
205												
210												
215												
220												
225												
RIVET HOLE DIAMETER (for reference)	Basic Tol H12	13.5	15.5	17.5	19.5	21.5	23.5	25.5	29.0	32.0	35.0	38.0

Note 1 : The nominal diameters in parenthesis are of second preference.
Note 2 : The preferred lengths are between the stepped lines.

Table 8.37 Snap head rivets
(As per IS: 2155)
All dimensions in mm

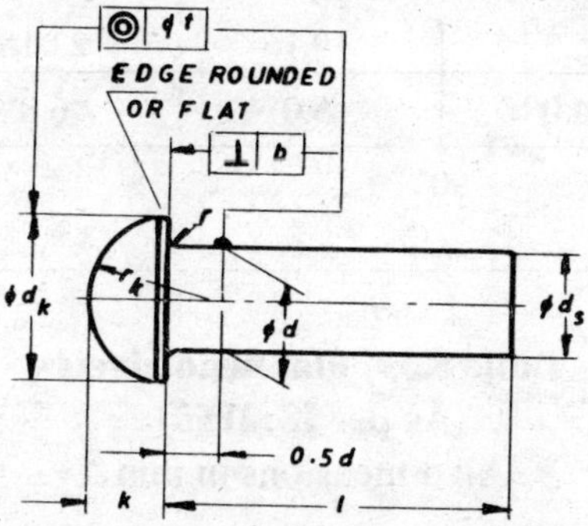

d *Nom* / *Max* / *Min*	6 6.15 5.85	8 8.15 7.85	10 10.3 9.7	12 12.3 11.7	(14) 14.3 13.7	16 16.3 15.7
d_s *Min*	5.82	7.76	9.4	11.3	13.2	15.2
d_k *Nom* / *Max* / *Min*	9.6 9.9 9.3	12.8 13.1 12.5	16.0 16.36 15.64	19.2 19.60 18.8	22.4 22.8 22.0	25.6 26.1 25.1
k *Nom* / *Max* / *Min*	4.2 4.44 3.96	5.6 5.84 5.36	7.0 7.29 6.71	8.4 8.69 8.11	9.8 10.09 9.51	11.2 11.55 10.85
r Max	0.3	0.4	0.5	0.6	0.7	0.8
r_k *M* ≈	5.7	7.5	8	9.5	11	13

Table 8.38 Flat countersunk head rivets
(As per IS: 2155)
All dimensions in mm

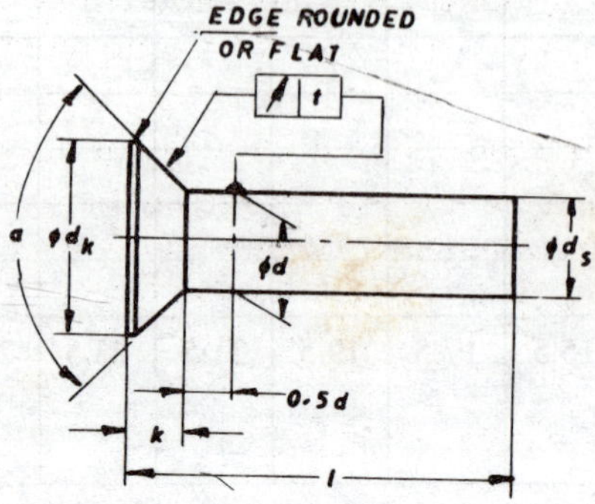

Nom d Max Min	6 6.15 5.85	8 8.15 7.85	10 10.3 9.7	12 12.3 11.7	(14) 14.3 13.7	16 16.3 15.7
d_s Min	5.82	7.76	9.4	11.3	13.2	15.2
Nom d_k Max Min	12 12 11.3	16 16 15.3	20 20 19.16	24 24 23.16	21 21 20.16	24 24 23.16
k Ref ≈	3.0	4.0	5.0	6.0	6.0	6.9
$\alpha + 5°$ 0	90°				60°	

Table 8.39 Flat head rivets
(As per IS:2155)
All dimensions in mm

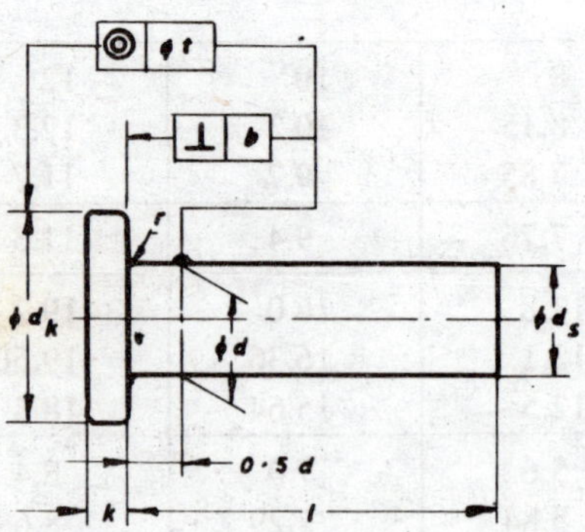

d	Nom	6	8	10	12	(14)	16
	Max	6.15	8.15	10.3	12.3	14.3	16.3
	Min	5.85	7.85	9.7	11.7	13.7	15.7
d_s	Min	5.82	7.76	9.4	11.3	13.2	15.2
d_k	Nom	12	16	20	24	28	32
	Max	12.0	16.0	20	24	28	32
	Min	11.3	15.3	19.16	23.16	27.16	31
k	Nom	1.5	2	2.5	3	3.5	4
	Max	1.9	2.5	3.0	3.6	4.1	4.6
	Min	1.5	2	2.5	3	3.5	4
r	Max	0.3	0.4	0.5	0.6	0.7	0.8

Table 8.40 Diameter-length combinations for cold forged rivets for hot closing

All dimensions in millimeters

l Tol $^{+0.5}_{0}$ For $d \leq 10$; $^{+1.0}_{0}$ For $d > 10$	Nominal Diameter d					
	6	*8*	*10*	*12*	*(14)*	*16*
12						
14						
16						
18						
20						
22						
24						
26						
28						
30						
32						
35						
38						
40						
42						
45						

48							
50							
55							
60							
65							
70							
75							
80							
85							
90							
95							
100							
105							
110							
RIVERT HOLE DIAMETER (for reference)	BASIC TOL H12	6.8	8.4	10.5	13	15	17

Note 1 : The nominal diameter in parenthesis is of second preference.
Note 2 : The preferred lengths are between the stepped lines

Table 8.41 Dimensions for cylindrical pins

All dimensions in millimeters

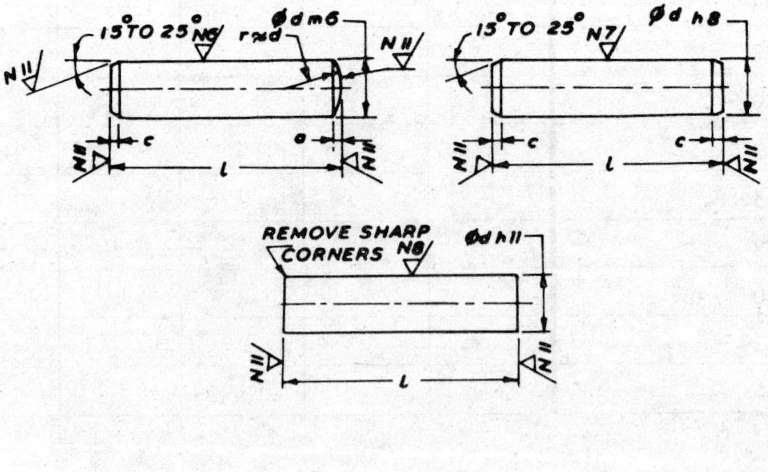

d m6/h8/h11 Nom	*1.5*	*2*	*2.5*	*3*	*4*	*5*	*6*	*8*	*10*	*12*	*16*	*20*	*25*	*30*	*40*	*50*
$a \approx$	0.2	0.25	0.3	0.4	0.5	0.63	0.8	1	1.2	1.6	2	2.5	3	4	5	6.3
$c \approx$	0.3	0.35	0.4	0.5	0.63	0.8	1.2	1.6	2	2.5	3	3.5	4	5	6.3	8

Material: Cylindrical pins conforming to the standard shall be manufactured from steel of grade C30 of IS: 5517-1972. "Steels for hardening and tempering" or any other suitable steel with a minimum tensile strength of 500 N/mm_2. At the request of purchaser, pins may also be supplied from free cutting steels conforming to IS:4431-1978. "Carbon and carbon-manganese free-cutting steels (*first revision*)."

Table 8.42 Dimensions of grooved taper pins

All dimensions in mm

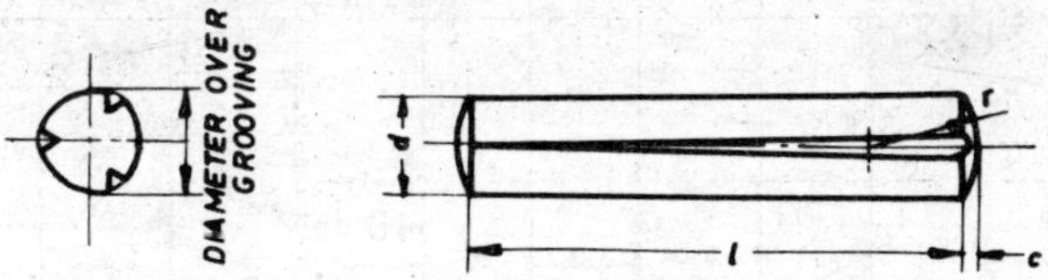

EXAMPLE OF APPLICATION

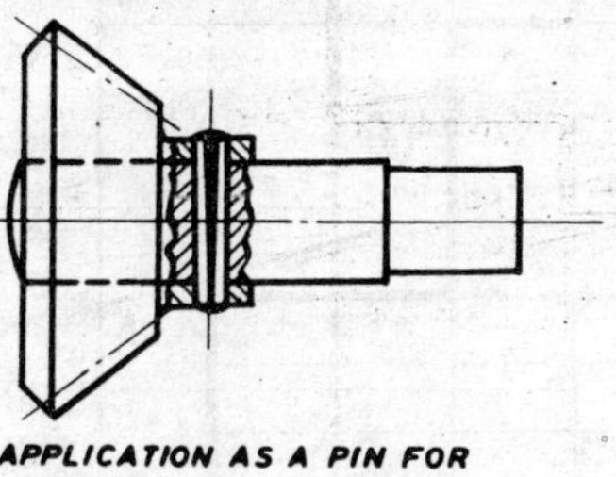

APPLICATION AS A PIN FOR SECURING A BEVEL GEAR

Nominal d	*1.5*	*2*	*2.5*	*3*	*4*	*5*	*6*	*8*	*10*	*12*	*(13)*	*14*	*16*	*20*	*25*
Diameter Tol	*h9*				*h11*										
C max	0.23	0.3	0.4	0.45	0.6	0.75	0.9	1.2	1.5	1.8	2	2	2.5	3	4
r Approx	1.6	2	2.5	3	4	5	6	8	10	12	12	16	16	20	25

Length l	Tol on l	Diameter over grooving														
4																
5																
6	+ 0.3	1.63			3.25											
8			2.20			4.30										
10							5.30									
12				2.70				6.30								
(14)					3.30											
16		1.60							8.35							
(18)						4.35				10.40						
20							5.35				12.40					
(22)																
24			2.15	2.65	3.25			6.35				13.45	14.45			
(26)									8.40					16.55		
28																
(30)						4.30				10.45						
32	+ 0.5				3.20		5.30				12.45					
36												13.50	14.50			
40								6.30						16.60		
(45)									8.35							
50						4.25										
(55)							5.25			10.40	12.40					
60															20.60	25.60
(65)								6.25				13.45	14.45			
70									8.30					16.55		
80	+ 0.8									10.35						
90									8.25							
100											12.30					
(110)										10.30		13.40	14.40			
120														16.50		
Tolerance on diameter over groving				+ 0.05		± 0.05					± 0.10					

Note : Sizes shown in brackets are of second preference.

8.5 MISCELLANEOUS SPRINGS

A spring is a machine element designed to store energy when deflected and to release the same when unloaded. On principle, any thing made of elastic material can show the "spring effect", that is, it deforms when force is applied thereby storing energy which is given up as it comes back to its original shape when the force is removed. However, for industrial applications, only certain materials with specific properties which are conducive to spring effect are used as spring materials, the foremost amongst them being the class of steels known as "spring steels". The prerequisites for such materials are: high strength, elastic limit and resilience, ability to retain the elasticity upto the maximum possible loading condition, and a high fatigue strength to avoid fatigue failure. For common purposes, high carbon spring steel wires with proper treatment and having an ultimate strength of around 3000 N/mm^2 are normally employed. Alloy steels having Si and Mn as well as Cr and V are also used. For springs subjected to corrosive surroundings, special types of stainless steels are in use. Besides, non-ferrous materials like bronze, brass and similar materials are also used in certain conditions.

Springs are classified as helical springs and flat springs. Helical springs are generally of three types: compressions springs, extension (or tension) springs and torsion springs. Flat springs are used as leaf springs and laminated springs commonly used in wagons and other types of vehicles and also in various equipments. Besides, there are other types of springs used in industries, such as, volute springs, ring springs, conical springs, disc springs and various combinations thereof.

Springs are virtually never drawn as true projections in working drawings. Instead, they are represented by approved, conventional schematic drawings. Some manufacturing concerns use standard printed proformas, carrying schematic drawing of the springs, on which simply the load characteristics, main dimensions, and conditions and other relevant data are entered. This way time and labour are saved.

The drawing representations and the relevant drawing data pertaining to common springs are discussed here. Calculation regarding springs are not included as they are beyond the scope of this book. Inclusion of the load diagram alongside the drawing of the spring will depend on its necessity and on the discretion of the designer. For ordinary purposes, it is enough to show the drawing of the spring with its dimensions. In Fig. 4.32, Section 4.4, the conventional representations of all kinds of springs for drawing making purposes have been shown. Symbolic representations of springs shown in that figure are generally used in general arrangement drawings. In working drawings, some designers prefer to show the spring data, *viz.*, outside diameter, mean diameter, inside diameter, wire diameter, free length, allowable deviations, spring rate or stiffness and other spring parameters in tabular form. This table is placed by the side of the spring drawing, as is done in case of gear drawings. A spring drawing should also contain a note regarding heat treatment and any other instructions, if necessary.

Compression spring : Compression spring and tension spring are the most common types of springs used in mechanical equipments. In the case of both the types of helical springs, spring steel wires up to 10 mm diameter are used for making the springs by the cold coiling method. For those exceeding 17 mm, hot coiling is used. For intermediate diameters, the manufacturing method will depend upon the stress involved, material and the application. In Fig. 8.4, the common end conditions of compression spring are

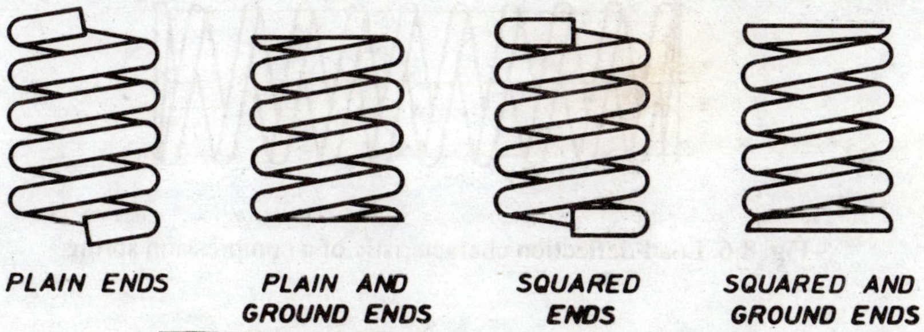

Fig. 8.4 End conditions of compression springs

shown. Figure 8.5 shows the end conditions of cold coiled and hot coiled springs. In Fig. 8.6, the compression spring diagram showing the load characteristics has been depicted. In the figure, the symbol "*D*" with the subscripts "*o*", "*i*" stand for the outside and the inside diameter of the coil respectively, while *D* without subscript stands for the mean diameter. Other parameters are: d = Diameter of wire, L (with the relevant subscripts) = Lengths of the spring, F (with the relevant subscripts) = Forces to which the spring are subjected, f (with subscripts) = Deflections of the spring corresponding to the forces. Thus, L_n is the length of loaded spring corresponding to the force F_n, the deflection being L_0 (free length) – L_n = f_n mm.

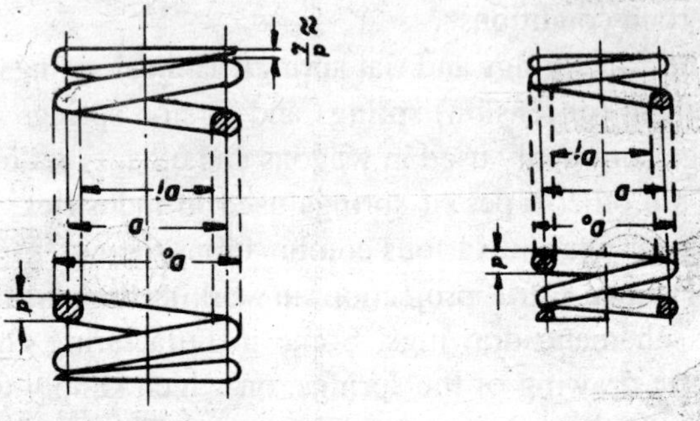

Fig. 8.5 End conditions of cold and hot coiled springs

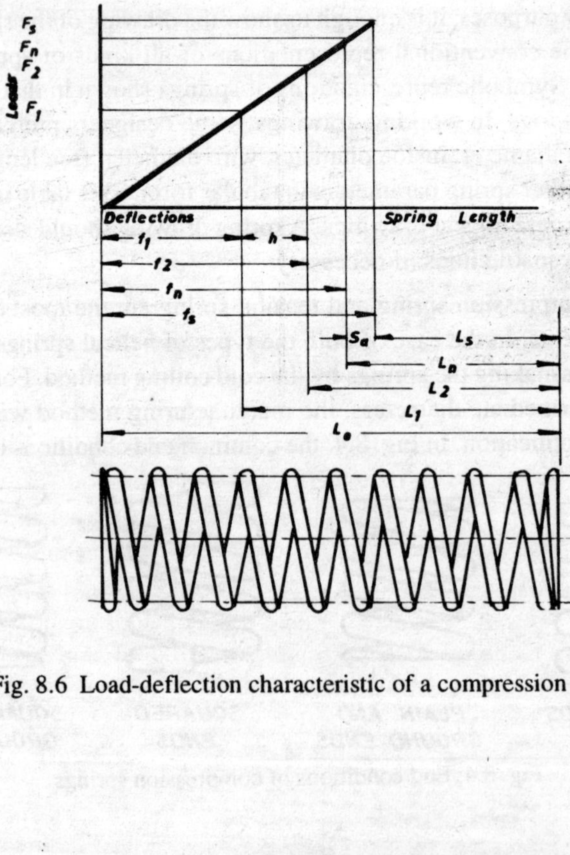

Fig. 8.6 Load-deflection characteristic of a compression spring

Tension spring: Helical tension springs are normally made by the cold coiling method with initial tension for wires upto 17 mm diameter, over which the springs are finish-tempered. Springs are normally coiled right-handed. Loops of different shapes are shown in Fig. 8.8. The force of the tension spring is transmitted through these loops. In Fig. 8.8, L_H is the distance of the inside edge of the loop from the spring body of which the inside diameter of coil is D_1.

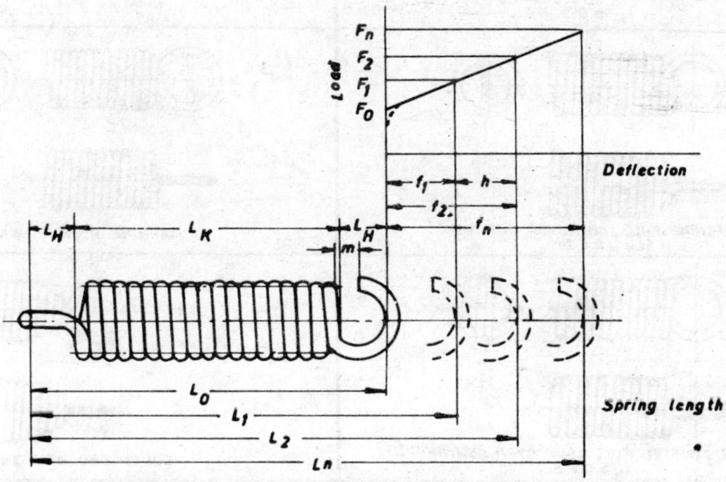

Fig. 8.7 Load-deflection characteristic of a tension spring

In Fig. 8.9 to 8.14, diagramatic representations of some of the common types of springs are shown.

8.6 CENTRE HOLES

Centre holes are generally provided at the ends of work-pieces to facilitate proper mounting of the job on lathe-centres and on other equipments to carry out machining operations or for other purposes such as, inspection, testing, measurement, control, etc. Four types of centre holes which are in common use are shown in Tables 8.43 and 8.44.

It is often required to indicate the appropriate centre holes in an engineering drawing and the following data are intended to provide practical guide-lines to the draughtsman to select them.

The angle of the conical portion of the centre hole is 60° in all the types, and is valid for work-pieces having a mass of upto 100 kg and where the cutting forces are small to medium in magnitude. For jobs with masses over 100 kg and where the cutting forces are large, the angle is usually 90°.

When the centre hole is intended to be cut off after the operation is over, then the corresponding raw-stock should at first be taken to be longer by an amount of "l_1" which is the cut-off length in each case. In cases where the centre holes are intended to be retained after operation, then the types *B* and *C* should be preferred. These types are protected against damage. In special cases, hardened plugs are inserted in centre holes for protection against damage when the jobs are not in use. In some centre holes, the cylindrical portions are threaded to accept fittings, extractor tools and other components. In particular, shaft ends of electrical machinery are provided with threaded centre holes.

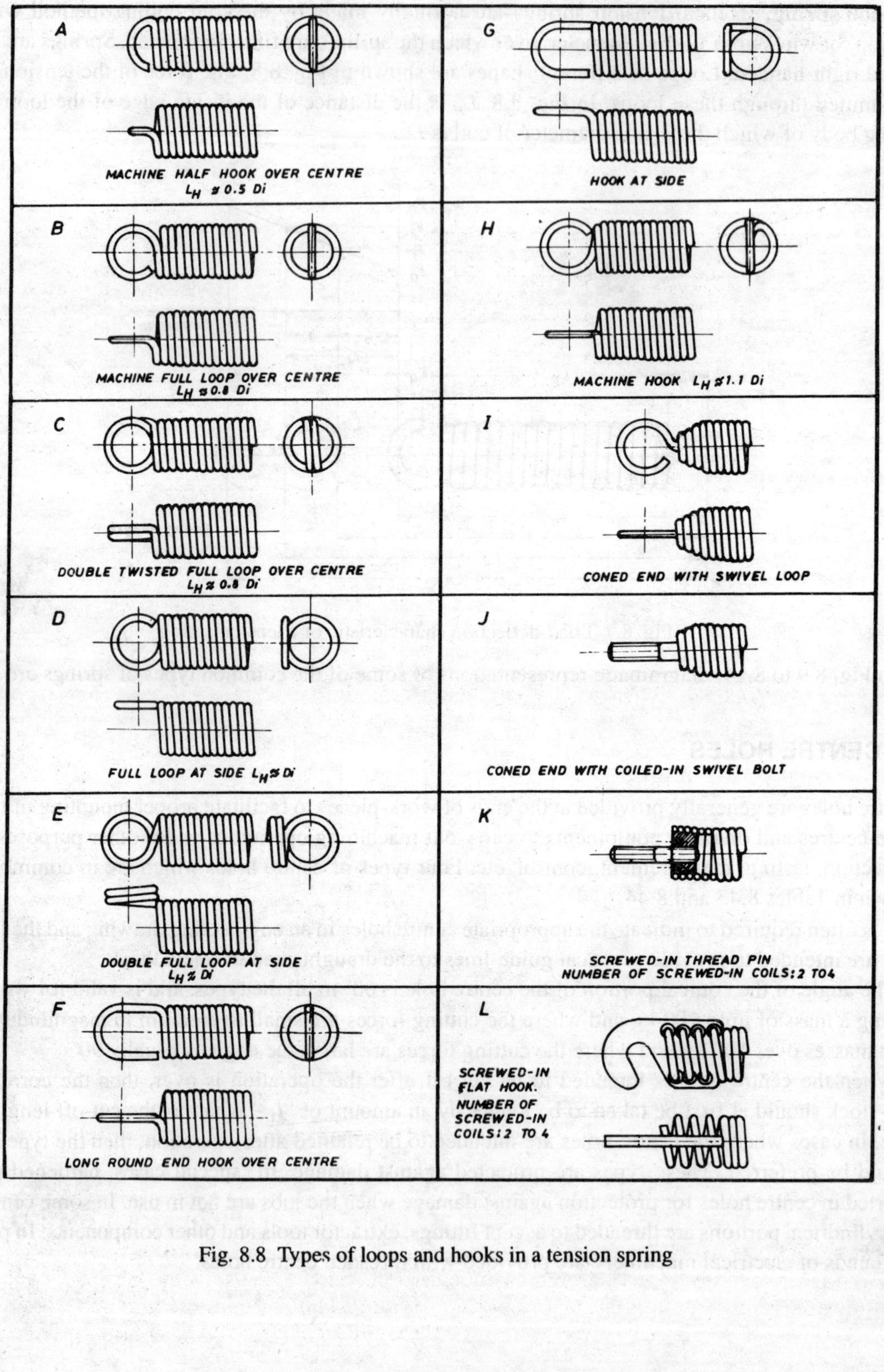

Fig. 8.8 Types of loops and hooks in a tension spring

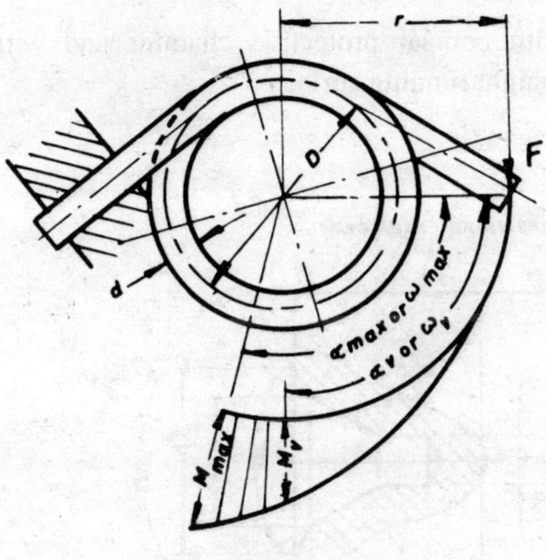

Fig. 8.9 Torsion spring

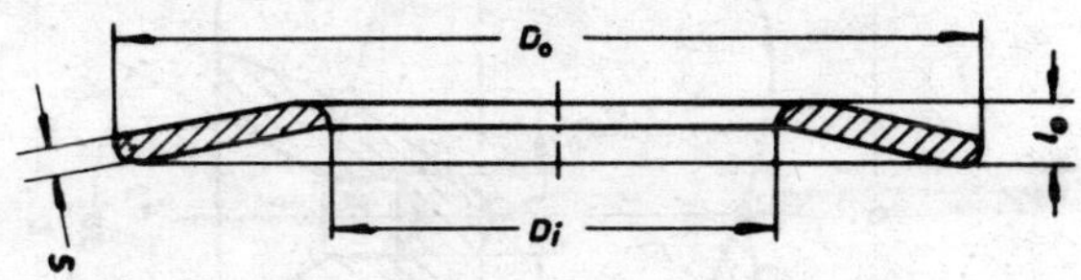

Fig. 8.10. Disc or Belleville spring

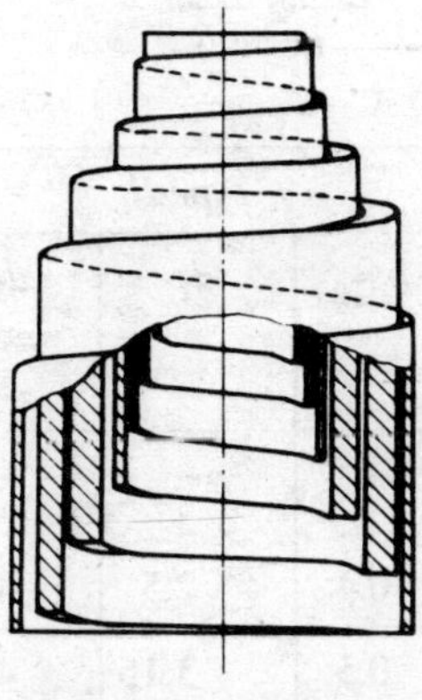

Fig. 8.11. Volute spring

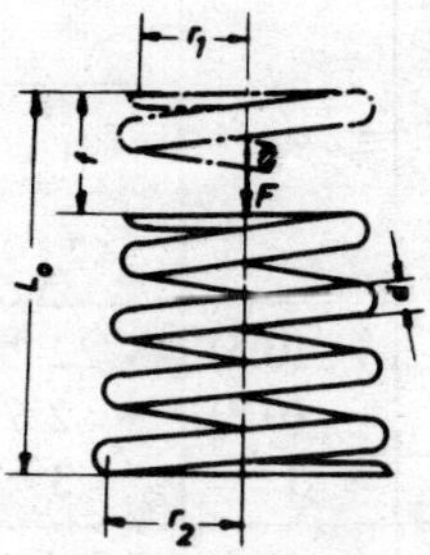

Fig. 8.12. Conical spring

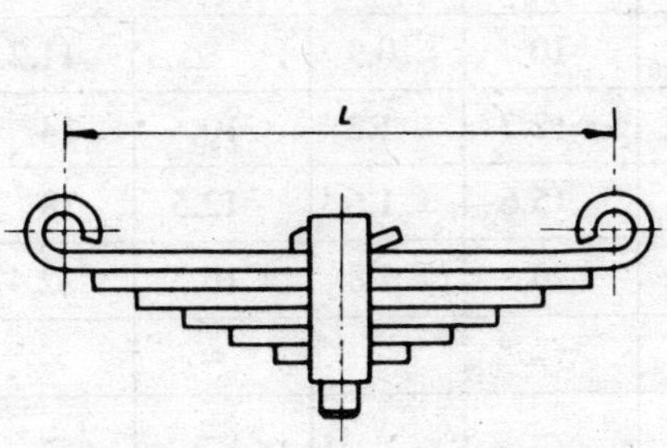

Fig. 8.13. Laminated leaf spring

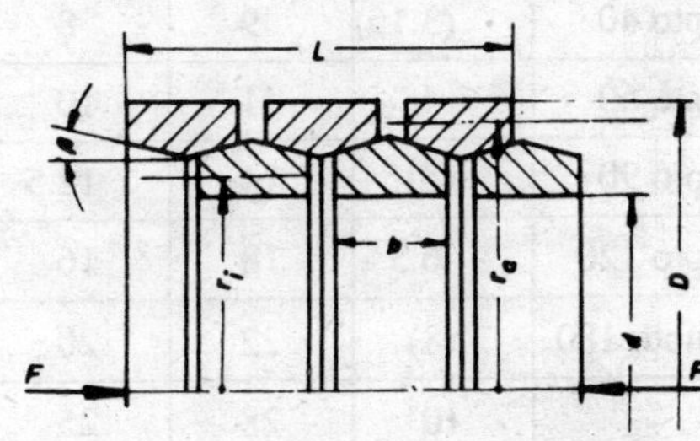

Fig. 8.14. Ring spring

Table 8.43 Data for centre holes, types A and B

Type A. Without protective chamfer and with straight running surfaces

Type B. With conical protective chamfer and with straight running surfaces.

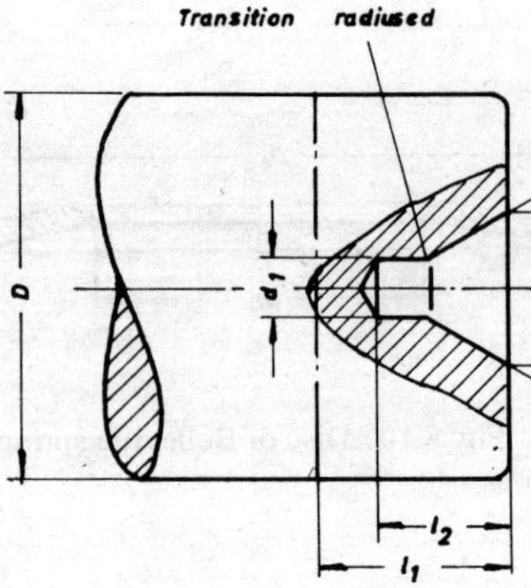

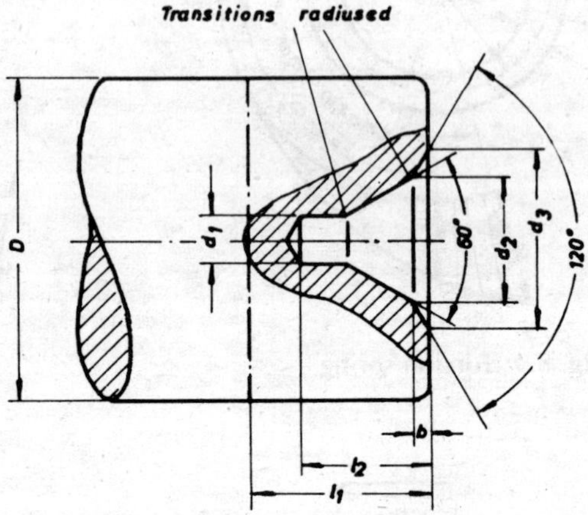

Diameter of work piece	d_1	Type A l_1	Type A d_2	Type A l_2 Minimum	Type B l_1	Type B $b \approx$	Type B d_2	Type B d_3	Type B l_2 Minimum
3 upto 9	(0.5)	2	1.6	1.4	–	–	–	–	–
	(0.8)	2.5	2	1.8	–	–	–	–	–
	1	3	2.5	2.2	3.5	0.4	2.5	4	2.6
Over 9 upto 12	(1.25)	4	3.15	2.8	4.5	0.5	3.15	4.75	3.3
Over 12 upto 15	1.6	5	4	3.5	5.5	0.5	4	5.6	4
Over 15 upto 20	(2)	6	5	4.5	6.6	0.6	5	7.1	5.1
Over 20 upto 30	2.5	7	6.3	5.5	8.3	0.8	6.3	9	6.3
Over 30 upto 40	(3.15)	9	8	7	10	0.9	8	11.2	7.9
Over 40 upto 60	4	11	10	9	12.7	1.2	10	14	10.2
Over 60 upto 90	(5)	14	12.5	11	15.6	1.6	12.5	18	12.6
Over 90 upto 120	6.3	18	16	14	20	1.8	16	22.4	15.8
Over 120 upto 180	(8)	22	20	18	–	–	–	–	–
Over 180	10	28	25	22	–	–	–	–	–

Table 8.44 Data for centre holes, types C and D

Type C. With protective chamfer of truncated conical form and with straight running surfaces.

Type D. Without protective chamfer and with convex running surface.

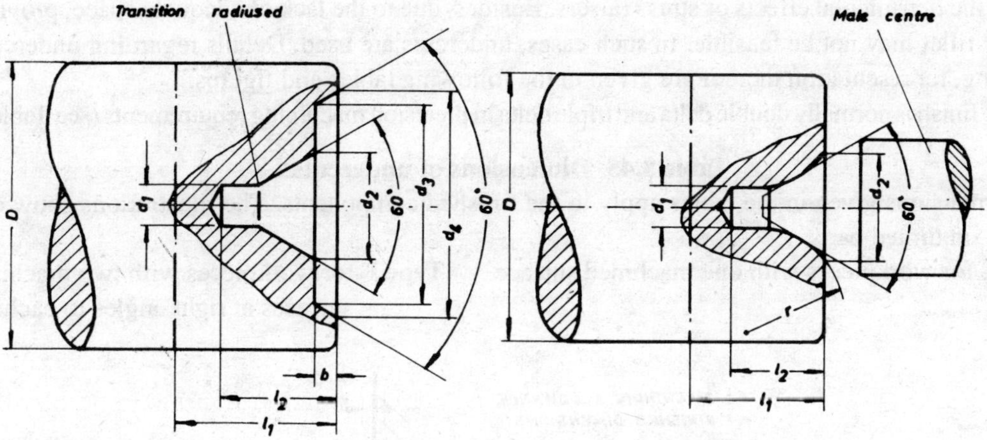

Diameters of work piece	d_1	Type C						Type D			
		l_1	$b \approx$	d_2	d_3	d_4	l_2 Minimum	l_1	d_2	r	l_2 Minimum
	(0.5)	–	–	–	–	–	–	2	1.6	2	1.4
3 upto 9	(0.8)	–	–	–	–	–	–	2.5	2	2.5	1.8
	1	3.6	0.6	2.5	5.3	6	2.8	3	2.5	3.15	2.2
Over 9 upto 12	(1.25)	4.7	0.7	3.15	6.3	7.1	3.5	4	3.15	4	2.8
Over 12 upto 15	1.6	6	0.9	4	7.5	8.5	4.4	5	4	5	3.5
Over 15 upto 20	2	7	0.9	5	9	10	5.4	6	5	6.3	4.5
Over 20 upto 30	2.5	8.6	1.1	6.3	11.2	12.5	6.6	7	6.3	8	5.5
Over 30 upto 40	(3.15)	10.7	1.7	8	14	16	8.7	9	8	10	7
Over 40 upto 60	4	13.2	1.7	10	18	20	10.7	11	10	12.5	9
Over 60 upto 90	(5)	16.3	2.3	12.5	22.4	25	13.3	14	12.5	16	11
Over 90 upto 120	6.3	21	3.0	16	28	31.5	17	18	16	20	14
Over 120 upto 180	8	26	3.9	20	35.5	40	21.9	22	20	25	18
Over 180	10	31.5	4.3	25	40	45	26.3	28	25	31.5	22

8.7 UNDERCUTS

Stepped shafts are generally provided with appropriate fillets at places where change of diameter takes place to alleviate the effect of stress concentration and other undesirable conditions. The standard dimensions of the fillets are given in Section 8.8. However, in special cases undercuts or relief grooves are provided at the positions of sudden changes in the cross-sections of the shafts. Such undercuts minimise the detrimental effects of stress raisers. Besides, due to the lack of adequate space, provision of a rounded fillet may not be feasible. In such cases, undercuts are used. Details regarding undercuts and the drawing, representation thereof are given in the following tables and figures.

Surface finish is normally double delta and triple delta in precision machining requirements (see Table 8.45).

Table 8.45 Dimensions of undercuts

The dimensions given in the tables apply to the finished components. The illustrations show outside undercuts on turned parts.

Type E for workpieces with one machined surface. **Type F** for workpieces with two machined surfaces at right angles to each other

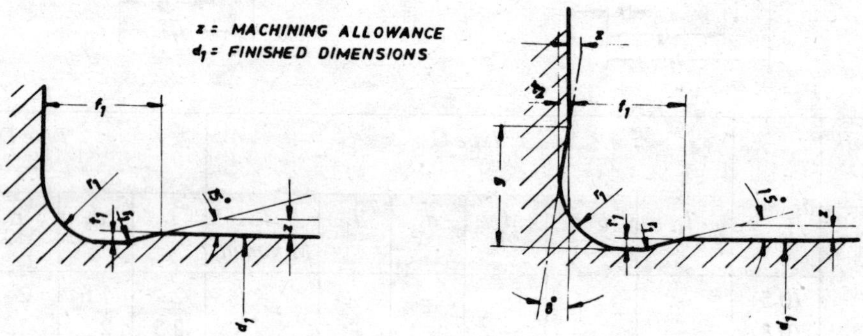

Dimensions in mm

r_1	t_1 + 0.1	f_1	$g \approx$	t_2 + 0.05	d_1 for normal duty conditions	d_1 with increased fatigue resistance
0.1	0.1	0.5	0.8	0.1	to 1.6	
0.2	0.1	1	0.9	0.1	above 1.6 to 3	–
0.4	0.2	2	1.1	0.1	above 3 to 10	
0.6	0.2	2	1.4	0.1	above 10 to 18	
0.6	0.3	2.5	2.1	0.2	above 18 to 80	–
1	0.4	4	3.2	0.3	above 80	
1	0.2	2.5	1.8	0.1		above 18 to 50
1.6	0.3	4	3.1	0.2		above 50 to 80
2.5	0.4	5	4.8	0.3	–	above 80 to 125
4	0.5	7	6.4	0.3		above 125

Table 8.46 Countersink in the companion part

Dimensions in mm

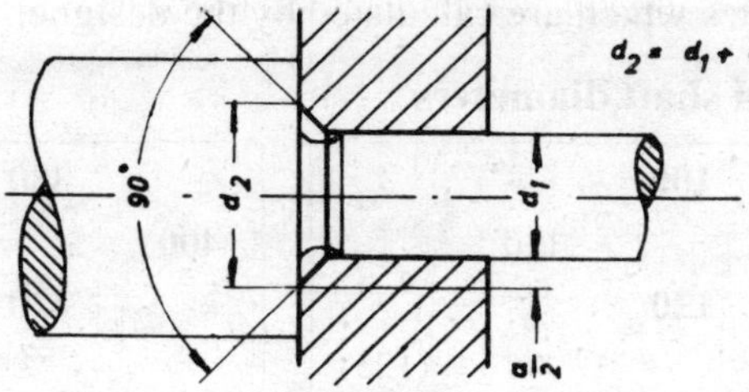

Undercut size $r_1 \times t_1$	a (min.) Type E	F
0.1 × 0.1	0	0
0.2 × 0.1	0.2	0
0.4 × 0.2	0.4	0
0.6 × 0.2	0.8	0.2
0.6 × 0.3	0.6	0
1 × 0.2	1.6	0.8
1 × 0.4	1.2	0
1.6 × 0.3	2.6	1.1
2.5 × 0.4	4.2	1.9
4 × 0.5	7	4.0

Undercuts shown in drawings can either be drawn in full, giving all the relevant dimensions, or they may be represented by simplified form with appropriate designation. This is illustrated in Fig. 8.15.

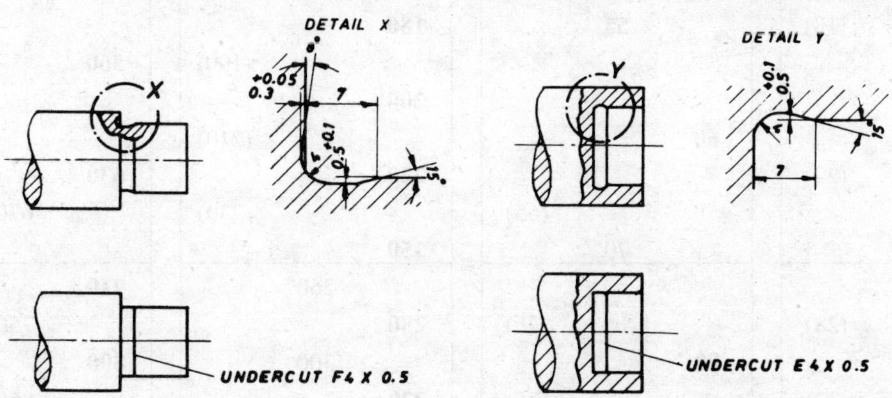

Fig. 8.15 Methods of representations of undercuts in drawings

Designation:

	Undercut type F of	Undercut type E of
	radius $r_1 = 4$ mm	radius $r_1 = 4$ mm
	depth $t_1 = 0.5$ mm	depth $t_1 = 0.5$ mm
	Undercut $F4 \times 0.5$	Undercut $E4 \times 0.5$

8.8 SHAFTS

Shafts are machine elements, the dimensions of which are determined as per the various design considerations involved. Hence shafts cannot be obtained as standardised items. Many diverse factors have to be taken into account while designing a shaft, the main considerations being the bending moment and the torque to which it is subjected. Stress concentration is another major design factor which determines the optimum shapes of the fillets in a stepped shaft.

Formulae and codes for shaft design are given in standard books on mechanical design. Standardised values of shaft diameters have been fixed for rationalisation and inventory control purposes. These are given in Table 8.47. Standard values of fillet radii, r, are given in Table 8.48. While making the drawing of a shaft, the draughtsman should stick to these standard dimensions given in the above Tables, unless special considerations call for ether non-standard values which are calculated by the designer.

Table 8.47 Standard shaft diameters

10	(38)	**100**	380
12	**40**	110	**400**
(14)	(42)	**120**	420
15	45	(125)	(440)
16	(48)	130	**450**
(17)	**50**	**140**	(460)
(18)	55	150	480
20	**60**	**160**	**500**
(22)	(65)	(170)	530
25	70	**180**	**560**
(28)	(75)	(190)	600
30	**80**	**200**	**630**
(32)	(85)	(210)	670
35	90	**220**	**710**
	(95)	(230)	750
		250	**800**
		260	850
		280	**900**
		300	950
		320	**1000**
		340	
		360	

Bracketed sizes shall not be preferred. Diameters in bold types shall be preferred.

Table 8.48 Fillet radii, *r*, for stepped shafts in torsion

D / *d*	*12*	*16*	20	25	30	40	50	60	80	*100*	*120*	*140*	*160*	*180*	200	220	250	280	*(300)*	320	*(340)*	360	*(380)*	400
10	0.5	1.0																						
12		1.0	1.5																					
14		0.5	1.0																					
16			1.0	1.5																				
18				1.5																				
20				1.5	2.0																			
22				1.0	1.5																			
25					1.5	2.0																		
28					1.0	2.0																		
30						2.0																		
(32)						1.5	2.5																	
35						1.5	2.5																	
38						1.0	2.5																	
40							2.5	3.0																
(42)							2.0	3.0																

45						1.5	3.0																	
(48)						1.0	2.5																	
50							2.5																	
55							1.5	4.0																
60								4.0																
(65)								3.5																
70								2.5	5.0															
(75)								1.5	5.0															
80									5.0	6.0														
(85)									3.5	6.0														
90									2.5	6.0														
(95)									2.0	5.5	6.5													
100										5.0	6.5													
(105)										3.5	6.5													
110										2.5	6.0	8.0												
120											5.0	8.0												
(125)											3.0	7.5	8.5											

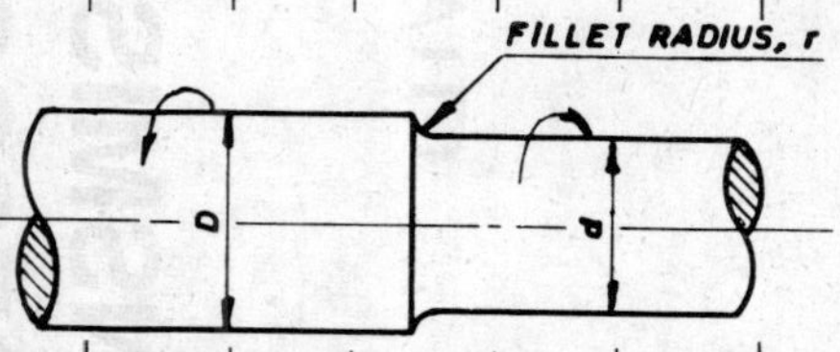

130											2.5	7.0	8.5											
140												5.0	8.5	10.0										
150												3.0	7.5	10.0										
160													5.5	9.5	11.0									
(170)													3.0	7.5	11.0									
180														5.5	9.5	12.0								
(190)														3.0	8.0	12.0								
200															5.5	11.0	14.0							
(210)															3.0	10.0	14.0							
220																8.0	13.0	16.0						
(230)																6.0	12.0	15.0						
240																5.0	11.0	15.0	17.0					
(250)																	7.5	13.0	16.0	17.0				
260																	5.5	11.0	14.0	16.0	17.0			
280																		5.5	10.0	13.0	16.0	18.0	20.0	
300																			6.0	10.0	15.0	18.0	20.0	

CHAPTER 9

COMMON TRANSMISSION ELEMENTS AND ACCESSORIES

9.1 TOOTHED GEARING

The subject of toothed gearing is a vast one and as such, only those aspects which are relevant from the drawing making point of view as well as some essential calculations which are required for the determination of dimensional and other parameters of different types of gearing will be discussed in this section.

In normal usage, gears are employed for transmission of torque and power between shafts. The driving and the driven shafts generally rotate at different speeds, measured in revolutions per minute (R.P.M.). In a gear box, two or more gears mounted on shafts and working together comprise a gear-drive system. A step-down gear-drive is one in which the output shaft rotates at a slower speed than the input shaft. A step-up gear-drive is one in which the output shaft rotates at a faster speed than the input shaft.

Gears operate in pairs. In a normal drive, the driving member is termed as the "pinion" and the driven member as the "gear". In general application, a gear-drive is usually a step-down one, so that the input member or the pinion rotates faster than the output member or the gear which rotates at a slower angular speed, depending on the transmission ratio. The torque delivered at the output end is thus greater than the torque received at the input end, and the system works as a torque converter. It will be noted here that the torque ratio is inversely proportional to the speed ratio, that is,

$$\frac{n_1}{n_2} = \frac{T_2}{T_1}$$

where n_1 and n_2 are the rotational speeds of the pinion and the gear respectively, and T_1 and T_2 are the input torque and the output torque respectively. Normally

$$n_1 > n_2\,,\ T_2 > T_1, \text{ and } z_2 > z_1,$$

where z_1 = Number of teeth of the pinion,

and z_2 = Number of teeth of the gear.

Classification of gears: Gears are classified as per the relation between the axes, shape of the solid on which the teeth are cut, curvature of the tooth trace and other special features. The general categorisation is given below:

Spur gears: In spur gears, the axes of mating components comprising the gear-drive are parallel. The reference or the pitch solid is a cylinder. The gear teeth are straight and parallel to the axes. A rack can be thought of as a spur gear of infinite diameter.

Helical gears: The parallel helical gears are similar to spur gears, except that the tooth traces are helical and not straight as in the case of spur gears. Helical gears have greater load carrying capacity, better contact relationship and quieter in operation. When the axes of a pair of helical gears are oriented at an angle relative to each other, they are called crossed helical gears.

Bevel gears: In bevel gearing, the axes are intersecting, the angle between the axes being usually 90°. The gear blank is a cone on which the teeth are developed. There are various types of bevel gears, the most common being the straight-sided bevel in which the teeth are straight which taper towards the cone apex from the big end, and the spiral bevel in which the tooth elements are curved in the shape of a spiral. Like helical gears, the spiral bevel gears have better load carrying capacity, better contact relation and quieter running *vis-a-vis* those of straight-sided bevels.

Worm and Worm-wheel: In worm-gearing, the axes are non-intersecting and the planes containing the axes are normally at right angles to each other. The tooth elements of both the components are helices.

Besides, there are numerous other gearing systems, *e.g.,* herringbone gears, Zerel bevels, Hypoid gears, crown or face gears, *etc*., and these are beyond the scope of this book. We will discuss in this section only those gearing systems which are in common usage.

Terminology: The nomenclatures used in this book will be as per IS: 2458 and IS: 2467. These specifications are mainly based on the ISO Recommendation No. 888 entitled "International Vocabulary of Gears", and ISO Recommendation R 701 entitled "International Gear Notation, Symbols for Geometrical Data".

Pitch circle: This is the circumference of an imaginary cylinder which rolls without slipping when in contact with another such cylinder as in friction drive. In any gear drive, the pitch circle diameter (P.C.D.) is the reference parameter from which other parameters are developed. The PCD is denoted by the letter "d" with proper subscripts, *e.g.*, d_1 for pinion, d_2 for gear, *etc.*

Tip circle: This is the outside circle of the gear and is denoted by "d_a" with the usual subscripts.

Root circle: This circle passes through the roots of the teeth and its diameter is denoted by "d_f".

Base circle: This is the circle from which the involute teeth are generated.

Addendum: It is the radial distance between the pitch circle and the tip circle, and is denoted by "h_a"

Dedendum: This is the radial distance between the pitch circle and the root circle, and is denoted by "h_f".

Module: It is defined as the ratio of the pitch diameter to the number of teeth of a gear. It is expressed in millimetres and is one of the major and determining parameters of a gear. In the FPS system, the diametral pitch denoted by "DP" is the reciprocal of module, but the unit of DP is inch^{-1}.

Pressure angle: It is the angle between a tooth profile and a radial line at its pitch point. The standard pressure angles are 14 1/2°, 20°, 25°, etc., the 20° angle being the most common pressure angle.

Figure 9.1 and 9.2 show the different gear parameters pictorially.

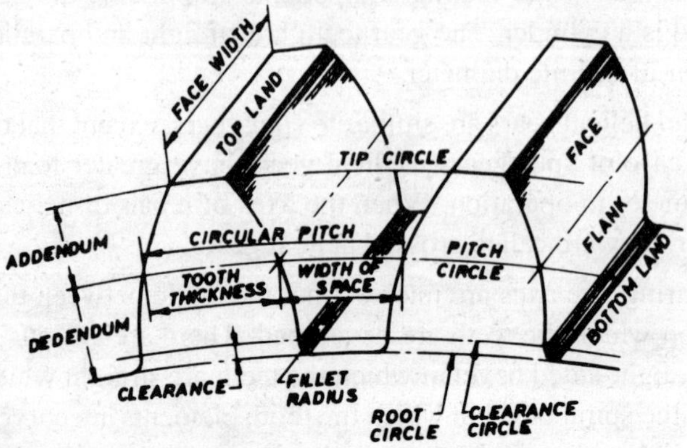

Fig. 9.1 Pictorial representation of gear tooth
Reproduced with permission from : Gitin M. Maitra, Handbook of Gear Design, Second Edition, Tata McGraw-Hill, New Delhi, 1994, Fig. 2.2, p. 2.3.

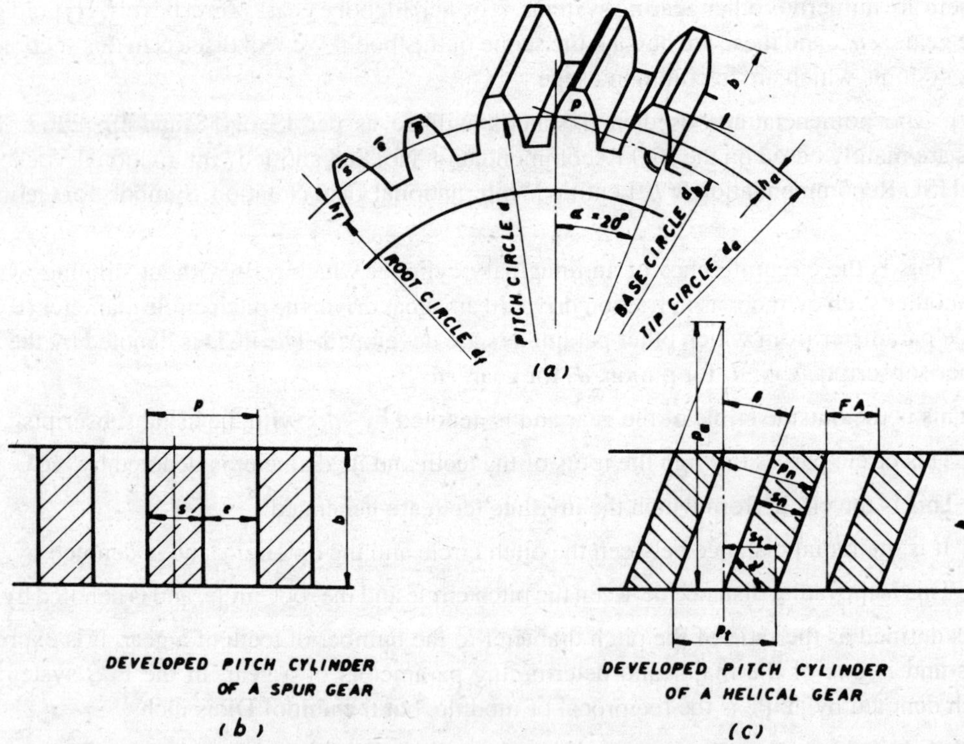

Fig. 9.2 Spur and helical gear parameters
Reproduced with permission from : Gitin M. Maitra, Handbook of Gear Design, Second Edition, Tata McGraw-Hill, New Delhi, 1994, Fig. 2.3, p. 2.3.

Spur gears: A pictorial view of two spur gears in mesh is shown in Fig. 9.3. Rack and pinion arrangement is shown in Fig. 9.4. As mentioned earlier, a rack can be thought of as a spur gear of infinite diameter. While gears have involute tooth profile, the rack tooth have straight sides as the involute curve becomes practically straight when the diameter approaches infinity.

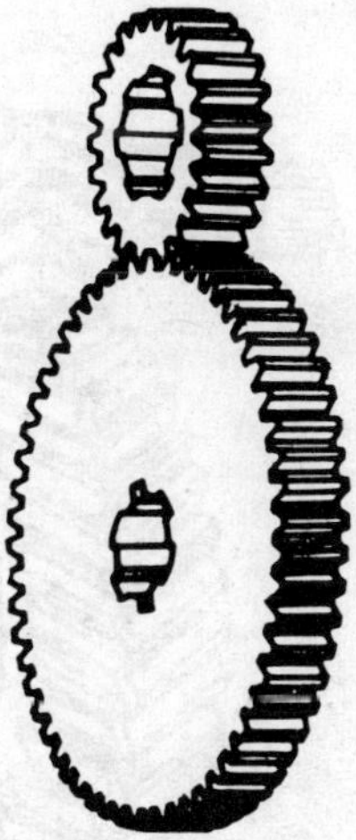

Fig. 9.3 Spur gears

Fig. 9.4 Rack and pinion

The tooth profile of a gear is as per the Basic Rack described in IS: 2535-1978. The dimensions for a tandard spur gearing is given in Table 9.1.

Table 9.1 Dimensions for standard spur gearing

Description	*Pinion*	*gear*
umber of teeth	z_1	z_2
itch circle diameter	$d_1 = z_1 m$	$d_2 = z_2 m$
ip circle diameter	$d_{a1} = d_1 + 2m$	$d_{a2} = d_2 + 2m$
oot circle diameter	$d_{f1} = d_1 - 2 \times 1.25$ m	$d_{f2} = d_2 - 2 \times 1.25\, m$
ooth thickness on pitch circle	$s = e = \frac{p}{2} = \frac{\pi.m}{2}$	As in pinion
entre distance	$a = \frac{d_1 + d_2}{2} = m\frac{z_1 + z_2}{2}$	

In the above Table, p represents the circular pitch ($p = \pi.m$) and m stands for the module.

kample: *Given* $z_1 = 21$, $z_2 = 73$, $m = 5$. *To find the parameters of the gear drive.*

lution: $d_1 = z_1 m = 21 \times m = 21 \times 5 = 105$ mm, $d_2 = z_2 m = 73 \times 5 = 365$ mm, $d_{al} = d_1 + 2m = 105 +$ $\times 5 = 115$ mm, $d_{a2} = d_2 + 2m = 365 + 2 \times 5 = 375$ mm. Centre distance = $(d_1 + d_2)/2 = 235$ mm.

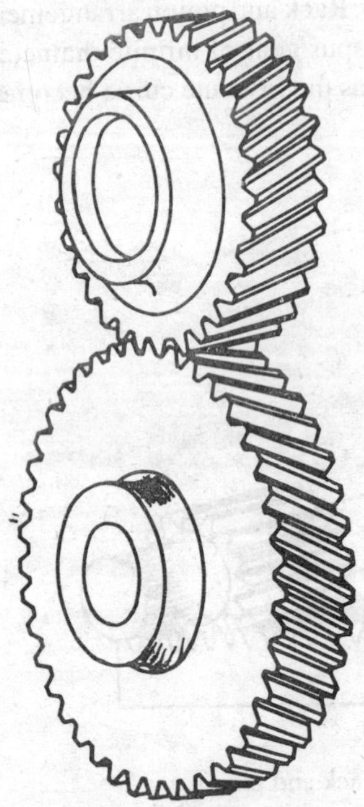

Fig. 9.5 Helical gears

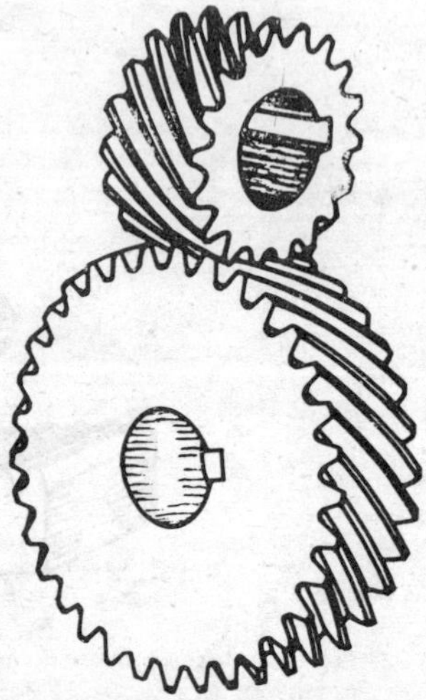

Fig. 9.6 Crossed helical gears

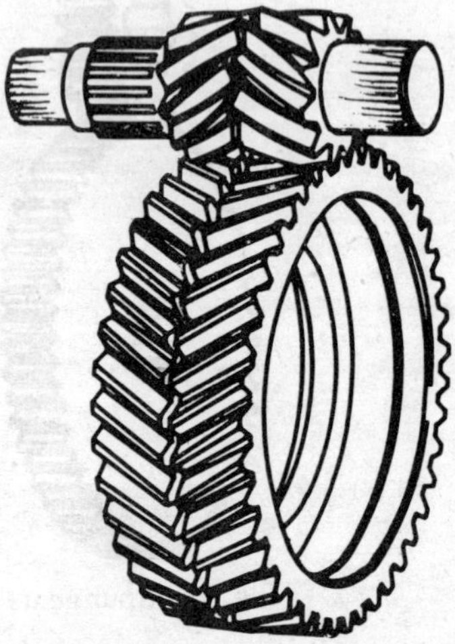

Fig. 9.7 Herringbone gears

Helical gears: As mentioned earlier, helical gears give smoother drive due to the progressive engag ment of gear teeth, carry greater load, have better contact relations and result in quieter operation. Crosse helical gears are used for non-parallel axes and since they have only point contact, they are not used fo load carrying purposes. Helical gears develop axial thrust. To eliminate this aspect, herringbone gears a used which nullify the thrust as they have both the right-handed helix and the left-handed helix on th same blank. Dimensions of a standard helical gear drive are given in Table 9.2.

Table 9.2 Dimensions for standard helical gearing

Description	*Pinion*	*Gear*
Number of teeth	z_1	z_2
Pitch circle diameter	$d_1 = z_1\, m/\cos\beta = z_1\, m_t$	$d_2 = z_2\, m/\cos\beta = z_2\, m_t$
Tip circle diameter	$d_{a1} = d_1 + 2\,m$	$d_{a2} = d_2 + 2\,m$
Root circle diameter	$d_{f1} = d_1 - 2 \times 1.25\,m$	$d_{f2} = d_2 - 2 \times 1.25\,m$
Centre distance	$a = (d_1 + d_2)/2$	

In helical gears, two kinds of modules are differentiated: the normal module in the normal plane and the transverse module in the transverse plane. They bear the following relation to each other:

$$m_t = \frac{m}{\cos\beta}$$

where m_t = The transverse module

and m = The normal module. The helix angle is represented by the Greek letter β.

Example : *Given* $z_1 = 21$, $z_2 = 73$, $m = 5$, $\beta = 9°$. *To find the parameters of the gear drive.*

Solution: $d_1 = z_1\, m/\cos\beta = 21 \times 5/\cos 9° = 105/0.98769 = 106.308$ mm

$d_2 = z_2\, m/\cos\beta = 73 \times 5/\cos 9° = 365/0.98769 = 369.549$ mm

$d_{al} = 106.308 + 2 \times 5 = 116.308,\ d_{a2} = 369.549 + 2 \times 5 = 379.549$ mm

$$a = \frac{(106.308 + 369.549)}{2} = 237.928 \text{ mm}$$

The informations concerning a gear should be furnished in a table adjoining the drawing of the gear. This table should be used for all general purposes. For special purposes, where additional data are to be furnished, tables suitably chosen may be used.

Table 9.3 Gear data

Sl. No	*Description*		*Data*
1.	No. of teeth	z	
2.	Module	m	
3.	Outside diameter	d_a	
4.	Pitch circle diameter	d	
5.	Pressure angle	α	
6.	Helix angle	β	
7.	Direction of helix		
8.	Profile correction	$x.m$	
9.	Centre distance in housing with tolerance	a	
10.	Quality of tolerance and zone		
11.	No. of teeth to be measured	z'	
12.	Tooth distance	w	
13.	No of teeth of mating gear		
14.	Drawing no. of mating gear		
15.	Tooth form		

Only those data which are of functional importance are to be given.

Sl. nos. 6 and 7 apply in case of helical gears only. They are, therefore, to be eliminated in case of spur gears.

Sl. no. 8 is eliminated in case of uncorrected gears.

Tooth form is as per IS:2535. If it is not, it should be indicated in Sl. No. 15.

Table 9.4 Calculation of bevel gears

Description	*Pinion*	*Gear*
Shaft angle	$\Sigma = \delta_1 + \delta_2$	
Pitch cone angle	$\tan \delta_1 = \dfrac{d_1}{d_2} = \dfrac{z_1}{z_2} = \dfrac{n_2}{n_1}$; $\delta_1 = \Sigma - \delta_2$	$\tan \delta_2 = \dfrac{d_2}{d_1} = \dfrac{z_2}{z_1} = \dfrac{n_1}{n_2}$; $\delta_2 = \Sigma - \delta_1$
Pitch circle diameter	$d_1 = z_1 \cdot m$	$d_2 = z_2 \cdot m$
Outside diameter	$d_{a1} = d_1 + 2 \cdot m \cdot \cos \delta_1$	$d_{a2} = d_2 + 2 \cdot m \cdot \cos \delta_2$
Addendum angle	$\tan \theta_{a1} = \dfrac{2 \cdot \sin \delta_1}{z_1}$; $\theta_{a1} = \theta_{a2}$	$\tan \theta_{a2} = \dfrac{2 \cdot \sin \delta_2}{z_2}$;
Blank cone angle	$\tan \delta_{a1} = \dfrac{z_1 + 2 \cos \delta_1}{Z_2 - 2 \sin \delta_1}$; $\delta_{a1} = \delta_1 + \theta_{a1}$	$\tan \delta_{a2} = \dfrac{z_2 + 2 \cos \delta_2}{z_1 - 2 \sin \delta_2}$; $\delta_{a2} = \delta_2 + \theta_{a2}$
Face width	$b\,(\max) \le \dfrac{m \cdot z_1}{6 \sin \delta_1}$	

We have seen that the spur and helical drives can be thought of as two pitch cylinders rolling against each other without slipping. Basically, bevel gears are analogous to a friction cone drive when the conical surface of one drives that of the other cone by friction.

For transmission of power through intersecting axes, the bevel gears are most commonly used systems. When two bevel gears mesh, their extended pitch cones meet at a common point called the apex. For straight-sided bevel gears, the parameters are specified as per the basic rack for bevel gears in IS: 5037. Some of the terms pertaining to the bevel gear drive are given below:

Pitch cones: These are defined as the cones which roll without slipping when they are in peripheral contact, and are analogous to the pitch cylinders of spur and helical gears.

Pitch cone angle: This is the angle subtended at the apex by the axis and the line which generates the pitch cone.

Shaft angle: This is the angle between the intersecting axes. This is denoted by Σ. It is equal to the sum of the pitch cone angles of the mating bevel gears. Normally, the shaft angle is 90°, but it can have also other values depending on the design. The parameters of bevel gears are apparent from Fig. 9.9. Since the tooth size decreases from the back end towards the apex and the tooth contour also varies accordingly, the terms such as addendum, dedendum, whole depth, PCD, outside circle diameter, etc, are all measured with reference to the large end of the teeth.

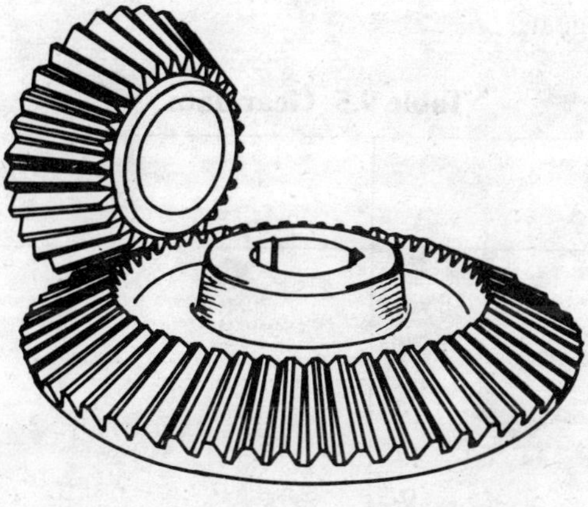

Fig. 9.8 Pictorial view of bevel gears

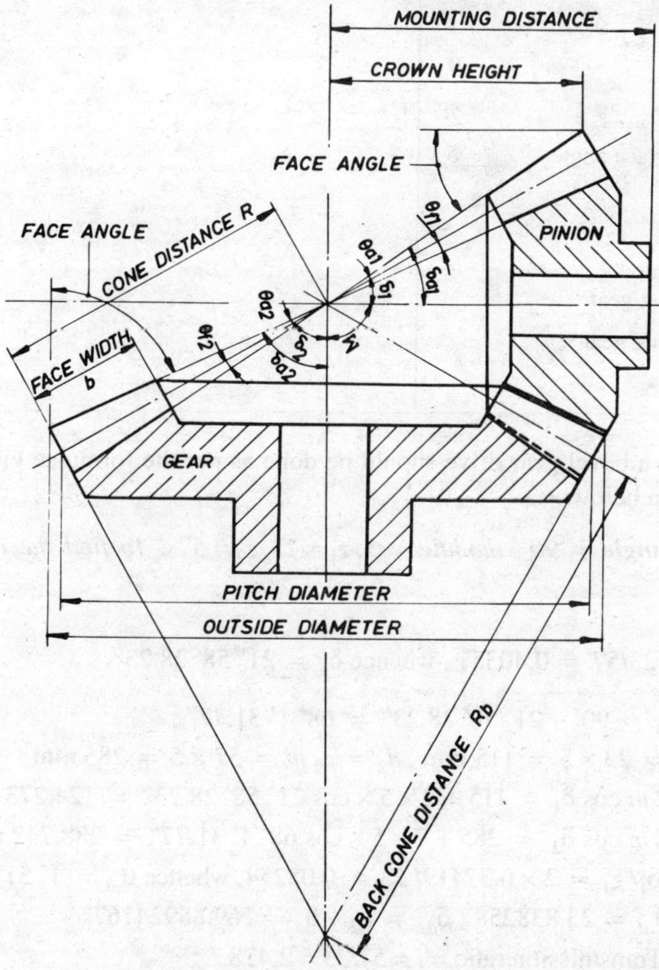

Fig. 9.9 Parameters of bevel gears

Reproduced with permission from : Gitin M. Maitra, Handbook of Gear Design, Second Edition, Tata McGraw-Hill, New Delhi, 1994, Fig. 5.3, p. 5.5.

The data for a bevel gear should be given in a tabular form as shown. The table should be given adjoining the actual drawing for the bevel gear.

Table 9.5 Gear data

Sl. No.	*Description*		*Data*
1.	No. of teeth	z	
2.	Module	m	
3.	Outside diameter	d_a	
4.	Pitch circle diameter	d	
5.	Pressure angle	α	
6.	Shaft angle	Σ	
7.	Pitch cone angle	δ	
8.	Addendum angle	θ_a	
9.	Whole depth of tooth	h	
10.	Blank cone angle or face angle	δ_a	
11.	Backlash		
12.	Quality of tolerance		
13.	No. of teeth of mating gear		
14.	Drawing no. of mating gear		
15.	Tooth form		

The basic calculations for a bevel gear drive should be done as per the formulae given in Table 9.4. A sample calculation is given below:

Example: *Given shaft angle* = *90°, module* = 5, $z_1 = 23$, $z_2 = 57$. *To find the relevant bevel gear parameters.*

Solution: $\tan \delta_1 = \dfrac{z_1}{z_2} = 23/57 = 0.40351$, whence $\delta_1 = 21°\,58'\,28.23''$.

$$\delta_2 = \Sigma - \delta_1 = 90° - 21°\,58'\,28.23'' = 68°\,1'\,31.77''$$

$$d_1 = z_1 m = 23 \times 5 = 115 \text{ mm},\ d_2 = z_2 m = 57 \times .5 = 285 \text{ mm}$$

$$d_{al} = d_1 + 2\,m \cos \delta_1 = 115 + 2 \times 5 \times \cos 21°\,58'\,28.23'' = .124.273 \text{ mm}$$

$$d_{a2} = d_2 + 2\,m \cos \delta_2 = 285 + 2 \times 5 \times \cos 68°\,1'\,31.77'' = 288.742 \text{ mm}$$

$$\tan \theta_{al} = 2 \sin \delta_1 / z_1 = 2 \times 0.37419/23 = 0.03254, \text{ whence } \theta_{al} = 1°\,51'\,49.5''$$

$$\delta_{al} = \delta_1 + \theta_{al} = 23.838258°,\ \delta_{a2} = \delta_2 + \theta_{a2} = 69.88924167°$$

$$\theta_{al} = \theta_{a2},\ \text{Transmission ratio} = i = 57/23 = 2.478.$$

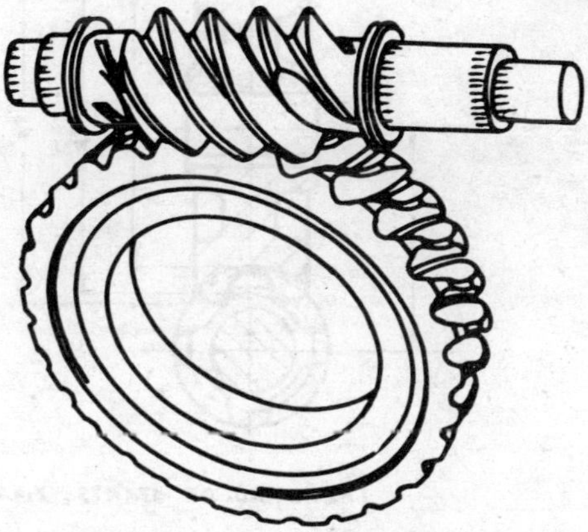

Fig. 9.10 Worm and Worm-wheel drive

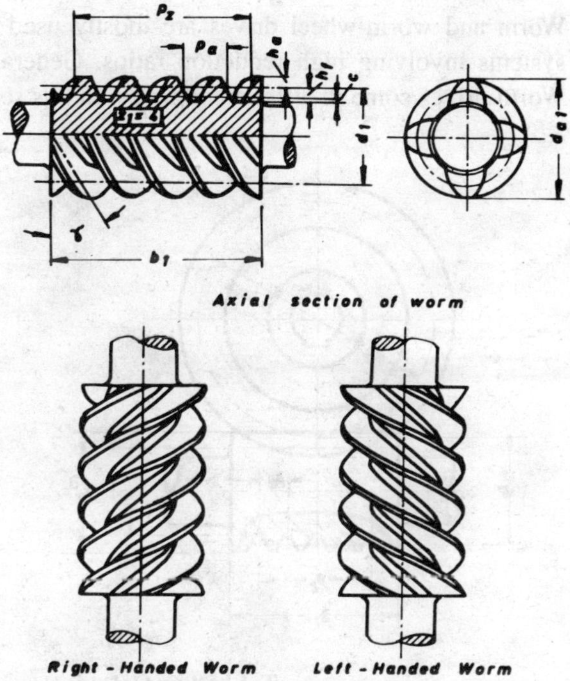

Fig. 9.11 Parameters of worm

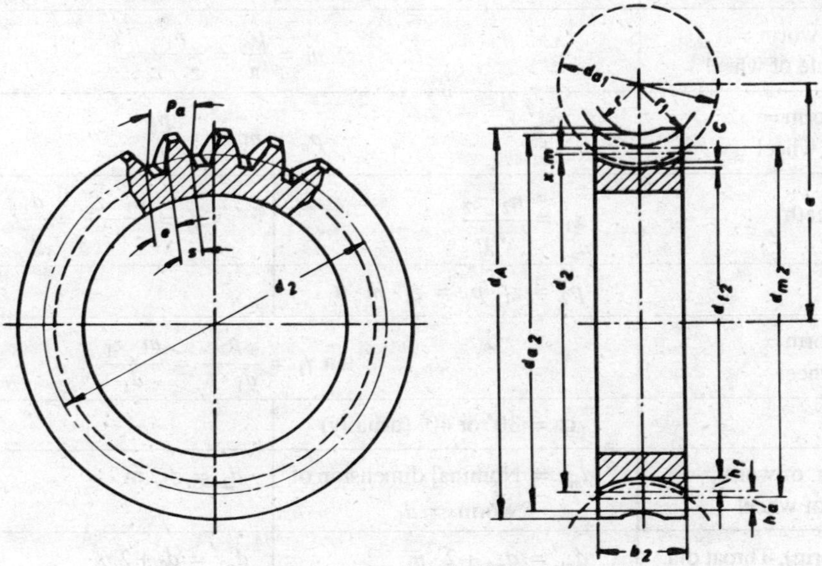

Fig. 9.12 Parameters of worm-wheel

Reproduced with permission from : Gitin M. Maitra, Handbook of Gear Design, Second Edition, Tata McGraw-Hill, New Delhi, 1994, Fig. 4.6, p. 4.8.

Worm and worm-wheel

Worm and worm-wheel drives are mostly used for non-parallel, non-intersecting, right-angle drive systems involving high reduction ratios. Generally, the worm is the driving component of the set. Worm-drives come in different varieties. In this section, the most commonly used type will considered.

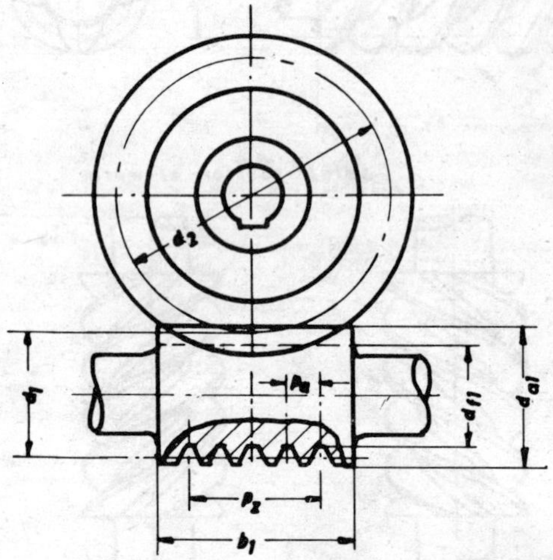

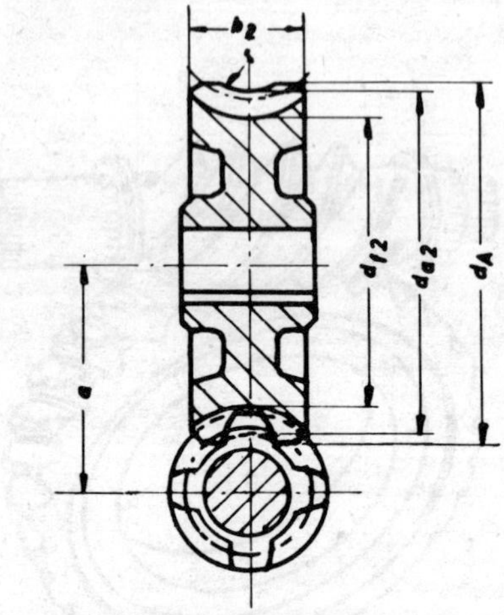

Table 9.6 Calculations of worm and worm-wheel

Description	*Worm*	*Worm-wheel*
Axial module of worm = Transverse module of wheel	$m = \frac{p_a}{\pi} = \frac{p_z}{z_1 \cdot \pi}$	
Axial pitch of worm = Circular pitch of wheel	$p_a = m \cdot \pi = \frac{p_z}{z_1}$	
No. of start (or teeth)	$z_1 = \frac{n_2 \cdot z_2}{n_1}$	$z_2 = \frac{n_1 \cdot z_1}{n_2} = \frac{d_2}{m}$
Lead	$p_2 = z_1 \cdot p_a = z_1 \cdot m \cdot \pi$	—
Lead angle of worm = Helix angle of wheel	$\tan \gamma_1 = \frac{p_2}{d_1 \cdot \pi} = \frac{m \cdot z_1}{d_1}$	
Flank angle	$\alpha = 30^\circ$ or 40° (usually)	
Middle circle dia. of worm, Pitch circle dia. of wheel	d_{m1} = Nominal dimension of worm = d_1	$d_2 = z_2 \cdot \text{m}$
Outside dia. (worm), Throat dia. (wheel)	$d_{a1} = d_{m1} + 2 \cdot m$	$d_{a2} = d_2 + 2.m$
Outside diameter	—	$d_A \approx d_{a2} + m$

Radius of wheel face	—	$r = \dfrac{d_{m1}}{2} - m = a - \dfrac{d_2}{2}$
Top clearance	$c = 0.167\,m$ to $0.3\,m$ Preferred value : $0.2\,m$	
Root diameter	$d_{f1} = d_{m1} - 2\,(m + c)$	$d_{f2} = d_2 - 2\,(m + c)$
Length of worm Face width of wheel	$b_1 = 2\,m\sqrt{z_2 + 1}$	$b_2 \approx 0.8\,dm_1$
Centre distance	$a = \dfrac{d_{m1} + d_2}{2}$	

Basic parameters of worm and worm-wheel : Referring to Figs. 9.11 and 9.12 the basic parameters are the following:

Number of starts of worm: This denotes the number of threads or teeth wound around the cylindrical body of the worm along its length, and carries the symbol z_1.

Middle circle diameter of worm: This denotes the nominal or reference dimension of the worm and is symbolised by d_1 or d_{m1}. It is analogous to the pitch circle diameter of a gear.

Axial module of worm: In a worm-drive, the axial module is the reference module on which all calculations are based and is denoted by m or sometimes by m_a.

Lead of worm: The lead of worm is defined as the distance, measured axially, between two consecutive points on the same worm-tooth when the tooth helix makes a complete turn around the worm-axis. It is denoted by p_z.

Lead angle of worm: It is the angle subtended between a tangent to the pitch helix and a plane normal to the worm-axis. This is also known as the middle lead angle of the worm and is denoted by γ or γ_m.

Pitch circle diameter of worm-wheel: This is given by

$$d_2 = m\,z_2,$$

where z_2 represents the number of teeth of worm-wheel.

Throat diameter of worm-wheel: This is given by

$$d_{a2} = d_2 + 2\,m.$$

Outside diameter of wheel: For better meshing conditions, the wheel envelopes the worm, thereby resulting in greater contact area. The outside diameter of wheel is thus larger than the throat diameter and is denoted by d_A.

Helix angle of wheel: This is equal in magnitude to the lead angle of the worm. Other parameters can be arrived at from Fig. 9.12 and Table 9.6.

Example: *Given $z_1 = 1$, $i = 52$, $m = 8$. To find the principal parameters of the worm-drive.*

Solution: Transmission ratio $i = z_2/z_1 = z_2/1 = 52, \quad \therefore z_2 = 52.$

$$p_a = \pi m = \pi \times 8 = 25.13, \; p_z = p_a z_1 = 25.13$$

As a guiding value, m can be taken to lie between

$$\frac{d_{m1}}{15} \text{ to } \frac{d_{m1}}{6},$$

with an average value of $d_{m1}/10$ or $0.1\, d_{m1}$. Hence,

$$d_{m1} \text{ (or } d_1) = 8 \times 10 = 80 \text{ mm}$$

$$\tan \gamma = \frac{m\, z_1}{d_1} = 8 \times 1/80 = 0.1, \text{ whence } \gamma = 5^\circ\, 42'\, 38.14''$$

$$d_2 = m\, z_2 = 8 \times 52 = 416 \text{ mm}.$$

(It is to be noted here that $d_{m2} = d_2 \pm 2\, xm$, where "x" represents the correction factor which is beyond the scope of this book. See later about gear correction).

$$d_{a1} = 80 + 2\, m = 80 + 16 = 96 \text{ mm}$$

$$d_{a2} = d_2 + 2\, m = 416 + 16 = 432 \text{ mm}$$

Other relevant data can be calculated using Table 9.6.

The data for worm and worm-wheel should be given in tabular form as shown in Table 9.7. The tables should be given adjoining the actual drawings of worm and worm-wheel respectively.

Table 9.7 Worm and worm-wheel data

Worm Data		
S. No	*Description*	*Data*
1.	No. of starts	z_1
2.	Module (axial)	m
3.	Outside diameter	d_{a1}
4.	Middle circle diameter	d_{m1}
5.	Pressure angle	α
6.	Lead angle	γ
7.	Direction of helix	
8.	Lead	p_z
9.	Axial Pitch	p_a

Worm-Wheel Data		
Sl. No.	*Description*	*Data*
1.	No. of teeth	z_2
2.	Module (transverse)	m
3.	Throat diameter	d_{a2}
4.	Pitch circle diameter	d_2
5.	Pressure angle	α
6.	Helix angle	γ
7.	Direction of helix	
8.	Profile correction	$x.m$
9.	Circular pitch	p_a

10.	Centre distance in housing with tolerance	a	
11.	Backlash		
12.	Quality of tolerance		
13.	No. of teeth at mating wheel		
14.	Drawing no. of mating wheel		
15.	Length of worm	b_1	

10.	Centre distance in housing with tolerance	a	
11.	Backlash		
12.	Quality of tolerance		
13.	Outside diameter	d_A	
14.	Drawing no. of mating worm		
15	Face width	b_2	

Sl. no. 8 (worm-wheel) is eliminated in case of uncorrected wheels.

Some aspects relevant to toothed gearing

Gear materials: Materials for gears should be carefully chosen. The satisfactory performance of a gear and its operational life are direct functions of the material it is made of and the treatment thereof. As a gear material, cast iron is used where the lead and the stress conditions are light in nature. Steel castings and structural steels are employed for light to medium duty gears. Case hardened steels and hardened and tempered steels are used when the stress conditions are severe. Alloy steels are used for exacting conditions and stainless steels are used against corrosion, heat, *etc.* worm-wheels are normally made of various types of bronzes. Non-metallic materials are used for noiseless running. In general, pinion should have a better material than gear since it has to work more than the gear for any particular time. Hardness of pinion also should be higher. The common gear materials are:

Cast iron: FG 200, FG: 260, IS: 210
Cast steel: Grade: 26 - 52, 30 - 57, IS: 1030
Structural steels: Fe 410, 490, 620, 590, IS:1570
Heat treatment steels: 45 C 8, 60 C 4, IS: 1570
Case hardening steels: 10 C 4, 15 C 4, 40 Cr 4, IS: 1570

Hardness of gears: To alleviate the detrimental effects of contact stress, the surfaces of gear teeth are often hardened. It decreases wear and increases the operational life of gears. Hardnesses of pinion and gear should roughly be related as

$$HB_{pinion} = HB_{gear} + 15' \text{ N/mm}^2. \qquad (\text{HB in N/mm}^2)$$

The above relation is not valid for gears of cast iron and for hardened gears.

Correction of gears: To satisfy certain conditions, the tooth profiles of gears are sometimes modified from the standard ones to suit the purpose. This is known as "profile correction" or simply "correction" of gears. Corrected gears are used for many reasons, the main factors being avoidance of undercutting, strengthening of gear teeth, adjustment as regards centre distance, etc. The relevant calculations are complicated and as such, the correction aspect is beyond the scope of this book.

Basic parameters of standard gear tooth: The basic parameters of gear tooth are standardised in IS: 2535, and are given in Table 9.1.

Tolerances of gear blanks: Before the blank goes for gear cutting, it has to be provided with proper tolerances. For this, refer Table 9.9. The tolerance to be assigned for the tip or outside diameter of the gear is to be taken from this table as per the quality of the gear and inserted in the gear drawing.

Engineering drawing of gears: An example of the working or manufacturing drawing of a gear is given in Fig. 9.13. The adjoining table contains the general data. The gear data tables have not been standardised and they vary from firm to firm. Some firms give the barest minimum details in the tables while other firms prefer to show elaborate data in their tables.

Table 9.8 shows the basic dimensions of gear tooth as per IS: 2535. The dedendum and the whole depth of tooth can be calculated by using the formulae.

Dedendum = 1.25 × Addendum, Whole depth of tooth = 2.25× Addendum in mm.

Table 9.8 Basic dimensions of standard gear tooth (Basic Rack as per IS:1535)

All dimensions in mm

Module	*Equivalent D.P*	*Circular pitch*	*Addendum*	*Dedendum*	*Whole depth*
0.30	64.667	0.943	0.30	0.375	0.675
0.40	63.500	1.257	0.40	0.500	0.900
0.50	50.800	1.571	0.50	0.625	1.125
0.60	42.300	1.885	0.60	0.750	1.350
0.70	36.286	2.199	0.70	0.875	1.575
0.80	31.750	2.513	0.80	1.000	1.800
0.90	28.222	2.827	0.90	1.125	2.025
1.00	25.400	3.142	1.00	1.250	2.250
1.25	20.320	3.927	1.25	1.562	2.812
1.50	16.933	4.712	1.50	1.875	3.375
1.75	14.514	5.498	1.75	2.188	3.938
2.00	12.700	6.283	2.00	2.500	4.500
2.25	11.289	7.069	2.25	2.812	5.062
2.50	10.160	7.854	2.50	3.125	5.625
2.75	9.236	8.639	2.75	3.438	6.188
3.00	8.466	9.425	3.00	3.750	6.750
3.25	7.816	10.218	3.25	4.062	7.312
3.50	7.257	10.996	3.50	4.375	7.875
3.75	6.773	11.781	3.75	4.688	8.438
4.00	6.350	12.566	4.00	5.000	9.000
4.50	5.644	14.137	4.50	5.625	10.125

5.00	5.080	15.708	5.00	6.250	11.250
5.50	4.618	17.279	5.50	6.875	12.375
6.00	4.233	18.850	6.00	7.500	13.500
6.50	3.008	20.428	6.50	8.125	14.625
7.00	3.628	21.991	7.00	8.750	15.750
8.00	3.175	25.132	8.00	10.000	18.000
9.00	2.822	28.274	9.00	11.250	20.250
10.00	2.540	31.416	10.00	12.500	22.500
11.00	2.309	34.558	11.00	13.750	24.750
12.00	2.117	37.699	12.00	15.000	27.000
13.00	1.954	40.841	13.00	16.250	29.250
14.00	1.814	43.982	14.00	17.500	31.500
15.00	1.603	47.124	15.00	18.750	33.750
16.00	1.587	50.266	16.00	20.000	36.000
18.00	1.411	56.549	18.00	22.500	40.500
20.00	1.270	62.832	20.00	25.000	45.000
22.00	1.155	69.115	22.00	27.500	49.500
24.00	1.058	75.398	24.00	30.000	54.000
27.00	0.941	84.823	27.00	33.000	60.750
30.00	0.847	94.248	30.00	37.500	67.500
33.00	0.770	103.673	33.00	41.250	74.250
36.00	0.706	113.097	36.00	45.000	81.000
39.00	0.651	122.522	39.00	48.750	87.750
42.00	0.605	131.947	42.00	52.500	94.500
45.00	0.564	141.372	45.00	56.250	101.250
50.00	0.508	157.080	50.00	62.500	112.500
55.00	0.462	172.788	55.00	68.750	123.750
60.00	0.423	188.486	60.00	75.000	135.000
65.00	0.391	204.204	65.00	81.250	146.250
70.00	0.363	219.911	70.00	87.500	157.500
75.00	0.339	235.619	75.00	93.750	168.750

Note: Equivalent D.P. is given for comparison only.

Table 9.9 Tolerance on gear blanks
(All values in microns except the diameter in mm)

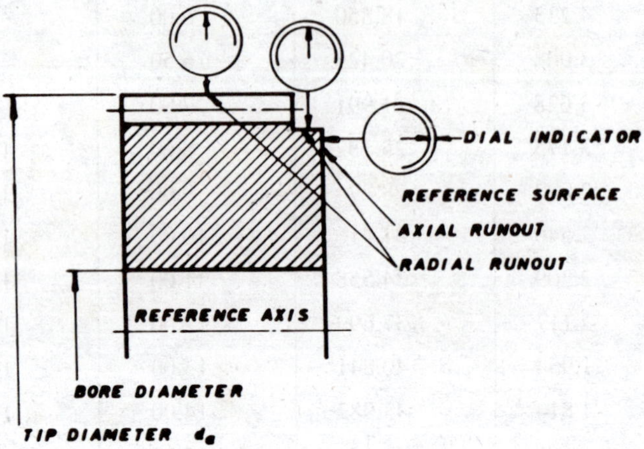

Quality	*3 4*	*5 6*	*7 8*	*9 10*	*11 12*
Bore siz[illegible]lerance	IT4 IT4	IT5 IT6	IT7 IT7	IT8 IT8	IT8 IT8
Shaft size tolerance	IT4 IT4	IT5 IT5	IT6 IT6	IT7 IT7	IT8 IT8
Radial runout of cylinder					
Radial runout of reference surface	$0.01d_a + 5$	$0.016d_a + 10$		$0.04d_a + 25$	$0.04d_a + 25$
Axial runout of reference surface			$0.025d_a + 15$		
Blank diameter d_a	*h7 h7*	*h7 h8*	*h8 h8*	*h9 h9*	*h11 h11*

9.2 FLAT BELT DRIVES

When the centre distance is large for gearing to transmit power and torque between shafts, the designer resorts to some sort of flexible connectors to attain the purpose. Such flexible connectors are: flat belts, Vee belts, roller chains and other similar power transmitting drives. However, if it is imperative that a certain velocity ratio must be strictly maintained under any condition, gearing has to be used, as flexible connectors are vulnerable to slippage due to the fact that frictional condition plays a large role in such drives, and belt drives are mainly friction drive systems.

Flat belt pulleys are usually made of cast iron, cast steel and mild steel fabrication as per design requirements. A variety of different materials are used for making flat belts. These include leather, fabric, rubber-impregnated fabric, and synthetic materials. In ideal condition, the velocity ratio is given by,

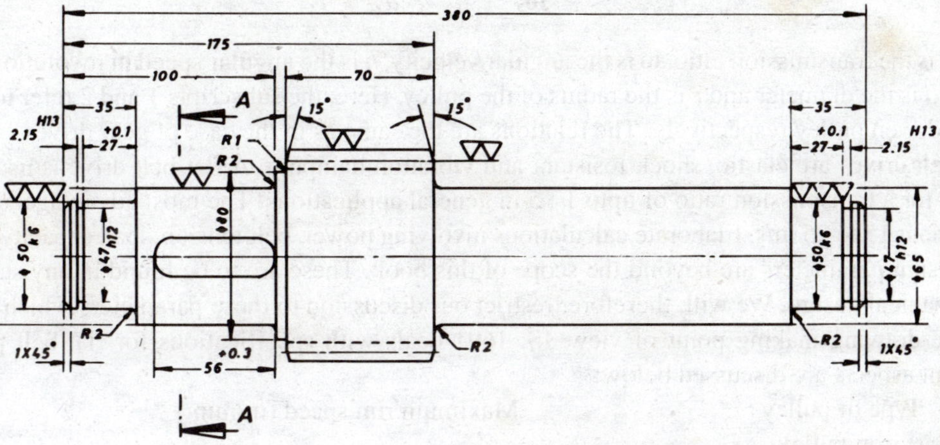

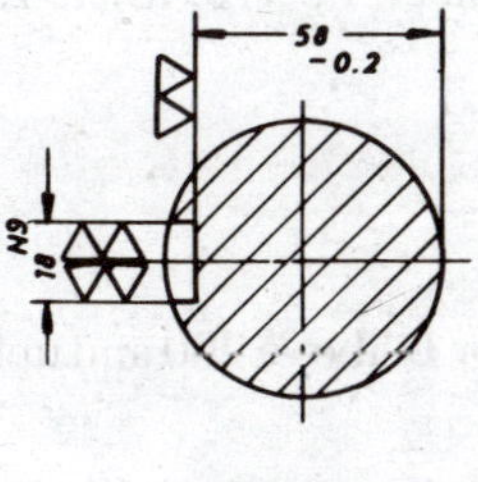

SECTION A-A

PINION DATA

SL.NO	DESCRIPTION		DATA
1	NO. OF TEETH	z	18
2	MODULE	m	5
3	OUTSIDE DIAMETER	d_a	102.42 h8
4	PITCH CIRCLE DIAMETER	d	90.88
5	PRESSURE ANGLE	α	20°
6	HELIX ANGLE	β	8°
7	DIRECTION OF HELIX		L.H.
8	PROFILE CORRECTION	$x.m$	+0.8
9	TOOTH FORM		AS PER IS:2535
10	QUALITY OF TOLERANCE		7 d S"
11	NO. OF TEETH TO BE MEASURED	z'	3
12	TOOTH DISTANCE	w	$38.745_{-0.078}^{-0.053}$
13	TOTAL COMPOSITE ERROR, DOUBLE FLANK	F''_i	0.083
14	NO. OF TEETH IN MATING GEAR		73
15	DRG. NO. OF MATING GEAR		DRG. NO. 2
16	CENTRE DISTANCE IN HOUSING WITH TOLERANCE	a	231.5 ± 0.034

MATERIAL: 40 Cr4

▽ (▽▽, ▽▽▽) SCALE 1:2

Fig. 9.13 Manufacturing drawing of a pinion shaft

Reproduced with permission from : Gitin M. Maitra, Handbook of Gear Design, Second Edition, Tata McGraw-Hill, New Delhi, 1994, Drawing 2.1, p. 2.149.

$$i = \frac{\omega_1}{\omega_2} = \frac{n_1}{n_2} = \frac{d_2}{d_1} = \frac{r_2}{r_1},$$

where, i is the transmission ratio, ω is the angular velocity, n is the angular speed in revolution per minute (RPM), d is the diameter and r is the radius of the pulley. Here, the subscripts 1 and 2 refer to the driving and the driven pulley respectively. The relations are the same as in the case of gear drives.

All belt drives are elastic, shock resistant and vibration damping. A flat belt drive is usually recommended for a transmission ratio of upto 1 : 5 in general applications. The most advantageous belt speed lies at about 15 to 25 m/s. Elaborate calculations involving power, belt tension, load capacity of belts and other design parameters are beyond the scope of this book. These are to be found in any standard book on mechanical design. We will, therefore, restrict our discussion to these parameters which are relevant from the drawing making point of view. IS: 1691 deals with specifications for flat belt pulleys. The important aspects are discussed below:

Type of pulley	Maximum rim speed (m/min.)
(a) Cast iron pulley	
(i) Solid	1500
(ii) Split	1000
(b) Mild steel pulley, Solid or split	1500

Material of pulley: Cast iron, FG 200 and FG 260, IS: 210; Mild steel, Fe 410, IS: 1570.

Tolerance on bore: Solid pulley: H 7 Split pulley: U 7

Nominal belt thickness (mm) as per IS: 1370 : 5, 6.5, 8, 10, 12.

Dimensional parameters of cast iron pulley

(i) Number of arms: (See Fig. 9.14) for D upto 200, web is used, for D above 200 and upto 450, number of arms, n is 4, for D above 450, n is 6.

(ii) Cross-section of arms: elliptical.

(iii) Radius of cross-section of arms: $r = 0.75\, b$.

(iv) Thickness of arms: $b = 2.94 \times \sqrt[3]{\frac{a \,.\, D}{4 \,.\, n}}$ for single belt

$= 2.94 \times \sqrt[3]{\frac{a \,.\, D}{2 \,.\, n}}$ for double belt

b_1 is tapered towards rim. Taper = 4 in 100.

(v) Minimum length, $l = 0.67\, a$ (l should never exceed a)

(vi) $\frac{d_1 - d_2}{2} = 0.412 \times \sqrt[3]{a \,.\, D} + 6$ mm for single belt

$= 0.529 \times \sqrt[3]{a \,.\, D} + 6$ mm for double belt

(vii) $r_1, = b/2$, $r_2 = b/2$, Rim thickness, t (without crowning = $(D/200) + 3$ mm for single belt

$= (D/200) + 6$ for double belt

In cast iron pulley, crowning on the rim is very important for alignment of the belt in running conditions. Fig. 9.14 shows the parameters and the usual designs of cast iron pulleys. Tables 9.10, 9.11, 9.12 and 9.13 show the relevant information concerning cast iron pulleys.

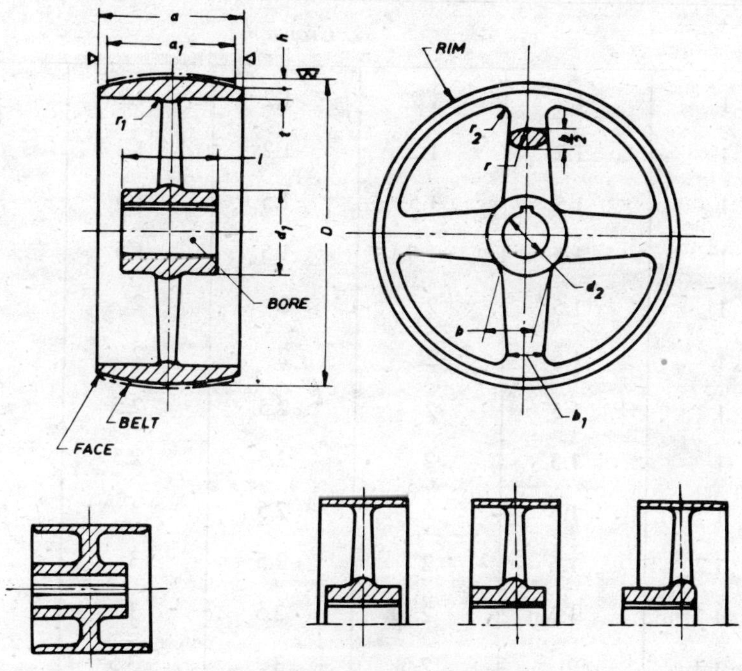

Fig. 9.14 Cast iron pulley

Reproduced with permission from : Gitin M. Maitra and L.V. Prasad, Handbook of Mechanical Design, Second Edition, Tata McGraw-Hill, New Delhi, 1995, Fig. 9.2, p. 9.27.

9.10 Crown of pulley

Crown of Pulleys of diameters from 40 mm to 355 mm inclusive

Diameter, D	*Crown, h*
40 to 112	0.3
125 and 140	0.4
160 and 180	0.5
200 and 220	0.6
250 and 280	0.8
315 and 355	1.0

Crown of Pulleys of diameters from 400 mm to 2000 mm

Diameter, D	Pulley width, a						
	125 and less	140 & 180	180 & 200	224 & 250	280 & 315	335	and larger
	Crown, h						
400	1	1.2	1.2	1.2	1.2	1.2	1.2
450	1	1.2	1.2	1.2	1.2	1.2	1.2
500	1	1.5	1.5	1.5	1.5	1.5	1.5
560	1	1.5	1.5	1.5	1.5	1.5	1.5
630	1	1.5	2	2	2	2	2
710	1	1.5	2	2	2	2	2
800	1	1.5	2	2.5	2.5	2.5	2.5
900	1	1.5	2	2.5	2.5	2.5	2.5
1000	1	1.5	2	2.5	3	3	3
1120	1.2	1.5	2	2.5	3	3	3.5
1250	1.2	1.5	2	2.5	3	3.5	4
1400	1.5	2	2.5	3	3.5	4	4
1600	1.5	2	2.5	3	3.5	4	5
1800	2	2.5	3	3.5	4	5	5
2000	2	2.5	3	3.5	4	5	5

Table 9.11 Relation between widths of pulley and belt

The pulleys should be wider than the belts by amounts shown below:

Belt width	*Pulleys to be wider than belt by*
Upto and including 125	13
From 125 upto and including 250	25
From 250 upto and including 375	38
From 375 upto and including 900	50

Table 9.12 Preferred diameters and tolerances on diameters

<table>
<tr><th>Nominal Diameter</th><th>Tolerance on Nominal Diameter</th><th>Nominal Diameter</th><th>Tolerance on Nominal Diameter</th></tr>
<tr><td>40</td><td>± 0.5</td><td>280</td><td rowspan="3">± 3.2</td></tr>
<tr><td>45</td><td rowspan="2">± 0.6</td><td>315</td></tr>
<tr><td>50</td><td>355</td></tr>
<tr><td>56</td><td rowspan="2">± 0.8</td><td>400</td><td rowspan="3">± 4.0</td></tr>
<tr><td>63</td><td>450</td></tr>
<tr><td>71</td><td rowspan="2">± 1.0</td><td>500</td></tr>
<tr><td>80</td><td>560</td><td rowspan="3">± 5.0</td></tr>
<tr><td>90</td><td rowspan="3">± 1.2</td><td>630</td></tr>
<tr><td>100</td><td>710</td></tr>
<tr><td>112</td><td>800</td><td rowspan="3">± 6.3</td></tr>
<tr><td>125</td><td rowspan="2">± 1.6</td><td>900</td></tr>
<tr><td>140</td><td>1000</td></tr>
<tr><td>160</td><td rowspan="3">± 2.0</td><td>1120</td><td rowspan="3">± 8.0</td></tr>
<tr><td>180</td><td>1250</td></tr>
<tr><td>200</td><td>1400</td></tr>
<tr><td>224</td><td rowspan="3">± 2.5</td><td>1600</td><td rowspan="3">± 10.0</td></tr>
<tr><td>250</td><td>1800</td></tr>
<tr><td></td><td>2000</td></tr>
</table>

Table 9.13 Preferred width of pulleys and tolerances on width

<table>
<tr><th>Width</th><th>Tolerances</th><th>Width</th><th>Tolerances</th></tr>
<tr><td>20</td><td rowspan="7">± 1</td><td>160</td><td rowspan="7">± 2</td></tr>
<tr><td>25</td><td>180</td></tr>
<tr><td>32</td><td>200</td></tr>
<tr><td>40</td><td>224</td></tr>
<tr><td>50</td><td>250</td></tr>
<tr><td>63</td><td>280</td></tr>
<tr><td>71</td><td>315</td></tr>
<tr><td>80</td><td rowspan="6">± 1.5</td><td>355</td><td rowspan="6">± 3</td></tr>
<tr><td>90</td><td>400</td></tr>
<tr><td>100</td><td>450</td></tr>
<tr><td>112</td><td>500</td></tr>
<tr><td>125</td><td>560</td></tr>
<tr><td>140</td><td>630</td></tr>
</table>

Mild steel pulley: In case of mild steel pulleys, these are usually supplied in split condition, *i.e.*, in two halves and the rim is flat. Single row of spokes are usual for pulleys upto 300 mm width. Hub length is normally half the width of face. The thickness of rim is 5 mm for all mild steel pulleys for sizes covered under Table 9.14. Fig. 9.15 shows the usual design of mild steel pulley and the belt orientations.

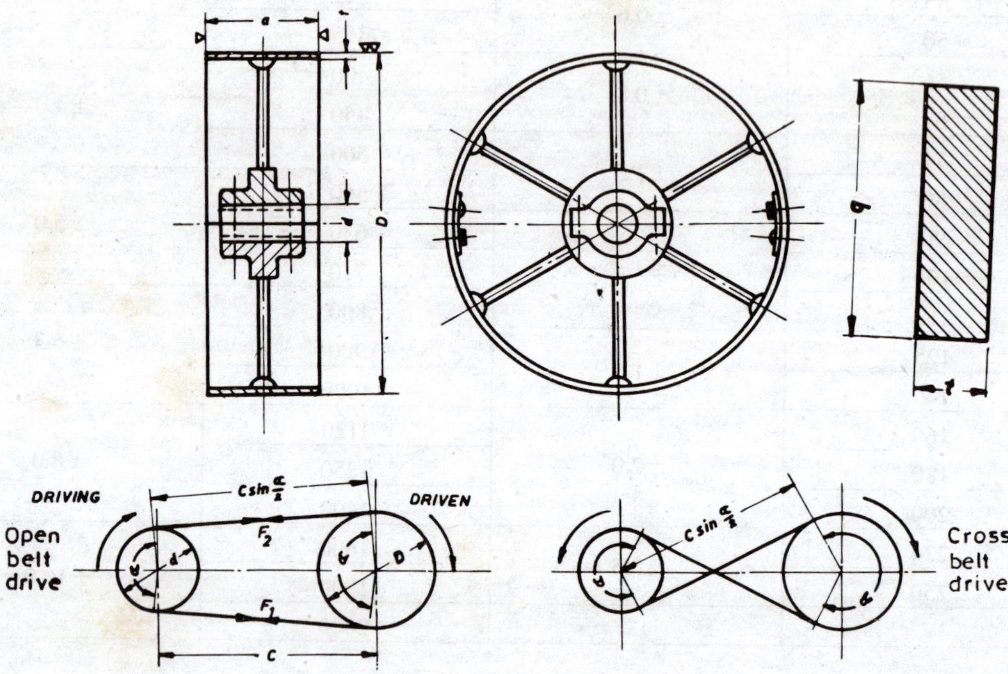

Fig. 9.15 Mild steel pulley

Table 9.14 Arms for mild steel pulleys

Diameter of pulley (mm)	*Number of arms*	*Diameter of arms (mm)*
200 to 500	6	19
560 to 710	8	19
800 to 1000	10	22
1120	12	22
1250	14	22
1400	16	22
1600	18	22
1800	18	22
2000	22	22

9.3 VEE BELT DRIVE

In a VEE belt drive, the belts run on pulleys having V-shaped grooves on the periphery. The drive consists of one or more endless belts which are standardised, bought out items, manufactured by reputed international manufacturers or their subsidiaries. Since the cross-section of belts, their lengths, size of grooves on the pulleys and other relevant data are fixed by standard specifications, the drawing office personnel are required to choose only those items after fixing the parameters by calculations involving power transmitted, torque, speed, and other factors. The catalogues of reputed belt manufacturers contains all the data relevant for calculation and final selection of these belts, and the reader is advised to get these catalogues for his use.

V-belt drive has largely replaced the conventional flat belt drive for its many inherent advantages. The drive is compact, quiet, resilient, flexible, elastic, simple in operation, and shock and vibration damping. Proper alignment in this drive is very important and some arrangement for adjusting the belt tension should be provided.

The *V*-belts drive the pulleys by wedge action. Because of the wedge shape of the *V*-belt, a gripping force greater than that in the case of a flat belt is developed between the belts and the grooves of the pulleys (see Fig. 9.16). Conditions being the same, a *V*-belt system can generally develop around 3 times the useful force transmitting capacity of a flat belt drive. The relevant informations normally required by a drawing maker are given in the paragraph, figures and tables which follow.

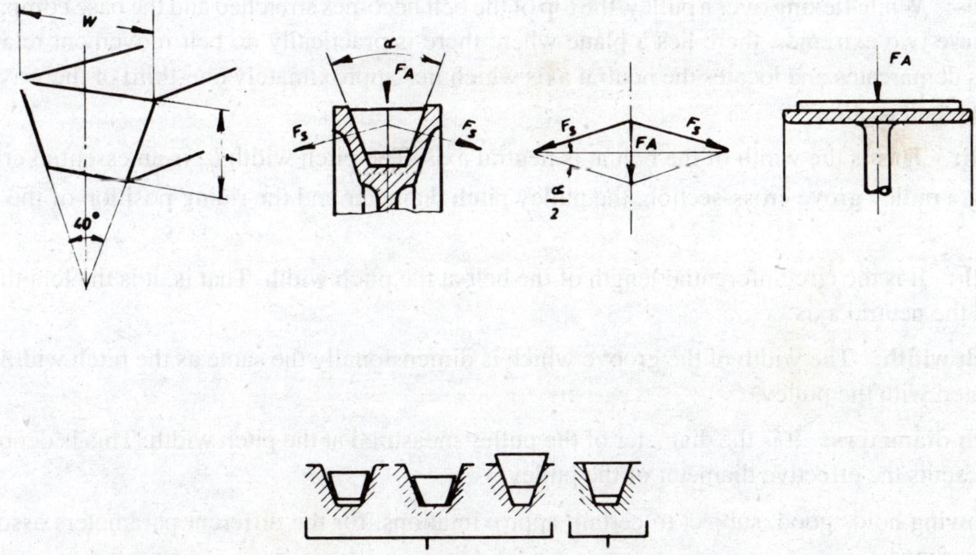

Fig. 9.16 V-belt transmission
Reproduced with permission from : Gitin M. Maitra and L.V. Prasad, Handbook of Mechanical Design, Second Edition, Tata McGraw-Hill, New Delhi, 1995, Fig. 9.1, p. 9.2.

Designation of *V*-belt: A *V*-belt is designated by the sectional symbol, nominal inside length and the relevant IS number. Thus, a *V*-belt of section *C*, nominal inside length of 3048 mm (120 inches) shall be designated as: *C* 3048/120, IS: 2494

Designation of *V*-pulley: A *V*-pulley of groove section *B* and pulley pitch diameter 250 mm shall be designated as: *V*-grooved pulley B 250, IS: 3142.

Table 9.15 Standard *V*-belt section

Belt cross-section	*Usual load of drive kW*	*Recommended minimum pulley pitch dia mm*	*Belt top width, W mm*	*Belt Thickness T mm*
A	0.75 - 5	75	13	8
B	2 - 15	125	17	11
C	7.5 - 75	200	22	14
D	22 - 149	350	32	19

Besides, section *E* is also in vogue in special cases.

Terminology: Some of the common nomenclatures associated with *V*-belt drive are given below:

Nominal inside length: This is the approximate length along the inside of a belt in untensioned condition. This is the length used to designate a *V*-belt.

Neutral axis: While flexing over a pulley, the top of the belt becomes stretched and the base compressed. Between these two extremes, there lies a plane where there is practically no belt movement relative to length. This demarcates and locates the neutral axis which lies approximately one-third of the way down the top of tne belt section.

Pitch width: This is the width of the belt at is neutral axis. The pitch width l_p, is an essential criterion for defining a pulley grove cross-section, the pulley pitch diameter and the riding position of the belt in the pulley.

Pitch length: It is the circumferential length of the belt at the pitch width. That is, it is the length of the belt around the neutral axis.

Pulley pitch width: The width of the groove which is dimensionally the same as the pitch width of the belt associated with the pulley.

Pulley pitch diameters: It is the diameter of the pulley measured at the pitch width. This is denoted by d_p and represents the effective diameter of the pulley.

The following holds good, subject to certain approximations, for the different parameters associated with pulley grooves:

$$l = (n - 1)\, e + 2f, \text{ where } n = \text{Number of grooves.}$$

$$b = 0.3\, l_p,\ e = 1.35\, l_p,\ f = 0.9\, l_p,\ h = 0.7\, l_p + 1.$$

Table 9.16 Dimensions of standard V-groove pulleys

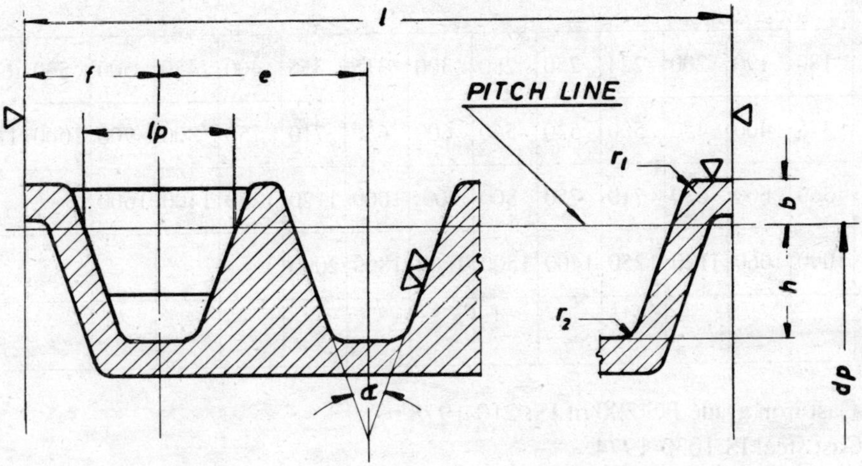

Groove cross section	l_p	b Min.	h Min.	e	t	α (± 0.5°)	Pitch diameter range d_p	g ± 0.13	r_1	r_2
A	11	3.3	8.7	15 ± 0.3	10^{+2}_{-1}	34°	75 to 125	13	1	1
						38°	126 and over	13.3		
B	14	4.2	10.8	19 ± 0.4	12.5^{+2}_{-1}	34°	125 to 200	16.6	1	1.5
						38°	201 and over	16.9		
C	19	5.7	14.3	25.5 ± 0.5	17^{+2}_{-1}	36°	200 to 300	22.7	1.5	2
						38°	301 and over	22.9		
D	27	8.1	19.9	37 ± 0.6	24^{+3}_{-1}	36°	355 to 500	32.3	2	3
						38°	501 and over	32.6		
E	32	9.6	23.4	44.5 ± 0.7	29^{+4}_{-1}	36°	500 to 630	38.2	2	4
						38°	631 and over	38.6		

Groove cross-section	Standardised pitch diameter d_p														
A	75	80	85	90	95	100	106	112	118	125	132	140	150	160	170
B	125	132	140	150	160	170	180	190	200	224	250	280	300	315	355
C	200	212	224	236	250	265	280	300	315	355	375	400	450	500	530
D	355	375	400	425	450	475	500	530	560	600	630	710	750	800	900
E															

Groove cross-section	*Standardised pitch diameter d_p*															
A	180	190	200	224	250	280	300	315	355	400	450	500	560	630	710	800
B	375	400	450	500	530	560	600	630	710	750	800	900	1000	1120	-	-
C	560	600	630	710	750	800	900	1000	1120	1250	1400	1600	-	-	-	-
D	1000	1060	1120	1250	1400	1500	1600	1800	2000	-	-	-	-	-	-	-
E														-		

Material: *I*-Cast iron grade FG 200 of IS: 210-1978
II-Cast steel IS:1030-1974

The value "*g*" relates to the medium width of the pulley groove, i.e., the groove width at the maximum or outside diameter of the pulley.

9.4 CHAIN DRIVES

For power transmission between shafts where the shafts are too far apart for gear-drive systems, some sort of positive chain drive system is usually resorted to. Since toothed wheels are used in such a system, there is no slippage. As power transmission chains, the roller chains and inverted tooth chains (also known as silent chains) are employed in general engineering purposes. Link chains are also used for power transmission, but they have their limitations. Since the roller-chain drive is mostly used in industrial applications, this type will be mainly discussed in this section as well as link chains which will be discussed briefly. Chains run on sprocket wheels. While sprocket wheels are design items, their dimensions depending on load and other determining factors, the chains are bought out items. The drawing of the contour of the sprocket teeth has been described in Section 2.6 as per the Indian Standard.

Roller chain drive: The roller chain consists of two parallel rows of inner and outer plates. Inside both the inner plates, bushes are press-fitted. Hollow rollers are mounted over the bushes with a clearance to enable the rollers to rotate freely. Inside the outer plates, pins are pressed which pass through the bushes with a clearance. The ends of pins are closed either by upsetting or by split pins after assembly. The ends of the chain are joined together by end links. A roller chain is usually recommended for continuous power transmission to alleviate the detrimental effects arising out of fatigue.

Bush chain drive: In a similar system, a bush chain is used. Construction is similar to roller chain except that there is no roller. As the roller is free to rotate over the bush in case of a roller chain drive, the rollers mesh with the sprocket with rolling friction whereas in case of bush chain, it operates with sliding friction which gives rise to more wear and tear. Both the types may be with single strand (single) or multi-strand (duplex, triplex, etc.). Roller chains and bush chains are shown in Fig. 9.17. Parts are heat treated to around 60 HRC. IS: 2403 gives the relevant dimensions as well as load data. Materials for chain: Fe 410, Fe 490, 45 C 8, Cr-Ni steel. Materials for sprockets: Cast iron, cast steel, Fe 620.

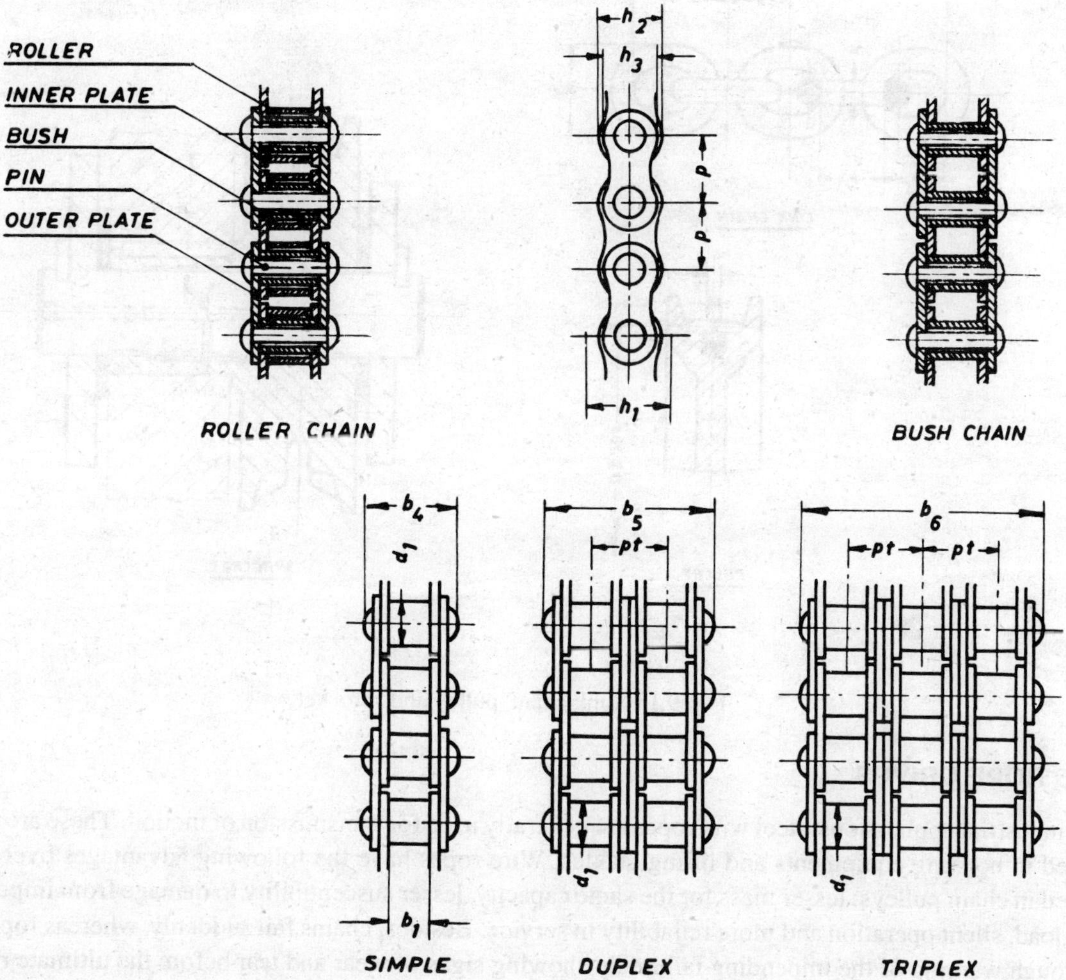

Fig. 9.17 Roller chains and bush chains

Link chain drive: Link chains are used in cranes, hoisting mechanisms and chain pulley blocks. In lifting tackles, round steel short link chains are mostly used which are standardised as per IS: 2429. Links are made by electric resistance welding or flash butt welding and also by hammer or forge welding. Chains are classified as per the pitch to diameter ratio, *e.g.*, short link ($p \leq 3\,d$) and long link ($p > 3\,d$). The pulleys or sheave for link chains are provided with grooved rims which correspond to the shape and form of the chain. Figure 9.18 shows the relevant components of a link chain drive.

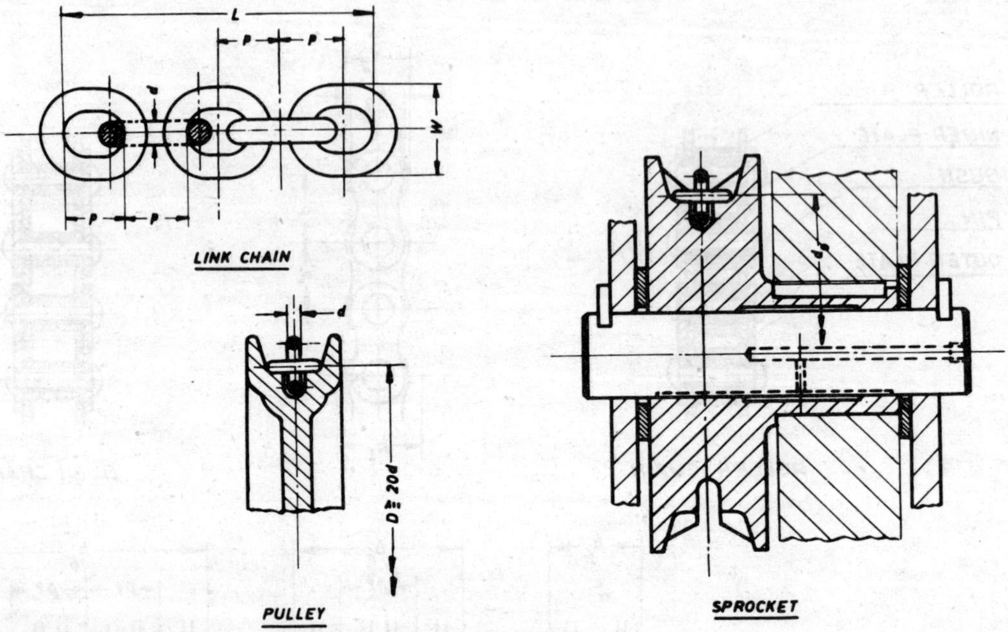

Fig. 9.18 Link chain, pulley and sprocket

9.5 ROPE DRIVES

In industrial applications, steel wire ropes are generally used for transmission of motion. These are mainly used in hoisting equipments and lifting tackles. Wire ropes have the following advantages over chains used in chain pulleys: lesser mass for the same capacity, lesser susceptibility to damage from impact type of load, silent operation and more reliability in service. Besides, chains fail suddenly, whereas ropes give enough warning of the impending failure by showing signs of wear and tear before the ultimate rupture. Flexibility is another desirable characteristic of wire ropes. The wires should conform to IS: 1835. Ropes are produced from steel wires, having special heat-treatment, and the ultimate tensile strength is in the range of 1750 to 2200 N/mm^2. IS: 3459 specifies small wire ropes of 2 to 7 mm diameter for general engineering purposes. Thicker wire ropes, conforming to IS: 2266, are mainly used for hoisting purposes. A selection of thicker wire ropes, which are in common use in hoisting mechanisms are given in Table 9.17.

Table 9.17 Data for thick wire ropes

Construction	Number of: Strands	Number of: Wire in each strand	Number of: Total wires	Tensile Designation of wires N/mm^2	Dia of Rope + 4% – 1%	Direction of Lay	Approximate Mass Kg/m: Fibre core	Approximate Mass Kg/m: Steel core	Minimum breaking load in kN: Fibre core	Minimum breaking load in kN: Steel core
6 × 19 = 114	6	12 6 1 19	114	1570	8	rh	0.22	–	31	–
					10	rh	0.35	–	4.9	–
6 × 51 = 222	6	18 12 6 1 37	222	1770	12	rh	0.5	–	7.6	–
					14	rh	0.68	–	10.4	–
					16	rh/lh	0.89	–	13.6	–
6 × 36 = 216	6	14 7 7 7 1 36	216	1770	20	rh/lh	1.52	1.67	21.8	25.7
					22	rh	1.84	2.02	28.7	31.0
					24	rh	2.19	2.41	34.2	36.9
					26	rh	2.57	2.83	40.1	43.3
					28	rh	2.98	3.28	46.6	50.3
					32	rh	3.89	4.28	60.8	65.7
					38	rh/lh	5.40	5.99	85.6	92.0
					40	rh	6.08	6.69	95.0	103.0

Construction of wire ropes: A wire rope basically consists of a number of strands laid in a helical manner around a metallic or non-metallic core, and each strand in turn is composed of a number of wires also laid in a similar way. A wire rope is generally designed by two numerals: the first indicates the number of strands and the second for the number of wires per strand, e.g., 6 × 7, 6 × 19, etc.

Specification of wire rope: While specifying a wire rope, the following data should be furnished: (a) Length of rope; (b) Nominal diameter of the wire rope; (c) Constructional details, such as, number of strands and number of wires in each strand; (d) Lay and hand, i.e., ordinary or regular lay and parallel or

Lang lay, and right-handed (*rh*) or left-handed (*lh*) lay; (e) Nominal breaking load of road; (f) Grade of steel used in rope, and (g) Particulars of ends and fittings.

Wire rope pulleys and drum: Wire rope pulleys or sheaves are generally made of cast steel as per IS: 1030 or mild steel as per IS: 226 or IS: 2062.

Rope drums are usually made of cast iron or are fabricated by welding. They are provided with helical grooves over which the rope is wound. The radius "*r*" and the pitch "*t*" of the groove for rope diameter "*d*" are given by:

$$r \approx 0.53 \times d, \text{ and } t = d + b$$

The groove dimensions are given in Table 9.18. These are to be read along with Fig. 9.19

Table 9.18 Dimensions of rope drum grooves

Rope diameter,	$d = 10$	13	16	19	22	27	33	40
	$r = 5.5$	7	8.5	10	12	14.5	17.5	21
	$b = 2$	2	2	3	3	4	4	5
	$a = 1$	1.5	2	2.5	3	3.5	4	5

The wall thickness of a cast iron drum can be calculated approximately by the empirical formula:

$$h = 0.02\,D + (6 \text{ to } 10) \text{ mm.}$$

Two designs of wire rope pulleys are shown in Fig. 9.19. One is with anti-friction bearing and the other with bush bearing. In Tables 9.19, 9.20, and 9.21, the dimensional values of both the types are given. These values which are tentative may be changed to meet any specific requirements. The nominal diameter by which a rope pulley is generally designated is the diameter at the bottom of the groove, that is, D_1. The value given in Table 9.21 are guiding dimensions only. They correspond to cases where ordinary roller bearings are used. If other anti-friction bearings are used, the dimensions along with their relevant tolerances will be fixed according to the bearing chosen, the axle diameter remaining the same. The tolerances corresponding to the bearings are to be taken from the bearing catalogues.

Table 9.19 Dimensions for pulleys with anti-friction bearings

Pulley Diameter, D_1	*Rope Dia*	D_5 H_8	D_6 h_6	D_7 M_7	D_8	$B3_{-0.2}$	B_4
250	9 - 16	50 - 55	60 - 65	95 - 100	125 - 130	75	87
280	10 - 18	50 - 60	60 - 70	95 - 100	125 - 130	75 - 83	87 - 95
315	11 - 19	50 - 70	60 - 80	95 - 110	125 - 140	75 - 86	87 - 98
355	12 - 22	50 - 80	60 - 90	95 - 125	125 - 165	75 - 96	87 - 108

400	14 - 24	60 - 90	70 - 110	100 - 150	135 - 190	83 - 106	95 - 118
450	15 - 27	60 - 90	70 - 110	100 - 150	135 - 190	83 - 106	95 - 118
500	17 - 31	70 - 110	80 - 130	110 - 180	145 - 225	86 - 126	98 - 138
560	19 - 34	80 - 140	90 - 160	125 - 200	165 - 245	96 - 126	108 - 138
630	22 - 39	90 - 150	110 - 170	150 - 215	190 - 265	106 - 126	118 - 138
710	24 - 44	110 - 170	130 - 190	180 - 240	225 - 290	106 - 126	120 - 138
800	27 - 48	120 - 180	140 - 200	190 - 250	235 - 300	106 - 126	120 - 140
900	31 - 54	130 - 200	150 - 220	190 - 270	235 - 320	106 - 131	120 - 145
1000	34 - 54	150 - 220	170 - 240	215 - 300	265 - 355	117 - 151	130 - 165
1120	40 - 54	150 - 240	170 - 260	215 - 320	265 - 375	117 - 151	130 - 165
1250	42 - 54	170 - 260	190 - 280	240 - 350	290 - 430	127 - 170	140 - 185

Table 9.20 Dimensions for pulleys fitted with bush bearings

D_1	*Rope dia.*	$D_5^{E_9}$	$D_6{}^{H7}_{r6}$	D_8	$B3_{-02}$
250	9 - 16	50	58	90	72
280	10 - 18	55	65	95	77
315	11 - 19	60	72	105	82
355	12 - 22	70	82	115	87
400	14 - 24	70	82	115	87
450	15 - 27	80	95	130	92
500	17 - 31	90	105	140	97
560	19 - 34	100 - 140	115 - 155	155 - 195	103
630	22 - 39	110 - 150	125 - 165	165 - 205	113 - 108
710	24 - 44	120 - 170	135 - 190	175 - 225	118 - 123
800	27 - 48	130 - 180	145 - 200	185 - 245	118 - 124
900	31 - 54	150 - 200	165 - 220	205 - 270	124 - 134
1000	34 - 54	160 - 220	180 - 250	225 - 300	134 - 144
1120	40 - 54	170 - 240	190 - 270	235 - 320	134 - 144
1250	42 - 54	180 - 260	200 - 290	245 - 345	149 - 169

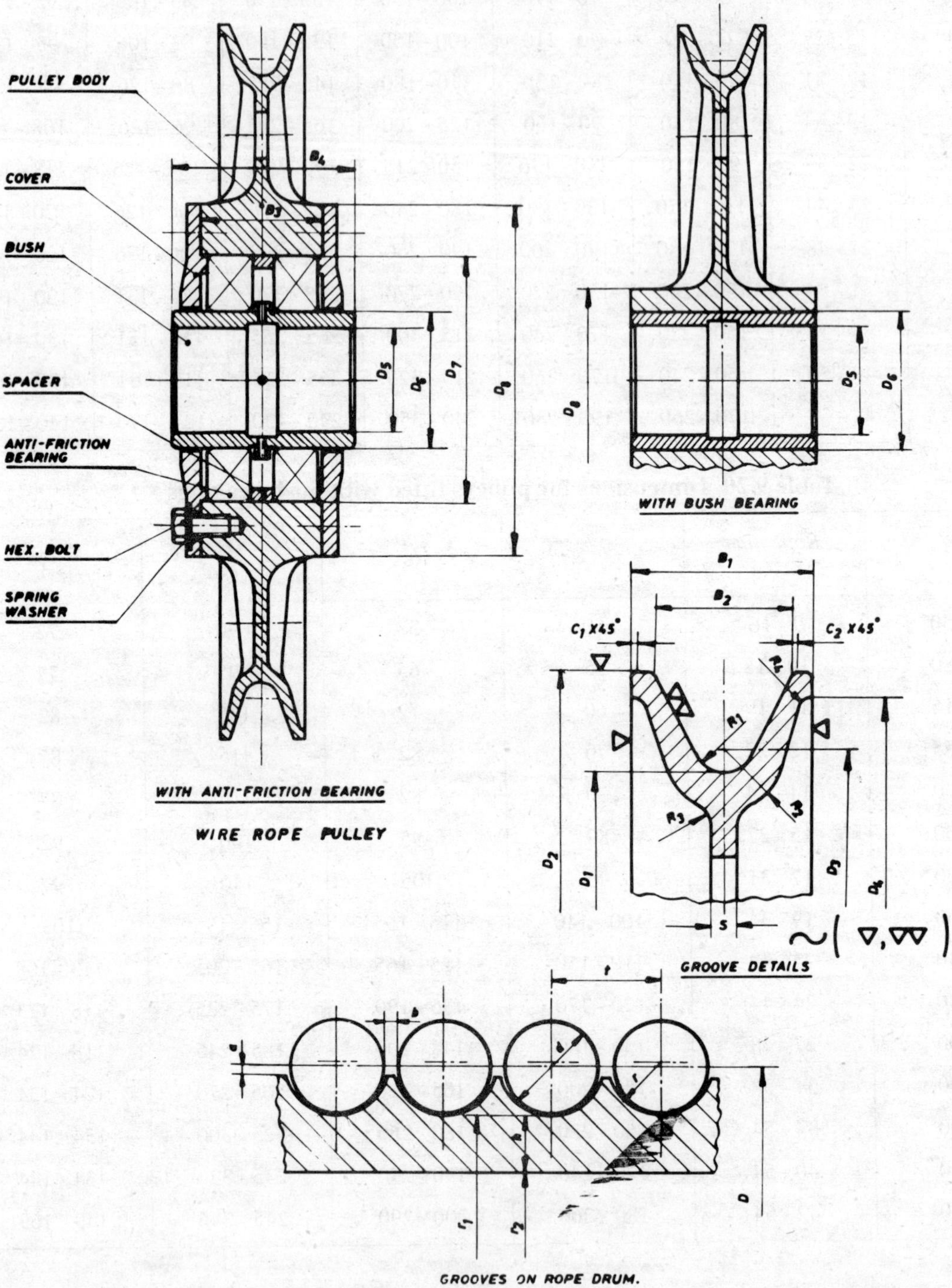

Fig. 9.19 Wire rope pulley and rope drum groove
Reproduced with permission from : Gitin M. Maitra and L.V. Prasad, Handbook of Mechanical Design, Second Edition, Tata McGraw-Hill, New Delhi, 1995, Fig. 9.4, p. 9.36.

Table 9.21 Dimensions common to both types of pulleys

D_1	D_2	D_3	D_4	B_1	B_2	R_1	R_2	R_3	R_4	S	C_1	C_2
250	290	259	280	36	27		12	5	3	5	1.5	1
280	325	290	315	39	31		13	6	3	6	1.5	1
315	365	328	355	43	34		14	7	3	6	1.5	1
355	415	370	403	50	41		16	9	3	7	1.5	1
400	465	415	453	55	44		17	10	3	8	2	1
450	520	467	508	60	49		19	12	3	9	2	1
500	575	524	561	65	53	$R_1 =$	22	12	3	10	2	1
560	640	588	626	70	56		25	15	3	11	2	1
630	720	654	705	75	65	$0.53 \times d$	25	18	3	12	2	1
710	810	740	795	85	73		30	20	3	14	2	1
800	910	830	894	95	80		30	20	3	16	2	1
900	1020	935	1004	105	88		35	20	4	18	2	1
1000	1135	1035	1119	115	94		35	25	4	20	2	2
1120	1255	1155	1239	115	94		35	25	4	22	2	2
1250	1385	1285	1369	115	94		35	25	4	22	2	2

9.6 COUPLINGS

A coupling is a device for connecting the ends of two shafts, so that the rotational motion of one shaft causes rotational motion of the other. There are various types of couplings in industrial applications which are broadly divided into two classes: rigid couplings and flexible couplings. A rigid coupling, also known generally as shaft coupling or flange coupling, is one of simplest and commonest forms of coupling. The hub portions of each coupling half are keyed to the ends of the shafts. The webs of the flange portions of the two halves are bolted together. The bolts prevent relative rotational motion between the halves and transmit the torque, motion and power from the driving to the driven shaft. Two types of rigid couplings are shown in Table 9.22. Type *A* represents the conventional type provided with tongue and recess for alignment purposes. An improved design is Type *B* where a split ring is provided in-between the two halves, so that when the ring is removed, each shaft becomes fully free of the other. Thus, enough clearance is obtained between the coupling halves, and no axial shifting of the coupling along the shafts is necessary to free the system, when necessary.

One inherent demerit of a rigid coupling is its vulnerability to misalignment. Also, it is incapable of absorbing shock and vibration. These defects are alleviated in a flexible coupling. Flexible couplings are the most commonly used contrivances for compensating the unavoidable errors in alignment of shafts which may occur at the time of mounting or may develop during the course of operation. These couplings are highly efficient and noiseless in operation. They take care of angular misalignment, misalignment connected with parallelity and other types of deviations.

There is a wide variety of flexible coupling designs to cater to different types of needs and they are marketed in various trade names, namely, Eupex, Periflex, Hadaflex, Elco, Bibby, Zapex, *etc.* Most of them essentially consists of two flanged members which are connected to each other by some sort of yielding arrangement. The yielding arrangement may be in the form of rubber bushes or inserts, tyres, coil springs, laminated bundles of leaf springs, steel strips bent in zig-zag fashion, *etc.* For coupling intended for low to medium duty, the flexibility may be imparted by inserts made of rubber, leather or fabric. For heavy duty, an all-metal type of flexible coupling is generally employed, *viz*., Bibby coupling, geared coupling, *etc*. In case of certain drives, the starting may be sudden or one machine may impose fluctuating or vibratory leads on the other. A flexible coupling provides a cushioning affect in such cases.

As mentioned earlier, there is a great variety of flexible couplings in use. In this section, parameters concerning three common types are given, namely, pin and bush type, tyre type and geared coupling.

The following points are of relevance in connection with the couplings:

1. The assignment of size numbers of the different types of couplings is done as per the numbering systems followed by the individual concerns manufacturing them. Thus, in some of pin and bush type of couplings, the size numbers are related to the quotient HP/100 RPM, in case of geared coupling to the maximum finish bore and so on.

2. In each case, H.P. stands for the metric Horse Power, and the torque is given in kgf. m, because most of the manufacturers still follow the metric system. To obtain the SI equivalents *i.e.*, kW and N m, refer the conversion tables in Section 10.6.

3. In Table 9.22 for flange couplings, the dimensions d_5, d_7, l_3, and t_2 belong to Type *C* which is not included here. Hence, these should be ignored.

4. Materials for the couplings are as follows:

Flange couplings: Cast iron as per IS: 210 for coupling halves.

Pin and bush type couplings: Gray cast iron in normal cases, cast iron with spheroidal graphite for special design.

Type coupling: Cast steel for coupling body, rubber with hardness 70 ± 5° IRH and tensile strength 20 N/mm^2 for tyre.

Geared coupling: Forged steel for mating gears, 45 C 8 (IS: 1570) for sleeves and hubs.

5. Misalignment: In general, the flexible coupling can take in angular misalignment of around 1 1/2 to 2° in normal designs. In case of geared couplings, an axial movement of around ± 0.5 to ± 2 mm and a maximum parallel misalignment of around ± 1.5 to ± 9 mm are generally permissible.

6. Tolerances: The tolerances for key-fitted couplings shall be as per the normal tolerances recommended for key-fitted components given in Section 8.3 or as per the tolerances given in company catalogues. For parallel key connections in case of pin and bush type couplings, the following fits are generally recommended:

Diameter (mm)	*Shaft end*	*Bore of coupling*
25 50 100	*m* 6 *k* 6 *n* 6	*H7*
50	*h* 6	*K7*
50	*h* 6	*M7*
All	*h* 8	*N7*

Table 9.22 Flange Coupling

FLANGE COUPLINGS

DIMENSIONS IN MM.

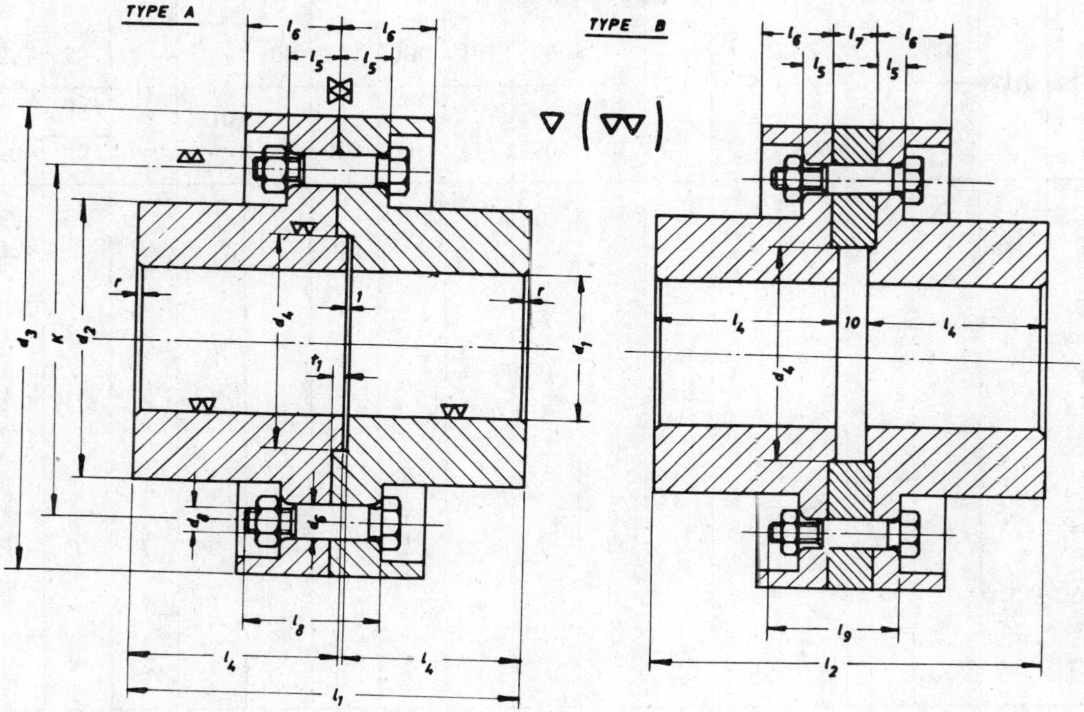

(Dimensions in mm)

d_1	d_2	d_3	d_4	d_5	d_6	d_7	k	l_1	l_2	l_3	l_4	l_5	l_6	l_7	Chamfer × 45°	t_1	t_2	Turned Hex. Bolt according to IS : 3640 with Hex Nul				Torqn transmitted	Speed r.p.m.
																			For A and C	Type B			
N7			H7/ h8		H7										r			d_8	l_8	l_9	Qty.	Kgf.m	max.
25	58	125	50	40	11	M10	90	101	110	117	50	16	31	16	1.6	3	8	M10	45	60	3	4.75	2120
30				45											2							9	
35	72	140	65	55			100	121	130	141	60				2		10					15	2000
40																						24.3	
45	95	150	75	65	11	M10	125	141	150	169	70	18	34	16	2	3	14	M10	50	65	3	36.5	1900
50															3							53	
55	110	180	90	75	13		140	171	180	203	85		37				16	M12		70	4	75	1800
60																						100	

70	130	200	100	85	13	M10	160	201	210	233	100	23	41	18	4	4	16	M12	60	80	6	175	1700
80	145	224	115	95			180	221	230	261	110						20				8	272	1600
90	164	250	135	105	17		200	241	250	281	120	30	54				20	M16	80	100	8	412	1500
100	180	280	150	120			224	261	270	301	130											600	1400
110	200	300	165	130	17	M10	250	281	290	329	140	33	60	13	5	4	24	M16	85	105	8	850	1320
125	225	335	185	145			280	311	320	359	155										10	1280	1250
140	250	375	210	160	21	M12	310	341	350	397	170	40	70				28	M20	105	125	10	1950	1180
160	290	425	240	180	25		350	401	410	457	200		75		6			M24	110			3070	1120
180	325	450	255	-	25	M12	380	451	460	-	225	45	80	20	-	5	-	M24	120	140	12	4620	1060
200	360	500	290	-			420	501	510	-	250				-		-				16	6300	1000
220	400	560	310	-	32		470	541	550	-	270	52	95		-		-	M30	140	160	14	8250	950
250	450	630	390	-			540	601	610	-	300				-		-				16	12000	900

Table 9.23 Pin and bush type couplings

Dimensions in mm

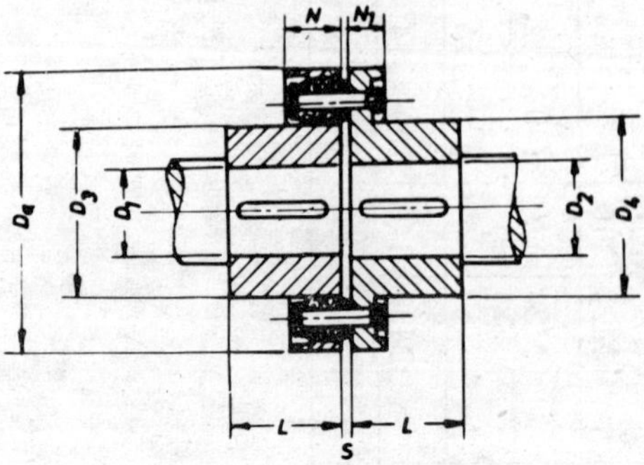

Size	Continuous rating under steady load		Max. speed RPM	BORE			D_a	D3	D4	L	N	N1	S
	HP/100 R P M	Torque kgf.m		Min Rough	Max. Finished								
					D_1	D_2							
0.06	0.6	4.3	7200	12	25	30	96	44	50	35	24	18	4 ± 2
0.16	1.6	11.5	6100	18	35	42	112	62	68	45	24	18	4 ± 2
0.63	6.3	45	4300	23	50	60	160	85	95	60	38	22	4 ± 2
1.4	14	100	3400	38	70	80	200	120	128	80	38	22	4 ± 2
2	20	144	3000	48	80	90	225	134	144	90	42	28	7 ± 3
4	40	280	2400	58	100	110	285	166	176	110	54	35	7 ± 3
8	80	580	1900	68	120	130	360	205	210	140	68	44	10 ± 4
11	110	790	1700	78	130	140	400	218	230	160	80	52	10 ± 4
16	160	1150	1500	88	140	160	450	240	260	180	80	52	10 ± 4
22	220	1580	1350	98	160	180	500	270	290	200	102	62	10 ± 4

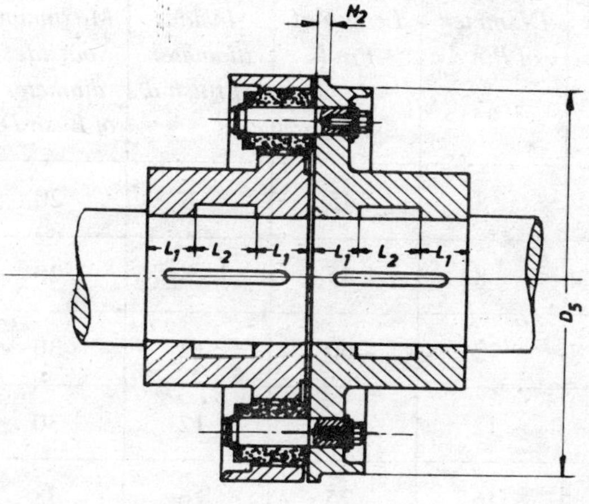

Size	Continuous rating under steady load		Max. speed RPM	Bore			Hub				L	Hub Recess		N	N_1	N_2	S
				Min.	Max. Finished												
	HP/100 RPM	Torque kgf.m		Rough	D_1	D_2	D_a	D_3	D_4	D_5		L_1	L_2				
45	450	3200	1050	100	200	200	630	340	340	580	240	-	-	122	90	22	12 ± 5
63	630	4500	950	110	190	190	710	320	320	660	260	-	-	145	100	25	12 ± 5
85	850	6100	850	125	160	160	800	270	270	750	290	-	-	145	100	25	12 ± 5
170	1700	12200	680	180	220	320	1000	370	520	940	350	-	-	164	110	30	12 ± 5
230	2300	16500	600	200	250	250	1120	425	425	1040	380	120	140	182	120	30	14 ± 6
450	4500	32000	490	300	350	400	1400	590	680	1310	460	140	180	204	140	35	14 ± 6

Details of pin and bush

Size	*PCD*	*No of Pin*	*Diameter of Pin d*	*Length of Pin l*	*Inside diameter of Bush d*	*Maximum outside diameter of Bush D_6*	*Length of Bush L_3*	*No. of Bellows of Bush*
0.06	68	8	8	45	8	20	22	1
0.16	84	10	8	45	8	20	22	1
0.63	118	9	12	60	12	30	34	1
1.4	156	12	12	60	12	30	34	1
2.0	174	11	16	75	16	38	42	1
4.0	218	11	20	95	20	48	52	1
8.0	276	11	25	120	25	60	66	1
11.0	298	10	30	140	30	75	80	1
16.0	345	11	30	140	30	75	80	1
22.0	375	10	35	170	35	90	100	1
45.0	480	10	45	200	45	110	120	2
63.0	540	10	50	225	50	125	140	2
85.0	628	11	50	225	50	125	140	2
170.0	810	13	55	255	55	140	160	2
230.0	905	13	60	280	60	160	180	2
450.0	1160	14	70	340	70	180	200	2

Note : Material : (a) Pins shall be of 45C8 or 50C4 (IS:1570), with hardness 1850 to 2000 N/mm^2 HB
(b) Bushes shall be of rubber with canvas impregnated shore hardness IRH 80.

Table 9.24 Tyre Type Couplings

Dimensions in mm

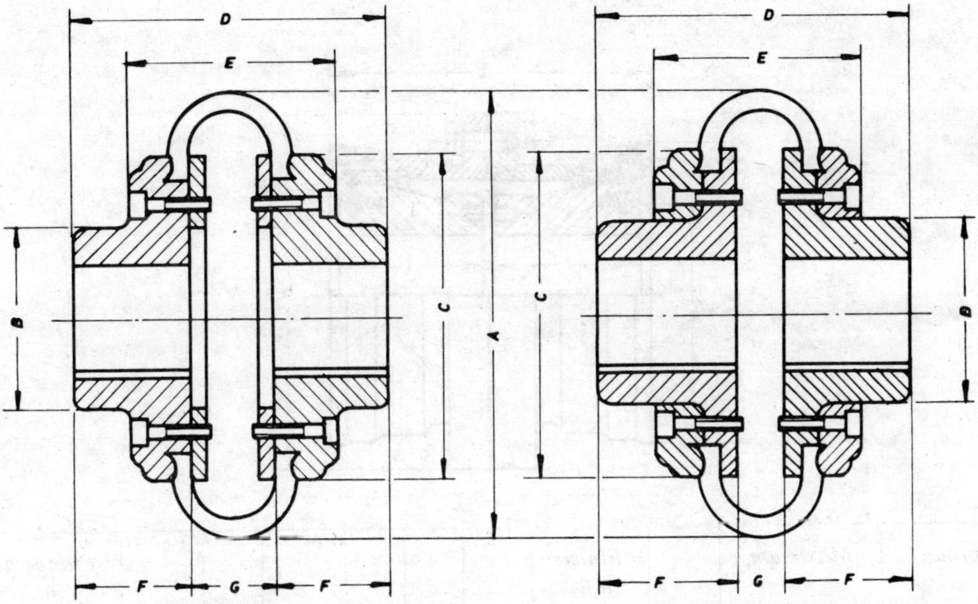

Size	*Continuous rating under steady load*		*Max. speed RPM*	*Bore*		*A*	*B*	*C*	*D*	*E*	*F*	*G*
	HP/100 RPM	*Torque kgf.m*		*Min. Rough*	*Max.*							
1	0.21	1.5	5500	12	24	108	43	80	108	52	40	28
2	0.42	3.0	5500	15	32	135	58	105	105	56	39	27
3	0.98	7.0	4200	17	44	178	71	133	146	83	53	40
4	2.1	15.0	3200	30	55	212	95	169	157	83	56	45
5	6.3	45.0	2600	38	90	277	145	225	205	132	76	53
6	11.0	80.0	2450	38	100	322	159	255	208	108	96	16
7	21.0	150.0	2000	45	125	370	203	310	250	125	115	20
8	52.0	400.0	1600	70	140	450	237	370	443	155	190	63
9	96.1	700.0	1250	70	145	550	260	420	358	224	142	74
10	172.8	1250.0	960	125	225	746	434	638	515	228	232	51

Table 9.25 Geared Coupling

Dimensions in mm

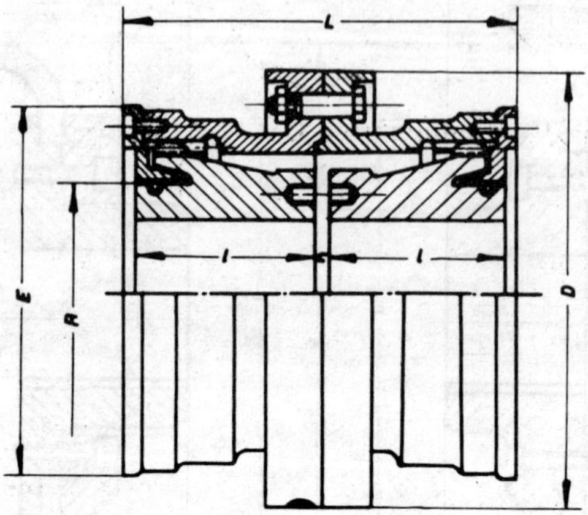

Size	Normal		Maximum		RPM Max	Finish Bore		D	L	l	E	c	R	Flange details		
	HP/RPM	Torque kgf.m	HP/RPM	Torque kgf m		Min.	Max.							PCD	No. of Holes	Fit Bolt Size
40	0.08	55	0.12	85	9500	18	40	160	135	60	123	5	60	126	12	M10
50	0.16	120	0.25	180	8500	28	50	183	155	70	145	5	75	147	14	M10
60	0.28	200	0.41	300	7500	30	60	215	183	80	165	6	90	175	14	M10
70	0.46	340	0.70	500	5700	32	70	230	198	90	183	6	100	190	14	M10
80	0.65	450	0.92	650	6000	35	80	255	223	100	204	6	120	215	14	M12
90	0.92	680	1.40	1000	5400	40	90	290	249	110	230	8	130	246	14	M14
100	1.30	950	2.00	1350	5300	58	100	300	269	120	250	8	140	252	14	M14

110	1.70	1250	2.62	1850	4750	68	110	330	289	130	270	8	155	282	14	M16
125	2.60	1850	3.85	2800	4250	78	125	355	330	150	305	10	175	300	14	M16
140	3.70	2500	5.45	3900	3750	88	140	410	365	165	335	10	200	352	14	M18
160	5.50	3900	8.00	5800	3350	125	160	460	421	190	380	12	230	402	16	M18
180	7.60	5200	11.00	8400	3000	145	180	505	481	220	425	12	260	443	16	M20
200	10.50	7500	15.50	11000	2650	160	200	570	537	245	475	14	290	504	16	M22
220	14.20	10000	21.00	15000	2500	185	220	610	593	270	520	16	325	536	16	M24
280	29.50	21000	43.00	32000	2000	245	280	785	751	340	670	22	415	703	24	M30
350	55.00	39500	82.00	58500	950	260	350	960	830	400	840	30	590	875	18	M42
400	78.00	56000	117.00	83700	875	340	400	1070	890	430	915	30	650	980	24	M48

9.7 JOURNAL BEARINGS AND ANTI-FRICTION BEARINGS

The term "bearing" means those machine elements which hold axles and shafts, and take the resultant forces acting thereon. Bearings are generally classified into sliding contact bearings and rolling contact bearings. The sliding contact bearings are categorized as: radial bearings which support rotating shafts and journals, thrust bearings which support axial loads on rotating components, and guide or slipper bearings which guide moving parts in a straight lines. The rolling contact bearings are also known as anti-friction bearings. These bearings are of many types and will be discussed later.

Journal bearings: Among the different types of sliding contact bearings, the radial bearings are the most common ones used in industrial applications. These bearings are also known as bush, sleeve and journal bearings. Both journal bearings and anti-friction bearings have their own peculiarities and fields of applications. The choice of bearing will depend upon the characteristics of the bearing and the scope of application. Journal bearings are cheaper, require less space, are quieter during operation, have greater rigidity and impact-damping properties. Their disadvantages are: higher frictional losses, particularly during starting, and more stringent lubrication requirements. They are also of shorter life comparatively.

Since the subject of bearing is a vast one, the discussion in this section will be mainly of general nature and will be confined only to those aspects which are commonly required for drawing making purposes. Assignment of proper fit between the hearing and the journal is a prime consideration. The following fits are generally recommended:

$$H\,7/g\,6,\ H\,7/f\,7,\ H\,8/f\,8,\ H\,7/e\,8,\ H\,7/d\,9,\ H\,7/d\,10.$$

The selection of fit will depend upon the degree of clearance required to meet the operational parameters. For simple bearings, bushes made of suitable bearing materials are press-fitted in bearing housings. The standardised bushes with their relevant data are given in Table 9.26.

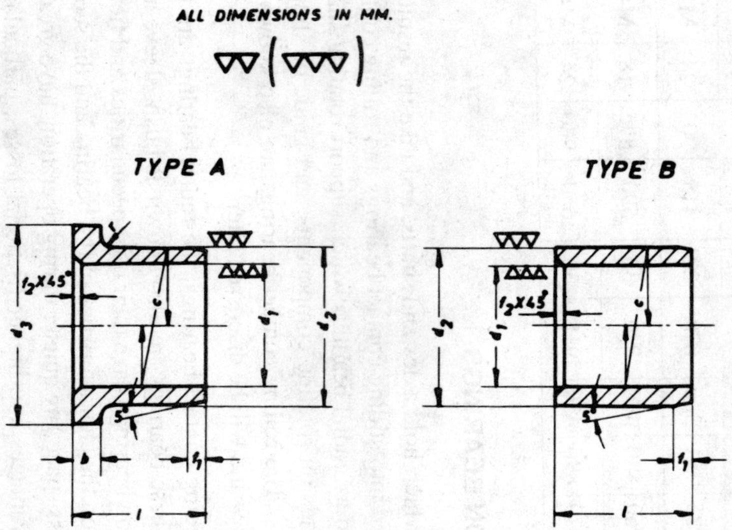

Reproduced with permission from : Gitin M. Maitra and L.V. Prasad, Handbook of Mechanical Design, Second Edition, Tata Mcgraw-Hill, New Delhi, 1995, Fig. 9.16. P 9.75.

Table 9.26 Dimensional data and tolerance for plain bushes

(All dimensions in mm)

<table>
<tr><td>d_1</td><td>10</td><td>12</td><td>14</td><td>16</td><td>18</td><td>20</td><td>22</td><td>25</td><td>28</td><td>30</td><td>32</td><td>35</td><td>36</td><td>38</td><td>40</td><td>42</td><td>45</td><td>48</td><td>50</td><td>52</td><td>56</td><td>60</td><td>63</td><td>65</td><td>68</td><td>70</td><td>72</td><td>75</td><td>80</td></tr>
<tr><td>d_2</td><td>16</td><td>18</td><td>20</td><td>22</td><td>25</td><td>26</td><td>28</td><td>32</td><td>36</td><td>38</td><td>40</td><td>45</td><td>45</td><td>48</td><td>50</td><td>52</td><td>56</td><td>58</td><td>60</td><td>63</td><td>68</td><td>72</td><td>75</td><td>78</td><td>80</td><td>82</td><td>85</td><td>90</td><td>95</td></tr>
<tr><td>d_3</td><td>20</td><td>22</td><td>25</td><td>28</td><td>30</td><td>32</td><td>35</td><td>40</td><td>44</td><td>46</td><td>48</td><td>50</td><td>52</td><td>55</td><td>58</td><td>60</td><td>64</td><td>66</td><td>68</td><td>70</td><td>75</td><td>84</td><td>86</td><td>90</td><td>92</td><td>95</td><td>98</td><td>105</td><td>110</td></tr>
<tr><td>b-0.2</td><td>3</td><td>4</td><td colspan="6">5</td><td colspan="4">6</td><td colspan="9">7</td><td colspan="6">8</td><td colspan="2">10</td></tr>
<tr><td>f_1</td><td colspan="5">1</td><td colspan="3">1.2</td><td colspan="4">1.6</td><td colspan="11">2</td><td colspan="5">2.5</td><td></td></tr>
<tr><td>f_2</td><td colspan="5">0.8</td><td colspan="4">1.2</td><td colspan="8">2</td><td colspan="4">2.5</td><td colspan="3">3</td><td colspan="5">4.5</td></tr>
</table>

r	0.2		0.4					0.6														0.8	
d_1	85	90	95	100	105	110	115	120	125	130	135	140	145	150	155	160	165	170	175	180	185	190	195	200
d_2	100	105	110	115	120	125	130	135	140	145	150	155	160	165	175	180	185	190	195	200	205	210	215	220
d_3	115	120	125	130	135	140	145	150	155	160	165	170	175	180	195	200	205	210	215	220	225	230	235	240
b-0.2	10																							
f_1	2.5												3.2											
f_2	4.5														5.5									
r	0.8														1									

Recommended ISO Tolerance for				
	d_1	Before Assembly	E9	C11
		Corresponding Tolerance for shaft	f7, e9	h9, h11
	d_2		r6	
	Hole for mounting the bush		H7	
Permissible deviation in concentricity			1/2 IT 8	

The bushes are provided with suitable grooves which are required to establish and maintain a film of lubricant between the mating surfaces and to provide sufficient cooling for the bearing system.

Some common forms of grooving with dimensions and some typical examples of application are shown in Table 9.27.

Table 9.27 Lubrication grooves and holes for bush bearings

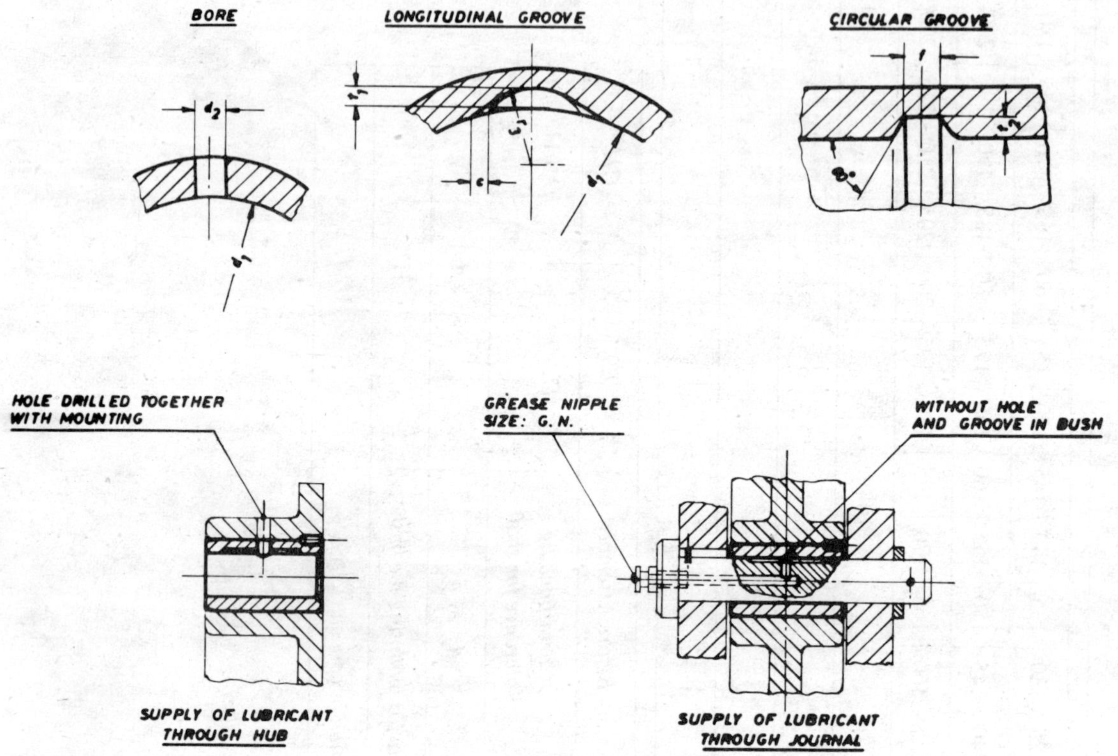

Diameter d_1	G.N.	$c \approx$	d_2	f	r_3	t_1	t_2
16 upto 25	$R\frac{1''}{4}$	0.8	3	2	2.5	0.8	0.8
Over 25 upto 38		1	3	2	4	1	1
Over 38 upto 56	$R\frac{3''}{8}$	1.2	4	2.6	6	1.3	1.2
Over 56 upto 72		1.5	6	4.3	8	1.8	1.5
Over 72 upto 100		2	6	4.3	12	2	1.8
Over 100 upto 150	$R\frac{1''}{2}$	3	8	7.5	20	2.5	2.2
Over 150 upto 200		4	10	11	28	3.5	2.7

Bearing bushes are sometimes provided with white metal linings. White metals are bearing metals which are also known as Babbitt metals. These metals are mainly tin and lead base alloys, and have excellent bearing properties. Their load carrying capacity is, however, lower when compared to other common bearing materials.

The white metal lined bushes are generally of two types: solid and split. These two types are shown in Tables 9.28 and 9.29.

Table 9.28 Dimensions of white metal lined bush, solid type

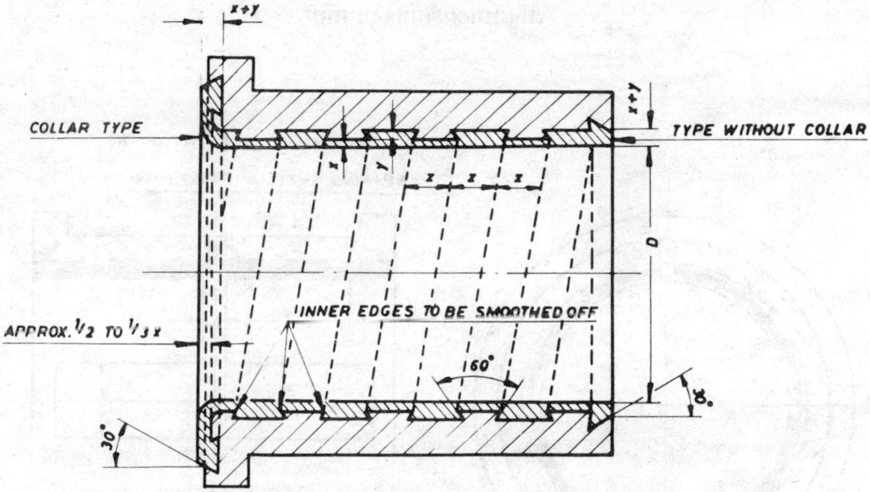

Steel or steel casting			
D	*X*	*Y*	*Z*
50-75	1.5		
76-100	2.5	2.5	10
101-200	2.5	2.5	12
201-300	3	2.5	14
301-400	3.5	2.5	16
401-500	4	3	18
501-600	5	3.5	20
> 600	6	4	25

Cast iron			
D	*X*		*Z*
50-100	2.5	2.5	10
101-200	3	2.5	12
201-300	3.5	2.5	14
301-400	4	3	16
401-500	5	3.5	18
500-600	6	4	20
> 600	7	4.5	25

The dimensions stated are minimum values

Reproduced with permission from : Gitin M. Maitra and L.V. Prasad, Handbook of Mechanical Design, Second Edition, Tata McGraw-Hill, New Delhi, 1995, Fig. 9.18. P 9.78.

Bearing bushes are mounted inside bearing housings or plummer blocks. These housings are generally of two kinds: the solid type and the split type. Sometimes the shafts are directly mounted on the housing, that is, without the bush.

The common material used for plummer blocks is cast iron, but cast steel is also used. Lubricant to the bearing housing is generally provided through a hole situated conveniently at the top portion of the housing. Lubrication may be effected through grease nipples. Some standard types of plummer blocks are shown in Table 9.29.

Table 9.29 Dimensions of white metal lined bush, split type

All dimensions in mm

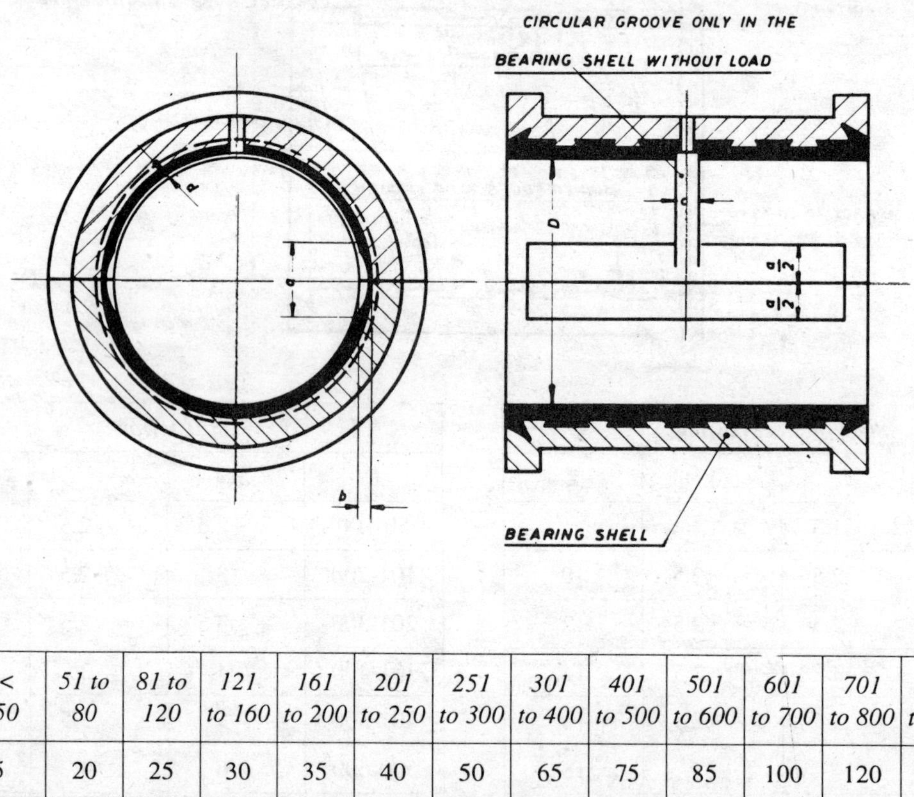

D	< 50	51 to 80	81 to 120	121 to 160	161 to 200	201 to 250	251 to 300	301 to 400	401 to 500	501 to 600	601 to 700	701 to 800	801 to 900	> 900
a	15	20	25	30	35	40	50	65	75	85	100	120	120	120
b	1,5	2	2,5	2,5	2.5	3	3	3.5	4	5	6	6	6	6
c	6	8	12	14	16	20	24	28	32	38	44	50	60	70
d	1	1,5	2	2	2	2,5	2,5	3	3	4	5	5	5	5

Note: The figures given are approximate values.

Table 9.30 Bearing housings, solid type and split type

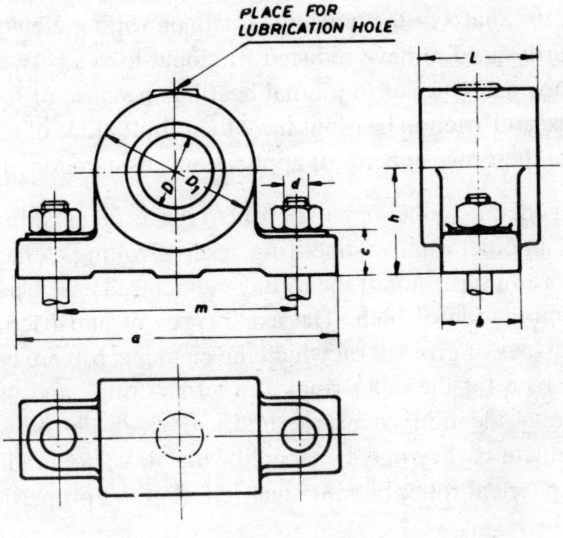

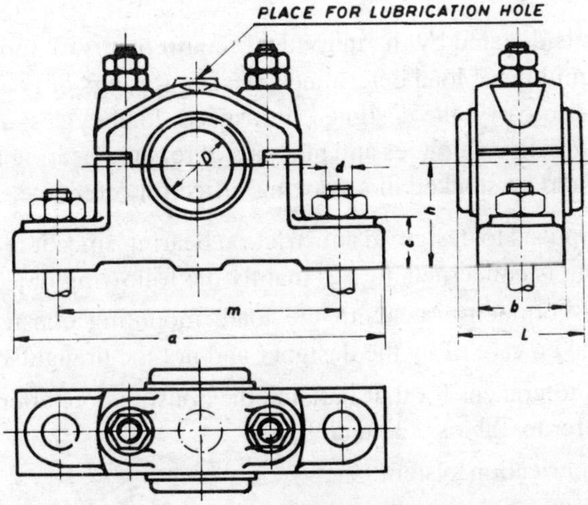

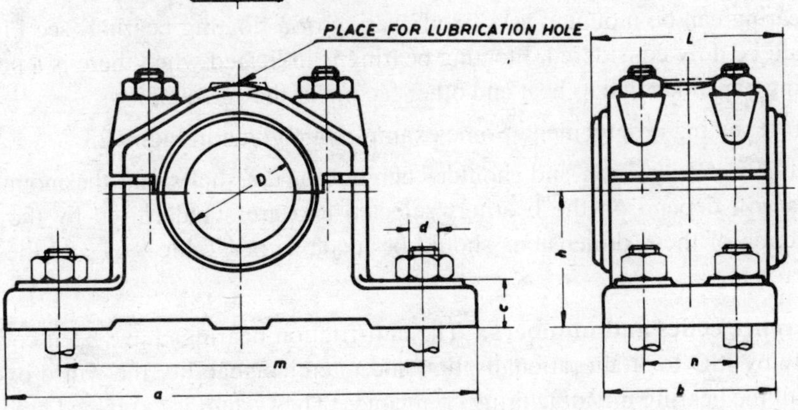

Anti-friction bearings

In the anti-friction bearings, the shafts or the housings rotate on rolling elements, such as, balls, rollers, needles, *etc.* They are primarily used to have reduced frictional losses between stationary and moving parts. Though prima facie they are superior to journal bearings because of lower coefficient of friction and lesser starting friction, the anti-friction bearings have their limitations of use, and both these bearings and the journal bearings have their own spheres of applications, as mentioned earlier.

Anti-friction bearings are generally either of ball or roller type as far as rolling elements are concerned. A typical bearing consists of an outer ring, an inner ring, a set of rolling elements (balls or rollers) and a cage or a separator which acts as a separator of the rolling elements. These bearings can carry pure radial or pure thrust leads or a combination of both. The usual types of anti-friction bearings are shown in Figs. 9.20 and 9.21. The pathways or grooves on which the elements roll are called races, *viz.*, inner race for the inner ring and outer race for the outer ring. The terms "ring" and "race" are sometimes used synonymously. In roller bearings, the rollers may be cylindrical which take only radial load with provision to take care of axial misalignment or they may be tapered which take both radial load and axial thrust. A still another variety, called spherical roller bearing, has self aligning properties to take care of angular misalignment.

The choice of bearing type is dictated by the individual requirements of application as well as by the nature of loads while the magnitude of load determines its size. Ratings of bearings are done according to the loads they are designed to carry, the designed life, speed, load cycles and other relevant factors. These are exhaustively dealt with in catalogues and manuals of reputed bearing manufacturers, *viz.*, SKF, FAG, Timken, *etc.*, which should be stocked in a drawing office for reference.

A drawing-maker is not supposed to design an anti-friction bearing since it is a bought-out item. So far as engineering drawing making is concerned, he has mainly the following objectives:

(i) Selection of bearing commensurate with the load, mounting constraints, and other factors. Usually, this task is taken care of by the designer and not the draughtsman.

(ii) Allocation of fits and tolerances for the shaft and the housing bore after the selection of bearing. In this connection, refer to Tables 9.32 and 9.33.

(iii) Provision of proper lubrication system.

(iv) A bearing can be mounted as a fixed bearing or a floating bearing (see Fig. 9.24). This aspect should be duly considered. Floating bearing is indicated when there is a possibility of the shaft getting expanded due to heat and other factors in the surroundings.

(v) Correct sealing arrangement. Some examples are given in Fig. 9.23.

(vi) Selection of fillet radii and shoulder heights for the shafts and the mountings. These dimensions will depend on the bearings selected and are standardised by the bearing companies. Selection of these dimensions should be made as per Table 9.31 and these should be strictly adhered to.

Bearing reference codes and numbers: The anti-friction bearings have been codified and numbered internationally by ISO to attain rationalisation and interchangeability the world over and the system is followed by all the bearing manufacturing companies. The system is explained below.

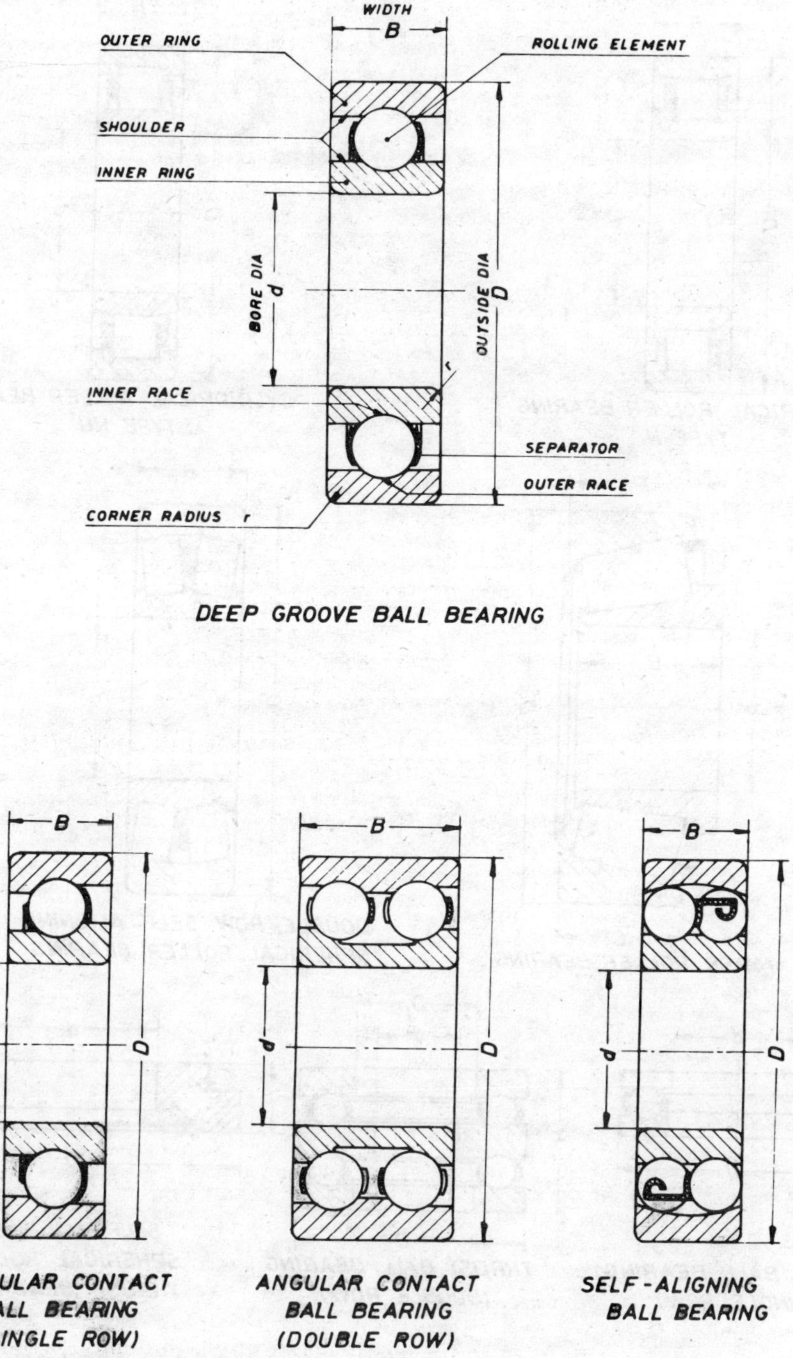

Fig. 9.20 Types of anti-friction bearings

Reproduced with permission from : Gitin M. Maitra and L.V. Prasad, Handbook of Mechanical Design, Second Edition, Tata McGraw-Hill, New Delhi, 1995, Fig. 9.24. P 9.87.

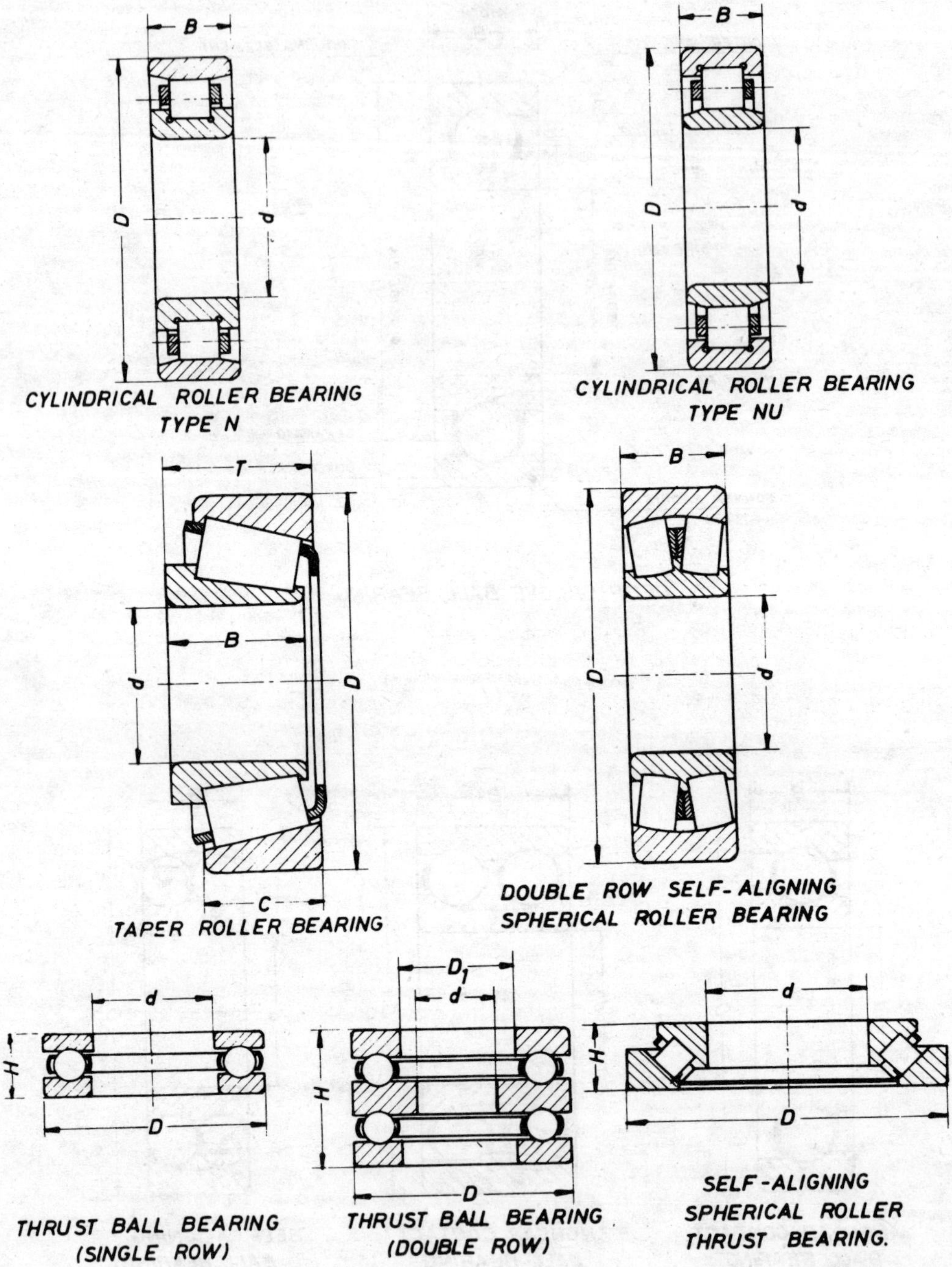

Fig. 9.21 Types of anti-friction bearings

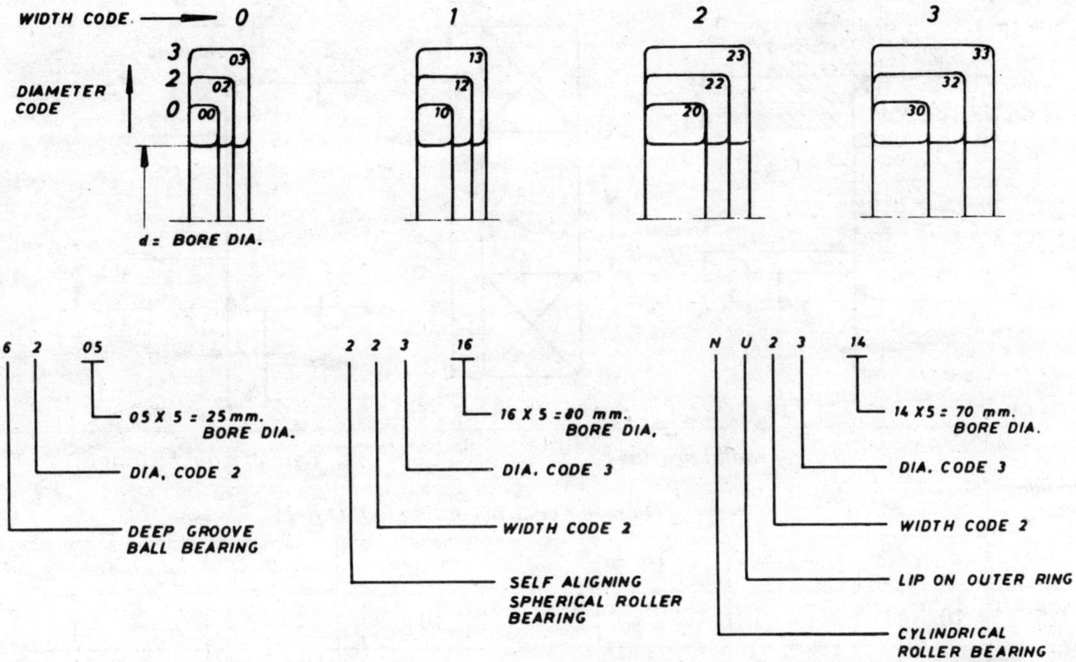

Fig. 9.22 Bearing reference codes and numbers

For any particular bore diameter, several outside diameters and widths are available. This is diagramatically represented in the top picture of Fig. 9.22. The diameter codes are: 8, 9, 0, 1, 2, 3 and 4, and the width codes are: 0, 1, 2, 3, 4, 5 and 6. The codes are in ascending order dimensionwise. The bore diameter remaining same, the proper bearing will be selected as per the design parameters and constraints, e.g., load, mounting space, *etc.*, conforming to the bearing code and number. The bearing reference number is made up in the following manner: the last two numerals refer to the diameter of the bore. In the range 20 mm to 480 mm, the bore can be calculated by multiplying these numerals by 5. Thus, in Fig. 9.22, $05 \times 5 = 25$ mm is the bore diameter of this particular bearing. The third numeral from the right refers to the diameter code. The fourth represents the width code in some cases or it may stand for the bearing type. Examples are given in the bottom picture of Fig. 9.22. Cylindrical roller bearings are now identified by the reference code letter *N*. Thus, by simply referring to the bearing number, one can know the relevant informations of that bearing.

Fillet radii and shoulder heights for shafts and housings: It has been emphasised before that the fillet radii of the shaft on which the bearing is mounted and of the housing in which the bearing is inserted are standardised by the bearing companies. While making a drawing of the shaft or the housing, only these prescribed radii should be entered in the drawing as applicable to the particular bearing in question.

Each bearing has a chamfer radius. This radius can be obtained from the bearing catalogues. These radii can also be determined by referring to Table 9.31, in which the chamfer radii "*r*" is shown as a function of the bore, type of bearing and the diameter code.

Table 9.31 Fillet radii and shoulder heights

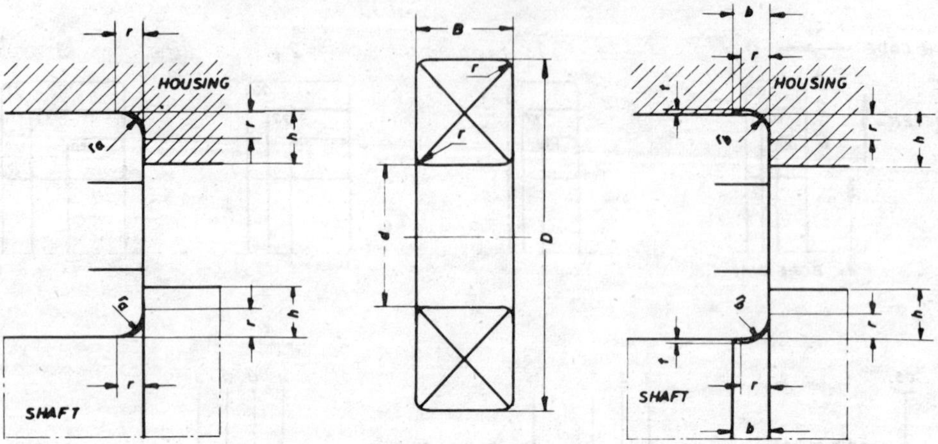

Chamfer Radius r	*Bore of radial bearings*				*Bore of axial bearings*			
	Diameter code according to ISO R-15							
	0	*2*	*3*	*4*	*0*	*2*	*3*	*4*
0.5	10 to 17	–	–	–	10 to 20	–	–	–
1.0	20 to 25	10 to 17	10	–	25 to 55	10 to 30	–	–
1.5	30 to 50	20 to 30	12 to 17	–	60 to 160	35 to 85	25 to 45	25 to 30
2.0	55 to 85	35 to 50	20 to 30	17 to 20	170 to 220	90 to 120	50 to 70	35 to 45
2.5	90 to 100	55 to 75	35 to 45	25 to 35	240 to 280	130 to 180	75 to 100	50 to 60
3.0	105 to 140	80 to 90	50 to 55	40 to 45	300 to 360	190 to 220	110	65 to 75
3.5	150 to 200	95 to 120	60 to 80	50 to 65	–	240 to 280	120 to 150	80 to 90
4.0	220 to 240	130 to 160	85 to 120	70 to 80	–	300 to 340	160 to 180	100 to 110
5.0	260 to 320	170 to 240	130 to 180	85 to 110	–	360 to 400	190 to 240	120 to 150
6.0	340 to 420	260 to 320	190 to 240	120 to 150	–	420 to 500	260 to 360	160 to 200
8.0	440 to 500	–	260 to 280	–	–	–	380 to 500	220 to 300
10.0	–	–	–	–	–	–	–	320 to 420
12.0	–	–	–	–	–	–	–	440 to 500

Reproduced with permission from : Gitin M. Maitra and L.V. Prasad, Handbook of Mechanical Design, Second Edition, Tata McGraw-Hill, New Delhi, 1995, Fig. 9.28. P 9.89.

Nominal Bearing Chamfer r	*Shoulder height for radial bearings*		*Fillet radius*	*Fillet with relief*		+
	h	h *Min.*	r_a *Max.*	t	r_b	b
0.2	0.8	0.5	0.1	–	–	–
0.3	0.8	0.6	0.1	–	–	–
0.4	1.0	0.8	0.2	–	–	–
0.5	1.0	1.0	0.3	–	–	–
0.8	2.0	1.5	0.5	–	–	–
1.0	2.5	2.0	0.6	–	–	–
1.5	3.0	2.5	1.0	0.2	1.3	2.0
2.0	3.5	3.0	1.0	0.3	1.5	2.4
2.5	4.5	3.5	1.5	0.4	2.0	3.2
3.0	5.0	4.0	2.0	0.5	2.5	4.0
3.5	6.0	4.5	2.0	0.5	2.5	4.0
4.0	7.0	5.5	2.5	0.5	3.0	4.7
5.0	9.0	7.5	3.0	0.5	4.0	5.9
6.0	11.0	9.0	4.0	0.6	5.0	7.4
8.0	14.0	11.0	5.0	0.6	6.0	8.6
10.0	18.0	14.0	6.0	0.6	7.0	10.0
12.0	22.0	17.0	8.0	0.6	9.0	12.0

After ascertaining the bearing chamfer radius "r", the next step is to take the value of the fillet radius "r_a" from the above Table. The shaft and the housing should be machined to this value which should be inserted in the drawing by the draughtsman, conforming to the bearing in question. If the fillet is provided with relief, then the value "r_b" should be used. It is most important that the correct values of r_a (or r_b) should be entered in the drawing for the steps in the shafts and the housings commensurate with the bearings selected. These values have been carefully standardised after considering such factors as stress concentration and other relevant factors.

The shoulder heights of the shafts and the housings are equally important. The bearing abuts against the shoulders. If the shoulder height is too small, there may be difficulty in holding this bearing. If it is too large, there is a chance of fouling the cage (separator) or the rolling elements against the shoulders, that is, the sides of the steps in the shaft or the housing.

Seals and lubrication: Anti-friction bearings are to be provided with adequate lubrication system to minimise friction and wear. Lubricants also help to dissipate heat which is generated during running. In bearing catalogues, the proper lubricants are indicated for each type of bearings. Under normal conditions, grease lubrication is generally resorted to. The choice of lubricants primarily depends on the temperature range and the operating speeds. Both oil and grease are used as the lubricating mediums, depending upon the design and the applications.

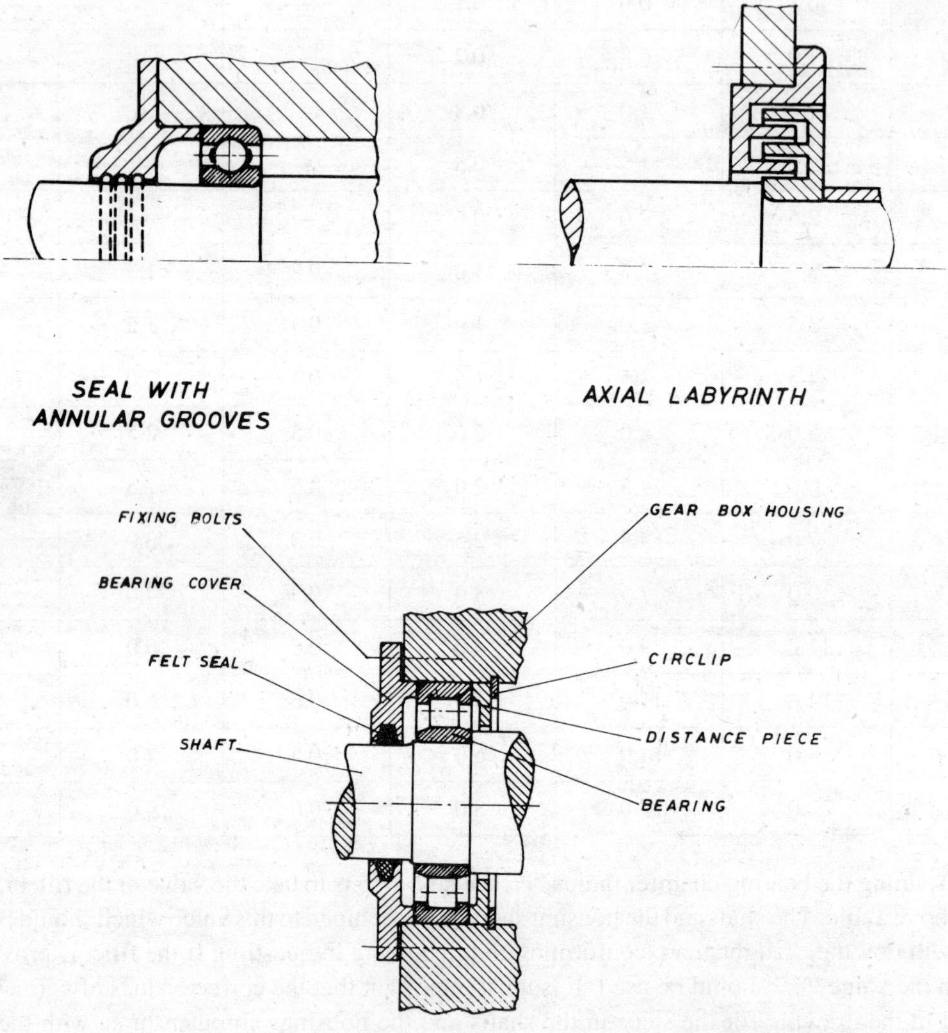

Fig. 9.23 Sealing arrangements in anti-friction bearings

In Fig. 9.23, some types of sealing arrangements, generally used in systems using anti-friction bearings, are shown. Seals are used to prevent leakage of lubricants and also to protect the bearings from the entry of dust, dirt and other undesirable particles to prevent seizure.

Two basic types of seals are commonly used: the non-rubbing type and the rubbing type. The non-rubbing seals are shown at the top portion of Fig. 9.23. The rubbing types are generally the rotary shaft oil seals and the felt seals. These seals have been dealt with in Section 9.8. Figure 9.23 also shows a typical bearing assembly.

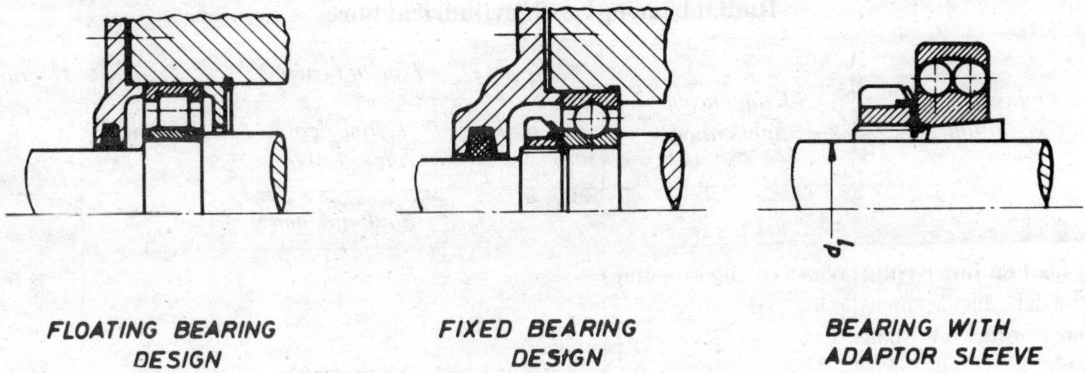

Fig. 9.24 Arrangements for bearing mounting

Mounting and location of bearings: The arrangements for bearing mounting should be carefully selected. Floating type of mounting is indicated when there has to be an axial play to take care of the expansion of the shaft or for some other factors. The type of floating bearing shown in Fig. 9.24 is a cylindrical roller bearing. This bearing is a detachable type, so that if the shaft expands, the inner ring glides over the rollers, thus affording relative movement between the inner ring (to which the shaft is fixed) and the rest of the bearing elements.

Fixed bearing is used when the expansion and other such factors are absent. Bearing with adapter sleeve is used when the bearing is to be fitted somewhere on the shaft where the bearing seat is quite away from the ends of the shaft.

Fits and tolerances for bearing fittings: The anti-friction bearings are manufactured to close tolerances. Depending on the bearing selected, the tolerances of the mountings, that is, the shaft on which the bearing is fitted and the housing in which the bearing is inserted, are internationally selected. These are given in Tables 9.32 and 9.33. The task of the draughtsman is only to choose the proper tolerances after consulting these tables.

The following points are of relevance while determining the tolerances:

Generally, the shaft is provided with a tolerance of ISO quality 5 to 6 and the housing bore with quality 6 to 7. When an adapter sleeve is used, the shaft is machined to a tolerance of h 9 or h 10. A most important factor for determining the tolerance is the orientation of the load on the rings. The following load conditions may arise:

1. **Circumferential load on the ring:** This occurs when
   (a) one ring rotates, the other ring as well as the direction of load is stationary, or
   (b) the ring is stationary but the other ring and the direction of load rotate at equal speeds.

2. Point load on the ring: This occurs when

(a) the ring in question and the direction of load are stationary, or

(b) the ring as well as the direction of load rotate at equal speeds while the other ring is stationary.

Table 9.32 Tolerances for solid steel shafts
Radial bearings with cylindrical bore

Load condition	*Examples of application*	*Type of bearing*			*Tolerance*
		Ball	*Cylindrical taper roller*	*Spherical roller*	
		Shaft dia. mm			
Point load on inner ring: Easy axial displacement of inner ring on shaft desirable.	Wheels on non-rotating axles	All diameters			g 6
-do- not necessary	Tension pulleys, rope sheaves				h 6
Circumferential load on inner rings: Light and variable load	Conveyors light duty gearboxes.	18 - 100 100 - 140	≤ 40 40 - 100	– –	*j* 6 *k* 6
Normal and heavy loads.	Electric motors, pumps I.C. engines	≤ 18 18 - 100 100 - 140 140 - 200 200 - 280	– ≤ 40 40 - 100 100 - 140 140 - 200	– ≤ 40 40 - 65 65 - 100 100 - 140	*j* 5 *k* 5 *m* 5 *m* 6 *n* 6
Very heavy loads and shock loads.	Rolling mills, heavy railway vehicles, etc.	– – –	50 - 140 140 - 200 > 200	50 - 100 100 - 140 > 140	*n* 6 *p* 6 *r* 6
High running accuracy with light loads.	Machine tools	≤ 18 18 - 100 100 - 200 –	– ≤ 40 40 - 140 140 - 200	– – – –	*h* 5 *j* 5 *k* 5 *m* 5

When the above bearings carry axial loads only, then the tolerance is *j* 6 for all diameters and for all kinds of application.

For radial bearings with taper bore and taper sleeve used for general engineering purposes, the tolerance on shaft in *h* 9/ IT 5 or *h* 10/IT 7. Here, IT 5 or IT 7 means that the deviation from circularity or the shaft must not exceed the quality 5 or 7 of the relevant nominal size.

Thrust bearings: Purely axial load - *h* 6 for all diameters. For combined radial loads and axial loads on spherical roller thrust bearings, (i) Stationary load on shaft washer - *j* 6 for $d \leq 250$ and *js* 6 for $d \geq 250$ (ii) Rotating load on shaft washer - *k* 6 for $d \leq 200$, *m* 6 for $d = 200$ to 400, and *n* 6 for $d \geq 400$.

Table 9.33 Tolerances for steel and cast iron housings & Radial bearings

Load condition	*Examples of application*	*Tolerance*
Circumferential load on the outer ring:		Solid housing:
Heavy and shock loads	Roller bearing wheel-hubs, big-end bearings.	P 7
Normal and heavy loads	Ball bearing wheel-hubs, crane wheels.	N 7
Light and variable loads	Conveyor rollers, rope sheaves, belt tension pulleys	M 7
Point lead on outer ring:		Solid or split housing:
Load of all kinds.	Railway axle boxes.	H 7
Normal light loads.	General engineering	H 8

When the direction of load is indeterminate, then for normal to heavy loads, the tolerance is K 7 for electric motors, pumps, crank-shaft bearings, etc. For heavy shock loads for electric traction motors, M 7 can be taken.

When accurate or silent running is desired, then the tolerances are:

Roller bearings for machine tools : K 6
Ball bearings for grinding spindles : J 6
Small electric motors : H 6

Thrust bearings:

Nature of load	*Bearing*	*Application*	*Tolerance*
Axial loads only	Thrust ball bearing.	General engineering purposes.	H 8
Combined loads	Spherical roller thrust bearing.		
Stationary load on housing Washer.		– do –	H 7
Rotating load on housing washer		– do –	H 7

Needle bearings : Needle bearings are mainly used in applications where the available radial space is limited, restricting the possibility of the mounting of other type of anti-friction bearings having larger sectional heights. These bearings have high load carrying capacity in relation to their sectional height. Both single-row and double-row designs are available. These bearings are furnished both with and without rings. Separators or cages hold the needles together in case of bearings without rings. Since the needle bearings are of separable type, the rings can be mounted independently. When the bearing is directly mounted on the shaft, i.e., in case of the design without the inner ring, the portion of shaft acting as the race for the needles should be hardened and ground, the surface hardness being around HRC 58 to 64. Where the needles run directly on the race, either on the shaft itself or on a sleeve fitted on the shaft and acting as the race, the following materials may be used for races :

Tool steel :	105 Cr 1 (IS : 1975)
Case hardening steel :	14 C6, 15 Cr 65 (IS : 1570)
Flame and induction hardening steels :	45 C 8, 50 C 4

The co-officiant of friction of needle bearing is around 0.0025.

Tolerances : The following tables show the tolerances normally used when needle bearings are fitted :

Table 9.34 Tolerances on housing and shaft for needle bearings

Tolerances for needle bearings without inner and outer races
(needles directly mounted on the shaft and housing)

Running Clearance: This depends on tolerance of needles, selected tolerance of race-ways and working temperature. Tolerance to be chosen as per operating condition.	*Tolerances on Housing*				
	G6	*H6*	*G6*	*G6*	*H6*
	Tolerances on Shaft				
	Upto 80 mm		*Over 80 upto 140 mm*	*over 140 mm*	
Smaller than Normal: Precision bearing seatings, impact loads oscillating movement.	*j5*	*h5*	*h5*	*h5*	–
Normal	*h5*	*g5*	*g5*	—	*f5*
Greater than Normal: High working temperatures, shaft deflections and errors in alignment of bearings.	*g6*	*f6*	*f6*	*f6*	–

Tolerances for needle bearings with outer race and with or without inner race:

Housing Tolerances

Working conditions		*ISO Limits*	*Typical Applications*
Peripheral load on outer race	Heavy loads with shocks	P 7	Idler gears
	Normal loads	N7	Rollers, crank drives
	Light variable loads	M7	Cable sheaves, pulleys
Intermediate direction of load	Heavy shock loads	M7	Pumps, eccentrics
	Normal loads	K7	Compressors
	Light Loads	J7	Compressors, crank shaft
Point load on outer race	Heavy and normal loads	J7	General engineering, transmission
	Normal loads with split housing	H7	General engineering
High precision mountings	Accurate running, low flexing under light loads	K6	Grinding, milling and drill spindles, steady

Shaft Tolerances

With Inner Race

Working Conditions		*Shaft Dia mm*	*ISO Limits*	*Typical Applications*
Point load on inner race	Light fit required	For all sizes	*g6*	All wheels on stationary axle, gear wheels, rollers, pulleys.
	Normal cases		*h6*	
	Increased accuracy		*h5*	
Peripheral load on inner race	Light loads	Upto 50	*j5*	Gearing, spindles, motors, pumps, compressors, cranks, general engineering.
	Normal loads	Upto 50	*k5*	
		From 50 - 150	*m5*	
		Over 150	*m6*	
	Heavy loads in difficult operating conditions, shock load	Upto 150	*m6*	
		Over 150	*n6*	

Without Inner Race

<table>
<tr><th colspan="2">Shaft Dia. (mm)</th><th colspan="3">Running Clearance</th></tr>
<tr><th>Over</th><th>To</th><th>Less than Normal</th><th>Normal</th><th>Greater than Normal</th></tr>
<tr><td>–</td><td>65</td><td rowspan="4">k5</td><td rowspan="2">h5</td><td>g6</td></tr>
<tr><td>65</td><td>80</td><td rowspan="2">f6</td></tr>
<tr><td>80</td><td>160</td><td rowspan="3">g5</td></tr>
<tr><td>160</td><td>180</td><td rowspan="4">e6</td></tr>
<tr><td>180</td><td>200</td><td rowspan="2">j5</td></tr>
<tr><td>200</td><td>250</td><td rowspan="2">f6</td></tr>
<tr><td>250</td><td>315</td><td>h5</td></tr>
<tr><td>315</td><td>400</td><td>g5</td><td></td><td>d6</td></tr>
</table>

9.8 SEALS AND SEALING RINGS

To prevent the loss of fluid, and to maintain the requisite pressure in hydraulic and pneumatic devices, seals, sealing rings, packings and gaskets of proper design are used. Gaskets used in pipe lines are covered in Section 6.8. In this section, common sealing devices are dealt with.

Seals are generally classified as per their application and material. Static seals are used in pressure vessels, cylinder heads in automobile engines, etc., and dynamic seals are used for preventing the flow of fluid in a sliding or rotating joint in the moving component of an equipment such as rotary pump. As regards materials, seals are classified into metallic and non-metallic seals. Lead, soft copper and aluminium are used for metallic seals, and leather, rubber, asbestos and other materials having similar properties are used for non-metallic seals.

Rotary shaft oil seals : Rotary shaft oil seals are widely used in machines and equipments for sealing rotating shafts to prevent the leakage of oil, entry of dust and other particles. These are mainly used for nominal pressure, the difference of pressure on either side being not more than around 0.04 N/mm^2 or MPa. It should be ensured that the lip of the seal is tight against mineral oil, grease and sometimes air.

The rotary shaft oil seals, used in general engineering applications, have been classified into three types in IS: 5129. These are :

Type A : *Rubber-cased seal.* It has a tight seating in the housing bore or mounting at all temperatures within the specified range, i.e., – 40° C to + 110°C. The rubber casing provides the tightness and prevents injury to the bore.

Type B : *Metal-cased seal.* Because of metal casing, the seal slides between the surfaces better than Type A while mounting or dismantling.

Type C : *Built-up seals for shaft diameter d_1 from 22 mm onwards.* The seal is used in special applications and is better protected against mechanical injury to the lip while mounting. In a rotary shaft oil seal, the sealing lip is spring-loaded which stretches when inserted over the shaft. Care should be taken to ensure that the lip is always turned towards the side to be sealed. In Fig. 9.25, an oil seal belonging to the Type A is illustrated. It also shows the relevant machining symbols, dimensions of bore of the mounting in relation to the width of the seal and the requisite tolerance for the shaft diameter, d_1. These data are also valid for other types. Table 9.35 contains the tolerance of the bore diameter, d_2. Seals are bought out items. For the dimensions of the seals, the reader may refer to IS: 5129. For drawing making purposes,

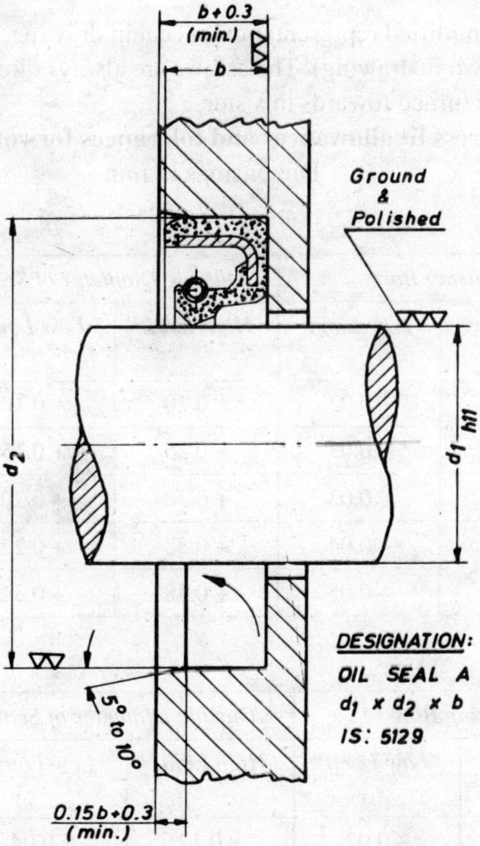

b Width of Seal	t_1 (0.15b + 0.3) Min.	t_2 (b + 0.3) Min.
7	1.35	7.3
8	1.50	8.3
9	1.65	9.3
10	1.80	10.3
12	2.10	12.3
15	2.55	15.3
20	3.30	20.3

Fig. 9.25 Rotary shaft oil seal

the draughtsman is required to enter the tolerances for mounting in the drawings which receive the seals. Hence, only the relevant mounting data are given. The portion of the shaft over which the seal is mounted should be ground, polished and hardened to around 60 HRC. Example of designation of a rotary shaft oil seal of type A for shaft diameter 25 mm, bore diameter of housing 40 mm and width 7 mm is :

Oil Seal A 25 × 40 × 7, IS: 5129.

Referring to Fig. 9.25 in a simplified representation of seal in drawing, two arrows are shown to indicate seal position (one arrow shown in drawing). The arrows are always directed towards the side to be sealed and the lip should always be turned towards this side.

Table 9.35 Press fit allowances and tolerances for rotary shaft oil seals

Dimensions in mm

Type A Seals

Nominal Bore Diameter of Housing d_2	*Housing Bore*		*Outside Diameter of Seal*		*Possible Press Fit Variation*	
	High Limit	*Low Limit*	*High Limit*	*Low Limit*	*Maximum interference*	*Minimum interference*
Up to 25	+ 0.03	– 0.03	+ 0.20	+ 0.10	0.23	0.07
25 to 55	+ 0.03	– 0.03	+ 0.25	+ 0.15	0.28	0.12
55 to 125	+ 0.03	– 0.03	+ 0.30	+ 0.20	0.33	0.17
125 to 200	+ 0.04	– 0.04	+ 0.38	+ 0.22	0.42	0.18
200 and above	+ 0.05	– 0.05	+ 0.48	+ 0.32	0.53	0.27

Type B and C Seals

Nominal Bore Diameter of Housing d_2	*Housing Bore*		*Outside Diameter of Seal*		*Possible Press Fit Variation*	
	High Limit	*Low Limit*	*High Limit*	*Low Limit*	*Maximum interference*	*Minimum interference*
Up to 50	Nominal	– 0.03	+ 0.12	+ 0.04	0.15	0.04
50 to 90	Nominal	– 0.03	+ 0.14	+ 0.06	0.17	0.06
90 to 115	+ 0.03	– 0.03	+ 0.18	+ 0.08	0.21	0.05
115 to 170	+ 0.03	– 0.03	+ 0.20	+ 0.10	0.23	0.07
170 to 215	+ 0.04	– 0.04	+ 0.23	+ 0.13	0.27	0.09
215 to 230	+ 0.04	– 0.04	+ 0.25	+ 0.15	0.29	0.11
230 and above	+ 0.04	– 0.04	+ 0.30	+ 0.20	0.34	0.16

Type A : Rubber-cased seal

Type B : Metal-cased seal

Type C : Built up seals for shaft diameter from 22 mm onwards.

O-ring: These are torus-shaped seals of circular cross-section used for static or dynamic applications. These rings are inserted in machined grooves. These rings are generally made of synthetic materials, namely,

Material	*Class*
Perbunan	A
Perbunan	B
Viton	C
Teflon	D

O-rings are designated by the inside diameter, chord thickness and material, viz.,

O-ring, 10/2.5, Viton

Figure 9.26 shows *O-ring* and the groove for mounting Tables 9.36 and 9.37 show the data relevant to *O*-ring and groove.

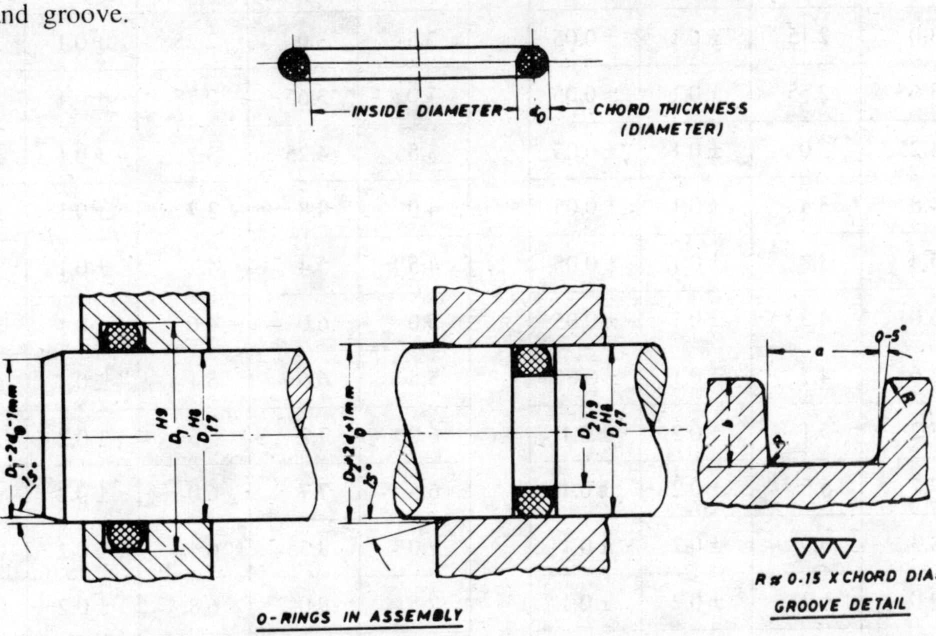

Fig. 9.26 O-Ring and groove

Table 9.36 Groove dimensions for *O*-rings
All dimensions in mm

Static sealing					*Dynamic sealing*				
Chord φ	*a*	*b*	*Tolerances*		*Chord* φ	*a*	*b*	*Tolerances*	
			a	*b*				*a*	*b*
1.0	1.2	0.8	± 0.1	± 0.05	1.0	1.2	0.8	± 0.1	± 0.05
1.5	1.8	1.25	± 0.1	± 0.05	1.5	1.8	1.3	± 0.1	± 0.05
2.0	2.4	1.7	± 0.1	± 0.05	2.0	2.4	1.8	± 0.1	± 0.05
2.5	3.0	2.15	± 0.1	± 0.05	2.5	3.0	2.25	± 0.1	± 0.05
3.0	3.65	2.55	± 0.1	± 0.05	3.0	3.65	2.75	± 0.1	± 0.05
3.5	4.25	3.0	± 0.1	± 0.05	3.5	4.25	3.2	± 0.1	± 0.05
4.0	4.8	3.4	± 0.1	± 0.05	4.0	4.8	3.7	± 0.1	± 0.05
4.5	5.4	3.8	± 0.1	± 0.05	4.5	5.4	4.2	± 0.1	± 0.05
5.0	6.0	4.3	± 0.1	± 0.05	5.0	6.0	4.6	± 0.1	± 0.05
5.5	6.6	4.7	± 0.1	± 0.05	5.5	6.6	5.1	± 0.1	± 0.05
6.0	7.2	5.1	± 0.2	± 0.1	6.0	7.2	5.5	± 0.2	± 0.1
6.5	7.7	5.5	± 0.2	± 0.1	6.5	7.7	6.0	± 0.2	± 0.1
7.0	8.3	5.9	± 0.2	± 0.1	7.0	8.35	6.4	± 0.2	± 0.1
7.5	9.0	6.3	± 0.2	± 0.1	7.5	9.0	6.8	± 0.2	± 0.1
8.0	9.6	6.8	± 0.2	± 0.1	8.0	9.6	7.3	± 0.2	± 0.1
8.5	10.2	7.2	± 0.2	± 0.1	8.5	10.2	7.7	± 0.2	± 0.1
9.0	10.8	7.7	± 0.2	± 0.1	9.0	10.8	8.3	± 0.2	± 0.1
9.5	11.4	8.1	± 0.2	± 0.1	9.5	11.4	8.7	± 0.2	± 0.1
10.0	12.0	8.5	± 0.2	± 0.1	10.0	12.0	9.2	± 0.2	± 0.1

a = *Width of groove*
b = *Depth of groove*

Table 9.37 Material properties of *O*-Rings

Material Class	*Hardness Share - A*	*Temperature Range °C*	*Tensile Strength N/mm²*	*Resistance Against*							*Field of Application*	*Remarks*
				Oil/Grease 100°C Without Additives	*Water 90°C Emulsion Oil + Water*	*Acid 20°C*	*Alkali 20°C*	*I S O Octane 20°C*	*Brake Fluid (Blue)*	*Ageing Weather Ozone*		
A	75 ± 3	-30 to + 130	10 to 20	*a*	*a – b*	*c*	*b*	*b – c*	*d*	*b*	High pressure hydraulic grease and lube oils general equipments, pumps water lines gas lines and compressors	Cheap, for normal use, standard quality
B	85 ± 3											
C	90 ± 3	– 25 to +200	10 to 15	*a*	*a – b*	*a – b*	*a*	*a*	*b – c*	*a*	Chemical equipments hydraulic oils with additives and all above applications	
D	98 ± 1	– 200 to + 200	20 to 35	*a*	*a*	*a*	a	*a*	*a*	*a*	Refrigerants, oxygen, aggressive fluids	Expensive, highly abrasionless, less elastic

a = Very good *b = Good* *c = Satisfactory* *d = Not usable*

Felt seal: Felt seals are usually employed in bearing covers for anti-friction bearings. The relevant data concerning felt seals and the mounting grooves are given in Table 9.38

Table 9.38 Felt seals and grooves

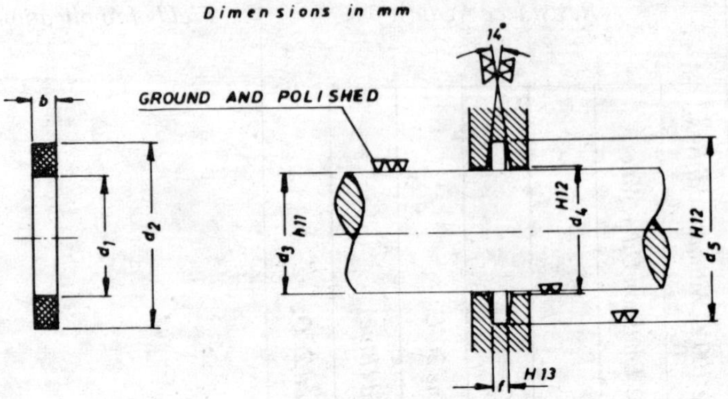

Sealing Ring No.	$d_1 = d_3$	b	d_2	d_4	d_5	f
Fi 5	20	4	30	21	31	3
Fi 6	25	5	37	26	38	4
Fi 7	30	5	42	31	43	4
Fi 8	35	5	47	36	48	4
Fi 9	40	5	52	41	53	4
Fi 10	45	5	57	46	58	4
Fi 11	50	6.5	66	51	67	5
Fi 12	55	6.5	71	56	72	5
Fi 13	60	6.5	76	61.5	77	5
Fi 15	65	6.5	81	66.5	82	5
Fi 16	70	7.5	88	71.5	89	6
Fi 17	75	7.5	93	76.5	94	6
Fi 18	80	7.5	98	81.5	99	6
Fi 19	85	7.5	103	86.5	104	6
Fi 20	90	8.5	110	92	111	7
Fi 21	95	8.5	115	97	116	7
Fi 22	100	10	124	102	125	8

Fi 24	110	10	134	112	135	8
Fi 26	115	10	139	117	140	8
Fi 28	125	11	153	127	154	9
Fi 30	135	11	163	137	164	9
FI 32	140	12	172	142	173	10
Fi 34	150	12	182	152	183	10
Fi 36	160	12	192	162	193	10
Fi 38	170	12	202	172	203	10
Fi 40	180	12	212	182	213	10

CHAPTER 10

MISCELLANEOUS TOPICS

10.1 ENGINEERING MATERIALS

Engineering material is a subject by itself, and so vast is the variety, scope of application, multifarious uses, ramifications of properties attainable by sophisticated treatments and effects thereof offered by the modern-day high technology that it is impossible to cover the subject in a comprehensive manner within the space available. Hence, a broad outline of the subject is being discussed here and general guidelines will be given as to the properties, typical uses and criteria of selection of common engineering materials which are of relevance to the draughtsmen and engineers for industrial drawing making purposes. Elsewhere in this book, details about the machine elements such as gears, springs, bolts, nuts, chains, etc., parts such as seals, gaskets, ropes, belts, *etc.*, as well as a host of other engineering items have been given. In all these cases, materials used specifically for those items have been mentioned and dealt with at appropriate places.

Since the performance of a part, its operational life and other inter-related aspects are direct functions of the material it is composed of, a drawing maker has to acquire enough knowledge of the materials commonly available besides the heat treatment thereof, machinability, strength and other allied aspects. As proper selection of the material is a pre-requisite for any component for efficient operation and reasonable life, this knowledge is imperative since the ultimate responsibility for specifying the appropriate material lies with him. Besides the functional and other requirements, the single most important factor in the determination of the proper material is the cost involved. This includes mainly the material cost and the production cost. Most often, the common materials serve the purpose, and a costly, special material is called for only in a special case where circumstance warrants it.

Among the engineering materials, common steels and cast irons are specified in most of the cases. Castings are resorted to where the shapes are complicated. For reasons of economy, fabricated designs are often used in lieu of cast parts when the number of parts to be made is small. Castings involve pattern making which entails extra cost. For higher strength and toughness, steel is used instead of cast iron. Normally different carbon steels are usually specified, but when bigger cross-sections combined with higher strength and hardness are the deciding factors, alloy steels are generally preferred. Case hardened steels are used where resistance to wear and contact pressure are the main design criteria. The different case hardening processes have been covered in Section 10.2. As spring materials, high silicon or silica-manganese steels are used. Stainless steels are used mainly for their anti-corrosive properties. For tool materials, high speed steels and tungsten carbide tips are extensively used. For parts requiring tensile strengths upto around 700 N/mm^2, carbon steels are generally used. For higher strengths (700 to 1600 N/mm^2), alloy steels are recommended. Common cast iron is suitable for applications upto around 300°C. Above 300°C, cast steel is indicated. For still higher temperatures, other special steels are used. Non-ferrous metals have their own spheres of application as given in Table 10.2.

Non-metals are extensively used as engineering materials. Such materials are ceramics, refractory materials, building materials, natural and synthetic rubbers, plastics and similar materials. Plastics are mainly covered in this Section since they have a wide field of application in mechanical design.

We shall now discuss the materials generally selected for common mechanical parts. A broad guideline will be given, the main emphasis being on the uses and applications of the parts *vis-a-vis* the materials of which the parts are made rather on the properties, chemical composition, etc., of the materials which are not of much relevance as far as drawing making is concerned. Thus, the practical, industrial drawing maker can choose the appropriate materials easily after going through the discussion given below, taking cognizance of the particular necessities involved. For the sake of convenience, the data are given in tabular form.

Table 10.1 Ferrous materials for common mechanical components

Material	*IS*	*IS Symbol*		*Typical applications*
		New	*Old*	
Ordinary Cast Iron	210	FG 200	Grade 20	All general engineering purposes, *e.g.*, housings, base and cover plates
		FG 260	Grade 25	Parts requiring higher strengths, *e.g.*, heavy machine parts, cylinders
Alloyed Cast Iron	3355	Gr. 2-E		Machine parts exposed to high temp., *e.g.*, grate bars, fire box parts.
Cast Iron with Spheroidal Graphite	1865	SG 500/7 SG 600/3		Gears with increased load capacity and resistance to contact stress
Ordinary Steel Casting	1030	Grade 20-40 Grade 25-45 Grade 26-52 Grade 27-54 Grade 30-57	Grade 1 Grade 2 Grade 3 Grade 4 Grade 5	Used for all types of general engineering purposes. Casting materials are classified in order of increasing strengths.
High Tensile Steel Castings	2644	CS 640 CS 700 CS 840 CS 1030 CS 1230	Grade 1 Grade 2 Grade 3 Grade 4 Grade 5	All applications where high tensile strengths are the design criteria and which are not covered by IS: 1030

Surface Hardening Steel Casting	2707	Grade 1 Grade 2		Applications where hardened surface combined with tough core are required for better wear resistance, *e.g.*, heavy gears, anvils, *etc*.
Steel Castings suitable for pressure and temp.	2856	Grade 1 Grade 2		Application for pressure containing parts and suitable for high temperature service.
Alloy steel castings	3038	Grade 1 to 7		For parts suitable for pressure and high temp. service.
Common Standard steels	1570	Fe 290 Fe 310 Fe 330 Fe 360 Fe 410 Fe 490 Fe 620 Fe 690	St 30 St 32 St 34 St 37 St 42 St 50 St 63 St 70	Structural steel sheets, tubes. Ordinary shafts, axles. General engineering purposes. Chemical pressure vessels. Structural steel for bridges, rolling stock, tubes, general engineering purposes. Structural steel for mines, machine parts. High tensile steel for axle and seamless tubes. Parts requiring high strength, gears.
Unalloyed Carbon Steels	1570	7 C4 10 C4 15 C4 35 C8	C 07 C 10 C 15 C 35 Mn 75	For cold shaping parts. For case carburized parts. For case carburized parts. For low stressed parts, fasteners.
Heat Treatable Steels	1570	40 C8 45 C8 60 C4	C 40 C 45 C 60	Shafts, spindles, studs, bolts, gears Shafts, keys, worms, gears. Shafts, medium duty gears.
Carbon and Carbon Manganese free cutting steels	1570	10 C8 S10 25 C12 Sl4 40 C10 S18 40 C15 S12	10 Sll 25 Mn1 Sl4 40 S18 40 Mn 2S 12	Parts suitable for cyaniding. Small heat treated parts, bolts, studs. Heat treated bolts, engine shafts, rods. Not suitable for high stress and wear. Heat treated shafts, axles, crank shafts.
Alloy steels (Hardened and tempered)	5517	20 C15 37 C15 35 Mn6 Mo 3 55 Cr 3 40 Cr 4 40 Ni 14 15 Cr 65 20 Mn Cr 1		Welded structures, crank shafts, levers Shafts, axles, connecting rods. Bolts, axles, studs, crank shafts. Shafts, rollers, cylinder liners, gears. Pinions, gears, axles, wear resisting parts. Parts requiring high toughness. Roller bearings, measuring gadgets and instruments. Gears, shafts.

Table 10.2 Non-ferrous metals for common applications

Material	IS	IS Symbol New	IS Symbol Old	Typical applications
Aluminium alloy	617	4520	A - 2	Piston and similar parts needing wear resistance and strength.
		4600	A - 6	Parts requiring medium strength and corrosion resistance.
		5500	A - 10	High strength, aircraft components.
		2280	A - 11	Parts for high strength and ductility.
		4652	A - 13	Chief alloy used for pistons of internal combustion engines. Low thermal expansion with good mechanical properties.
		2285	A - 14	Known as "Y" alloy. For strength at elevated temperatures, piston, cylinder heads.
		4300	A - 18	Widely used for electrical appliances.
		4420	A - 24	Used for die-cast components.
Gun metal	306		Grade 1	Worm wheels, sliding bearings, bushes.
Phospher bronze	28		PBZ 10	Most common material for worm wheels, part requiring hard,
			Grade 2	tough wear resistance material. Bearing bushes and liners.
Tin lead bronze	318		Grade 3	Hot rolling mill bearings.
Brass	232		Grade 3	Castings for armatures, fittings, housings.
Bearing metals	25		Grade 5	For bearings for good sliding and load carrying properties.
			Grade 84	Parts for good casting and soldering properties, high speed bearings.

Table 10.3 Non-metals for common applications

Material	Application
PVC (hard)	Pipes, plates, acid containers, gaskets.
PVC (soft)	Hoses, guards, pipes.
Polyamides (Nylon, Perlon)	Bearings, wearing surfaces, gears.
Polystyrene	Insulators, chemical equipments.
Polyacrylates (Plexiglass)	Extruded tubings, pipes.
PTFE (Teflon)	General purpose application. Extreme temperature parts.
Polyethylene	Packing materials, abrasion resistant parts.
Cellulose nitrate	Hand grips, hoses, plates.
Silicone	For high heat resistant electro-structural parts.
Synthetic rubbers	Marketed by various trade names, *e.g.*, buna, perbunan, *etc.*, used for seals (neoprene), hoses, insulators, *etc.*, Better heat resistance than natural rubber.

Table 10.4 Density of common materials (g/cm^3)

Material	Density	Material	Density	Material	Density
Steel (aver. value)	7.85	Bronze	8.6 - 8.8	PVC	1.3 - 1.4
Aluminium alloys	2.5 - 3.0	Brass	8.2 - 8.6	Rubber	1.0 - 2.0
Copper alloys	7.5 - 9.0	PTFE	2.1 - 2.3	Polyamides	1.3

10.2 HEAT TREATMENT AND HARDNESS

Heat treatment is a general term which includes a number of processes to achieve specific properties of metals. Among the metals, steel is the metal which is most commonly used for engineering purposes. While designing or draughting, the drawing maker is often required to specify the appropriate heat treatment to be imparted to a part for proper manufacturing and subsequent functioning of the part. Hence, a broad knowledge of the common heat treatment processes is necessary.

From the point of availability and economic viability, the choice of engineering materials is often limited to those which are commonly used. Engineering components are, by and large, made of carbon steels and alloy steels. Plain carbon steel is primarily an alloy of iron and carbon. Though mainly the carbon content determines the properties of a given steel, it is the form rather than the amount of carbon present which is the more important factor in determining the properties of the steel. Hence, the form in which the carbon is present and the pattern of its distribution in the steel are the main criteria for heat treatment considerations. Thus, the heat treatment of steel is mainly concerned with altering the form and distribution of carbon. Since the heat treatment processes can take a variety of types, the main consideration for choice is based on the properties desired in the end product. Common heat treatment methods are : annealing, normalizing, hardening, quenching, tempering, patenting, austempering, martempering, *etc*. The various hardening methods will be separately discussed later in this section, as a drawing office man is more concerned with the hardness of materials. The other common processes are briefly discussed below :

Annealing: The main purpose of annealing is to soften the steel. Softening changes the micro-structure which in turn results in altered mechanical properties. Annealing can have many modified forms, each with specific name and purpose, such as, soft annealing, blue annealing, bright annealing, etc. In common annealing process, the part is heated to a specific temperature in a controlled atmosphere in a furnace. It is held for a prescribed period, ensuring uniform temperature throughout, followed by slow cooling. Since the rate of cooling is the governing factor which determines the final properties of the part, it is this aspect of the cycle which is more important. Generally, this is accomplished by putting off the furnace heat and the part is allowed to cool inside and with the furnace. As annealing makes the parts softer, it renders the parts more amenable to cold shaping by dies, presses, *etc*., without developing cracks and other defects, and it also increases the ease of machining.

Normalizing: Though frequently used as synonymous terms, annealing and normalizing are quite distinct and different processes. As the name indicates, it normalizes the grain structure of the part after it is worked upon. In both heating and cooling and the rates thereof, normalizing differs from annealing. In normalizing process, as compared to annealing, the heating temperature is higher, the period of heating is shorter, and the part is cooled in still air after removing it from the furnace. Since this is tantamount to a sort of quenching as the cooling period is much quicker, normalizing results in higher strength values as compared to annealing. Normalizing is sometimes specified for forgings for grain refinement and improved machinability. The method is faster than annealing, but because of shorter heating period, the structure and properties are not so uniform as in the case of an annealed end product.

Stress relieving : This is a similar process used for relieving stresses in cast or work-hardened pieces. It increases dimensional stability and makes the castings amenable for machining to close dimensional tolerances.

Quenching: Quenching is a cooling operation. This is done by immersing the heated piece in a quenching bath which may contain water, oil, brine or other liquids, depending on the specifications. A part can also be air-quenched.

Tempering : Due to quenching, parts become hardened. Stresses are developed which are relieved by reheating the parts following the quench. This operation is generally termed as tempering. Tempering helps in the recovery of a degree of ductility and toughness of the steel lost due to quenching. Thus, tempering results in a reduction in hardness and increases the toughness of the part. By suitably manipulating the tempering process, the desired strength/toughness or strength/hardness ratio can be achieved.

Hardness : Allocation of proper hardness values on surfaces of components, wherever required, is a prime prerequisite of an engineering drawing. In cases where a part is subjected to abrasion and wear, the requisite hardness ensures prolonged life of the component, when it is imparted to the wearing surfaces. Increased hardness of a component results in increased strength. As an example, the case of a gear box having full-hardened gears can be cited. A gear-set having normally hardened gears of Brinell Hardness Number (HB) 2000 N/mm^2 can be replaced by a set having full-hardened gears of HB 6000 N/mm^2, both sets developing the same power, but the second set weighing only one-eighth of the first one.

Since steel is the most commonly used engineering material, its properties as regards hardness will be discussed at length. The property that determines the ability of a particular steel to harden right through has been termed as hardenability. As far as mechanical properties are concerned, fully hardened and tempered steels exhibit the best combination of those properties. Hence, the main criterion for selection from the point of view of mechanical properties is the hardenability. The rate of heat abstraction in quenching is a deciding factor while hardening. For larger sizes and for higher tensile strength, it is necessary to use steel which hardens effectively even when this rate is slower. This is achieved by the use of alloy steels. Thus, by the use of alloy steels, it becomes possible to impart effective hardness in sizes larger than those which could be hardened in case of carbon steels.

A steel component can be fully hardened or only its outer surfaces can be hardened, leaving the inner core comparatively soft. The former process is known as through hardening and the latter as case hardening.

Through hardening is desirable from the standpoint of overall strength. A steel is strongest when it is hardest. There is a, more or less, direct relationship between the ultimate tensile strength (UTS) and hardness mainly in case of carbon steels. This is given by

$$\text{UTS} \approx 0.36 \text{ HB},$$

both parameters having the same unit, viz., kgf./mm^2 or N/mm^2.
For annealed Cr. - Ni steel, UTS $\approx$ 0.34 HB.

Through hardening is generally obtained by the quenching process. Since a lot of complicated metallurgical considerations are involved which are beyond the scope of this book, the process is not further elaborated.

Case hardening is a surface hardening process. Through hardening is a result of direct quenching. The main purpose of case hardening is to have a hard surface of the part inside which there will be a strong, tough core. This is a combination which is much desired in many engineering applications. A surface hardened gear will have teeth which are resistant to abrasion while retaining a tough interior portion or core which has resilience to take care of shock, impact and similar forces. There are a number of case hardening processes commonly in vogue which are discussed below :

Case carburising : This process is taken as synonymous to case hardening in general, though it is only a type of the many case hardening methods available. Case carburising is generally resorted to for low carbon steels, the carbon content being 0.25% or lower. The part to be hardened is packed in a container having carburizing compound rich in carbon content. The surfaces of the part soaks carbon, thus increasing their carbon content. A case enriched in carbon is formed all around the surfaces, and the depth of this case should be carefully controlled by specified procedures to impart the desired effects to the part. Case carburizing is normally recommended for case depths upto 2 mm.

Flame hardening : In cases where thin hard surfaces will serve the purpose and where the carbon content of steel is high, hardening by case carburizing cannot be done. This is then achieved by several methods in the shop, the flame hardening being the most common. The skin of the part is heated by flame from oxy-acetylene torch or some similar device. It is then quenched to have the desired hardening effect. Steel used is generally plain carbon (0.45% C) or a very mildly alloyed steel. Gear teeth are usually flame hardened. Long shafts and large work pieces are case hardened by this process. Flame hardening is also done when small area of the piece is to be case hardened and where higher dimensional accuracy is required.

Induction hardening : This is a similar process. Heat is developed by electrical induction by high frequency alternating current followed by quenching to achieve the desired hardness.

Nitriding: In this process, the steel part is treated in ammonia, producing nitrides, which act as the hardening agents. A very thin but extremely hard and wear resistant case is produced .

Cyaniding: Steel is treated in a bath containing upto 50% cyanide. Hardness obtained is high. Relations between the UTS and hardnesses are given in Appendix-F. These values can also be read off from Fig. 10.1.

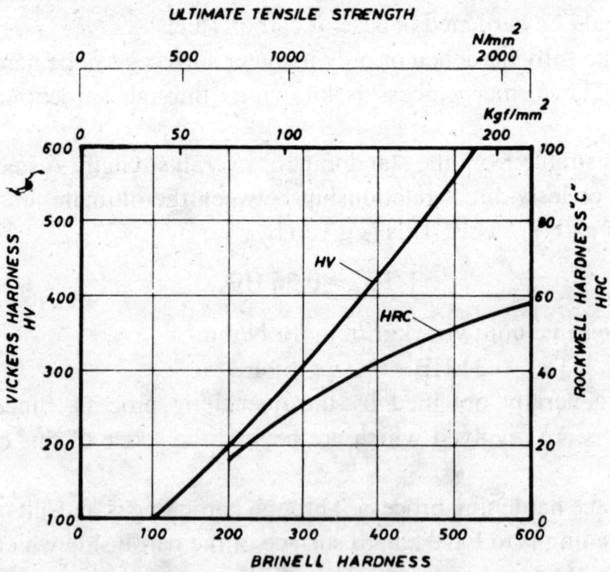

Fig. 10.1 Relation between hardnesses and UTS.

In Fig. 10.1 above, the unit of HB is kgf/mm^2

10.3 MANUFACTURING PROCESSES

As applied to engineering and technology, the term manufacturing means the production of a component or an equipment or any allied product from engineering materials by using systematised methods, so that a satisfactory product ensures conforming to the specifications laid down for the product. In a successful manufacturing organisation, all functions are co-ordinated. Though not directly involved in the processes of manufacturing, personnel connected with drawing making in an engineering drawing office play a key role in the selection of manufacturing processes of a particular end product, the sequences thereof and in the determination of any innovations or constructive suggestions to manufacture better, cheaper and novel products in this highly competitive market. Hence, it is imperative that the draughtsmen and the engineers should have a basic knowledge, if not a comprehensive knowledge in certain cases, of the manufacturing processes and various methods involved in producing engineering parts. Many a time an apparently sound design, when it is considered in the manufacturing plant from various angle connected with productions, has been found to be impossible to produce or the cost involved may be prohibitive. In other words, lack of sufficient and relevant knowledge has made the design unproducible. This lacuna must be overcome by imbibing enough knowledge regarding manufacturing processes involved besides the functionability of the part, its cost, material and other aspects. A thorough knowledge of common engineering materials is a fundamental prerequisite and as such, this subject has been dealt with elaborately in a separate section (see Section 10.1). In this section, the common engineering manufacturing processes and allied subjects as well as the pertinent factors will be briefly discussed.

Casting: Any machine component in which metal is used has to have a required shape which should conform to the relevant design drawing. To this end, several methods are employed, namely, casting, forging, machining and a host of other allied processes. A relatively uncomplicated method of shaping metals is afforded by the various casting processes. A casting is produced by pouring metal into a cavity which conforms to the desired form of the part. This cavity or mould has been made prior to casting with the help of a pattern, generally wooden, which has the shape or form of the part. In majority of the cases, sand of some type is used to produce the moulds necessary, and hence sand is the basic foundry material. The most commonly used method is known as the green sand moulding process. By using modern technology, castings of all conceivable shapes involving most of the metals used in engineering can be made. For making foundry patterns, a variety of materials such as , wood, metal, plaster, plastics, *etc.*, are in vogue. After studying the drawing made at the design office and while making the pattern, the pattern maker makes allowance of shrinkage or contraction of the metal, provides for extra metal for surfaces which require subsequent machining and also provides the necessary draft or taper for easing the casting procedures. In Section 4.3, guidelines for producing better castings are given, showing better casting designs and avoiding common pitfalls. Since sand casting is the method most predominantly used for flexibility and economy, and since no other process for shaping metal offers so much latitude, freedom of expression and other advantages, it is prudent for the designers and the draughtsmen to acquire knowledge of casting methods to make workable design resulting in economical casting. They should also frequently consult the foundry people in order to produce satisfactory end products.

Though sand casting is the process primarily used in the industry, there are several casting methods in which sand plays no part. Sandless castings have the following advantages: better surface finish, greater mechanical strength, finer dimensional accuracy, speedier production, lesser losses, lower machining costs and finishing costs. Among the various sandless casting methods, the centrifugal casting processes are best suited for parts of cylindrical shape. The centrifugal castings are produced in moulds rotating about their own axes. Metal moulds are generally used. The process assures sound, fine-grained structure of the metal, and any slag or other inclusions can be easily removed by subsequent machining operations. Die casting is another method which is very much in use. In this method, molten metal under pressure is forced in metal dies. It has all the usual advantages of other sandless casting methods. Die castings are made almost exclusively in case of non-ferrous alloys. Instead of moulds, dies are used. Hence, the name. The dies are composed of at least two parts which open up for the removal of the casting. The necessary pressure to force the metal into the cavity of the (closed) dies while casting is provided by compressed air or by some hydraulic system. Besides, other sandless casting methods include permanent mould or gravity die casting, injection moulding (used primarily for plastic materials), vacuum die casting, and other processes.

Powder metallurgy : Powder metallurgy is a process by which metallic and semi-metallic parts can be produced by pressing metallic or semi-metallic powder or mixture thereof under high pressure in a suitable die and then sintering or heating the product, so that the particles coalesce, resulting in a satisfactory end product. The advantages of this process are : it makes possible the blending of disparate metal powders having widely different melting points and characteristics, namely, copper, silver, tungsten, *etc.*, and lets the end product to be shaped successfully which would have otherwise not been possible by resorting to processes like casting, hot working, *etc.*, it produces good surface finish, wastage of raw material is alleviated or altogether eliminated, it can make the end product to obtain the properties of alloys made from the constituent elements, and it is an economically viable process in many cases. One of the most widely used items produced by powder metallurgy is the so-called "oil-less" bearing. Those bush bearings have oil impregnated inside the minute pores left after pressing the powdered bearing materials, *e.g.*, brass, gun metal, *etc.*, while shaping the bushes. When the shaft rotates inside such a bearing, it draws out oil which lubricates the mating surfaces.

Hot shaping : Almost all engineering metallic products can be formed by one or the other methods of hot shaping. Though steel and wrought iron have the best known capability to be hot-worked, the non-ferrous metals are also now widely subjected to hot-forming methods. The various hot shaping processes are: hammering (manual or mechanical), forging, hot rolling, extrusion, and other methods. Hammering a suitably heated piece by applying repeated blows is a very common method of shaping work-pieces. If pressure is applied in lieu of repeated blows, it is called pressing or press forging. Hot rolling is resorted to when production requirements involving great tonnages of structural sections, *e.g.*, beams, channels, angles, bars as well as plates and sheets are to be met. Ingots or semi-finished products are pressed through rolls having the suitable contours or "passes" in rolling mills. The end products are then subjected to further processing like shearing, heat treatment (if necessary), coiling in case of hot-rolled sheets, *etc*. Forging methods comprise hot-working by hammer, press and rolling, the choice being governed by such considerations as size, design and production aspects. Individual forgings are made by hammering the piece on an anvil. This method, known as black-smith forging, is restricted to small parts. Large forgings are produced by steam hammers and similar machines. Drop forging, also known as impact die forging, is a process where forging is made in a die by impact in which drop hammers are employed to supply the impact needed. In the extrusion process, non-ferrous alloys are hot-formed by forcing them through dies under high pressures. A multitude of shapes can, thus, be obtained by this process. Extruded end products are bars, tubulars, etc. Aluminium alloys are mainly extruded, though this process is also used for other non-ferrous alloys and sometimes for steel as well with suitable arrangements and dies.

Cold shaping: Metals which have been previously mechanically worked upon are sometimes cold worked to meet certain specifications. Cold worked products are generally sheet metals, wires, tubing and allied items. This type of process changes the mechanical properties of metals. Through cold working, hot-formed materials are finished at room temperature. The two principal methods employed are cold rolling in cold rolling mills and cold drawing. Cold working improves surface and dimensional finish, increases yield and tensile strengths as well as hardness and endurance limit. However, ductility is lowered and for better machinability or for cold shaping in dies and in presses to conform to certain desired shapes and contours, such as in auto-body sheets, annealing is done which makes the sheets amenable to smooth shaping by press-tools, *etc*., and the formation of cracks at bonds, undesirable warping, *etc*. are eliminated.

Machining processes : Machining processes provide the means to produce the desired surface finishes, dimensions and forms of the work-pieces. An extensive variety of machine tools are available to achieve these objectives. In each of these machines, metal is removed through the formation of chips. Chips are formed by cutting tools as well as by grinding wheels. With the advent of high speed steel (HSS) tools and later, tungsten carbide tipped tools in almost all categories of cutting tools, the machining processes in machine tools have been greatly diversified and enhanced. The basic machine tool is the engine lathe which is the most common machine tool to be found in any workshop. In a lathe, a single-point tool can be used to meet specific requirements after suitably grinding it to any desired shape. A lathe is an extremely flexible contrivance in the sense that it can be used for just about any type of machining operation with suitable arrangements. Primary functions are turning, facing, boring threading. Secondary functions are drilling, centering, reaming, parting, knurling, chamferring, and also the functions such as gear cutting, *etc.,* with suitable extra gadgets fitted to the lathe. Turret lathes are innovations of the ordinary lathes and offer the production of interchangeable parts, greater and speedier production as well as increased precision. In a milling machine, the cutting is performed by a rotating cutting tool called a milling cutter. A milling cutter can have multiple teeth or a single tooth. In a milling process, the cutting is done by combined, synchronised relative motions between the rotating cutter and the job. A milling machine is a versatile machine which can be used for a variety of operations, such as, straight-milling, gear cutting, drilling, boring, slotting and a host of other operations.

Besides the above, a number of other machine tools are used in a workshop, namely, the shaping machine which is generally used for producing flat surfaces, drilling and boring machines, grinding machines, breaching, honing and lapping machines, planing machines, slotting machines, and an array of other common or precision machine tools.

Jigs and fixtures : Though not concerned with manufacturing processes directly, jigs and fixtures are essential auxiliary contrivances to help production in a machine shop. The design of jigs and fixtures is a subject by itself and only the rudimentary conceptual ideas will be dealt with here.

An engineering component can be manufactured in a number of ways, and the cost involved is the most important criterion for the proper selection of the manufacturing method. If parts are required to be made in large quantities and in the shortest possible time, obviously some sort of device has to be used on which the work-pieces can be quickly clamped and positioned in correct relation to the cutting tools, so that time is saved before the actual machining takes place and accurate guiding of the tool is ensured. A device of this sort is known as a jig or fixture.

A jig is a device which holds the job in a positive manner and guides the tool for machining. The job is generally held in the jig by some sort of clamping arrangement, but the jig itself is not fixed to the machine table, but is held by hand. When a large number of machining operations of repetitive type like drilling, reaming, tapping, *etc.*, are to be executed quickly, jigs are the best devices to be used.

A fixture, on the other hand, is a device which is fixed to the machines bed by clamping in such a manner that the job is in the right position vis-a-vis the cutter. Unlike the jig, a fixture does not guide the cutter. For this purpose, a setting gauge is often employed to enable the initial setting of the job in relation to the cutter so that the correct setting is quickly and easily done before machining begins. For large scale milling, turning and grinding, fixtures are often used.

Though a precise distinction as to the definitions of jigs and fixtures exists, the two terms are quite often used loosely and synonymously. In Table 10.5, the jig bushes as per IS: 666 used for drilling operations are shown. In Fig. 10.2, an example of a fixture is illustrated. Example of a typical drilling jig arrangement is shown in Fig. 10.3.

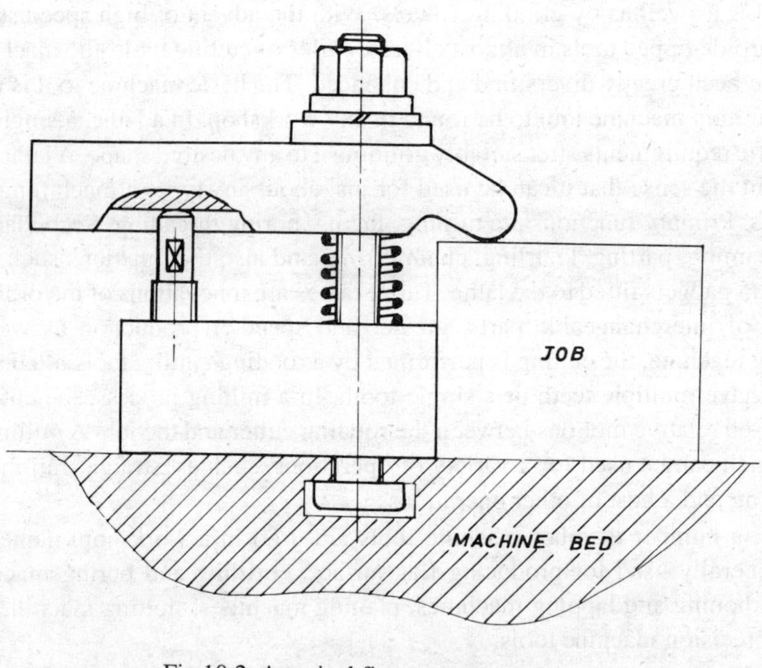

Fig.10.2 A typical fixture arrangement
Based on *Fachzeichnen Metalltechnik, Grundstufe,* 1984, Fig. No. 83/3. P. 83,
Verlag Europa - Lehrmittel, Wuppertal, Germany.

As indicated earlier, jig design is a vast arena. Readers who want to delve deep into its intricacies are advised to study many excellent books specialised on the subject which are available.

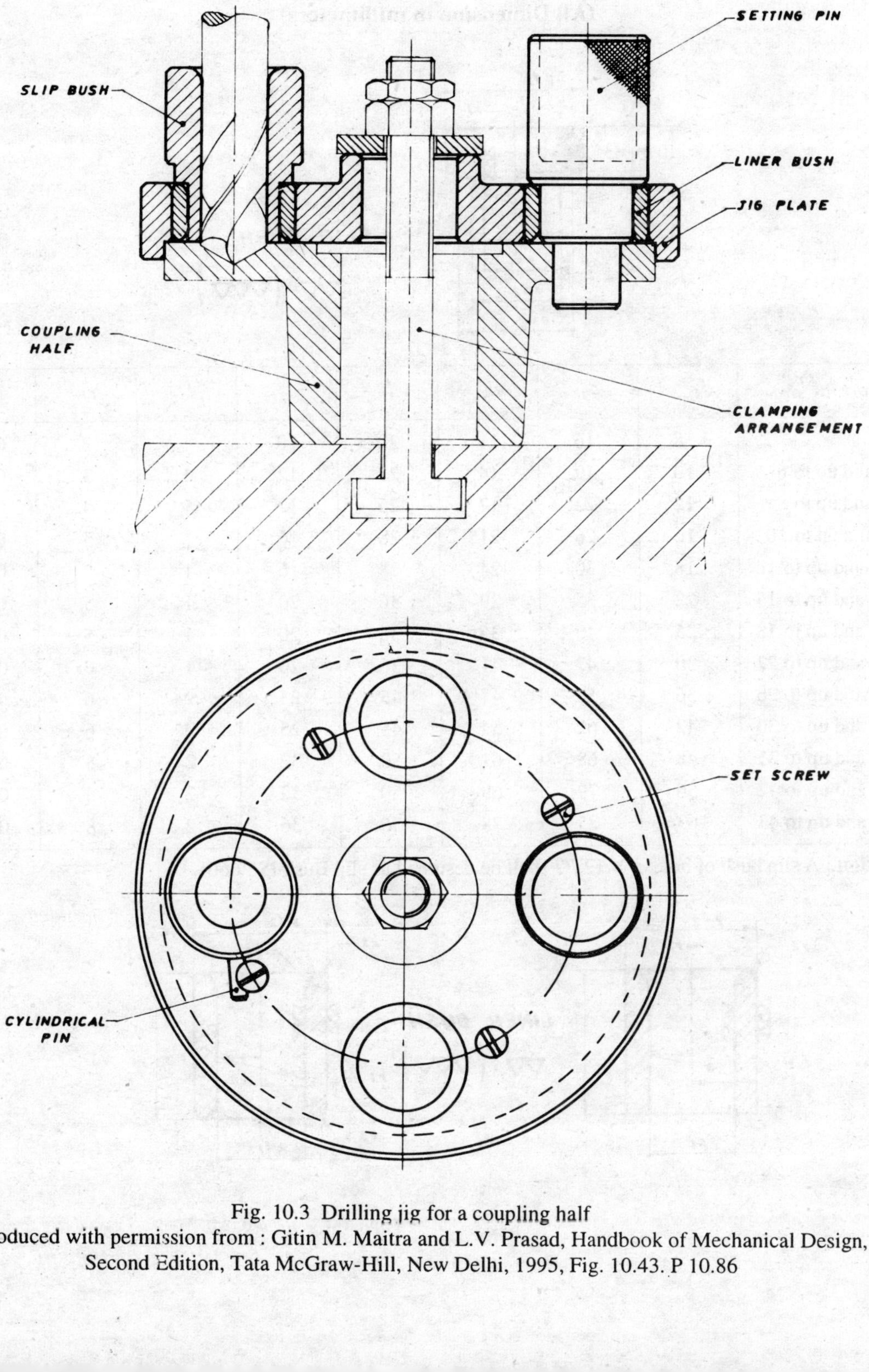

Fig. 10.3 Drilling jig for a coupling half
Reproduced with permission from : Gitin M. Maitra and L.V. Prasad, Handbook of Mechanical Design, Second Edition, Tata McGraw-Hill, New Delhi, 1995, Fig. 10.43. P 10.86

Table 10.5 Dimensions of slip bushes and liner bushes
(All Dimension in millimeters)

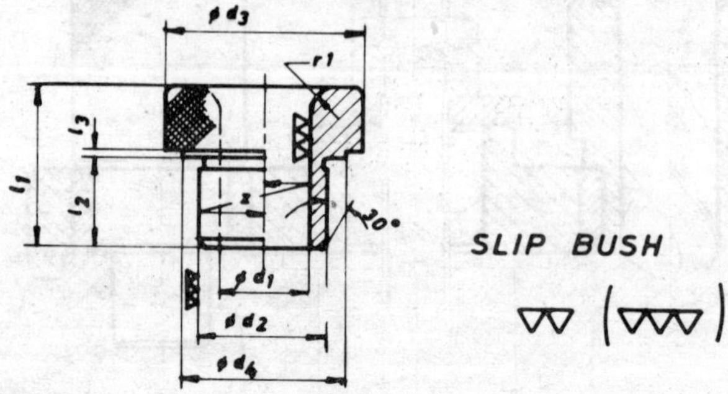

Bore d_1^{F7}	d_{2m6}	d_3	d_4	l_1	l_2	l_3	r_1	z
Up to 4	8	16	11	20	10	1	3	0.01
Over 4 and up to 6	10	19	14	22	12	1	3	0.01
Over 6 and up to 8	12	22	17	25	12	1	4	0.01
Over 8 and up to 10	16	26	21	28	16	1	5	0.01
Over 10 and up to 12	18	30	24	28	16	1	5	0.01
Over 12 and up to 15	22	35	29	36	20	1	5	0.01
Over 15 and up to 18	25	40	35	36	20	1	5	0.01
Over 18 and up to 22	30	47	41	36	20	1	6	0.01
Over 22 and up to 26	36	56	47	45	25	2	6	0.02
Over 26 and up to 30	42	62	54	45	25	2	6	0.02
Over 30 and up to 35	48	68	61	50	32	2	8	0.02
Over 35 and up to 42	56	79	69	50	32	2	8	0.02
Over 42 and up to 43	63	85	78	56	36	2	8	0.02

Designation : A slip bush of bore $d_1 = 12\ F7$ shall be designed as slip Bush IS : 666

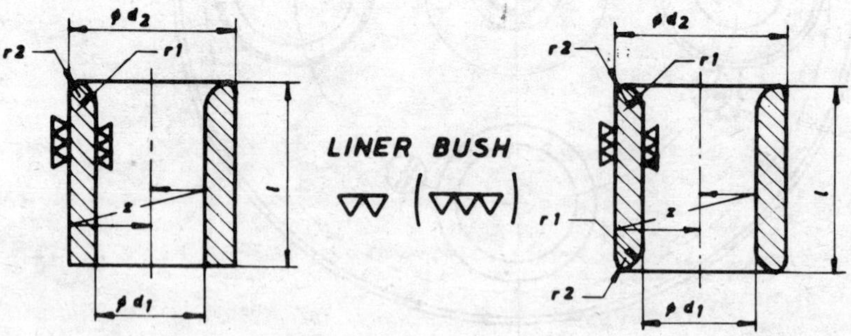

Bore d_1 *F7*	*l* *Short*	*Long*	d_2 *n6*	r_1	r_2	*z*
Up to 1	6	9	3	1.2	0.2	0.005
Over 1 and up to 1.8	6	9	4	1.2	0.2	0.005
Over 1.8 and up to 2.6	6	9	5	1.2	0.3	0.005
Over 2.6 and up to 3.3	8	12	6	1.6	0.3	0.005
Over 3.3 and up to 4	8	12	7	1.6	0.4	0.005
Over 4 and up to 5	8	12	8	2.0	0.4	0.005
Over 5 and up to 6	10	16	10	2.0	0.4	0.01
Over 6 and up to 8	10	16	12	2.0	0.6	0.01
Over 8 and up to 10	12	20	16	2.5	0.8	0.01
Over 10 and up to 12	12	20	18	2.5	0.8	0.01
Over 12 and up to 15	16	28	22	4	0.8	0.01
Over 15 and up to 18	16	28	25	4	0.8	0.01
Over 18 and up to 22	20	36	30	6	0.8	0.01
Over 22 and up to 26	20	36	36	6	1.0	0.02
Over 26 and up to 30	20	36	42	6	1.0	0.02
Over 30 and up to 35	25	45	48	8	1.0	0.02
Over 35 and up to 42	25	45	56	8	1.0	0.02
Over 42 and up to 48	32	56	63	8	1.6	0.02
Over 48 and up to 55	32	56	70	8	1.6	0.02
Over 55 and up to 63	36	70	80	8	1.6	0.02

Designation

A liner bush of Type *A* with bore $d_1 = 12^{F7}$ and length $l = 20$ shall be designated as

Liner Bush *A* 12 × 20 *IS*: 666

10.4 SPARE PART DRAWING

In a running factory, a draughtsman or an engineer is often called for making the manufacturing drawing of a part or an engineering component which belongs to an equipment installed in a plant unit. Such a drawing may be required for indenting purpose, for floating enquires or it may be needed for preventive or routine maintenance of the equipment. Sometimes a drawing may have to be made in a hurry in case of breakdown of some sort. These drawings, called spare part drawings, are also made to meet the requirements of engineering stores, stock control and inventory control departments of the factory, so that parts which have already been made according to these drawings and which have subsequently been stocked in the store can be made available "off the shelf" when an immediate need arises. Besides, in cases where the indigenous or the foreign supplier has not provided the purchaser with the shop drawing of a component of an equipment installed in the factory, either due to the absence of any contractual obligations between the client and the supplier or for preparatory reasons, it is always advisable to have the spare part drawings made from internal resources of the factory, so that these drawings are always handy whenever required.

A spare part drawing is made in the same way as in the case of an ordinary drawing. In this connection, some useful guidelines are given below:

1. A part of which the drawing is to be made often comes in damaged or in broken condition. In such cases, the drawing maker has to make intelligent guess as to the possible configuration of the missing portion. If it is a symmetrical part, such imagination is easy. Otherwise, it is advisable to visit the plant unit to see the running condition which may help him to have an idea as to the whole part or he may also get an identical part in the factory store.
2. Measuring instruments such as Vernier calipers of sizes 150 mm and 300 mm or more, ordinary inside and outside calipers, thread pitch gauges of different types of threads, *e.g.*, metric, fine, pipe, Whitworth, various American standard thread profiles, radius gauges and steel scales, all properly calibrated and checked for their accuracy, should be kept handy for use in the shop floor where sketches for spare part drawings are mostly made. A pliable rod made of lead is very useful for copying curves of the part such as impeller blade and similar objects. Accurate measurements of the part, both linear and angular, should be made to ensure correctness of the drawings.
3. If a plane surface has a complicated configuration, this surface may be lightly smeared with some dirty oily substance. A white sheet of paper, often the back side of a used print of requisite size, is then spread over the surface and pressed against it. The resulting impression is a mirror-image of the surface from which features such as grooves, bolt circle diameters, location of odd contours, etc., can be measured off.
4. To assign the proper tolerance, the mating part should be taken into consideration to determine the appropriate fit. Usually, a quality from 6 to 8 suffices for ordinary requirements of tolerance. The zone of tolerance will, of course, depend upon the fit.
5. As for surface finish, one delta (∇) is sufficient for non-mating machined surfaces, two delta (∇) for toleranced surfaces and three delta (∇∇∇) for mountings of anti-friction bearings. Ground and polished surfaces should be specially indicated.
6. As for material specification, ordinary cast iron, cast steel or rolled steels or non-ferrous metals, as the case may be, are usually specified for normal cases from cost considerations. However, in special cases, drillings are taken from the part and analysed in the chemical laboratory to determine the exact composition of the material and also the heat treatment thereof, if any. If the exact material is not available, as it mostly happens in case of foreign materials, then the nearest equivalent material is specified as per IS: 1870 (latest version) which shows in Indian equivalents vis-a-vis the foreign steels, or as per IS: 1570 (latest version) which contains designation, material composition, strength properties and other relevant data of various kinds of indigenous steels.

7. If a spare part for which a drawing has been made is to be ordered for bulk manufacture, it is always advisable to make one piece first for trial. If the trial piece suits the operational conditions and maintenance parameters, the accuracy of the drawing is vindicated, and bulk order can be placed. If it fails, the cause of the failure can be investigated and errors can be pin-pointed. An accurate spare part drawing can then be made from the knowledge thus obtained.
8. For making spare part drawings of gears and allied components, special care is to be taken while measuring gear parameters. Centre distance in the mounting should be accurately measured as this is the basis of calculations in a gear drive. Module can be arrived at after measuring the whole depth by a depth gauge and then comparing it with the standard chart. Outside diameter of a gear having odd number of teeth can be ascertained by mounting the gear on a mandrel in a lathe or a similar machine and then taking the help of a dial indicator whose stylus traces the tips of the teeth. In any case, elaborate calculations are to be made to match the measured values, especially in case of corrected gears. Gears should preferably be replaced in pairs as changing one mating component of the gear pair only may result in faulty operation. In case of helical gears and worm-wheel, the helix angle at the tip circle can be roughly determined by first applying dirty oil or industrial blue on the top land of the teeth, rolling the gear on a white sheet of paper and then measuring off the angle from the impression thus obtained. The final adjusted parameters are to be entered in the Gear Data table adjacent to the gear drawing as shown in Fig. 9.13.
9. For line diagram of piping, isometric sketches can be made on diamond shaped graph papers to convey three dimensional visual impression.
10. While rendering the sketch into actual drawing, only essential views and sections should be drawn. In case of fabrication drawing, the sequence should preferably be given. Cast or welded pieces should carry the notes, instructions and other information as regards cleanliness, stress-relieving, radiographic ultra-sonic tests, etc., if required.

10.5 MENSURATION

Mensuration deals with the determinations of linear dimensions, areas, volumes and other allied data of geometrical features. By using mensuration formulae, unknown data of figures, *e.g.*, lengths, areas or volumes can be calculated in terms of their known dimensions.

Areas of common geometrical figures are given in Table 10.6. Volumes as well as other data concerning common solids are given in Table 10.7.

Incorporation of values of mass (or weight, as it used to be previously referred to) of individual parts in as engineering drawing is essential for the completeness of the drawing. These values are mainly required for material and cost estimation purposes, but they are useful data for other purposes as well.

Mass can be calculated by multiplying the volume of the solid by the density of the material of the solid. For steel, the density is generally taken as 7.85 gramms per cubic centimetre. The densities of common engineering materials are given in Table 10.4 of Section 10.1

Table 10.6 Areas of common geometrical figures (e_x, e_y, etc. refer to the distances of the controidal axes).

Figure	*Area*	*Figure*	*Area*
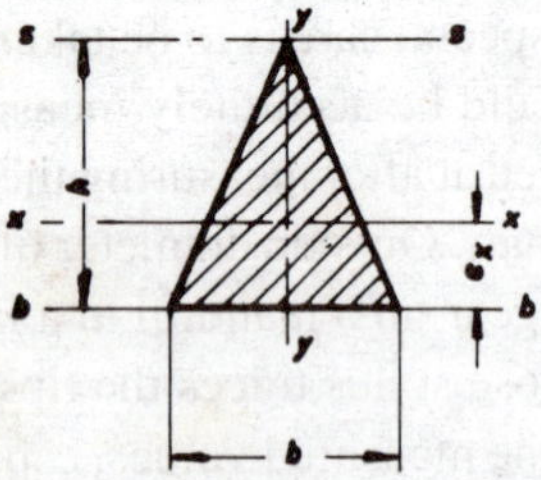 Triangle	$\frac{bh}{2}$	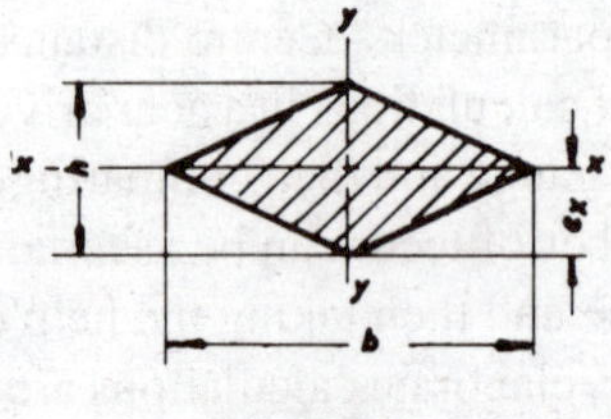 Rhombus	$\frac{bh}{2}$
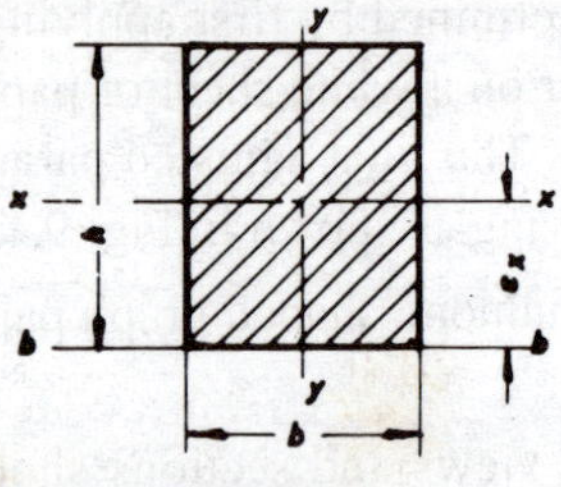 Rectangle	bh	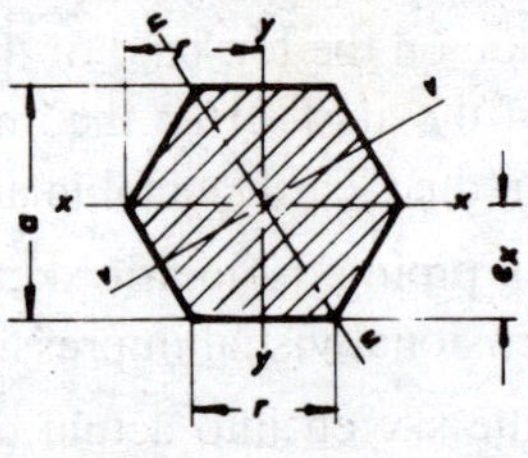 Hexagon	$=\sqrt{\frac{3}{2}}\,a^2$ $=\frac{3\sqrt{3}}{2}\,r^2$
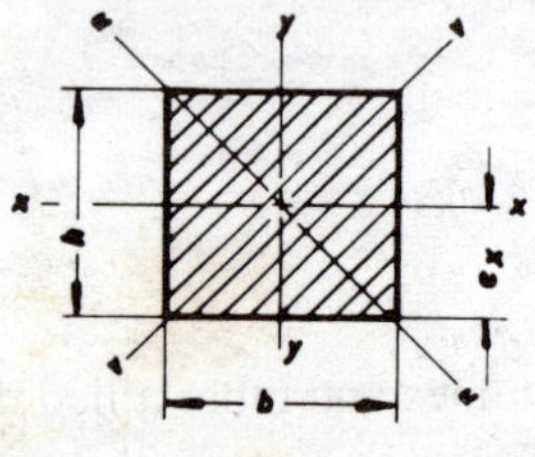 Square	h^2		
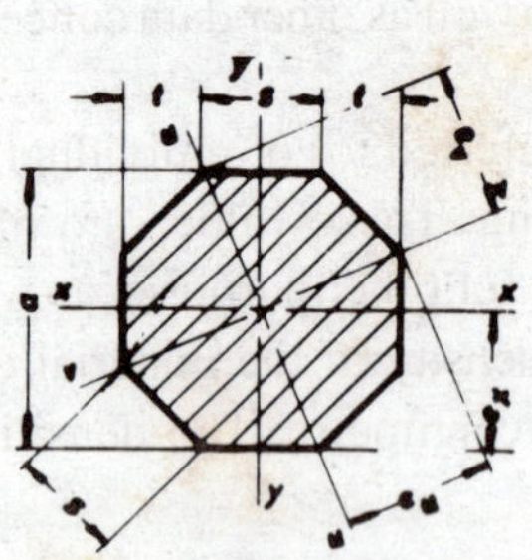 Octagon	$0.8284\,a^2$	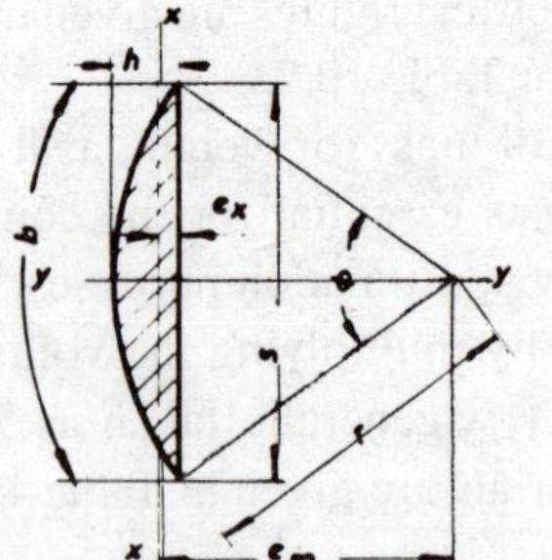 Segment	$\frac{r^2}{2}\left(\frac{\pi\phi^\circ}{180}-\sin\phi\right)$ $=\frac{r(b-s)+sh}{2}$ $r=\frac{s^2}{8h}+\frac{h}{2}$ $b=r\pi\frac{\phi^\circ}{180}$ $\tan\phi/2 = s/2\,(r-h)$

Figure	*Area*	*Figure*	*Area*
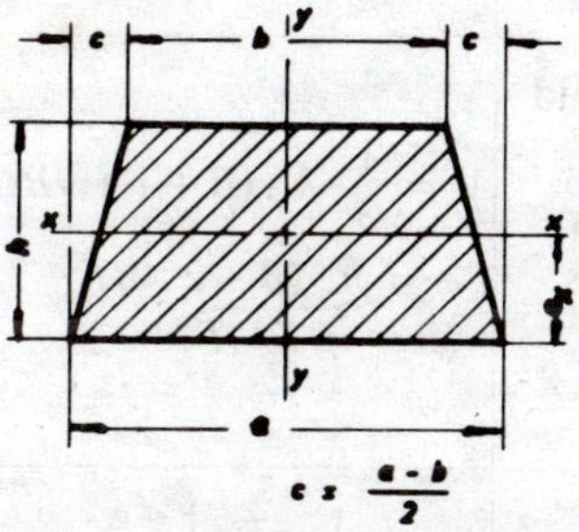 Trapezium	$\frac{h}{2}(a+b)$	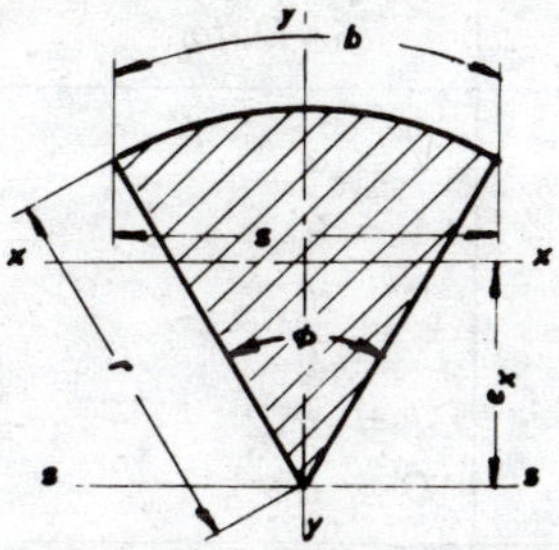 Sector	$\frac{br}{2} = \frac{\phi^\circ}{360} r^2 \pi$ $b = r\pi \frac{\phi^\circ}{180}$ $\phi = \frac{180}{\pi} \frac{b}{r}$
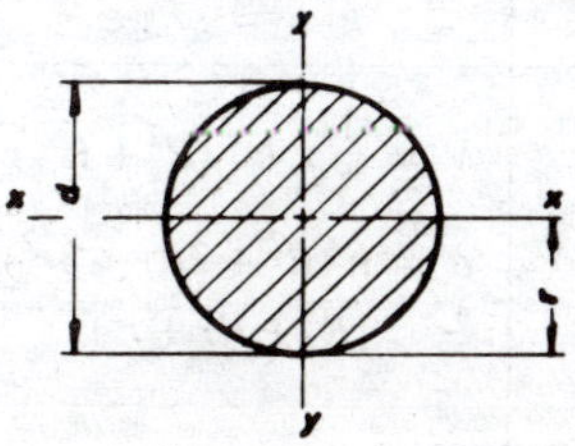 Circle	$\pi r^2 = \frac{\pi d^2}{4}$	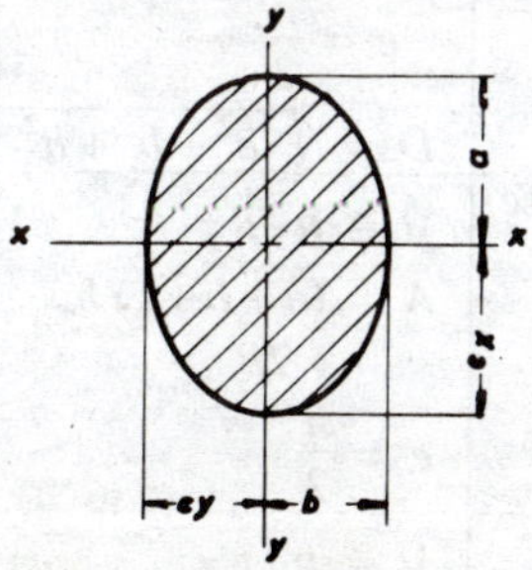 Ellipse	πab
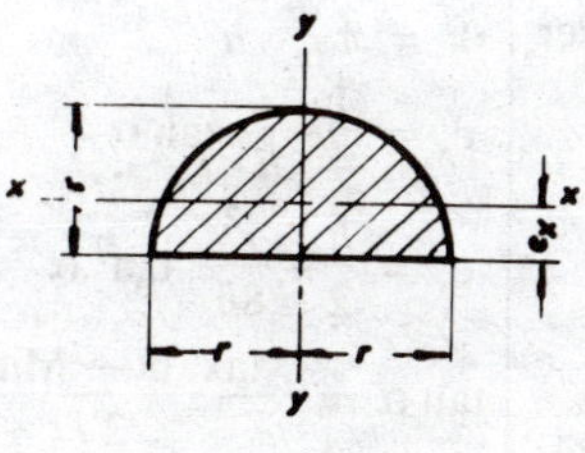 Semicircle	$\frac{\pi}{2} r^2$ $= 1.5708\, r^2$	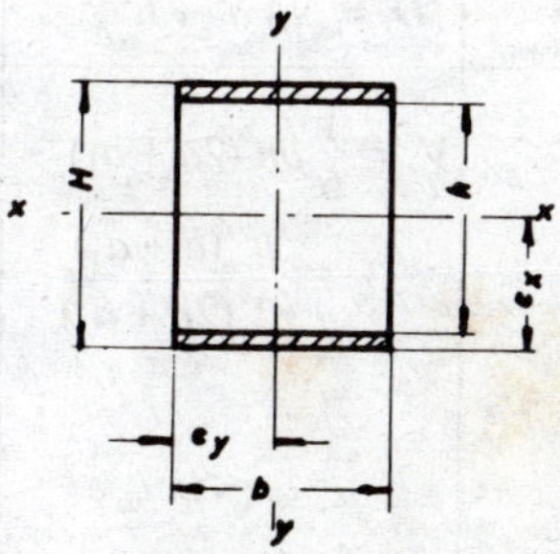 Rectangle	$(H-h)\, b$

Reproduced with permission from : Gitin M. Maitra and L.V. Prasad, Handbook of Mechanical Design, Second Edition, Tata McGraw-Hill, New Delhi, 1995, Table, 2.6. P 2.21.

Table 10.7. Volumes, surface areas and other data of common solids

V = Volume A = Total area B = Area at base T = Area At top

e = Distance of C.G. from base

Name of Solid	*Data*	*Name of Solid*	*Data*
1. Cube 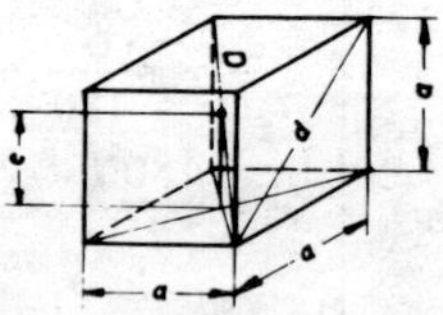	$V = a^3$ $A = 6 . a^2$ $e = \frac{a}{2}$ $d = a\sqrt{2}$ $D = a\sqrt{3}$	7. Frustum of Pyramid 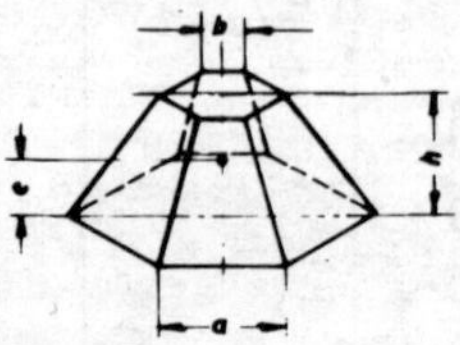	$V = \frac{1}{3} \cdot h \cdot (B + T + \sqrt{B\,T})$ $e = h \cdot (B + 2\sqrt{BT} + 3\,T)/4$ $(B + \sqrt{BT + T})$
2. Prism with rectangular base 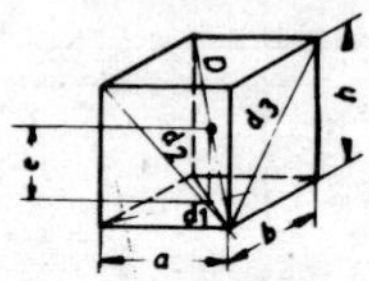	$V + a \cdot b \cdot h$ $A = 2\,(ab + ah + bh)$ $e = \frac{h}{2}$ $D = \sqrt{\sqrt{a^2 + b^2 + h^2}}$	8. Cone 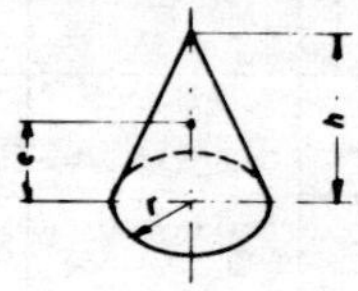	$V = \frac{1}{3} \cdot \pi \cdot r^2 \cdot h$ $A = \pi\, r \cdot (r + \sqrt{r^2 + h^2})$ $e = \frac{h}{4}$
3. Prism with triangular base 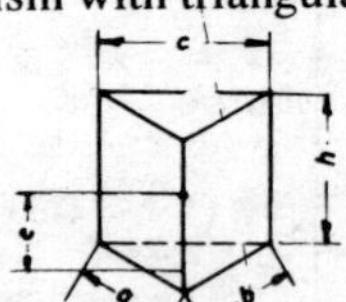	$V = B \cdot h$ $A = (a + b + c)\,h + 2B$ $e = \frac{h}{2}$	9. Frustum of cone 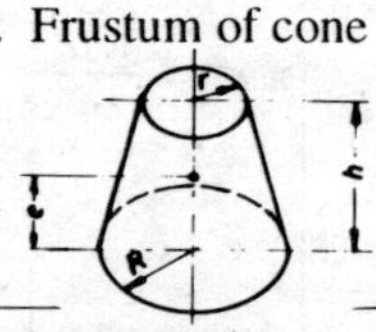	$V = \frac{1}{3}\pi\, h \cdot (R^2 + R \cdot r + r^2)$ $e = h\,(R^2 + 2Rr + 3r^2)/4$ $(R^2 + Rr + r^2)$
4. Prism with hexagonal base 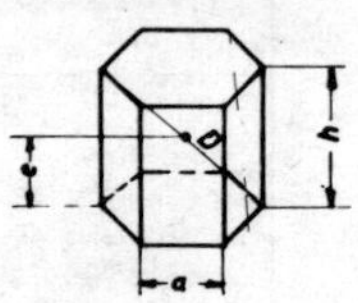	$V = B \cdot h$ $\approx 2.598\, a^2 \cdot h$ $A \approx 5.1963\, a^2 + 6 \cdot a \cdot h$ $e = h/2$ $D = \sqrt{4\,a^2 + h^2}$	10. Cylinder 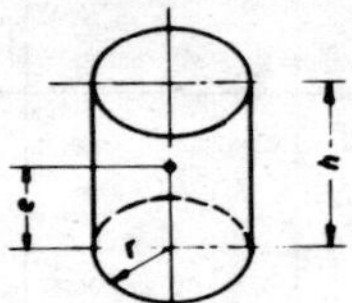	$V = \pi\, r^2 \cdot h$ $A = 2\,\pi \cdot r \cdot (h + r)$ $e = \frac{h}{2}$
5. Wedge 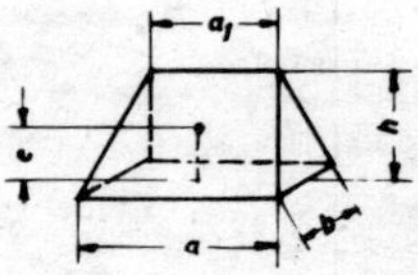	$V = \frac{1}{6}\, bh\,(2a + a_1)$ $e = \frac{h}{2}\,\frac{(a + a_1)}{(2a + a_1)}$	11. Portion of cylinder 	$V = \pi\, r^2 \cdot h$ $e_x = \frac{1}{4} \cdot \frac{r}{h} \cdot \tan\alpha$ $e_y = \frac{h}{2} + \frac{r^2}{8h} . \tan^2\alpha$ $\tan\alpha = \frac{\text{Max. ht.} - \text{Min. ht.}}{2\,r}$
6. Pyramid 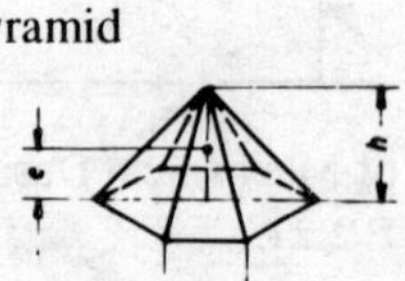	$V = \frac{1}{3}\, B \cdot h$ $e = \frac{h}{4}$	12. Sphere 	$V = \frac{4}{3} . \pi\, R^3$ $A = 4\pi \cdot R^2$ e = Centre of sphere

Name of Solid	*Data*	*Name of Solid*	*Data*
13. Hemi sphere 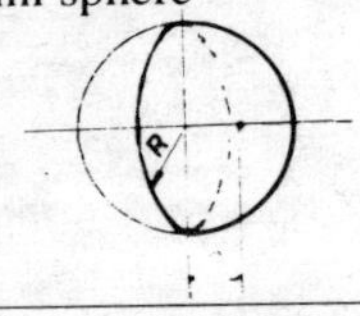	$V = \frac{2}{3} \cdot \pi \cdot R^2$ $A = 3\pi R^2$ $e = \frac{3}{8} \cdot R$	19. Torus 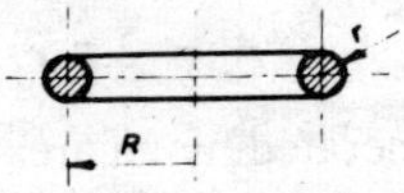	$V = 2\pi^2 \cdot R \cdot r^2$ $A = 4\pi^2 R \cdot r$
14. Spherical Segment 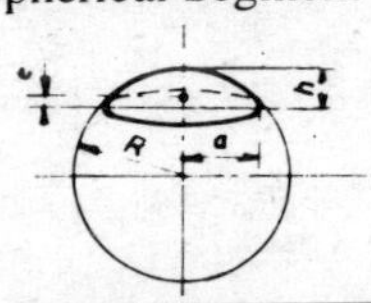	$V = \pi h^2 \left(R - \frac{h}{3}\right)$ $A = 2\pi R h + B$ $e = \frac{3(2R-h)^2}{4(3R-h)}$	20. Ellipsoid 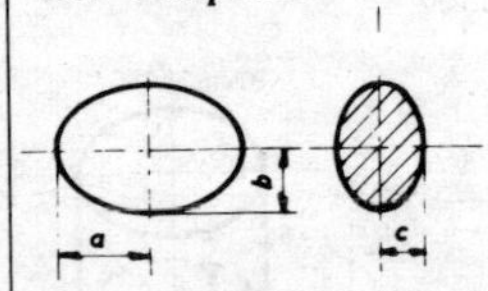	$V = \frac{4}{3}\pi a \cdot b \cdot c$
15. Spherical Sector 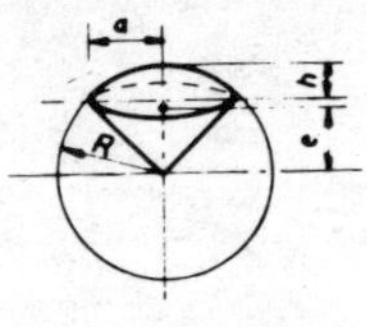	$V = \frac{2}{3} \cdot \pi R^2 \cdot h$ $A = \pi R (2h + a)$ $e = \frac{3}{8}(2R - h)$ $a = \sqrt{h(2R-h)}$	21. Paraboloid 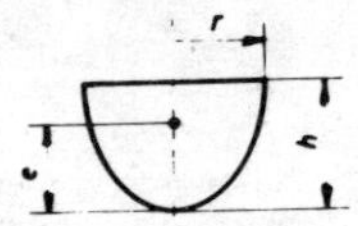	$V = \frac{1}{2}\pi r^2 \cdot h$ $A = \frac{\pi r}{6h^2}[(4h^2 + r^2)^{3/2}$ $- r^3] + T$ $e = 2/3 \cdot h$
16. Spherical Zone 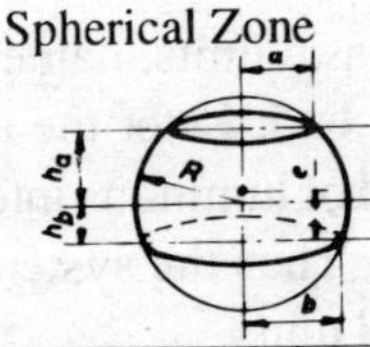	$V = \frac{1}{6}\pi h (3a^2$ $+ 3b^2 + h^2)$ $A = 2\pi R \cdot h$ $+ T + B.$	22. Segment of Paraboloid 	$V = \frac{1}{2}\pi (r^2 + r_1^2)\, h_1$
17. Spherical Wedge 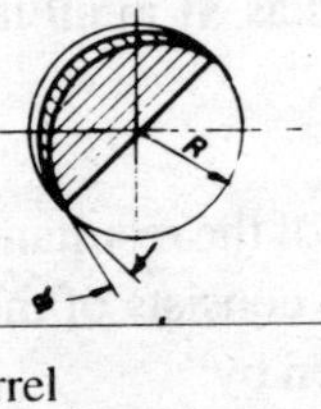	$V = \frac{\phi}{360} \cdot \frac{4\pi r^3}{3}$ $A = \frac{\phi}{360} \cdot 4\pi r^2$ ϕ = Centre angle in degrees.	23. Frustum of Elliptical Cone 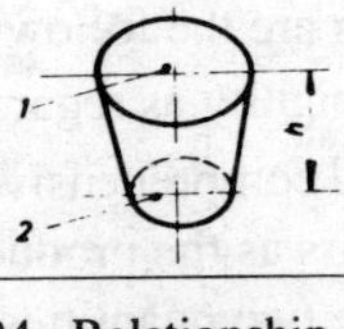	$V = \pi \frac{h}{6}[(2a_1 + a_2)\, b_1$ $+ (2a_2 + a_1)\, b_2]$ $a_1 \cdot a_2$ and b_1, b_2 are semi major and minor axes.
18. Barrel 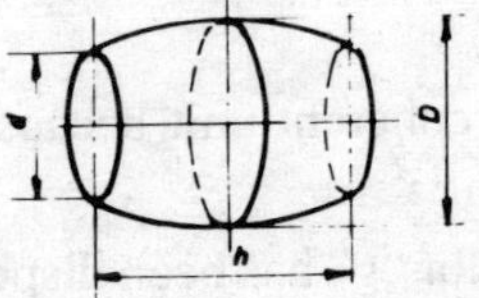	$V \approx$ $\frac{1}{12}\pi h \cdot (2D^2 + d^2)$	24. Relationship of Certain Solids 	The following volume relationship holds good : Cone : Paraboloid : sphere: cylinder $= \frac{1}{3} : \frac{1}{2} : \frac{2}{3} : 1$

Reproduced with permission from : Gitin M. Maitra and L.V. Prasad, Handbook of Mechanical Design, Second Edition, Tata McGraw-Hill, New Delhi, 1995, Table, 1.15. P 1.32.

Rule of Pappus or Guldinus

1. The area generated by rotating a line about an axis is equal to its length multiplied by the path of its centre of gravity. Here.

$$A = 2\pi\tilde{x} \cdot \text{length} = 2\pi R \cdot l.$$

2. Similarly, volume generated by an area.

$$V = 2\pi\tilde{x} \cdot \text{Area} = 2\pi R \cdot a \cdot b.$$

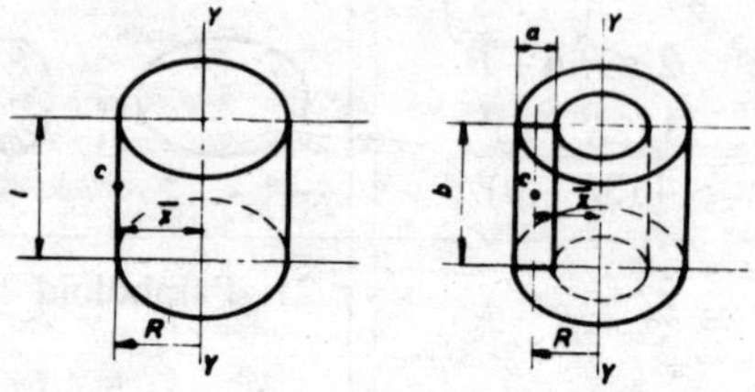

10.6 SI UNIT SYSTEM

The SI is an international system of units which developed initially from three basic units, namely, the unit of length or metre, the units of mass or kilogram and the unit of time or second. Later the unit of electric current or Ampere was added followed by unit of temperature or Kelvin and of luminous intensity, candela. The unit came to be used for the amount of substance was called a mole. Thus the system now consists of seven base units and two supplementary units and a number of derived units.

This international system of units is in fact an extension and refinement of the traditional metric system of units, and is known as : Système International d'Unités, which is abbreviated as SI in all languages. The salient features of this system are the following:

1. It dispenses with the ambiguity as regards the unit of force and mass.
2. It is rational, coherent and comprehensive. In a coherent system, the unit of the resultant quantity of two units which appears as the product or quotient of these two units consists of the units of the individual units of the two quantities. Thus, the unit of power is given by:

   1 Watt = 1 Joule/1 sec = 1 Newton × 1 m/sec.

   The metric Horse Power which is equal to 75 m × kg/sec is not a coherent unit because the coefficient 75 is involved. The coherent system simplifies calculation.
3. The SI is an absolute system and not a gravitational one. Thus the factor "g" has been dispensed with. The unit of force is Newton which is given by:

   $1\text{ N} = 1\text{ kg.m/sec}^2$.
4. The SI is based on decimal arithmetic. The units of different sizes are found by multiplying or dividing a single base value by powers of 10. Thus, changes from one size to another can be made by simply adding zeros or shifting the decimal points.

5. Since both the technologists and the scientists use the same units of basic quantities like energy, force, power, etc., they can work more closely and without confusion arising out of multiplicity of units in the different disciplines.

In India, the SI systems are being increasingly used in schools, colleges, industries and other fields. The Indian Standard Specifications are now practically all in SI units. Tables of SI units which are of relevance in engineering mainly are given in subsequent paragraphs.

Table 10.8 Base units of SI system

Sl. No.	*Quantity*	*Unit*	*Symbol*
1.	Length	Metre	m
2.	Mass	Kilogram	kg
3.	Time	Second	s
4.	Electric current	Ampere	A
5.	Thermodynamic temperature	Kelvin	K
6.	Luminous intensity	Candela	cd
7.	Amount of substance	Mole	mol

These units are defined below:

1. The metre is the length equal to 1, 650, 763.73 wave lengths in vacuum of the radiation corresponding to the transition between the energy levels 2 p_{10} and $5d_5$ of the krypton-86 atom.
2. The kilogram is the mass of the international proto-type of the kilogram which is in the custody of the Bureau International des Poids et Mesures (*BIPM*) at Serves, near Paris.
3. The second is the duration of 9, 192, 631, 770 periods of the radiation corresponding to the transition between the two hyperfine levels of the ground state of the caesium 133 atom.
4. The ampere is the constant current which, if maintained in two straight parallel conductors of infinite length, of negligible circular cross-section, and placed at a distance of 1 metre apart in a vacuum, would produce between the conductors a force equal to 2×10^{-7} newton per metre of length.
5. The Kelvin, unit of thermodynamic temperature, is the fraction 1/273.16 of the thermodynamic temperature of the triplex point of water.
6. The candela is the luminous intensity in the perpendicular direction of a surface of $\frac{1}{600,000}$ square metre of a black body at the temperature of freezing plantinum under a pressure of 101, 325 newtons per square metre.
7. The mole is the amount of substance of a system which contains as many elementary entities as there are atoms in 0.012 kilogram of carbon 12.

Table Supplementary Units

The SI has two supplementary units of plane angle and solid angle:

1. The radian in the plane angle which, having its vertex of the centre of a circle cuts off a length on the circumference at the circle equal to the radius of the circle.

2. The steradian is the solid angle which, having its vertex at the centre of a sphere, cuts off an area on the surface of the sphere equal to that at a square with sides of length equal to the radius of the sphere.

Derived units with special unit symbol

The SI is augmented by derived units. Special names have been adopted for some of the derived SI units, together with special unit symbols, which are given below.

The symbols given by the side of nomenclatures used in the Quantity column below correspond to the recommendations of the ISO. It is preferred that these symbols are used in equations and formulae to effect overall uniformity in scientific and technical writings. For example, such symbol or force is "F" and is no longer P or Q or K or any other letter or symbol.

Table 10.9 ISO symbols and definitions

Quantity	*Unit*	*Definition*
Force, *F*	N	1 Newton is that force which, when applied to a body having a mass of 1 kg, gives it an acceleration of 1 m per sec squared. 1 N = 1 kg.m/s^2.
Work, *W* Energy, *E* Quantity of heat, *Q*	J	1 Joule is the work done when the point of application of a force of 1 Newton is displaced through a distance of 1 metre in the direction of the force, 1 J = 1 N.M = 1 W.s
Power, *P*	W	1 Watt is one Joule per sec. 1 W = 1 J/s.
Pressure, *p* Normal stress, σ Shear stress, τ	Pa	1 Pascal is the pressure equally exerted on an area of 1 m^2 by a force of 1 Newton acting vertically, 1 Pa = N/m^2.
Frequency, *f*	Hz	1 Hertz is equal to the frequency of one cycle per second. 1 Hz = 1 c/s

Table 10.10 Derived units with complex name

Some derived SI units with complex names are in use

Sl. No.	*Quantity*		*SI unit*	*Symbol*
1.	Area	*A*	Square metre	m^2
2.	Volume	*V*	Cubic metre	m^3
3.	Density (mass density)	ρ	Kilogram per cubic metre	kg/m^3
4.	Velocity	*v*	Metre per second	m/s
5.	Angular velocity	ω	Radians per second	rad/s
6.	Acceleration	*a*	Metre per second squared	m/s^2
7.	Angular acceleration	α	Radian per second squared	rad/s^2

Decimal fractions and multiples of SI units along with the relevant prefixes are given in the following table. These are approved factors by which the base units are to be multiplied to arrive at a higher or a smaller new unit convenient for the purpose of calculation:

Table 10.11 Multiples of SI-Units

Fraction	*Prefix*	*Symbol*	*Multiple*	*Prefix*	*Symbol*
10^{-1}	deci	d	10^{1}	deka	d_a
10^{-2}	centi	c	10^{2}	hecto	h
10^{-3}	milli	m	10^{3}	kilo	k
10^{-6}	micro	μ	10^{6}	mega	M
10^{-9}	nano	n	10^{9}	giga	G
10^{-12}	pico	p	10^{12}	tera	T
10^{-15}	femto	f	–	–	–
10^{-18}	atto	a	–	–	–

Compound prefixes should not be used, namely, 10^{-9} metre is represented by: 1 nm, and not 1 mμm.

The equivalent values or the conversion factors of some important units in common usage are given in the following Table. In each table, the SI units are written within thick boxes:

Table 10.12 Conversion factors for units of force

Units of force	*Newton - N*	*Kilogramme-force=kgf=kp*	*dyne*
1 N	1	≈ 0.102	10^{5}
1 kgf	≈ 9.81	1	$\approx 9.81 \times 10^{5}$
1 dyne	10^{-5}	$\approx 1.02 \times 10^{-6}$	1

Table 10.13 Conversion factors for units of work, energy and torque

Units of work energy and torque	*Joule*	*Kilowatt-hour*	*Newton-metre*	*erg*	*kgf.m*
1 Joule = 1 J = 1 Watt-sec = 1 W.s	1	0.277778×10^{-6}	1	10^{7}	≈ 0.102
1 kilowatt-hour = 1 kwh	3.6×10^{6}	1	3.6×10^{6}	3.6×10^{13}	$\approx 0.367 \times 10^{6}$
1 Newton-metre = 1 N.m 1 Volt-ampere-sec. = 1 V.A.s	1	0.277778×10^{-6}	1	10^{7}	≈ 0.102
1 erg	10^{-7}	0.277778×10^{-13}	10^{-7}	1	$\approx 0.102 \times 10^{-7}$
1 kilogramme-force metre = 1 Kgf.m	≈ 9.81	2.72407×10^{-6}	≈ 9.81	$\approx 9.81 \times 10^{7}$	1

Table 10.14 Conversion factors for units of stress

Units of stress	$Pa = \frac{N}{m^2}$	$MPa = \frac{N}{mm^2}$	$\frac{N}{cm^2}$	$\frac{kgf}{mm^2}$	$\frac{kgf}{cm^2}$
1 Pascal = 1 Pa = 1 N/m^2	1	10^{-6}	10^{-4}	$\approx 0.102 \times 10^{-6}$	$\approx 0.102 \times 10^{-4}$
1 Megapascal = 1 MPa = 1 N/mm^2	10^6	1	10^2	≈ 0.102	≈ 10.2
1 N/cm^2	10^4	10^{-2}	1	$\approx 0.102 \times 10^{-2}$	≈ 0.102
1 kgf/mm^2	$\approx 9.81 \times 10^6$	≈ 9.81	$\approx 9.81 \times 10^2$	1	10^2
1 kgf/cm^2	$\approx 9.81 \times 10^4$	$\approx 9.81 \times 10^{-2}$	≈ 9.81	10^{-2}	1

Table 10.15 Conversion factors for units of pressure

Units of stress	$Pa = \frac{N}{m^2}$	*Bar*	$\frac{kgf}{cm^2}$	*Atm*	*Torr*
1 Pascal = 1Pa = 1 N/m^2	1	10^{-5}	$\approx 0.102 \times 10^{-4}$	0.98692×10^{-5}	0.75006×10^{-2}
1 Bar = 0.1 MPa = 10^6 dyne/cm^2	10^5	1	1.01972	0.98692	≈ 750
1 Technical atmosphere = 1 at = 1 kgf/cm^2	$\approx 9.81 \times 10^4$	$\approx 9.81 \times 10^{-1}$	1	9.67841×10^{-1}	≈ 736
1 Physical atmosphere = 1 atm	$\approx 1.01 \times 10^5$	1.01325	1.03325	1	≈ 760
1 Torr = 1 mm of mercury	$\approx 1.33 \times 10^2$	1.333224×10^{-3}	1.35951×10^{-3}	1.31579×10^{-3}	1

Table 10.16 Conversion factors for units of power and energy rate

Units of power and energy rate	*Watt*	*Kilowatt*	$\frac{\text{Newton–metre}}{\text{Second}}$	$\frac{\text{kgf.m}}{\text{Second}}$	*Metric Horse Power PS*
1 Watt = 1 W = 1 Joule/sec. = 1 J/s	1	10^{-3}	1	≈ 0.102	$\approx 1.36 \times 10^{-3}$
1 Kilowatt = 1 kW	10^3	1	10^3	≈ 102	≈ 1.36
$1 \frac{\text{Newton metre}}{\text{Second}}$ = 1 N.m/s	1	10^{-3}	1	≈ 0.102	$\approx 1.36 \times 10^{-3}$

$1 = \dfrac{\text{Kilogramme - force.m}}{\text{Second}}$ = 1 kgf.m/Sec.	≈ 9.81	$\approx 9.81 \times 10^{-3}$	≈ 9.81	1	$\dfrac{1}{75}$
1 Metric Horse Power = 1 PS	≈ 735.5	≈ 0.7355	≈ 735.5	75	1

Table 10.17 Conversion Factors

The equivalent values of other units in SI units are given below:

Sl. No.	*Quantity*	*Unit*	*Equivalent*
1.	Length	1 angstrom	10^{-10} m
		1 inch	0.0254 m
		1 foot	0.3048 m
		1 yard	0.9144 m
		1 mile	1.60934 km
		1 nautical mile, international	1.85318 km
		1 chain	20.12 m
		1 engineer, chain	30.48 m
2.	Area	1 square inch	645.16 mm^2
		1 square foot	0.092903 m^2
		1 square yard	0.836127 m^2
		1 square mile	2.58999 km^2
		1 acre	4047 m^2
3.	Velocity	1 inch per second	25.4 mm/s
		1 foot per second	0.3048 m/s
		1 foot per minute	0.00508 m/s
		1 mile per hour	1.609 km/h
4.	Acceleration	1 foot per second squared	0.3048 m/s^2
5.	Mass	1 pound	0.45359237 kg.
		1 slug	14.5939 kg
6.	Density	1 pound per cubic inch	2.76799×10^4 kg m^{-3}
		1 pound per cubic foot	16.0185 kg.m^{-3}
7.	Force	1dyne	10^{-5} N
		1 poundal	0.138255 N
		1 pound-force	4.44822 N
		1 kilogram-force	9.80655 N
8.	Pressure	1 atmosphere (Physical)	101.325 kNm^{-2}
		1 pound force per square inch	6.89476 kNm^{-2}
		1 torr	133.322 Nm^{-2}

9.	Energy	1 erg 1 calorie (IT) 1 calorie (15°C) 1 cakoric (thurmo chemical) 1 Btu 1 foot poundal 1 foot pound-force	10^{-7} J 4.1868 J 4.1855 J 4.184 J 1055.06 J 0.042149 J 1.35582 J
10.	Power	1 hp (horse power) 1 metric hp (PS)	745.7 W 735.5 W
11.	Volume	1 litre 1 cubic inch 1 cubic foot 1 UK gallon	10^{-3} m^3 1.63871×10^{-5} m^3 0.0283168 m^3 0.004546092 m^3

10.7 PREFERRED NUMBERS AND PREFERRED SIZES

Preferred numbers are numbers which have been standardised for the purpose of rationalisation. The need for such standardisation stems from the fact that a rational approach is necessary to limit the number and the variety of sizes of products of all types. It stands to reason, therefore, that some sort of standardised values should be nationally and universally adopted. This will help to avoid multiplicity of product sizes, simplify stock and inventory control, limit the sizes of tools and gauges used in the engineering shops, assure interchangeability, bring order in selecting series of items, standardise ratings of machineries and bring about other advantageous features of such system.

The basis on which the preferred members are selected is the geometrical progression. The geometrical progression has several advantages over the arithmetical progression. A geometrical series provides small steps for small numbers and large steps for large numbers. An arrangement like that meets best the most requirements in the practical field and also conforms to the mode of variations found in nature, *e.g.*, organic growth, radioactive decay and other natural laws.

In a geometrical series, each term is larger than the preceding one by a fixed percentage. The preferred nembers are classified into five principal series, namely, R 5, R 10, R 20, R 40 and R 80, where the numeral indicates the particular root of 10 on which the series is based. After careful considerations and agreement at international level, the following system for preferred numbers has evolved:

The first number is 10 in any series and the other numbers in the series are obtained by multiplying or dividing the first number by the constant factor relevant to the series and repeating the process with each ensuing number. For example, in the case of the R 5 series, the constant factor is:

$$\sqrt[5]{10} = 1.5849... \approx 1.6$$

In the range of 10 to 100 in this series, the preferred numbers are:

$$10$$
$$10 \times 1.6 = 16$$
$$16 \times 1.6 \approx 25$$

$$25 \times 1.6 = 40$$
$$40 \times 1.6 \approx 63$$
$$63 \times 1.6 \approx 100$$

Numbers of R5 series are, therefore, around 60% apart. In a similar way, we get the numbers of the other series thus:

$$R\,10 = \sqrt[10]{10} = 1.2589 \approx 1.25$$
$$R\,20 = \sqrt[20]{10} = 1.1220 \approx 1.12$$

and so on for the other series. Tables of preferred numbers and sizes are given in the following tables.

Table 10.18 Preferred Numbers

Basic numbers				*Exact numbers*	*Variation of the basic numbers from the exact number %*	*Mantissas*
R5	*R10*	*R20*	*R40*			
1.00	1.00	1.00	1.00	1.0050	0	000
			1.06	1.0593	+ 0.07	025
		1.12	1.12	1.1220	– 0.18	050
			1.18	1.1885	– 0.71	075
	1.25	1.25	1.25	1.2589	– 0.71	100
			1.32	1.3335	– 1.01	125
		1.40	1.40	1.4125	– 0.88	150
			1.50	1.4962	+ 0.25	175
160	1.60	1.60	1.60	1.5849	+ 0.95	200
			1.70	1.6788	+ 1.26	225
		1.80	1.80	1.7783	+ 1.22	250
			1.90	1.8836	+ 0.87	275
	2.00	2.00	2.00	1.9953	+ 0.24	300
			2.12	2.1135	+0.31	325
		2.24	2.24	2.2387	+ 0.06	350
			2.36	2.3714	– 0.48	375

2.50	2.50	2.50	2.50	2.5119	– 0.47	400
			2.65	2.6607	– 0.40	425
		2.80	2.80	2.8184	– 0.65	450
			3.00	2.9854	+ 0.49	475
	3.15	3.15	3.15	3.1623	– 0.39	500
			3.35	3.3497	+ 0.01	525
		3.55	3.55	3.5481	+ 0.05	550
			3.75	3.2584	–0.22	575
4.00	4.00	4.00	4.00	3.9811	+ 0.47	600
			4.25	4.2170	+ 0.78	625
		4.50	4.50	4.4668	+ 0.74	650
			4.75	4.7315	+ 0.39	675
	5.00	5.00	5.00	5.0119	– 0.24	700
			5.30	5.3088	– 0.17	725
		5.60	5.60	5.6234	– 0.42	750
			6.00	5.9566	+ 0.73	775
6.30	6.30	6.30	6.30	6.3096	– 0.15	800
			6.70	6.6834	+ 0.25	825
		7.10	7.10	7.0795	+ 0.29	850
			7.50	7.4989	+ 0.01	875
	8.00	8.00	8.00	7.9433	+ 0.71	900
			8.50	8.4140	+ 1.02	925
		9.00	9.00	8.9125	+ 0.98	950
			9.50	9.4406	+ 0.63	975

Table 10.19 Preferred sizes

0.1	1	10	100				370
			105				375
	11	11	110			38	380
			115				390
0.12	1.2	12	120	0.4	4.0	40	400
			125				410
		13	130			42	420
			135				430
	1.4	14	140			44	440
			145		4.5	45	450
	1.5	15	150			46	460
			155				470
0.16	1.6	16	160			48	480
			165				490
		17	170	0.5	5	50	500
			175			52	520
	1.8	18	180			53	530
			185		5.5	55	550
		19	190			56	560
			195			58	580
0.2	2	20	200	0.6	6	60	600
		21	210			62	
	2.2	22	220			63	630
		23	230			65	650
		24	240			67	670
0.25	2.5	25	250			68	
		26	260		7	70	700
			270			71	710
	2.8	28	280			72	

			290			75	750
0.3	3	30	300			78	
			310	0.8	8	80	800
			315			82	
	3.2	32	320			85	850
			330			88	
		34	340		9	90	900
	3.5	35	350			92	
			355			95	950
		36	360			98	

10.8 COMMON MATHEMATICAL FORMULAE

While making engineering drawings, it is often required to make mathematical calculations to determine drawing parameters. To this end, the common algebraic and particularly the trigonometric formulae are very useful. In the following paragraphs, such formulae and data which are generally of relevance for drawing making purposes are given.

Algebraic expansion and factors

$$(a \pm b)^2 = a^2 \pm 2ab + b^2$$

$$(a \pm b)^3 = a^3 \pm 3a^2 b + 3ab^2 \pm b^3$$

$$(a + b + c)^2 = a^2 + b^2 + c^2 + 2ab + 2ac + 2bc$$

$$a^2 - b^2 = (a - b)(a + b)$$

$$a^3 - b^3 = (a - b)(a^2 + ab + b^2)$$

$$a^3 + b^3 = (a + b)(a^2 - ab + b^2)$$

$$a^4 + a^2b^2 + b^4 = (a^2 + ab + b^2)(a^2 - ab + b^2)$$

Quadratic Equations

An equation in the form

$$ax^2 + bx + c = 0 \qquad (a \neq 0)$$

is known as a quadratic equation. The general solution of the above equation is given by :

$$x = \frac{-b \pm \sqrt{b^2 - 4ac}}{2a}$$

Ratio of proportion

If $a : b = c : d$ or $\frac{a}{b} = \frac{c}{d}$, then ad $= bc$ and $\frac{a}{c} = \frac{b}{d}$

If $\frac{a}{b} = \frac{c}{d} = \frac{e}{f} = ... = k$, then $k = \ , \frac{a+c+e+.....}{b+d+f+....} = \frac{ra+sc+te+\dots}{rb+sd+tf+\dots}$

Trigonometric formulae

$$\sin^2 A + \cos^2 A = 1 \qquad \tan A = \frac{\sin A}{\cos A} = \frac{1}{\cot A}$$

$$\cot A = \frac{\cos A}{\sin A} = \frac{1}{\tan A} \qquad \sec A = \frac{1}{\cos A} \qquad \text{cosec } A = \frac{1}{\sin A}$$

$$\sin A = \sqrt{1-\cos^2 A} = \frac{\tan A}{\sqrt{1+\tan^2 A}}$$

$$\cos A = \sqrt{1-\sin^2 A} = \frac{\cot A}{\sqrt{1+\cot^2 A}}$$

$$\sin (A+B) = \sin A \cos B + \cos A \sin B$$

$$\sin (A-B) = \sin A \cos B - \cos A \sin B$$

$$\cos (A+B) = \cos A \cos B - \sin A \sin B$$

$$\cos (A-B) = \cos A \cos B + \sin A \sin B$$

$$\tan (A+B) = \frac{\tan A + \tan B}{1 - \tan A \tan B} \qquad \tan (A-B) = \frac{\tan A - \tan B}{1 + \tan A \tan B}$$

$$\cot (A+B) = \frac{\cot A \cot B - 1}{\cot B + \cot A} \qquad \cot (A-B) = \frac{\cot A \cot B + 1}{\cot B - \cot A}$$

$$\tan A + \tan B = \frac{\sin (A+B)}{\cos A \cos B} \qquad \tan A - \tan B = \frac{\sin (A-B)}{\cos A \cos B}$$

$$\cot A + \cot B = \frac{\sin (B+A)}{\sin A \sin B} \qquad \cot A - \cot B = \frac{\sin (B-A)}{\sin A \sin B}$$

$$\sin^2 A - \sin^2 B = \cos^2 B - \cos^2 A = \sin (A+B) \sin (A-B)$$

$$\cos^2 A - \sin^2 B = \cos^2 B - \sin^2 A = \cos (A+B) \cos (A-B)$$

$$\sin A \sin B = \frac{1}{2} \cos (A-B) - \frac{1}{2} \cos (A+B)$$

$$\cos A \cos B = \frac{1}{2} \cos (A-B) + \frac{1}{2} \cos (A+B)$$

$$\tan A \tan B = \frac{\tan A + \tan B}{\cot A + \cot B}$$

$$\cot A \cot B = \frac{\cot A + \cot B}{\tan A + \tan B}$$

$$\sin A = 2\sin\frac{1}{2}A\cos\frac{1}{2}A. \quad \sin 2A = 2\sin A\cos A$$

$$\sin A\cos B = \frac{1}{2}\sin(A+B) + \frac{1}{2}\sin(A-B)$$

$$\cos 2A = \cos^2 A - \sin^2 A = 1 - 2\sin^2 A = 2\cos^2 A - 1$$

$$\tan 2A = \frac{2\tan A}{1-\tan^2 A} = \frac{2}{\cot A - \tan A}$$

$$\cot 2A = \frac{\cot^2 A - 1}{2\cot A} = \frac{\cot A - \tan A}{2}$$

$$\sin A = \frac{2\tan\frac{1}{2}A}{1+\tan^2\frac{1}{2}A}, \quad \cos A = \frac{1-\tan^2\frac{1}{2}A}{1+\tan^2\frac{1}{2}A}$$

$$2\sin^2 A = 1 - \cos 2A, \quad 2\cos^2 A = 1 + \cos 2A$$

Functions of angles in any quadrant in terms of angles in first quadrant

	$-\alpha$	$90° \pm \alpha$	$180° \pm \alpha$	$270° \pm \alpha$	$n(360°) \pm \alpha$
sin	$-\sin\alpha$	$+\cos\alpha$	$\mp\sin\alpha$	$-\cos\alpha$	$\pm\sin\alpha$
cos	$+\cos\alpha$	$\mp\sin\alpha$	$-\cos\alpha$	$\pm\sin\alpha$	$+\cos\alpha$
tan	$-\tan\alpha$	$\mp\cot\alpha$	$\pm\tan\alpha$	$\mp\cot\alpha$	$\pm\tan\alpha$
cot	$-\cot\alpha$	$\mp\tan\alpha$	$\pm\cot\alpha$	$\mp\tan\alpha$	$\pm\cot\alpha$
sec	$+\sec\alpha$	$\mp\text{cosec}\,\alpha$	$-\sec\alpha$	$\pm\text{cosec}\,\alpha$	$+\sec\alpha$
cosec	$-\text{cosec}\,\alpha$	$+\sec\alpha$	$\mp\text{cosec}\,\alpha$	$-\sec\alpha$	$\pm\text{cosec}\,\alpha$

n denotes any integer.

Signs of the Finctions

Quadrant	*sin*	*cos*	*tan*	*cot*	*sec*	*csec*
I	+	+	+	+	+	+
II	+	–	–	–	–	+
III	–	–	+	+	–	–
IV	–	+	–	–	+	–

Functions of 0°, 30°, 45°, 60°, 90°, 180°, 270°, 360°

	0°	30°	45°	60°	90°	180°	270°	360°
sin	0	1/2	$\sqrt{2}/2$	$\sqrt{3}/2$	1	0	–1	0
cos	1	$\sqrt{3}/2$	$\sqrt{2}/2$	1/2	0	–1	0	1
tan	0	$1/\sqrt{3}$	1	$\sqrt{3}$	∞	0	∞	0

APPENDICES

APPENDIX - A Tables of Dimensional Toleranaces

TOLERANCE VALUES IN µm

Tolerances for Nominal Dimension 450mm pictorially represented (as an example)

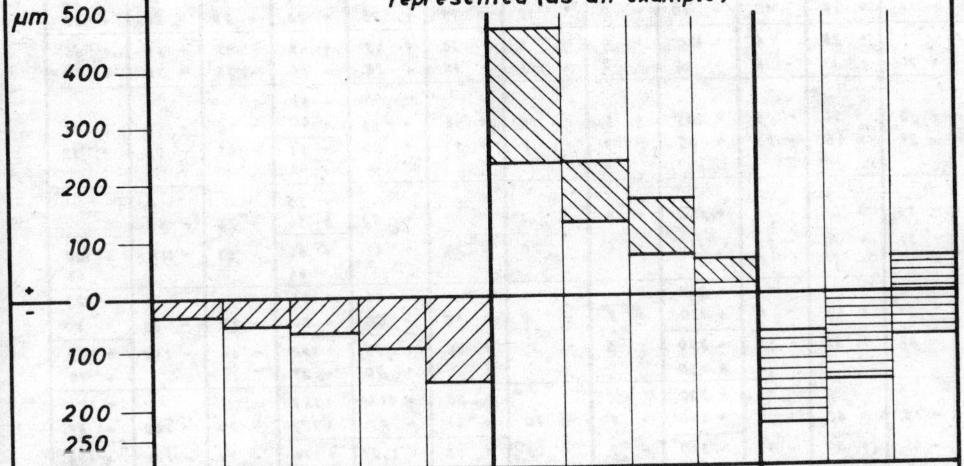

RANGE OF NOMINAL DIMENSION mm	SHAFT					BORE				KEYWAY		
	h5	h6	h7	h8	h9	D10	E8	F8	G6	P9	N9	JS9
1 TO 3	0 - 4	0 - 6	0 - 10	0 - 14	0 - 25	+ 60 + 20	+ 28 + 14	+ 20 + 6	+ 8 + 2	- 6 - 31	- 4 - 29	+12,5 -12.5
OVER 3 UPTO 6	0 - 5	0 - 8	0 - 12	0 - 18	0 - 30	+ 78 + 30	+ 38 + 20	+ 28 + 10	+ 12 + 4	- 12 - 42	- 0 - 30	+ 15 - 15
OVER 6 UPTO 10	0 - 6	0 - 9	0 - 15	0 - 22	0 - 36	+ 98 + 40	+ 47 + 25	+ 35 + 13	+ 14 + 5	- 15 - 51	0 - 36	+ 18 - 18
OVER 10 UPTO 18	0 - 8	0 - 11	0 - 18	0 - 27	0 - 43	+ 120 + 50	+ 59 + 32	+ 43 + 16	+ 17 + 6	- 18 - 61	0 - 43	+21.5 -21.5
OVER 18 UPTO 30	0 - 9	0 - 13	0 - 21	0 - 33	0 - 52	+149 + 65	+ 73 + 40	+ 53 + 20	+ 20 + 7	- 22 - 74	0 - 52	+ 26 - 26
OVER 30 UPTO 50	0 - 11	0 - 16	0 - 25	0 - 39	0 - 62	+ 180 + 80	+ 89 + 50	+ 64 + 25	+ 25 + 9	- 25 - 88	0 - 62	+ 31 - 31
OVER 50 UPTO 80	0 - 13	0 - 19	0 - 30	0 - 46	0 - 74	+ 220 + 100	+106 + 60	+ 76 + 30	+ 29 + 10	- 32 -106	0 - 74	+ 37 - 37
OVER 80 UPTO 120	0 - 15	0 - 22	0 - 35	0 - 54	0 - 87	+ 260 + 120	+126 + 72	+ 90 + 36	+ 34 + 12	- 37 -124	0 - 87	+43.5 -43.5
OVER 120 UPTO 180	0 - 18	0 - 25	0 - 40	0 - 63	0 -100	+ 305 + 145	+148 + 85	+106 + 43	+ 39 + 14	- 43 -143	0 - 100	+ 50 - 50
OVER 180 UPTO 250	0 - 20	0 - 29	0 - 46	0 - 72	0 - 115	+ 355 + 170	+172 +100	+122 + 50	+ 44 + 15	- 50 - 165	0 - 115	+57.5 -57.5
OVER 250 UPTO 315	0 - 23	0 - 32	0 - 52	0 - 81	0 -130	+ 400 + 190	+ 191 + 110	+137 + 56	+ 49 + 17	- 56 -186	0 -130	+ 65 - 65
OVER 315 UPTO 400	0 - 25	0 - 36	0 - 57	0 - 89	0 -140	+ 440 + 210	+ 214 + 125	+ 151 + 62	+ 54 + 18	- 62 -202	0 -140	+ 70 - 70
OVER 400 UPTO 500	0 - 27	0 - 40	0 - 63	0 - 97	0 - 155	+ 480 + 230	+232 +135	+165 + 68	+ 60 + 20	- 68 -223	0 -155	+77.5 -77.5

RANGE OF NOMINAL DIMENSION mm	BORE				SHAFT							
	P6	F7	N7	C11	j5	g6	r6	u7	x7	f8	d9	u9
1 TO 3	− 6 − 12	+ 16 + 6	− 4 − 14	+ 120 + 60	+ 2 − 2	− 2 − 8	+ 16 + 10	+ 28 + 18	+ 30 + 20	− 6 − 20	− 20 − 45	—
OVER 3 UPTO 6	− 9 − 17	+ 22 + 10	− 4 − 16	+ 145 + 70	+ 3 − 2	− 4 − 12	+ 23 + 15	+ 35 + 23	+ 40 + 28	− 10 − 28	− 30 − 60	+ 53 + 23
OVER 6 UPTO 10	− 12 − 21	+ 28 + 13	− 4 − 19	+ 170 + 80	+ 4 − 2	− 5 − 14	+ 28 + 19	+ 43 + 28	+ 49 + 34	− 13 − 35	− 40 − 76	+ 64 + 28
OVER 10 UPTO 14	− 15	+ 34	− 5	+ 205	+ 5	− 6	+ 34	+ 51	+ 58 + 40	− 16	− 50	+ 76
OVER 14 UPTO 18	− 26	+ 16	− 23	+ 95	− 3	− 17	+ 23	+ 33	+ 63 + 45	− 43	− 93	+ 33
OVER 18 UPTO 24	− 18	+ 41	− 7	+ 240	+ 5	− 7	+ 41	+ 62	+ 75 + 54	− 20	− 65	—
OVER 24 UPTO 30	− 31	+ 20	− 28	+ 110	− 4	− 20	+ 28	+ 41	+ 85 + 64	− 53	− 117	+ 100 48
OVER 30 UPTO 40	− 21	+ 50	− 8	+ 280 + 120	+ 6	− 9	+ 50	+ 85 + 60	+ 105 + 80	− 25	− 80	+ 122 + 60
OVER 40 UPTO 50	− 37	+ 25	− 33	+ 290 + 130	− 5	− 25	+ 34	+ 95 + 70	+ 122 + 97	− 64	− 142	+ 132 + 70
OVER 50 UPTO 65	− 25	+ 60	− 9	+ 330 + 140	+ 6	− 10	+ 60 + 41	+ 117 + 87	+ 152 + 122	− 30	− 100	+ 161 + 87
OVER 65 UPTO 80	− 45	+ 30	− 39	+ 340 + 150	− 7	− 29	+ 62 + 43	+ 132 + 102	+ 176 + 146	− 76	− 174	+ 176 + 102
OVER 80 UPTO 100	− 30	+ 71	− 10	+ 390 + 170	+ 6	− 12	+ 73 + 51	+ 159 + 124	+ 213 + 178	− 36	− 120	+ 211 + 124
OVER 100 UPTO 120	− 52	+ 36	− 45	+ 400 + 180	− 9	− 34	+ 76 + 54	+ 175 + 144	+ 245 + 210	− 90	− 207	+ 231 + 144
OVER 120 UPTO 140				+ 450 + 200			+ 88 + 63	+ 210 + 170	+ 288 + 248			+ 270 + 170
OVER 140 UPTO 160	− 36 − 61	+ 83 + 43	− 12 − 52	+ 460 + 210	+ 7 − 11	− 14 − 39	+ 90 + 65	+ 230 + 190	+ 320 + 280	− 43 − 106	− 145 − 245	+ 290 + 190
OVER 160 UPTO 180				+ 480 + 230			+ 93 + 68	+ 250 + 210	—			+ 310 + 210
OVER 180 UPTO 200				+ 530 + 240			+ 106 + 77	+ 282 + 236				+ 351 + 236
OVER 200 UPTO 225	− 41 − 70	+ 96 + 50	− 14 − 60	+ 550 + 260	+ 7 − 13	− 15 − 44	+ 109 + 80	+ 304 + 258	—	− 50 − 122	− 170 − 285	+ 373 + 258
OVER 225 UPTO 250				+ 570 + 280			+ 113 + 84	+ 330 + 284				+ 399 + 284
OVER 250 UPTO 280	− 47	+ 108	− 14	+ 620 + 300	+ 7	− 17	+ 126 + 94	+ 367 + 315	—	− 56	− 190	+ 445 + 315
OVER 280 UPTO 315	− 79	+ 56	− 66	+ 650 + 330	− 16	− 49	+ 130 + 98	+ 402 + 350		− 137	− 320	+ 480 + 350
OVER 315 UPTO 355	− 51	+ 119	− 16	+ 720 + 360	+ 7	− 18	+ 144 + 108	+ 447 + 390	—	− 62	− 210	+ 530 + 390
OVER 355 UPTO 400	− 87	+ 62	− 73	+ 760 + 400	− 18	− 54	+ 150 + 114	+ 492 + 435		− 151	− 350	+ 575 + 435
OVER 400 UPTO 450	− 55	+ 131	− 17	+ 840 + 440	+ 7	− 20	+ 166 + 126	+ 553 + 490	—	− 68	− 230	+ 645 + 490
OVER 450 UPTO 500	− 95	+ 68	− 80	+ 880 + 480	− 20	− 60	+ 172 + 132	+ 603 + 540		− 165	− 385	+ 695 + 540

RANGE OF NOMINAL DIMENSION mm	BORE D9	BORE E6	BORE E9	BORE H5	BORE H6	BORE H9	BORE H10	BORE H12	BORE H13	BORE H14	SHAFT u8	SHAFT x8
1 TO 3	+45 / +20	+20 / +14	+39 / +14	+4 / 0	+6 / 0	+25 / 0	+40 / 0	+100 / 0	+140 / 0	+250 / 0	+32 / +18	+34 / +20
OVER 3 UPTO 6	+60 / +30	+28 / +20	+50 / +20	+5 / 0	+8 / 0	+30 / 0	+48 / 0	+120 / 0	+180 / 0	+300 / 0	+41 / +23	+46 / +28
OVER 6 UPTO 10	+76 / +40	+34 / +25	+61 / +25	+6 / 0	+9 / 0	+36 / 0	+58 / 0	+150 / 0	+220 / 0	+360 / 0	+50 / +28	+56 / +34
OVER 10 UPTO 14	+93 / +50	+43 / +32	+75 / +32	+8 / 0	+11 / 0	+43 / 0	+70 / 0	+180 / 0	+270 / 0	+430 / 0	+60 / +33	+67 / +40
OVER 14 UPTO 18												+72 / +45
OVER 18 UPTO 24	+117 / +65	+53 / +40	+92 / +40	+9 / 0	+13 / 0	+52 / 0	+84 / 0	+210 / 0	+330 / 0	+520 / 0	+74 / +41	+87 / +54
OVER 24 UPTO 30											+81 / +48	+97 / +64
OVER 30 UPTO 40	+142 / +80	+66 / +50	+112 / +50	+11 / 0	+16 / 0	+62 / 0	+100 / 0	+250 / 0	+390 / 0	+620 / 0	+99 / +60	+119 / +80
OVER 40 UPTO 50											+109 / +70	+136 / +97
OVER 50 UPTO 65	+174 / +100	+79 / +60	+134 / +60	+13 / 0	+19 / 0	+74 / 0	+120 / 0	+300 / 0	+460 / 0	+740 / 0	+133 / +87	+168 / +122
OVER 65 UPTO 80											+148 / +102	+192 / +146
OVER 80 UPTO 100	+207 / +120	+94 / +72	+159 / +72	+15 / 0	+22 / 0	+87 / 0	+140 / 0	+350 / 0	+540 / 0	+870 / 0	+178 / +124	+232 / +178
OVER 100 UPTO 120											+198 / +144	+264 / +210
OVER 120 UPTO 140	+245 / +145	+110 / +85	+185 / +85	+18 / 0	+25 / 0	+100 / 0	+160 / 0	+400 / 0	+630 / 0	+1000 / 0	+233 / +170	+311 / +248
OVER 140 UPTO 160											+253 / +190	+343 / +280
OVER 160 UPTO 180											+273 / +210	+273 / +310
OVER 180 UPTO 200	+285 / +170	+129 / +100	+215 / +100	+20 / 0	+29 / 0	+115 / 0	+185 / 0	+460 / 0	+720 / 0	+1150 / 0	+308 / +236	+422 / +350
OVER 200 UPTO 225											+330 / +258	+457 / +385
OVER 225 UPTO 250											+356 / +284	+497 / +425
OVER 250 UPTO 280	+320 / +190	+142 / +110	+240 / +110	+23 / 0	+32 / 0	+130 / 0	+210 / 0	+520 / 0	+810 / 0	+1300 / 0	+396 / +315	+556 / +475
OVER 280 UPTO 315											+431 / +350	+606 / +525
OVER 315 UPTO 355	+350 / +210	+161 / +125	+265 / +125	+25	+36 / 0	+140 / 0	+230 / 0	+570 / 0	+890 / 0	+1400 / 0	+479 / +390	+697 / +590
OVER 355 UPTO 400											+524 / +435	—
OVER 400 UPTO 450	+385 / +230	+175 / +135	+290 / +135	+27 / 0	+40 / 0	+155 / 0	+250 / 0	+630 / 0	+970 / 0	+1550 / 0	+587 / +490	—
OVER 450 UPTO 500											+637 / +540	—

RANGE OF NOMINAL DIMENSION, mm	BORE			SHAFT								
	H7	H8	H11	j6	k6	m6	n6	p6	f7	e8	h11	a11
1 TO 3	+ 10 0	+ 14 0	+ 60 0	+ 4 - 2	+ 6 0	+ 8 + 2	+ 10 + 4	+ 12 + 6	- 6 - 16	- 14 - 28	0 - 60	- 270 - 330
OVER 3 UPTO 6	+ 12 0	+ 18 0	+ 75 0	+ 6 - 2	+ 9 + 1	+ 12 + 4	+ 16 + 8	+ 20 + 12	- 10 - 22	20 - 38	0 - 75	- 270 - 345
OVER 6 UPTO 10	+ 15 0	+ 22 0	+ 90 0	+ 7 - 2	+ 10 + 1	+ 15 + 6	+ 19 + 10	+ 24 + 15	- 13 - 28	- 25 - 47	0 - 90	- 280 - 370
OVER 10 UPTO 18	+ 18 0	+ 27 0	+ 110 0	+ 8 - 3	+ 12 + 1	+ 18 + 7	+ 23 + 12	+ 29 + 18	- 16 - 34	- 32 - 59	0 - 110	- 290 - 400
OVER 18 UPTO 30	+ 21 0	+ 33 0	+ 130 0	+ 9 - 4	+ 15 + 2	+ 21 + 8	+ 28 + 15	+ 35 + 22	- 20 - 41	- 40 - 73	0 - 130	- 300 - 430
OVER 30 UPTO 40	+ 25 0	+ 39 0	+ 160 0	+ 11 - 5	+ 18 + 2	+ 25 + 9	+ 33 + 17	+ 42 + 26	- 25 - 50	- 50 - 89	0 - 160	- 310 - 470
OVER 40 UPTO 50	+ 25 0	+ 39 0	+ 160 0	+ 11 - 5	+ 18 + 2	+ 25 + 9	+ 33 + 17	+ 42 + 26	- 25 - 50	- 50 - 89	0 - 160	- 320 - 480
OVER 50 UPTO 65	+ 30 0	+ 46 0	+ 190 0	+ 12 - 7	+ 21 + 2	+ 30 + 11	+ 39 + 20	+ 51 + 32	- 30 - 60	- 60 - 106	0 - 190	- 340 - 530
OVER 65 UPTO 80	+ 30 0	+ 46 0	+ 190 0	+ 12 - 7	+ 21 + 2	+ 30 + 11	+ 39 + 20	+ 51 + 32	- 30 - 60	- 60 - 106	0 - 190	- 360 - 550
OVER 80 UPTO 100	+ 35 0	+ 54 0	+ 220 0	+ 13 - 9	+ 25 + 3	+ 35 + 13	+ 45 + 23	+ 59 + 37	- 36 - 71	- 72 - 126	0 - 220	- 380 - 600
OVER 100 UPTO 120	+ 35 0	+ 54 0	+ 220 0	+ 13 - 9	+ 25 + 3	+ 35 + 13	+ 45 + 23	+ 68 + 43	- 36 - 71	- 72 - 126	0 - 220	- 410 - 630
OVER 120 UPTO 140	+ 40 0	+ 63 0	+ 250 0	+ 14 - 11	+ 28 + 3	+ 40 + 15	+ 52 + 27	+ 68 + 43	- 43 - 83	- 85 - 148	0 - 250	- 460 - 710
OVER 140 UPTO 160	+ 40 0	+ 63 0	+ 250 0	+ 14 - 11	+ 28 + 3	+ 40 + 15	+ 52 + 27	+ 68 + 43	- 43 - 83	- 85 - 148	0 - 250	- 520 - 770
OVER 160 UPTO 180	+ 40 0	+ 63 0	+ 250 0	+ 14 - 11	+ 28 + 3	+ 40 + 15	+ 52 + 27	+ 68 + 43	- 43 - 83	- 85 - 148	0 - 250	- 580 - 830
OVER 180 UPTO 200	+ 46 0	+ 72 0	+ 290 0	+ 16 - 13	+ 33 + 4	+ 46 + 17	+ 60 + 31	+ 79 + 50	- 50 - 96	- 100 - 172	0 - 290	- 660 - 950
OVER 200 UPTO 225	+ 46 0	+ 72 0	+ 290 0	+ 16 - 13	+ 33 + 4	+ 46 + 17	+ 60 + 31	+ 79 + 50	- 50 - 96	- 100 - 172	0 - 290	- 740 - 1030
OVER 225 UPTO 250	+ 46 0	+ 72 0	+ 290 0	+ 16 - 13	+ 33 + 4	+ 46 + 17	+ 60 + 31	+ 79 + 50	- 50 - 96	- 100 - 172	0 - 290	- 820 - 1110
OVER 250 UPTO 280	+ 52 0	+ 81 0	+ 320 0	+ 16 - 16	+ 36 + 4	+ 52 + 20	+ 66 + 34	+ 88 + 56	- 56 - 108	- 110 - 191	0 - 320	- 920 - 1240
OVER 280 UPTO 315	+ 52 0	+ 81 0	+ 320 0	+ 16 - 16	+ 36 + 4	+ 52 + 20	+ 66 + 34	+ 88 + 56	- 56 - 108	- 110 - 191	- 0 - 320	- 1050 - 1370
OVER 315 UPTO 355	+ 57 0	+ 89 0	+ 360 0	+ 18 - 18	+ 40 + 4	+ 57 + 21	+ 73 + 37	+ 98 + 62	- 62 - 119	- 125 - 214	0 - 360	- 1200 - 1560
OVER 355 UPTO 400	+ 57 0	+ 89 0	+ 360 0	+ 18 - 18	+ 40 + 4	+ 57 + 21	+ 73 + 37	+ 98 + 62	- 62 - 119	- 125 - 214	0 - 360	- 1350 - 1710
OVER 400 UPTO 450	+ 63 0	+ 97 0	+ 400 0	+ 20 - 20	+ 45 + 5	+ 63 + 23	+ 80 + 40	+ 108 + 68	- 68 - 131	- 135 - 232	0 - 400	- 1500 - 1900
OVER 450 UPTO 500	+ 63 0	+ 97 0	+ 400 0	+ 20 - 20	+ 45 + 5	+ 63 + 23	+ 80 + 40	+ 108 + 68	- 68 - 131	- 135 - 232	0 - 400	- 1650 - 2050

APPENDIX - B Tables of General Tolerances on Iron and Steel Castings

The deviations specified in this Appendix relate to untoleraned dimensions of unmachined iron and steel castings in all cases.

Tables B1 to B6 are meant for steel castings as per IS: 4897 - 1976. Classes 1 and 2 are relevant for castings involving machine moulding and hand moulding respectively.

In each ease, the Nominal Dimension is the dimension prescribed on the production drawing to which deviation is to be applied. The Overall Dimension is the largest measurable dimension of a casting on a particular part of the casting in which the Nominal Dimension is located. (See Fig. B1).

Table B1 Permissible deviation in untoleranced dimensios for steel castings (class of accuracy 1)

All dimensions in millimetres

Nominal Dimension of Raw Casting		*Overall Dimensions*											
Over	*Upto*	*Up to 18*	*Over 18 Up to 30*	*Over 30 Up to 80*	*Over 80 Up to 180*	*Over 180 Up to 315*	*Over 315 Up to 500*	*Over 500 Up to 800*	*Over 800 Up to 1250*	*Over 1250 Up to 2000*	*Over 2000 Up to 3150*	*Over 3150 Up to 5000*	*Over 5000 Up to 8000*
(1)	*(2)*	*(3)*	*(4)*	*(5)*	*(6)*	*(7)*	*(8)*	*(9)*	*(10)*	*(11)*	*(12)*	*(13)*	*(14)*
	6	± 0.6	± 0.8	± 0.8	± 0.8	± 1.0	± 1.5	± 1.5	± 2.0	± 2.5	± 3.0	± 4.0	± 4.0
6	10	± 0.8	± 0.8	± 0.8	± 1.0	± 1.5	± 1.5	± 2.0	± 2.5	± 3.5	± 4.0	± 4.0	± 5.0
10	18	± 0.8	± 1.0	± 1.2	± 1.5	± 1.5	± 2.0	± 2.5	± 3.5	± 4.0	± 4.0	± 5.0	± 5.0
18	30	–	± 1.2	± 1.5	± 1.5	± 2.0	± 2.5	± 3.5	± 4.0	± 4.5	± 5.0	± 5.0	± 5.0
30	80	–	–	± 1.5	± 2.0	± 2.5	± 3.0	± 3.5	± 4.0	± 4.5	± 5.0	± 5.0	± 5.0
80	180	–	–	–	± 2.5	± 3.0	± 3.5	± 4.0	± 4.5	± 5.0	± 5.0	± 5.5	± 6.0
180	315	–	–	–	–	± 3.0	± 3.5	± 4.0	± 4.5	± 5.0	± 5.5	± 6.0	± 6.5
315	500	–	–	–	–	–	± 4.0	± 4.5	± 5.0	± 5.0	± 6.0	± 6.0	± 7.0
500	800	–	–	–	–	–	–	± 5.0	± 5.0	± 5.5	± 6.0	± 6.0	± 7.0
800	1250	–	–	–	–	–	–	–	± 5.5	± 6.0	± 6.0	± 7.0	± 8.0
1250	2000	–	–	–	–	–	–	–	–	± 7.0	± 7.0	± 8.0	± 9.0
2000	3150	–	–	–	–	–	–	–	–	–	± 8.0	± 9.0	± 10.0
3150	5000	–	–	–	–	–	–	–	–	–	–	± 10.0	± 11.0
5000	8000	–	–	–	–	–	–	–	–	–	–	–	± 12.0

Table B2 Permissible deviation in untoleranced dimensions for steel casting (class of accuracy 2)

All dimensions in millimetres

Nominal Dimension of Rough Castings		*Overall Dimensions*												
Over	*Up to*	*Up to 18*	*Over 18 Up to 30*	*Over 30 Up to 80*	*Over 80 Up to 180*	*Over 180 Up to 315*	*Over 315 Up to 500*	*Over 500 Up to 800*	*Over 800 Up to 1250*	*Over 1250 Up to 2000*	*Over 2000 Up to 3150*	*Over 3150 Up to 5000*	*Over 5000 Up to 8000*	*Over 8000*
(1)	*(2)*	*(3)*	*(4)*	*(5)*	*(6)*	*(7)*	*(8)*	*(9)*	*(10)*	*(11)*	*(12)*	*(13)*	*(14)*	*(15)*
–	6	± 1.0	± 1.0	± 1.5	± 1.5	± 1.5	± 2.0	± 2.5	± 3.5	± 4.0	± 5.0	± 6.0	± 7.0	± 7.0
6	10	± 1.0	± 1.0	± 1.5	± 1.5	± 2.0.	± 2.5	± 3.5	± 4.0	± 5.0	± 6.0	± 6.0	± 7.0	± 8.0
10	18	± 1.0	± 1.5	± 1.5	± 2.0	± 2.5	± 3.5	± 4.0	± 5.0	± 6.0	± 6.0	± 7.0	± 8.0	± 9.0
18	30	–	± 1.5	± 2.0	± 2.5	± 3.0	± 4.0	± 5.0	± 6.0	± 7.0	± 7.0	± 8.0	± 9.0	± 10.0
30	80	–	–	± 2.5	± 3.0	± 3.5	± 4.0	± 5.0	± 6.0	± 8.0	± 8.0	± 9.0	± 10.0	± 10.0
80	180	–	–	–	± 3.5	± 4.0	± 5.0	± 6.0	± 7.0	± 8.0	± 8.0	± 9.0	± 10.0	± 11.0
180	315	–	–	–	–	± 4.5	± 5.0	± 6.0	± 7.0	± 8.0	± 9.0	± 10.0	± 11.0	± 11.0
315	500	–	–	–	–	–	± 6.0	± 7.0	± 8.0	± 8.0	± 9.0	± 10.0	± 11.0	± 12.0
500	800	–	–	–	–	–	–	± 7.0	± 8.0	± 9.0	± 10.0	± 11.0	± 12.0	± 12.0
800	1250	–	–	–	–	–	–	–	± 9.0	± 9.0	± 10.0	± 11.0	± 12.0	± 12.0
1250	2000	–	–	–	–	–	–	–	–	± 10.0	± 11.0	± 12.0	± 12.0	± 14.0
2000	3150	–	–	–	–	–	–	–	–	–	± 12.0	± 13.0	± 14.0	± 16.0
3150	5000	–	–	–	–	–	–	–	–	–	–	± 15.0	± 16.0	± 18.0

Table B3 Permissible deviations in untoleranced wall thickness of steel castings

All dimensions in millimetres

Overall Dimension of Castings, Max	*Nominal Thickness of Castings*	*Permissible Deviation*	
		Class 1	*Class 2*
Up to 500	Upto 10	± 1.5	± 2.0
	Over 10 to 30	± 2.0	± 3.0
	Over 30 to 50	± 3.0	± 3.5
	Over 50 to 80	± 3.0	± 4.0
	Over 80 to 120	± 3.5	± 4.5
Over 500 to 1250	Upto 10	± 2.0	± 3.0
	Over 10 to 30	± 2.5	± 3.5
	Over 30 to 50	± 3.0	± 4.0
	Over 50 to 80	± 3.5	± 4.5
	Over 80 to 120	± 4.0	± 5.0

Over 1250 to 2500	Up to 18	± 3.0	± 4.0
	Over 18 to 30	± 3.5	± 4.5
	Over 30 to 50	± 4.0	± 5.0
	Over 50 to 80	± 4.5	± 5.5
	Over 80 to 120	± 5.0	± 6.0
Over 2500 to 4000	Up to 18	± 4.0	± 5.0
	Over 18 to 30	± 4.5	± 5.5
	Over 30 to 50	± 5.0	± 6.0
	Over 50 to 80	± 5.5	± 6.5
	Over 80 to 120	± 5.5	± 7.0
Over 4000	Up to 18	–	± 6.0
	Over 18 to 30	–	± 6.5
	Over 30 to 50	–	± 7.0
	Over 50 to 80	–	± 7.0
	Over 80 to 120	–	± 8.0

Table B4 Permissible deviations on mass of steel castings

Nominal Mass of Casting kg	*Permissible Deviation on Mass, Percent*	
	Class 1	*Class 2*
Up to 80	± 6.0	±7.0
Over to 80 to 500	± 5.5	± 6.5
Over 500 to 2000	± 4.5	± 6.0
Over 2000	–	± 5.5

Table B5 Permissible deviations on centre-to-centre distance between cored holes, bosses, etc.

All dimensions in millimeteres

Drawing Dimension	*Tolerance*	
	Class 1	*Class 2*
Up to 200	± 1.5	± 2.5
Over 200 to 500	± 2.0	± 3.5
Over 500 to 1000	± 2.5	± 4.0
Over 1000 to 1700	± 3.5	± 5.0
Over 1700 to 2500	± 4.5	± 6.0
Over 2500		± 7.0

Table B6 Corner radii

Corner radii (inside) minimum values, mm	*Nominal Wall Thickness mm*								
	Upto 30	*Over 30 to 50*	*Over 50 to 80*	*Over 80 to 120*	*Over 120 to 180*	*Over 180 to 250*	*Over 250 to 315*	*Over 315 to 400*	*Over 400 to 500*
	10	15	20	30	50	70	90	120	150

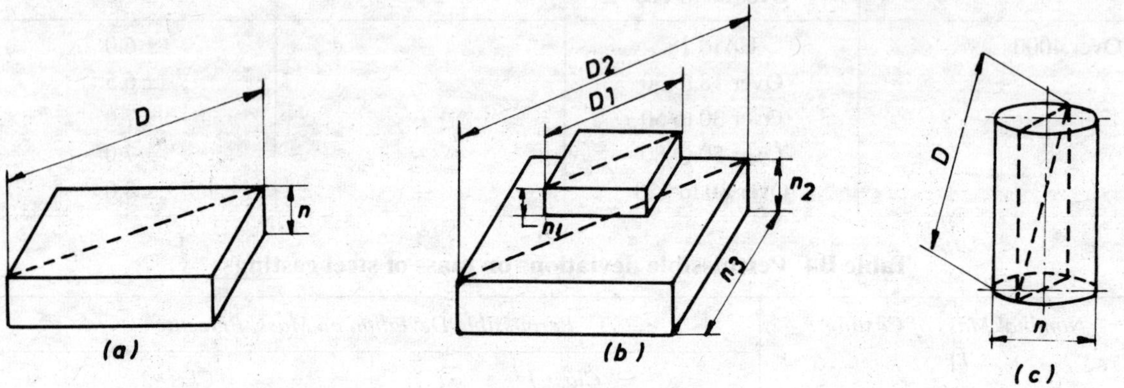

Fig. B1 Nominal and overall dimensions of castings

(a) Nominal dimension is "n" and overall dimension is "D".

(b) Nominal dimension is "n_1" and overall dimension is "D_1".
Nominal dimension is "n_2" and overall dimension is "D_2".
Nominal dimension is "n_3" and overall dimension is "D_3".

(c) Nominal dimension is "n" and overall dimension is "D".
Nominal dimension is "D" and overall dimension is also "D".

Tables B7 to B13 show the allowable deviations for untoleranced dimensions of grey iron castings. The values are as per IS: 5519. The deviations are classified into the folllowing three categories :

Class 1 : For castings for mass production.

Class 2 : For castings for batch production.

Class 3 : For castings for piece production.

Table B7 Permissible deviations on untoleranced dimensions of grey iron castings, class 1

All Dimensions in millimetres

Maximum Overall Dimension of casting	*Nominal Dimension*									
	Upto 6	*Over 6 to 10*	*Over 10 to 18*	*Over 18 to 30*	*Over 30 to 80*	*Over 80 to 180*	*Over 180 to 315*	*Over 315 to 500*	*Over 500 to 800*	*Over 800 to 1250*
	±	±	±	±	±	±	±	±	±	±
Up to 18	0.5	0.5	0.5	–	–	–	–	–	–	–
Over 18 to 30	0.5	0.5	0.6	0.8	–	–	–	–	–	–
Over 30 to 80	0.5	0.5	0.8	1.0	1.2	–	–	–	–	–
Over 80 to 180	0.6	0.6	1.0	1.2	1.5	1.5	–	–	–	–
Over 180 to 315	0.8	1.0	1.2	1.5	1.5	2.0	2.0	–	–	–
Over 315 to 500	1.0	1.2	1.2	1.5	2.0	2.0	2.5	2.5	–	–
Over 500 to 800	1.2	1.5	1.5	2.0	2.0	2.5	2.5	2.5	3.0	–
Over 800 to 1250	1.5	2.0	2.0	2.5	2.5	2.5	2.5	3.0	3.0	3.5

Table B8 Permissible deviations on untoleranced dimensions of grey iron castings, class 2

All Dimensions in millimetres

	Nominal Dimensions													
Maximum Overall Dimension of Casting	*Upto 6*	*Over 6 to 10*	*Over 10 to 18*	*Over 18 to 30*	*Over 30 to 80*	*Over 80 to 180*	*Over 180 to 315*	*Over 315 to 500*	*Over 500 to 800*	*Over 800 to 1250*	*Over 1250 to 2000*	*Over 2000 to 3150*	*Over 3150 to 5000*	*Over 5000 to 8000*
	±	±	±	±	±	±	±	±	±	±	±	±	±	±
Up to 18	0.6	0.8	0.8	–	–	–	–	–	–	–	–	–	–	–
Over 18 to 30	0.8	0.8	1.0	1.2	–	–	–	–	–	–	–	–	–	–
Over 30 to 80	0.8	0.8	1.2	1.5	1.5	–	–	–	–	–	–	–	–	–
Over 80 to 180	0.8	1.0	1.5	1.5	2.0	2.5	–	–	–	–	–	–	–	–

Over 180 to 315	1.0	1.5	1.5	2.0	3.0	3.0	3.0	–	–	–	–	–	–	–
Over 315 to 500	1.5	1.5	2.0	2.5	3.5	3.5	3.5	4.0	–	–	–	–	–	–
Over 500 to 800	1.5	2.0	2.5	3.5	4.0	4.0	4.0	4.0	4.5	5.0	–	–	–	–
Over 800 to 1250	2.0	2.5	3.5	4.0	4.5	4.5	4.5	5.0	5.0	5.5	–	–	–	–
Over 1250 to 2000	2.5	3.5	4.0	4.5	5.0	5.0	5.0	5.0	5.5	5.5	6.0	7.0	–	–
Over 2000 to 3150	3.0	3.5	4.0	5.0	5.0	5.0	5.5	6.0	6.0	6.0	7.0	8.0	–	–
Over 3150 to 5000	4.0	4.0	5.0	5.0	5.0	5.5	6.0	6.0	6.0	7.0	8.0	9.0	10	–
Over 5000 to 8000	4.0	5.0	5.0	5.0	6.0	6.0	6.5	7.0	7.0	8.0	9.0	10	11	12

Table B9 Permissible deviations on untoleranced dimensions of grey iron castings, class 3

All Dimensions in millimetres

Maximum Overall Dimension of Casting	*Nominal Dimensions*														
	Upto 6	*Over 6 to 10*	*Over 10 to 18*	*Over 18 to 30*	*Over 30 to 80*	*Over 80 to 180*	*Over 180 to 315*	*Over 315 to 500*	*Over 500 to 800*	*Over 800 to 1250*	*Over 1250 to 2000*	*Over 2000 to 3150*	*Over 3150 to 5000*	*Over 5000 to 8000*	*Over 8000*
	±	±	±	±	±	±	±	±	±	±	±	±	±	±	±
Up to 18	0.8	1.0	1.0	–	–	–	–	–	–	–	–	–	–	–	–
Over 18 to 30	1.0	1.0	1.2	1.5	–	–	–	–	–	–	–	–	–	–	–
Over 30 to 80	1.2	1.2	1.5	2.0	2.5	–	–	–	–	–	–	–	–	–	–
Over 80 to 180	1.2	1.5	2.0	2.5	3.0	3.5	–	–	–	–	–	–	–	–	–
Over 180 to 315	1.5	2.0	2.5	3.0	3.5	4.0	4.5	–	–	–	–	–	–	–	–
Over 315 to 500	2.0	2.5	3.5	4.0	4.0	5.0	5.0	6.0	–	–	–	–	–	–	–
Over 500 to 800	2.5	3.5	4.0	5.0	5.0	6.0	6.0	7.0	7.0	–	–	–	–	–	–
Over 800 to 1250	3.5	4.0	5.0	6.0	6.0	7.0	7.0	8.0	8.0	9.0	–	–	–	–	–
Over 1250 to 2000	4.0	5.0	6.0	7.0	8.0	8.0	8.0	8.0	9.0	9.0	10.0	–	–	–	–
Over 2000 to 3150	5.0	6.0	6.0	7.0	8.0	8.0	9.0	9.0	10.0	10.0	11.0	12.0	–	–	–
Over 3150 to 5000	6.0	6.0	7.0	8.0	9.0	9.0	10	10	11	11.0	12	13	15	–	–
Over 5000 to 8000	7.0	7.0	8.0	9.0	10	10	11	11	12	12	12	14	16	18	–
Over 8000	7.0	8.0	9.0	10	10	11	11	12	12	12	14	16	18	19	20

Table B10 Permissible deviations on untoleranced wall thickness of grey iron castings

All Dimensions in millimetres

Maximum Overall Dimension of Casting	*Thickness of Casting*	*Permissible Deviation*	
		Class 1	*Classes 2 and 3*
		±	±
Up to 500	Up to 10	0.8	1.0
	Over 10 to 30	1.0	1.5
	Over 30 to 50	1.5	2.0
	Over 50 to 80	2.0	2.5
	Over 80 to 120	2.5	3.0
Over 500 to 1250	Up to 10	1.0	1.5
	Over 10 to 30	1.5	2.0
	Over 30 to 50	2.0	2.5
	Over 50 to 80	2.5	3.0
	Over 80 to 120	3.0	3.5
Over 1250 to 2000	Up to 18	2.0	2.5
	Over 18 to 30	2.5	3.0
	Over 30 to 50	3.0	3.5
	Over 50 to 80	3.5	4.0
	Over 80 to 120	4.0	4.5
Over 2000 to 4000	Up to 18	2.0	2.5
	Over 18 to 30	2.5	3.0
	Over 30 to 50	3.0	3.5
	Over 50 to 80	3.5	4.0
	Over 80 to 120	4.0	4.5
Over 4000	Up to 18	2.0	2.5
	Over 18 to 30	2.5	3.0
	Over 30 to 50	3.5	4.0
	Over 50 to 80	4.0	5.0
	Over 80 to 120	4.5	6.0

Table B11 Corner radii

All Dimensions in millimetres

	Nominal Wall Thickness								
	Up to 30	*Over 30 to 50*	*Over 50 to 80*	*Over 80 to 120*	*Over 120 to 180*	*Over 180 to 250*	*Over 250 to 315*	*over 315 to 400*	*Over 400 to 500*
Corner Radii (Inside) Minimum Values	10	15	20	30	50	70	90	120	150

Table B12 Deviation in mass of grey iron castings

Nominal Mass of Casting, kg	*Permissible Deviation in Mass (Percent)*		
	Class 1	*Class 2*	*Class 3*
Up to 100	5	8	10
Over 100 to 1000	4	6	10
Over 1000	3	5	8

Table B13 Deviations in centre to centre distance between cored holes, bosses, etc. of grey iron castings

Nominal dimension mm	*Deviation mm*
upto 500	± 3
over 500	± 5

Table B14 to B17 relate to deviations for untoleranced dimensions of malleable iron castings as per IS: 8349 - 1977. The following classifications are used:

Class 1: Diviations in case of improved moulding procsses for precision work,

Class 2: Deviations when machine moulding processes are involved.

Class 3: Deviations for hand moulding processes.

Table B14 Deviations on untoleranced length, width, height and centre-to-centre distance on malleable iron castings (class 1)

All dimensions in millimetres

	Range of Nominal Dimensions								
	Up to 18	*Over 18 to 30*	*Over 30 to 50*	*Over 50 to 80*	*Over 80 to 120*	*Over 120 to 200*	*Over 200 to 315*	*Over 315 to 500*	*Over 500 to 800*
Outside deminsion	+ 0.9	+ 1.1	+ 1.3	+ 1.5	+ 1.8	+ 2.1	+ 2.6	+ 3.1	+ 3.8
	– 0.7	– 0.8	– 1	– 1.2	– 1.4	– 1.9	– 2.1	– 2.5	– 3
Inside dimension and centre-to-centre distance	+ 0.7	+ 0.8	+ 1	+ 1.2	+ 1.4	+ 1.9	+ 2.1	+ 2.5	+ 3
	– 0.9	– 1.1	– 1.3	– 1.5	– 1.8	– 2.1	– 2.6	– 3.2	– 3.8

Table B15 Deviations on untoleranced length, width, height and cenure-to-centre distance of malleable iron castings (classes 2 and 3)

All dimensions in millimetres

		Range of nominal dimensions				
Tolerance Class		*Upto 18*	*Over 18 to 50*	*Over 50 to 180*	*Over 180 to 500*	*Over 500 to 800*
2	Outside demension	+ 1.2 – 1	– 2 – 1.5	+ 4 – 2.5	+ 6 – 3.5	+ 8 – 5
	Inside dimension and centre-to-centre distance	+ 1 – 1.2	+ 1.5 – 2	+ 2.5 – 4	+ 3.5 – 6	+ 5 – 8
3	Outside demension	+ 2 – 1	+ 3 – 2	+ 5 – 3.5	+ 7 – 5	+ 9 –7
	Inside dimension and centre-to-centre distance	+ 1 – 2	+ 2 –3	+ 3.5 – 5	+ 5 – 7	+ 7 – 9

Table B16 Deviations on untoleranced wall thickness of malleable iron castings (class 1)

All Dimensions in millimetres

Range of Nominal Dimensions				
Upto 6	*Over 6 to 10*	*Over 10 to 18*	*Over 18 to 30*	*30 to 50*
+ 0.9	+ 1.1	+ 1.4	+ 1.7	+ 2
– 0.8	– 0.9	– 1.1	– 1.3	– 1.6

Table B17 Deviations on untoleranced wall thickness of malleable iron castings (classes 2 and 3)

All dimensions in millimetres

	Range of Nominal Dimensions			
Tolerance Class	*Upto 6*	*Over 6 to 18*	*Over 18 to 50*	*Over 50 to 120*
2	± 1	± 2	± 2.5	± 3.5
3	± 1.5	± 2.5	± 3.5	± 4.5

Tables B 18 and B 19 concern deviations for untoleranced dimensions of spheroidal or nodular graphite iron castings as per IS: 8350 - 1977. The deviations are categorized into the following classes :

Class 1: Castings involving machine moulding.

Class 2: Castings involving hand moulding.

Table B18 Deviations on untoleranced length, width and height of spheroidal or nodular graphite iron castings

All dimensions in millimetres

Tolerance Class		*Range of Nominal dimensions*				
		Upto 18	*Over 18 to 50*	*Over 50 to 180*	*Over 180 to 500*	*Over 500 to 1250*
1	Outside dimension	+1.2 –1	+2 –1.5	+4 –2.5	+5 –3.5	+8 –5
	Inside dimension	+1 –1.2	+1.5 –2	+2.5 –4	+3.5 –5	+5 –8
2	Outside dimension	+2 –1	+3 –2	+5 –3.5	+7 –5	+9 –7
	Inside dimension	+1 –2	+2 –3	+3.5 –5	+5 –7	+7 –9

Table B19 Deviations on untoleranced wall thickness of spheroidal or nodular graphite iron castings

All dimensions in millimetres

Tolerance Class	*Range of Nominal Dimensions*			
	Upto 6	*Over 6 to 18*	*Over 18 to 50*	*Over 50 to 120*
1	± 1	± 2	± 2.5	± 3.5
2	± 1.5	± 2.5	± 3.5	± 4.5

APPENDIX - C Tables of Common Fractions, Decimals and Constants

n	$\sqrt{n}$	$\sqrt[3]{n}$	n	$\sqrt{n}$	$\sqrt[3]{n}$
0.01	0.10000	0.21544	0.25	0.50000	0.62996
0.02	0.14142	0.27144	0.30	0.54772	0.66943
0.03	0.17321	0.31072	0.40	0.63246	0.73681
0.04	0.20000	0.34200	0.50	0.70711	0.79370
0.05	0.22361	0.36840	0.60	0.77460	0.84343
0.06	0.24495	0.39149	0.70	0.83666	0.88790
0.07	0.26458	0.41213	0.80	0.89443	0.92832
0.08	0.28284	0.43089	0.90	0.94868	0.96549
0.09	0.30000	0.44814			
0.10	0.31623	0.46416			
0.20	0.44721	0.58480			

DECIMAL EQUIVALENTS OF COMMON FRACTIONS

1/64 = 0.015625	1/2 = 16/32 = 32/64 = 0.50
1/32 = 2/64 = 0.03125	33/64 = .515625
3/64 = .046875	17/32 = 34/64 = .53125
1/16 = 2/32 = 4/64 = 0.0625	35/64 = 546875
5/64 = .078125	9/16 = 18/32 = 36/64 = .5625
3/32 = 6/64 = .09375	37/64 = .571325
7/64 = .109375	19/32 = 38/64 = .59375
1/8 = 4/32 = 8/64 = 0.125	
9/.64 = .140625	39/64 = .609375
5/32 = 10/64 = .15625	41/64 = .640625
11/64 = .171875	21/32 = 42/64 = .65625
3/16 = 6/32 = 12/64 = .1875	43/64 = .671875
13/64 = .203125	11/16 = 22/32 = 44/64 = .6875

7/32 = 14/64 = .21875

15/64 = .234375

1/4 = 8/32 = 16/64 = 0.25

17/64 = .265625

9/32 = 18/64 = .28125

19/64 = .296875

5/16 = 10/32 = 20/64 = .3125

21/64 = .328125

11/32 = 22/64 = .34375

23/64 = .359375

3/8 = 12/32 = 24/64 = 0.375

25/64 = 0.390625

13/32 = 26/64 = .40625

27/64 = .421875

7/16 = 14/32 = 28/64 = .4375,

15/16 = 30/32 = 60/64 = .9375

29/64 = .45312

45/64 = .703125

23/32 = 46/64 = .71875

47/64 = .734375

3/4 = 24/32 = 48/64 = 0.75

49/64 = .765625

25/32 = 50/64 = .78125

51/64 = .796875

13/16 = 26/32 = 52/64 = .8125

53/64 = .828125

27/32 = 54/64 = .84375

55/64 = .859375

7/8 = 28/32 = 56/64 = 0.875

57/64 = .890625

29/32 = 58/64 = .90625

59/64 = .921875

15/32 = 30/64 = .46825

63/64 = .984375

Common Multiples and Functions of π

n	*Value*
π	3,1415927
2π	6,2831853
3π	9,4247780
4π	12,5663706
5π	15,7079633
6π	18,8495559
7π	21,9911486
8π	25,1327412
9π	28,2743339
π^2	9,8696044
π^3	31,0062767
π^4	97,4090910
$\pi : 2$	1,5707963
$\pi : 3$	1,0471976
$\pi : 4$	0,7853982
$\pi : 5$	0,6283185
$\pi : 6$	0,5235988
$\pi : 7$	0,4487990
$\pi : 8$	0,3926991
$\pi : 9$	0,3490659
$\pi : 12$	0,2617994
$\pi : 16$	0,1963495
$\pi : 32$	0,0981748
$\pi : 64$	0,0490874
$\pi : 90$	0,0349066
$\pi : 180$	0,0174533
$\pi : 360$	0,0087266
$1 : \pi$	0,3183099
$2 : \pi$	0,6366198
$3 : \pi$	0,9549296
$4 : \pi$	1,2732395
$5 : \pi$	1,5915494
$6 : \pi$	1,9098593
$7 : \pi$	2,2281692
$8 : \pi$	2,5464791
$9 : \pi$	2,8647890
$16 : \pi$	5,0929582
$64 : \pi$	20,3718327
$180 : \pi$	57,2957795
$360 : \pi$	114,5915590
$1 : \pi^2$	0,1013212
$1 : \pi^3$	0,0322515
$\log \pi$	0,4971499
$\log \pi^2$	0,9942997
$\log \pi^3$	1,4914496
$\log \sqrt{\pi}$	0,2485749

n	*Value*
$1 : \pi^4$	0,0102660
$\sqrt{\pi}$	1,7724539
$\sqrt{2\pi}$	2,5066283
$\sqrt[3]{\pi}$	1,4645919
$\sqrt[3]{2\pi}$	1,8452701
$\sqrt[3]{3\pi}$	2,1123070
$\sqrt[3]{4\pi}$	2,3248947
$\sqrt[3]{\pi^2}$	2,1450294
$\pi \sqrt{\pi}$	5,5683280
$\pi \sqrt[3]{\pi}$	4,6011511
$\pi \sqrt[3]{\pi^2}$	6,7388086
$4\pi^2$	39,4784176
$4 : \pi^2$	0,4052847
$\pi^2 : 4$	2,4674011
$16 : \pi$	5,0929582
$16 : \pi^2$	1,6211389
$\pi^2 : 16$	0,6168503
$\pi \sqrt{2}$	4,4428829
$\pi \sqrt{3}$	5,4413981
$\pi : \sqrt{2}$	2,2214415
$\pi : \sqrt{3}$	1,8137994
$2 \sqrt{\pi}$	3,5449077
$3 \sqrt{\pi}$	5,3173615
$1 : \sqrt{\pi}$	0,5641896
$2 : \sqrt{\pi}$	1,1283792
$\log \sqrt[3]{\pi}$	0,1657166
$\log 1 : \pi$	0,5028502-1
$\log 1 : \pi^2$	0,0057003-1

n	*Value*
$3 : \sqrt{\pi}$	1,6925687
$\sqrt{\pi : 2}$	1,2533141
$\sqrt{\pi : 3}$	1,0233267
$\sqrt{\pi : 4}$	0,8862270
$\sqrt{1 : \pi}$	0,5641896
$\sqrt{2 : \pi}$	0,7978846
$\sqrt{3 : \pi}$	0,9772050
$\sqrt{4 : \pi}$	1,1283792
$\sqrt{90 : \pi}$	5,3523723
$\sqrt[3]{\pi : 2}$	1,1624474
$\sqrt[3]{\pi : 3}$	1,0154913
$\sqrt[3]{\pi : 4}$	0,9226351
$\sqrt[3]{1 : \pi}$	0,6827841
$\sqrt[3]{2 : \pi}$	0,8602540
$\sqrt[3]{3 : \pi}$	0,9847450
$1 : \sqrt[3]{\pi}$	0,6827840
$\pi \sqrt{g}$	9,8397570
$\pi \sqrt{2g}$	13,9155179
$\pi^2 : g$	1,0060759
$\pi : \sqrt{g}$	1,0030333
$\pi : \sqrt{2g}$	0,7092516
$\log 1 : \pi^3$	0,5085504-2
$\log \sqrt{1 : \pi}$	0,7514251-1
$\log \sqrt[3]{1 : \pi}$	0,8342834-1

APPENDIX D : Tables of Common Structural Sections, Plates and Sheets

In this table and subsequent tables, a selection from structural sections, plates and sheets commonly used in mechanical industries is presented.

Beams

Size 100 to 600 belong to Indian Standard Medium Beam (ISMB).
Size 150 to 200 belong to Indian Standard Heavy Beam (ISHB).
Designation of an ISMB of size 250 and 1000 long :

ISMB 250 × 1000 LG

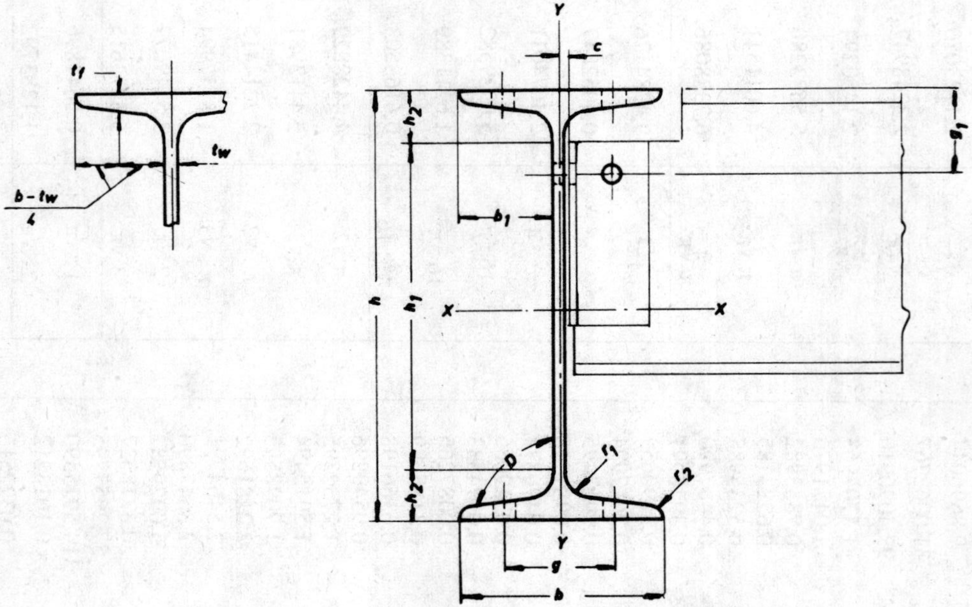

Size	Mass	Sectional area a	Depth of sec-tion h	Width of flange b	Thickness of flange t_f	Thickness of web t_w	Moments of inertia		Radii of gyration		Moduli of section		Radius at root r_1	Radius at toe r_2	Slope D degrees	Connection Details						Max. Size of flange rivet/bolt
							I_{xx}	I_{yy}	r_{xx}	r_{yy}	Z_{xx}	Z_{yy}				h_1	h_2	b_1	c	g	g_1 min	
mm	kg/m	cm^2	mm	mm	mm	mm	cm^4	cm^4	cm	cm	cm^3	cm3	mm	mm		mm	mm	mm	mm	mm	mm	mm
100	11.5	14.6	100	75	7.2	4.0	257.5	40.8	4.20	1.67	51.5	10.9	9.0	4.5	98	65.0	17.50	35.5	3.50	35	55	12
125	13.0	16.6	125	75	7.6	4.4	449.0	43.7	5.20	1.62	71.8	11.7	9.0	4.5	98	89.2	17.90	35.3	3.70	35	55	12
150	14.9	19.0	150	80	7.6	4.8	726.4	52.6	6.18	1.66	96.9	13.1	9.0	4.5	98	113.9	18.05	37.6	3.90	40	55	12
175	19.3	24.62	175	90	8.6	5.5	1272.0	85.0	7.19	1.86	145.4	18.9	10.0	5.0	98	134.5	20.25	42.25	4.25	50	55	12
200	25.4	32.33	200	100	10.8	5.7	2235.4	150.0	8.32	2.15	223.5	30.0	11.0	5.5	98	152.7	23.65	47.15	4.35	55	60	16
225	31.2	39.72	225	110	11.8	6.5	3441.8	218.3	9.31	2.34	305.9	39.7	12.0	6.0	98	173.3	25.85	51.75	4.75	60	60	20
250	37.3	47.55	250	125	12.5	6.9	5131.6	334.5	10.39	2.65	410.5	53.5	13.0	6.5	98	194.1	27.95	59.05	4.95	65	65	22
300	44.2	56.26	300	140	12.4	7.5	8603.6	453.9	12.37	2.84	573.6	64.8	14.0	7.0	98	241.5	29.25	66.25	5.25	80	65	22
350	52.4	66.71	350	140	14.2	8.1	13630.3	537.7	14.29	2.84	778.9	76.8	14.0	7.0	98	288.0	31.00	65.95	5.55	80	65	22
400	61.6	78.46	400	140	16.0	8.9	20458.4	622.1	16.15	2.82	1022.9	88.9	14.0	7.0	98	334.4	32.80	65.55	5.95	80	70	22
450	72.4	92.27	450	150	17.4	9.4	30390.8	834.0	18.15	3.01	1350.7	111.2	15.0	7.5	98	379.2	35.40	70.30	6.20	90	70	22
500	86.9	110.74	500	180	17.2	10.2	45218.3	1369.8	20.21	3.52	1808.7	152.2	17.0	8.5	98	424.1	37.95	84.90	6.60	100	75	28
600	122.6	156.21	600	210	20.8	12.0	91813.0	2651.0	24.24	4.12	3060.4	252.5	20.0	10.0	98	509.7	45.15	99.00	7.50	100	80	25,32
150	27.1	34.48	150	150	9.0	5.4	1455.6	431.7	5.50	3.54	194.1	57.6	8.0	4.0	94	112.0	19.0	72.30	4.20	90	55	22
200	37.3	47.54	200	200	9.0	6.1	3608.6	967.1	8.71	4.51	360.8	96.7	9.0	4.5	94	158.4	20.8	96.95	4.55	100	55	22.32

Channels

All channels given in the table below belong to Indian Standard Medium Channel (ISMC).
Designation of an ISMC of size 125 and 800 long :

ISMC 125 × 800 LG.

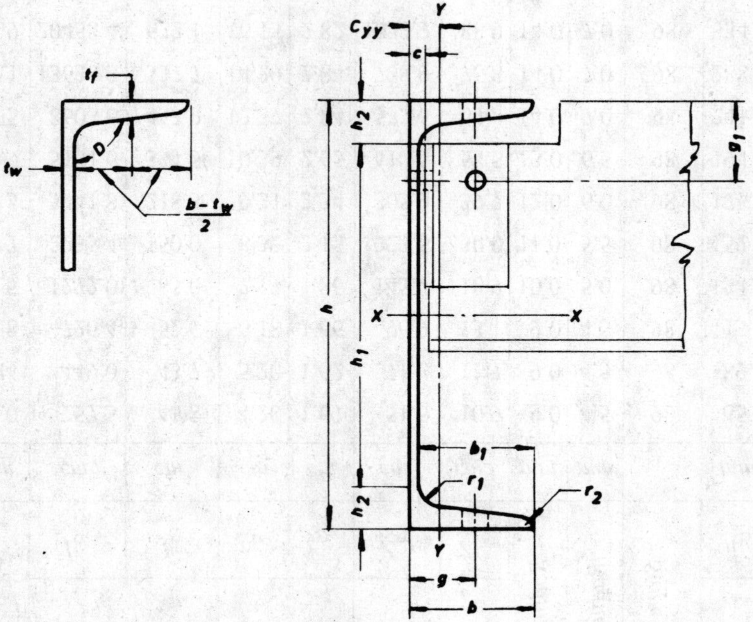

Size	Mass	Sectional area a	Depth of section h	Width of flange b	Thickness of flange t_f	Thickness of web t_w	Centre of gravity C_{yy}	Moments of inertia		Radii of gyration		Moduli of section		Radius of root r_1	Radius of root r_2	Slope D degrees	Connection Details						Max. Size of flange rivet/bolt
								I_{xx}	I_{yy}	I_{xx}	I_{yy}	Z_{xx}	Z_{yy}				h_1	h_2	$b_1/2$	c	g	g_1 min	
mm	kg/m	cm^2	mm	mm	mm	mm	cm	cm^4	cm^4	cm	cm	cm^3	cm^3	mm	mm		mm	mm	mm	mm	mm	mm	mm
75	6.8	8.67	75	40	7.3	4.4	1.31	760	12.6	2.96	1.21	20.3	4.7	8.5	4.5	96	41.4	16.8	17.8	5.9	21	-	12
100	9.2	11.70	100	50	7.5	4.7	1.53	186.7	25.9	4.00	1.49	37.3	7.5	9.0	4.5	96	64.0	18.0	22.7	6.2	28	50	16
125	12.7	16.19	125	65	8.1	5.0	1.94	416.4	59.9	5.07	1.92	66.6	13.1	9.5	5.0	96	85.4	19.8	30.0	6.5	35	55	22
150	16.4	20.88	150	75	9.0	5.4	2.22	779.4	102.3	6.11	2.21	103.9	19.4	10.0	5.0	96	106.7	21.7	34.8	6.9	40	55	25
175	19.1	24.38	175	75	10.2	5.7	2.20	1223.3	121.0	7.08	2.23	139.8	22.8	10.5	5.5	96	128.4	23.3	34.7	7.2	40	55	25
200	22.1	28.21	200	75	11.4	6.1	2.17	1819.3	140.4	8.03	2.23	181.9	26.3	11.0	5.5	96	150.2	24.9	34.5	7.6	40	60	25
225	25.9	33.01	225	80	12.4	6.4	2.30	2694.6	187.2	9.03	2.38	239.5	32.8	12.0	6.0	96	170.9	27.1	36.8	7.9	45	60	25
250	30.4	38.67	250	80	14.1	7.1	2.30	3816.8	219.1	9.94	2.38	305.3	38.4	12.0	6.0	96	192.5	28.7	36.5	8.6	45	65	25
300	35.8	45.64	300	90	13.6	7.6	2.36	6362.6	310.8	11.81	2.61	424.2	46.8	13.0	6.5	96	240.7	29.6	41.2	9.1	50	65	28
400	49.4	62.93	400	100	15.3	8.6	2.42	15082.8	504.8	15.48	2.83	754.1	65.6	15.0	7.5	96	332.8	33.6	45.7	10.1	60	70	28

Equal Angles

Designation of an Equal Angle of size 80 × 80, thickness 12 and 600 long :

ISA 8080 × 12 × 600 LG.

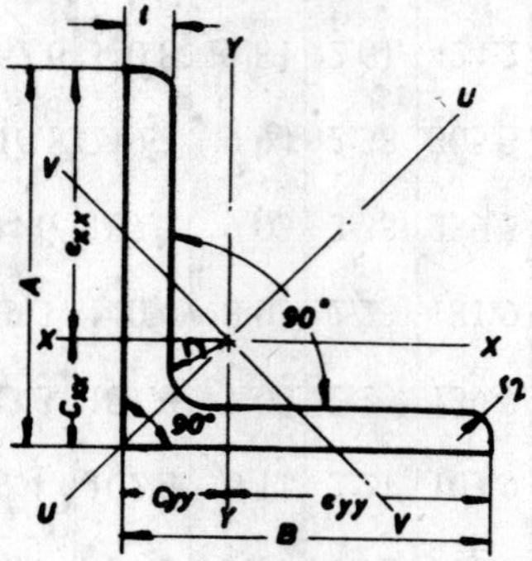

Size	Thick-ness t	Sectional area a	Mass	Centre of gravity	Distance of extreme fibre	Moments of Inertia			Radii of Gyration			Moduli of section Z_{xx} =	Radius at root	Radius at toe	Product of inertia
A × B mm mm	mm	cm^2	kg/m	C_{xx} = C_{yy} cm	e_{xx} = e_{yy} cm	$I_{xx} = I_{yy}$ cm^4	I_{uu} cm^4	I_{vv} cm^4	$r_{xx} = r_{yy}$ cm	r_{uu} cm	r_{vv} cm	Z_{yy} cm^3	r_1 mm	r_2 mm	I_{xx} cm^4
35 × 35	5.0	3.27	2.6	1.04	2.46	3.5	5.6	1.5	1.04	1.31	0.67	1.4	5.0	3.0	2.1
40 × 40	5.0	3.78	3.0	1.16	2.84	5.4	8.6	2.2	1.20	1.51	0.77	1.9	5.5	3.0	3.2
	6.0	4.47	3.5	1.20	2.80	6.3	10.0	2.6	1.19	1.50	0.77	2.3	”	”	3.7
45 × 45	5.0	4.28	3.4	1.29	3.21	7.9	12.6	3.2	1.36	1.72	0.87	2.5	”	”	4.6
	6.0	5.07	4.0	1.33	3.17	9.2	14.6	3.8	1.35	1.70	0.87	2.9	”	”	5.4

65 × 65	6.0	7.44	5.8	1.81	4.69	29.1	46.5	11.7	1.98	2.50	1.26	6.2	6.5	4.5	17.2
	8.0	9.76	7.7	1.89	4.61	37.4	59.5	15.3	1.96	2.47	1.25	8.1	”	”	22.0
75 × 75	6.0	8.66	6.8	2.06	5.44	45.7	73.1	18.4	2.30	2.91	1.46	8.4	7.0	4.5	27.0
	8.0	11.38	8.9	2.14	5.36	59.0	94.1	24.0	2.28	2.88	1.45	11.0	”	”	34.8
	10.0	14.02	11.0	2.22	5.28	71.4	113.3	29.4	2.26	2.84	1.45	13.5	”	”	41.7
80 × 80	6.0	9.29	7.3	2.18	5.82	56.0	89.6	22.5	2.46	3.11	1.56	9.6	8.0	4.5	33.0
	8.0	12.21	9.6	2.27	5.73	72.5	115.6	29.4	2.44	3.08	1.55	12.6	”	”	42.7
	12.0	17.81	14.0	2.42	5.58	101.9	161.4	42.4	2.39	3.01	1.54	18.3	”	”	59.7
90 × 90	6.0	10.47	8.2	2.42	6.58	80.1	128.1	32.0	2.77	3.50	1.75	12.2	8.5	5.5	47.2
	8.0	13.79	10.8	2.51	6.49	104.2	166.4	42.0	2.75	3.47	1.75	16.0	”	”	61.5
	10.0	17.03	13.4	2.59	6.41	126.7	201.9	51.6	2.73	3.44	1.74	19.8	”	”	74.5
	12.0	20.19	15.8	2.66	6.34	147.9	234.9	60.9	2.71	3.41	1.74	23.3	”	”	86.5
100 × 100	6.0	11.67	9.2	2.67	7.33	111.3	178.1	44.5	3.09	3.91	1.95	15.2	8.5	5.5	65.7
	8.0	15.39	12.1	2.76	7.24	145.1	231.8	58.4	3.07	3.88	1.95	20.0	”	”	85.8
	10.0	19.03	14.9	2.84	7.16	177.0	282.2	71.8	3.05	3.85	1.94	24.7	”	”	104.4
	12.0	22.59	17.7	2.92	7.08	207.0	329.3	84.7	3.03	3.82	1.94	29.2	”	”	121.6
110 × 110	8.0	17.02	13.4	3.00	8.00	195.0	311.7	78.2	3.38	4.28	2.14	24.4	10.0	6.0	115.1
	10.0	21.06	16.5	3.08	7.92	238.4	380.5	96.3	3.36	4.25	2.14	30.1	”	”	140.6
130 × 130	10.0	25.06	19.7	3.58	9.42	402.7	643.4	162.1	4.01	5.07	2.54	42.7	”	”	238.3
	12.0	29.87	23.4	3.66	9.34	473.8	755.9	191.8	3.99	5.03	2.54	50.7	”	”	279.9
150 × 150	10.0	29.03	22.8	4.06	10.94	622.4	995.4	249.4	4.63	5.86	2.93	56.9	12.0	8.0	368.2
	12.0	34.59	27.2	4.14	10.86	735.4	1174.8	296.0	4.61	5.83	2.93	67.7	12.0	8.0	435.0
	16.0	45.70	35.9	4.29	10.71	949.0	1510.0	391.0	4.56	5.74	2.93	88.7	16.0	8.0	
200 × 200	16.0	61.80	48.5	5.52	14.48	2340.0	3740.0	943.0	6.15	7.78	3.91	162.0	18.0	9.0	
	20.0	76.4	59.9	5.68	14.32	2850.0	4540.0	1160.0	6.11	7.72	3.89	199.0	18.0	9.0	

Unequal Angles

Designation of an Unequal Angle of size 100 × 75, thickness 8 and 350 long :

ISA 10075 × 8 × 350 LG

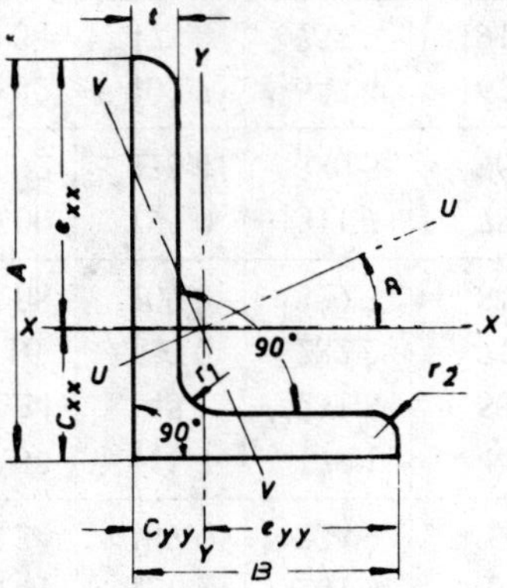

Size	*Thic-kne-ss*	*Secti-onal area*	*Mass*	*Centre of gravity*		*Distance of exteme fibre*		*Moments of inertia*				*Radii of Gyration*				*Moduli of section*		*Posi-tion of axis*	*Rad-ius at root*	*Rad-ius at toe*	*Pro-duct of inertia*
A × B	*t*	*a*		C_{xx}	C_{yy}	*exx*	*eyy*	I_{xx}	I_{yy}	I_{uu}	I_{vv}	r_{xx}	r_{yy}	r_{uu}	r_{vv}	Z_{xx}	Z_{yy}		r_1	r2	I_{xy}
mm mm	*mm*	cm^2	*kg/m*	*cm*	*cm*	*cm*	*cm*	cm^4	cm^4	cm^4	*cm*	*cm*	*cm*	*cm*	*cm*	cm^3	cm^3	tan − α	*mm*	*mm*	cm^4
45 × 30	5.0	3.52	2.8	1.51	0.77	2.99	2.23	6.9	2.4	7.9	1.4	1.40	0.83	1.50	0.63	2.3	1.1	0.43	5.0	3.0	2.3
	6.0	4.16	3.3	1.55	0.81	2.95	2.19	8.0	2.8	9.2	1.7	1.39	0.82	1.49	0.63	2.7	1.3	0.42	"	"	2.7

75 × 50	6.0	7.16	5.6	2.44	1.20	5.06	3.80	40.3	14.3	46.4	8.2	2.37	1.41	2.55	1.07	8.0	3.8	0.44	6.5	4.0	13.9
	8.0	9.38	7.4	2.52	1.28	4.98	3.72	51.8	18.3	59.4	10.6	2.35	1.40	2.52	1.06	10.4	4.9	0.43	”	”	17.7
80 × 50	10.0	12.02	9.4	2.81	1.32	5.19	3.68	74.7	22.1	83.3	13.5	2.49	1.36	2.63	1.06	14.4	6.0	0.38	7.0	4.5	22.9
90 × 60	6.0	8.65	6.8	2.87	1.39	6.13	4.61	70.6	25.2	81.5	14.3	2.86	1.71	3.07	1.28	11.5	5.5	0.44	7.5	5.0	24.5
100 × 75	8.0	13.36	10.5	3.10	1.87	6.90	5.63	131.6	63.3	161.3	33.5	3.14	2.18	3.48	1.59	19.1	11.2	0.55	8.5	6.0	53.4
	10.0	16.50	13.0	3.19	1.95	6.81	5.55	160.4	76.9	196.1	41.2	3.12	2.16	3.45	1.58	23.6	13.8	0.55	”	”	64.7
125 × 75	8.0	15.38	12.1	4.15	1.68	8.35	5.82	245.5	67.2	272.8	40.0	4.00	2.09	4.21	1.61	29.4	11.5	0.36	9.0	6.0	74.0
	10.0	19.02	14.9	4.74	1.76	8.26	5.74	300.3	81.6	332.9	49.1	3.97	2.07	4.18	1.61	36.3	14.2	0.36	”	”	89.9
150 × 75	10.0	21.56	16.9	5.32	1.61	9.68	5.89	499.1	85.3	529.8	54.5	4.81	1.99	4.96	1.59	51.6	14.5	0.26	10.0	6.0	116.2
150 × 115	10.0	25.52	20.0	4.55	2.82	10.45	8.68	573.3	293.4	714.3	152.4	4.74	3.39	5.29	2.44	54.9	33.8	0.58	11.0	7.5	241.0
	12.0	30.38	23.8	4.64	2.90	10.36	8.60	676.5	345.3	841.4	180.4	4.72	3.37	5.26	2.44	65.3	40.2	0.58	”	”	283.6

Rivet Gauge Distances in Legs of Angles

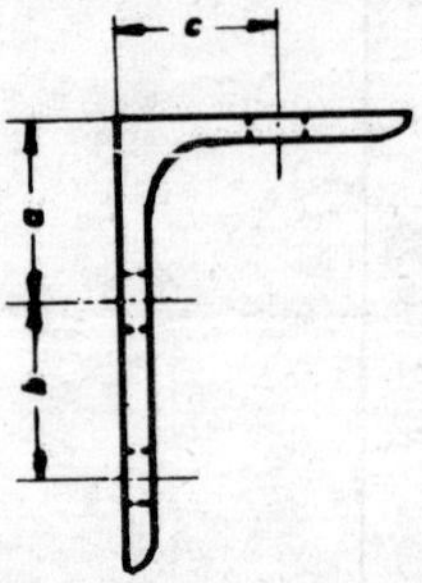

Leg Size	Double Row of Rivets		Single Row of Rivets	Maximum Rivet Size for Double Row of Rivets
	a	*b*	C	
200	75	85	115	27
150	55	65	90	22
130	50	55	80	20
125	45	55	75	20
115	45	50	70	12
110	45	45	65	12
100	40	40	60	12
95	–	–	55	–
90	–	–	50	–
80	–	–	45	–
75	–	–	40	–
70	–	–	40	–
65	–	–	35	–
60	–	–	35	–
55	–	–	30	–
50	–	–	28	–
45	–	–	25	–
40	–	–	21	–
35	–	–	19	–
30	–	–	17	–
25	–	–	15	–
20	–	–	12	–

Tee Bars

Designation of an Indian Standard Normal Tee bar (ISNT) of size 60 :

ISNT 60

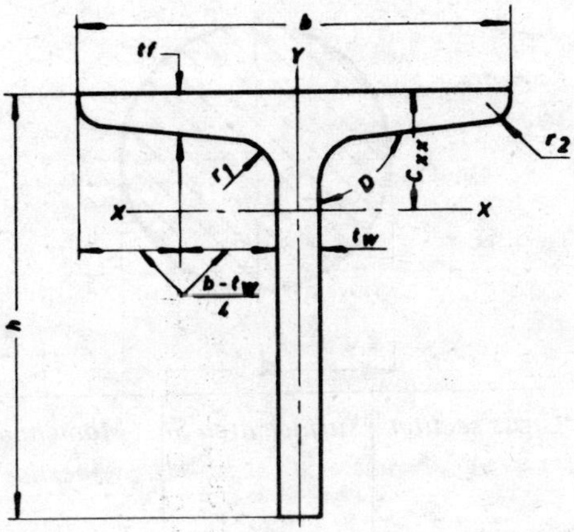

Connection Details

Size mm	*Max. size of flange rivet/bolt d mm*	*Rivet/bolt centre dist. g mm*
50	6	30
60	6	30
70	12	35

Size	*Mass*	*Sectional area*	*Depth of section*	*Width of flange*	*Thickness of flange*	*Thickness of web*	*Centre of gravity*	*Moment of inertia*		*Radii of gyration*		*Moduli of section*		*Radius at root*	*Radius at toe*	*Slope of flange*
		a	*h*	*b*	t_f	t_w	C_{xx}	I_{xx}	I_{yy}	r_{xx}	r_{yy}	Z_{xxx}	Z_{yy}			
mm	*kg/m*	cm^2	*mm*	*mm*	*mm*	*mm*	*cm*	cm^4	cm^4	*cm*	*cm*	cm_3	cm^3	r_1 *mm*	r_2 *mm*	*D Degrees*
50	4.5	5.70	50	50	6.0	6.0	1.44	12.7	5.9	1.50	1.02	3.6	2.4	6	4	91
60	5.4	6.90	60	60	6.0	6.0	1.67	22.5	10.1	1.81	1.21	5.2	3.7	6.5	4.5	91
75	15.3	19.49	75	150	9.0	8.4	1.62	96.2	230.2	2.22	3.44	16.4	30.1	8.0	4.0	94

Rounds

Designation of a hot rolled round of 40 mm diameter :

Round 40

Dimensions in mm

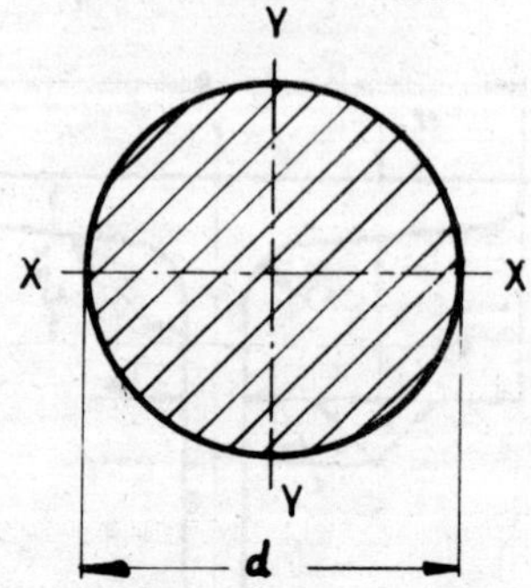

Diameter	*Mass*	*Cross section A*	*Surface area S*	*Moment of inertia*	*Radius of gyration*	*Modulus of section*
d	*kg/m*	*cm*^{2}	*cm*^{2}*/m*	$I_{xx} = I_{yy}$ *cm*^{4}	$r_{xx} = r_{yy}$ *cm*	$Z_{xx} = Z_{yy}$ *cm*^{3}
12	0.9	1.13	377	0.1018	0.30	0.1696
14	1.2	1.54	440	0.1886	0.35	0.2694
16	1.6	2.01	503	0.3217	0.40	0.4021
18	2.0	2.54	565	0.5153	0.45	0.5726
20	2.5	3.14	628	0.7854	0.50	0.7854
22	3.0	3.80	691	1.1499	0.55	1.045
25	3.8	4.91	785	1.9175	0.63	1.534
28	4.8	6.16	880	3.0172	0.70	2.155
32	6.3	8.04	1010	5.1472	0.80	3.217
36	8.0	10.2	1130	8.2448	0.90	4.580
40	9.9	12.6	1260	12.5664	1.00	6.283
45	12.5	15.9	1410	20.1289	1.13	8.946
50	15.4	19.6	1570	30.6796	1.25	12.272
56	19.3	24.6	1760	48.2750	1.40	17.241
63	24.5	31.2	1980	77.3272	1.57	24.548
80	39.5	50.3	2510	201.0619	2.00	50.265
100	61.7	78.5	3140	490.8738	2.50	98.175

Flats

Designation of a hot rolled flat of 50 mm breadth and 16 mm thickness :

Flat 50 × 16

Dimensions in mm

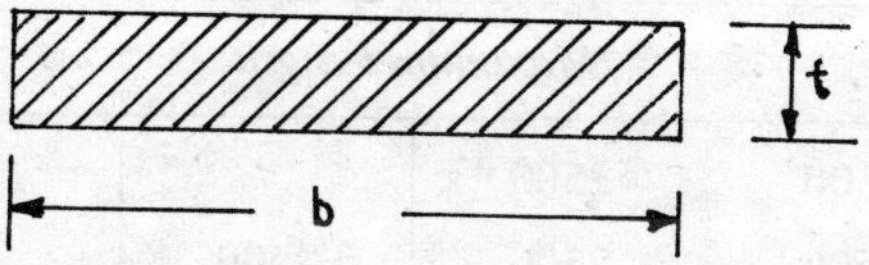

Breadth b	Thickness t	Mass	Sectional area a	Moments of inertia		Radii of gyration		Moduli of section	
			a	I_{xx}	I_{yy}	r_{xx}	r_{yy}	Z_{xx}	Z_{yy}
mm	mm	kg/m	cm^2	cm^4	cm^4	cm	cm	cm^3	cm^3
45	6	2.1	2.7	0.081	4.556	0.173	1.299	0.270	2.025
45	10	3.5	4.5	0.375	7.594	0.289	1.299	0.750	3.375
45	12	4.2	5.4	0.648	9.113	0.346	1.299	1.080	4.050
50	6	2.4	3.0	0.090	6.250	0.173	1.443	0.300	2.500
50	8	3.1	4.0	0.213	8.333	0.231	1.443	0.533	3.333
50	10	3.9	5.0	0.417	10.417	0.289	1.443	0.834	4.167
50	12	4.7	6.0	0.720	12.500	0.346	1.443	1.200	5.000
50	16	6.3	8.0	1.707	16.667	0.462	1.443	2.134	6.667
65	8	4.1	5.2	0.277	18.308	0.231	1.876	0.693	5.633
65	10	5.1	6.5	0.542	22.885	0.289	1.876	1.084	7.042
65	12	6.1	7.8	0.936	27.463	0.346	1.876	1.560	8.450
75	8	4.7	6.0	0.320	28.125	0.231	2.165	0.800	7.500
75	10	5.9	7.5	0.625	35.156	0.289	2.165	1.250	9.375
75	12	7.1	9.0	1.080	42.188	0.346	2.165	1.800	11.250
75	20	11.8	15.0	5.000	70.313	0.577	2.165	5.000	18.750
100	10	7.8	10.0	0.833	83.333	0.289	2.890	1.666	16.667
100	12	9.4	12.0	1.440	100.000	0.346	2.890	2.400	20.000
100	16	12.6	16.0	3.413	133.333	0.462	2.890	4.266	26.667
100	20	15.7	20.0	6.666	166.666	0.577	2.890	6.666	33.333
200	10	15.7	20.0	1.667	666.666	0.289	5.774	3.334	66.667
200	12	18.8	24.0	2.880	800.000	0.346	5.774	4.800	80.000
200	25	39.2	50.0	26.042	1666.666	0.722	5.774	20.834	166.667
250	12	23.6	30.0	3.600	1562.500	0.346	7.217	6.800	80.000
250	20	39.2	50.0	16.667	2604.166	0.577	7.217	16.667	208.333
250	25	49.1	62.5	32.552	3255.208	0.722	7.217	26.042	260.417
300	12	28.3	36.0	4.320	2700.000	0.346	8.660	7.200	180.000

Sheets

Dimensions in mm

Cold Rolled Sheets

Thickness	*Width*						*Mass per sq. metre kg*
	750	*900*	*1000*	*1100*	*1250*	*1400*	
	Maximum Length						
0.63	2500	2500	2500	-	-	-	4.95
0.80	2500	2500	2500	2500	-	-	6.28
1.00	3200	3200	3200	3200	3200	-	7.85
1.25	-	3200	3200	3200	3200	3200	9.81
1.60	-	4000	4000	4000	4000	4000	12.56

Hot Rolled Sheets

Thickness	*Width*						*Mass per sq. metre kg*
	900	*1000*	*1100*	*1250*	*1400*	*1500*	
	Maximum Length						
2.00	4000	4000					15.70
2.50	4000	4000	4000	4000	4000		19.63
3.15	4000	4000	4000	4000	4000	4000	24.73
3.55	10000	10000	10000	10000	10000	10000	27.87
4.00	10000	10000	10000	10000	10000	10000	31.40

Plates

Dimensions in mm

Light Plate

Thickness	*Width*						*Mass per sq. metre kg*
	900	*1000*	*1100*	*1250*	*1400*	*1500*	
	Maximum Length						
5.00	10000	10000	10000	10000	10000	10000	39.25
6.00	10000	10000	10000	10000	10000	10000	47.10

Sheared Heavy Plates (Rimming or Semi-killed quality)

Thickness	*Width*							*Mass per sq. metre kg*
	1000	*1200*	*1600*	*1800*	*2000*	*2200*	*2500*	
	Maximum Length							
8	10000	12500	10000	10000	10000	11000	12500	62.80
10	12500	12500	10000	11000	12500	12500	12500	78.50
12	12500	12500	12500	12500	12500	12500	12500	94.20
14	12500	12500	12500	12500	12500	12500	12500	109.90
16	12500	12500	12500	12500	12500	12500	12500	125.60
18	12500	12500	12500	12500	12500	12500	12500	141.30
20	11000	11000	12500	12500	12500	12500	11000	157.00
22	3600	3600	12500	12500	12500	11000	10000	172.70
25	3600	3600	12500	12500	11000	10000	9000	196.25
28	3600	3600	12500	11000	9000	8000	7100	219.80
32	3600	3600	10000	9000	8000	7100	6300	251.20
36	3600	3600	9000	8000	7100	6300	5600	282.60

Sheared Heavy Plates (Killed quality)

Thickness	*Width*							*Mass per sq. metre kg*
	1000	*1250*	*1600*	*1800*	*2000*	*2200*	*2500*	
	Maximum Length							
8	9000	11000	9000	10000	9000	11000	11000	62.80
10	11000	11000	9000	11000	11000	12500	11000	78.50
12	12500	12500	12500	12500	12500	12500	12500	94.20
14	12500	12500	12500	11000	12500	11000	12500	109.90
16	11000	12500	11000	12500	11000	12500	12500	125.60
18	12500	12500	12500	11000	12500	11000	12500	141.30
20	12500	11000	11000	12500	12500	12500	11000	157.00
22	3600	3600	12500	12500	11000	10000	9000	172.70
25	3600	3600	12500	12500	11000	10000	9000	196.25
28	3600	3600	12500	11000	9000	8000	7100	219.80
32	3600	3600	10000	9000	8000	7100	6300	251.20
36	3600	3600	9000	8000	7100	6300	5600	282.60

Unsheared Heavy Plates (Rimming or Semi-killed)

Thickness	*Width*					*Mass per sq. metre kg*
	1600	*1800*	*2000*	*2200*	*2500*	
	Maximum Length					
40	11000	10000	9000	8000	7100	314.00
45	10000	9000	8000	7100	6300	353.25
50	9000	8000	7100	6300	5600	392.50
56	8000	7100	6300	5600	5000	439.60
63	7100	6300	5600	5000	4500	494.55

Unsheared Heavy Plates (Killed and Special Quality Steel)

Thickness	*Width*					*Mass per sq. metre kg*
	1600	*1800*	*2000*	*2200*	*2500*	
	Maximum Length					
40	9000	8000	7100	6300	5600	314.00
45	8000	7100	6300	5600	5000	353.25
50	7100	6300	5600	5000	4500	392.50
56	6300	5600	5000	4500	5000	439.60
63	5600	5000	4500	5000		494.55

APPENDIX - E Tables of Mass of Flats and Bars

Mass of Steel Bars

In this table, size indicates the diameter in case of rounds and a side in case of squares. Density of steel has been assumed as 7.85 g per cubic cm. In case the mass is required for other metals, the figures for steel are to be multiplied by the following factors :

Material	*Factor (approx)*	*Material*	*Factor (approx)*
Cast iron	0.926	Aluminum	0.344
Coppor	1.132	Aluminum alloy	0.363
Brass	1.082		

Size (mm)	*Round (kg/m)*	*Square (kg/m)*
10	0.617	0.785
20	2.466	3.140
30	5.549	7.065
40	9.865	12.56
50	15.413	19.625
60	22.195	28.260
70	30.210	38.465
80	39.458	50.240
90	49.939	63.585
100	61.654	78.500
150	138.721	176.625
200	246.615	314.000
250	385.336	490.625
300	554.883	706.500
350	755.258	961.625
400	986.460	1256.000
450	1248.488	1589.625
500	1541.344	1962.500

Mass of flats-kg/metre

Width mm	*Thickness mm*									
	1	*2*	*3*	*4*	*5*	*6*	*7*	*8*	*9*	*10*
10	0.079	0.157	0.235	0.314	0.393	0.471	0.550	0.628	0.707	0.785
20	0.157	0.314	0.471	0.628	0.785	0.942	1.099	1.256	1.413	1.570
30	0.236	0.471	0.707	0.943	1.178	1.413	1.648	1.884	2.120	2.355
40	0.314	0.628	0.942	1.256	1.570	1.884	2.198	2.512	2.826	3.140
50	0.393	0.785	1.178	1.570	1.963	2.355	2.748	3.140	3.533	3.925
60	0.471	0.942	1.413	1.884	2.355	2.826	3.297	3.768	4.239	4.710
70	0.550	1.099	1.649	2.198	2.748	3.297	3.847	4.396	4.946	5.495
80	0.628	1.256	1.884	2.512	3.140	3.768	4.396	5.024	5.652	6.280
90	0.707	1.413	2.120	2.826	3.533	4.239	4.946	5.652	6.359	7.065
100	0.785	1.570	2.355	3.140	3.925	4.710	5.495	6.280	7.065	7.850
125	0.981	1.962	2.944	3.925	4.906	5.888	6.869	7.850	8.831	9.813
150	1.177	2.355	3.533	4.710	5.888	7.066	8.243	9.420	10.598	11.775
175	1.374	2.748	4.121	5.495	6.868	8.243	9.616	10.990	12.364	13.738
200	1.570	3.140	4.710	6.280	7.850	9.420	10.990	12.56	14.130	15.700
225	1.766	3.533	5.299	7.065	8.831	10.598	12.364	14.130	16.896	17.663
250	1.963	3.925	5.888	7.850	9.813	11.776	13.738	15.700	17.662	19.626
275	2.159	4.318	6.476	8.635	10.794	12.950	15.111	17.270	19.428	21.587
300	2.355	4.710	7.065	9.420	11.775	14.130	16.485	18.840	21.195	23.550

APPENDIX - F Tables of Strength Versus Hardness Numbers

Comparison of tensile strength of steels versus hardness given in these tables must be regarded as giving no more than a general indication of the relationship.

Strength and hardness values are given in the conventional metric system since this system is still largely in vogue for assigning hardness values. To convert the values in SI units, refer section 10.6.

Tensile Strength	*Brinell hardness*	*Rockwell hardness*		*Vickers hardness*	*Scleroscope hardness*
Kgf/mm²	*Kgf/mm²* HB	*C-Scale Diamond Cone P-100 Kgf HRC*	*B-Scale Steel Ball 1/16 Inch ϕ P = 100 Kgf HRB*	*P ≥ 5 kgf.* HV	*Model-C* Sh
31	88	–	45	88	–
32	91	–	48	91	–
33	93	–	50	93	–
34	96	–	53	96	–
35	99	–	55	99	–
36	102	–	58	102	–
37	105	–	60	105	–
38	108	–	62	108	–
39	111	–	64	111	–
40	115	–	66	115	–
41	118	–	68	118	–
42	121	–	70	121	–
43	124	–	71	124	–
44	127	–	73	127	–
45	130	–	74	130	–
46	133	–	76	133	–
47	137	–	77	137	–
48	139	–	78	139	–
49	143	–	80	143	–
50	145	–	81	145	–
51	149	–	82	149	–
52	152	–	83	152	–
53	154	–	84	154	–
54	158	–	85	158	–
55	161	–	86	161	–

56	164	–	87	164	–
57	167	–	87	167	–
58	170	–	88	170	–
59	172	–	89	172	–
60	175	–	89	175	–
61	179	–	90	179	–
62	182	–	91	182	–
63	184	–	92	184	–
64	187	–	93	187	–
65	191	–	93	191	–
66	193	–	-	193	–
67	195	–	94	195	–
68	198	–	95	198	–
69	202	–	95	202	–
70	204	–	96	204	–
71	207	–	96	207	–
72	211	–	97	211	–
73	213	–	97	213	–
74	217	–	98	217	–
75	219	–	98	219	–
76	222	–	99	222	–
77	226	–	99	226	–
78	229	19	–	229	–
79	231		–	231	–
80	234	20	–	234	–
81	236	21	–	236	36
82	241	21	–	241	21
83	244	22	–	244	37
84	246	–	–	246	–
85	249	23	–	249	–
86	252	23	–	252	38
87	255	24	–	255	–
88	257	24	–	257	–
89	260	25	–	260	39
90	263	25	–	263	–
91	266	–	–	266	40

92	269	26	–	269	–
93	272	–	–	272	41
94	275	27	–	275	–
95	278	27	–	278	42
96	282	28	–	282	–
97	285	28	–	285	–
98	288	–	–	288	43
100	292	29	–	292	–
101	295	–	–	295	44
102	298	30	–	298	–
103	302	–	–	302	45
105	306	31	–	306	–
106	309	–	–	309	46
107	313	–	–	313	–
109	317	32	–	317	–
110	321	–	–	321	47
112	325	33	–	325	48
113	329	34	–	329	–
114	333	–	–	333	–
116	337	–	–	337	49
117	341	35	–	341	–
119	345	–	–	345	50
120	350	36	–	350	51
122	354	–	–	354	–
123	359	37	–	360	52
125	363	–	–	365	53
127	368	38	–	370	–
128	373	–	–	376	54
130	378	39	–	382	55
132	383	40	–	388	–
134	388	–	–	395	56
135	393	41	–	401	57
137	398	–	–	407	58
139	404	42	–	415	–
141	409	–	–	422	59
143	415	43	–	430	60

145	420	44	–	436	61
147	426	–	–	444	62
150	432	45	–	453	–
–	–	–	–	455	63
–	–	–	–	460	–
–	–	46	–	465	64
–	–	–	–	470	64.5
–	–	–	–	475	65
–	–	47	–	480	65.5
–	–	–	–	485	66
–	–	–	–	490	66.5
–	–	48	–	495	67
–	–	–	–	500	67.5
–	–	–	–	505	68
–	–	49	–	510	–
–	–	–	–	515	68.5
–	–	–	–	520	69
–	–	50	–	525	69.5
–	–	–	–	530	70
–	–	–	–	535	70.5
–	–	51	–	540	71
–	–	–	–	545	71.5
–	–	–	–	550	72
–	–	52	–	560	73
–	–	–	–	570	74
–	–	53	–	580	75
–	–	–	–	590	75.5
–	–	54	–	600	76.5
–	–	55	–	610	77.5
–	–	–	–	620	78.5
–	–	56	–	630	79
–	–	–	–	640	80
–	–	57	–	650	81
–	–	–	–	660	82
–	–	58	–	670	82.5
–	–	–	–	680	83.5

–	–	59	–	690	84.5
–	–	–	–	700	85
–	–	60	–	710	86
–	–	–	–	720	87
–	–	–	–	730	87.5
–	–	61	–	740	88.5
–	–	–	–	750	89.5
–	–	62	–	760	90
–	–	63	–	780	92
–	–	64	–	800	93.5
–	–	–	–	820	95
–	–	65	–	840	96.5
–	–	66	–	860	98
–	–	–	–	880	99.5
–	–	67	–	900	101
–	–	–	–	920	102.5
–	–	68	–	940	103.5

BIBLIOGRAPHY

1. ASME Handbook, Engineering Tables, McGraw Hill, New York, U.S.A.
2. Boettcher u. Forberg, Technisches Zeichnen, B.G. Teubner, Stuttgart, Germany.
3. DIN 406, 3141.
4. DIN Taschenbuch Nr. 2, Deutsches Institut fuer Normung e.v. Berlin, Germany.
5. DUBBEL Taschenbuch fuer den Maschinenbau, Vol. 1, Springer Verlag, Berlin, Germany.
6. Fachzeichnen Metalltechnik, Grundstufe, Verlag Europe - Lehrmittel, Wuppertal, Germany.
7. Faires and Keown, Mechanism, McGraw-Hill, New York, U.S.A.
8. Haenchen u. Decker, Neue Festigkeitsberechnung fuer den Maschinenbau, 3rd edition, Carl Hanser Verlag, Munich, Germany.
9. Hand buch fuer Technische Zeichner u. Teilkonstrukteure, Fried-rich u. Voigt, Technik - Tabellen Verlag, Darmstadt, Germany.
10. Hoischen H., Technisches Zeichnen, Verlag W. Girardet, Essen, Germany.
11. Huette des Ingenieurs Taschenbuch, Vol. II A, Verlag von Wilhelm Ernst u. Sohn, Berlin, Germany.
12. IS: 210, 554, 696, 1030, 1364, 1365, 1366, 1367, 1570, 1835, 1865, 1870, 1975, 2102, 2266, 2269, 2403, 2429, 2494, 2535, 2644, 2707, 2856, 3038, 3142, 3355, 3459, 4897, 5129, 5517, 5519, 8000, 8349, 8350, 10702, 10711, 10713, 10714, 10719.
13. ISO 128 — 1982T, R 129, R 273, R 286, R 1101, 1302-1978, 5261-1981, 5455-1979, 5457-1980.
14. Luft, H. Maschinen elemente, Technik Tabellen Verlag, Darmstadt, Germany.
15. Machinery's Handbook, Industrial Press, New York, U.S.A.
16. Maitra, Gitin M., Handbook of Gear Design, 2nd edition, Tata McGraw-Hill Co Ltd., New Delhi, 1995.
17. Maitra, Gitin M. and L.V. Prasad, Handbook of Mechanical Design, 2nd Edition, Tata McGraw-Hill Co. Ltd., New Delhi, 1995.
18. Niemann, G., Maschinenelemente, Springer Verlag, Berlin, Germany.
19. Shigley, J.E., Mechanical Engineering Design, McGraw-Hill, New York, U.S.A.
20. Tabellenbuch fuer Metallverarbeitende Berufe, Verlag Europa Lehrmittel, Wuppertal, Germany.

INDEX